AF617876

Operator Theory: Advances and Applications
Vol. 117

Editor:
I. Gohberg

Editorial Office:
School of Mathematical Sciences
Tel Aviv University
Ramat Aviv, Israel

Editorial Board:
J. Arazy (Haifa)
A. Atzmon (Tel Aviv)
J. A. Ball (Blacksburg)
A. Ben-Artzi (Tel Aviv)
H. Bercovici (Bloomington)
A. Böttcher (Chemnitz)
L. de Branges (West Lafayette)
K. Clancey (Athens, USA)
L. A. Coburn (Buffalo)
K. R. Davidson (Waterloo, Ontario)
R. G. Douglas (Stony Brook)
H. Dym (Rehovot)
A. Dynin (Columbus)
P. A. Fillmore (Halifax)
C. Foias (Bloomington)
P. A. Fuhrmann (Beer Sheva)
S. Goldberg (College Park)
B. Gramsch (Mainz)
G. Heinig (Chemnitz)
J. A. Helton (La Jolla)
M.A. Kaashoek (Amsterdam)
T. Kailath (Stanford)
H.G. Kaper (Argonne)
S.T. Kuroda (Tokyo)
P. Lancaster (Calgary)
L.E. Lerer (Haifa)
E. Meister (Darmstadt)
B. Mityagin (Columbus)
V. V. Peller (Manhattan, Kansas)
J. D. Pincus (Stony Brook)
M. Rosenblum (Charlottesville)
J. Rovnyak (Charlottesville)
D. E. Sarason (Berkeley)
H. Upmeier (Marburg)
S. M. Verduyn-Lunel (Amsterdam)
D. Voiculescu (Berkeley)
H. Widom (Santa Cruz)
D. Xia (Nashville)
D. Yafaev (Rennes)

Honorary and Advisory Editorial Board:
P. R. Halmos (Santa Clara)
P. D. Lax (New York)
M. S. Livsic (Beer Sheva)

Differential Operators and Related Topics

Proceedings of the Mark Krein International Conference on Operator Theory and Applications, Odessa, Ukraine, August 18–22, 1997
Volume I

V. M. Adamyan
I. Gohberg
M. Gorbachuk
V. Gorbachuk
M. A. Kaashoek
H. Langer
G. Popov
Editors

Springer Basel AG

Editors:
V.M. Adamyan
Department of Theoretical Physics
University of Odessa
270026 Odessa
Ukraine

I. Gohberg
Department of Mathematical Sciences
Raymond and Beverly Sackler
Faculty of Exact Sciences
Tel Aviv University
69978 Ramat Aviv
Israel

M. Gorbachuk and V. Gorbachuk
Institute of Mathematics
National Academy of Sciences of Ukraine
Kyiv, Ukraine

M.A. Kaashoek
Department of Mathematics
Vrije Universiteit
De Boelelaan 1081a
1081 HV Amsterdam
The Netherlands

H. Langer
Department of Mathematics
Technical University of Vienna
Wiedner Hauptstrasse 8–10/1411
1040 Vienna
Austria

G. Popov
Institute of Mathematics, Economics and Mechanics
Odessa State University
270057 Dvoryanskaya str. 2
Odessa
Ukraine

1991 Mathematics Subject Classification 47-06

A CIP catalogue record for this book is available from the
Library of Congress, Washington D.C., USA

Deutsche Bibliothek Cataloging-in-Publication Data

Mark Krein International Conference on Operator Theory and Applications <1997, Odessa>: Proceedings of the Mark Krein International Conference on Operator Theory and Applications : Odessa, Ukraine, August 18 - 22, 1997 / V. M. Adamyan .... ed.. - **Springer Basel AG**
ISBN 978-3-7643-6287-4 **ISBN 978-3-0348-8403-7 (eBook)**
DOI 10.1007/978-3-0348-8403-7
Vol. 1. Differential operators and related topics. - 2000
(Operator theory ; Vol. 117)
ISBN 978-3-7643-6287-4

This work is subject to copyright. All rights are reserved, whether the whole or part of the material is concerned, specifically the rights of translation, reprinting, re-use of illustrations, recitation, broadcasting, reproduction on micro[illegible] or in other ways, and storage in data banks. For any kind of use permission of the copyright owner must be obtained.

© 2000 Springer Basel AG
Originally published by Birkhäuser Verlag Basel - Boston - Berlin in 2000

Printed on acid-free paper produced from chlorine-free pulp. TCF ∞
Cover design: Heinz Hiltbrunner, Basel

ISBN 978-3-7643-6287-4

Preface

The present book is the first of the two volume Proceedings of the Mark Krein International Conference on Operator Theory and Applications. This conference, which was dedicated to the 90th Anniversary of the prominent mathematician Mark Krein, was held in Odessa, Ukraine from 18–22 August, 1997. The conference focused on the main ideas, methods, results, and achievements of M.G. Krein.

This first volume is devoted to the theory of differential operators and related topics. It opens with a description of the conference, biographical material and a number of survey papers about the work of M.G. Krein. The main part of the book consists of original research papers presenting the state of the art in the area of differential operators.

The second volume of these proceedings, entitled Operator Theory and related Topics, concerns the other aspects of the conference. The two volumes will be of interest to a wide-range of readership in pure and applied mathematics, physics and engineering sciences.

Table of Contents

Picture of M.G. Krein

Operator Theory:
Advances and Applications, Vol. 117
© 2000 Birkhäuser Verlag Basel/Switzerland

About the Mark Krein International Conference

The Mark Krein International Conference on Operator Theory and Applications was held in Odessa, Ukraine, and took place between August 18–22, 1997. This Conference was devoted to the 90th Anniversary of the prominent mathematician Mark Krein who has made fundamental contributions to mathematical analysis (understood in a very broad sense), algebra, and mechanics. His investigations are distinguished by an amazing combination of the finest analytic and algebraic techniques and geometric methods, deep intrinsic unity, and interrelation between classical and modern problems raised by the development of both pure and applied mathematics.

The Conference focused on the main ideas and methods of works of M. Krein, his achievements in various fields of mathematics, and their applications and further development. As a matter of fact, in each of the areas mentioned above M. Krein's works had a substantial influence. As for functional analysis, it would not be an exaggeration to say that, being a source of inspiration for many mathematicians, these works determined to a large extent the present-day status of this branch of mathematics.

More than 150 mathematicians from Austria, Armenia, Byelorussia, Finland, France, Germany, Israel, Italy, Japan, Moldova, Netherlands, Poland, Romania, Russia, Ukraine, USA, Venezuela took part in the Conference.

The opening session of the Conference contained the following addresses about M.G. Krein, his life, his seminars, and his influence in mathematics and on mathematicians:

1) *I. Gohberg* (Israel) (presented by *M.A. Kaashoek* (Netherlands)) Mark Grigorievich Krein (a short biography);
2) *G. Popov* (Ukraine) On Mark G. Krein seminars;
3) *C. Sadosky* (USA) How we followed some leads from M.G. Krein;
4) *V. Sizov* (Ukraine) (presented by *Yu. Vorobjov* (Ukraine)) The ship hydrodynamics.

The following mathematicians gave 45-minute plenary survey lectures:

1) *Yu. Berezansky* (Ukraine) The works of M.G. Krein on eigenfunction expansions for self-adjoint operators and their applications and development;
2) *H. Dym* (Israel) M.G. Krein and prediction theory;
3) *M. Gorbachuk and V. Gorbachuk* (Ukraine) M.G. Krein and extension theory of symmetric operators. Theory of entire operators;

4) *H. Langer* (Austria) Spectral functions of self-adjoint operators in Krein spaces;
5) *R. Mennicken* (Germany) Spectral theory of systems of differential operators of mixed order and applications;
6) *A. Nudelman* (Ukraine) On M.G. Krein's contribution to the moment problem;
7) *Ł. Sakhnovich* (Ukraine) Works by M.G. Krein on inverse problems;
8) *L. Vainerman* (Ukraine) On M.G. Krein's works on the representation theory and harmonic analysis on topological groups.

About 130 twenty-five minute contributions were presented in the sections:

- Operator Theory;
- Functional Analysis, Groups and Semigroups;
- Integral Operators;
- Applications of Operator Theory to Function Theory;
- Inverse Problems;
- Mechanics;
- Spectral Theory;
- Probability and Krein Strings;
- Random Operators;
- Differential Operators;
- Convex Sets;
- Moment Problem;
- Extension Theory;
- Operator Theory and System Theory;
- Spectral Theory of Differential Equations;
- Operator Theory and Operator Equations;
- Applications of Operator Theory.

The Conference was organized by the Institute of Mathematics of the National Academy of Sciences of Ukraine and the Institute of Mathematics, Economics and Mechanics of the Odessa Mechnikov University. The Organizing Committee consisted of Yu. Berezansky (Chair, Ukraine) and V. Adamyan, V. Kruglov, A. Samoilenko (Deputy Chairs, Ukraine).

The scientific program was a responsibility of the Scientific Program Committee whose members were: S. Albeverio (Germany), T. Ando (Japan), M. Birman (Russia), M. Cotlar (Venezuela), Yu. Daletsky (Ukraine), C. Foias (USA), I. Gohberg (Israel), M. Gorbachuk (Ukraine), B. Helton (USA), M.A. Kaashoek (Netherlands), S. Krein (Russia), O. Ladyzhenskaya (Russia), H. Langer (Austria), P. Lancaster (Canada), P. Lax (USA), M. Livsic (Israel), V. Marchenko (Ukraine), R. Mennicken (Germany), S. Mkhitarjan (Armenia), L. Nirenberg (USA), C. Sadosky (USA), I. Skrypnik (Ukraine), V. Yakubovich (Russia).

Other preparations for the Conference were in the hands of the Working Committee consisting of: S. Andronati, D. Arov, Yu. Chapovsky, V. Gorbachuk, A. Kochubei, A. Nudelman, V. Ostrovsky, V. Pivovarchik, Yu. Procerov, V. Reut, L. Sakhnovich, S. Shumihin, Yu. Vorobyov, N. Whitefield (Ukraine).

The main sources of funds for the Conference were provided by its sponsors:

- Ministry of Education of Ukraine;
- National Academy of Sciences of Ukraine;
- The Executive Committee of Odessa City Council of People Deputies;
- International Association for Promotion of Cooperation with Scientist from the New Independent States of the Former Soviet Union (INTAS);
- Odessa Mechnikov University.

A book exhibition was open throughout the Conference.

On 19th of August a memorial plate in honour of M.G. Krein was displayed on the house where this brilliant mathematician has spent the main part of his life. His daughter, Irma Krein, took part in the event and also addressed the participants in the opening session of the conference.

The above is a picture of the memorial plate in honour of M.G. Krein on the house 14 Arteoma Street in Odessa, where he lived. The plate was inaugurated during the conference. The text on the plate is in Ukrainian and its translation is as follows:

In this building
from 1928 to 1989
lived and worked
the outstanding mathematician
Marko Grigorovich Krein

Operator Theory:
Advances and Applications, Vol. 117
© 2000 Birkhäuser Verlag Basel/Switzerland

Mark Grigorevich Krein (A Short Biography)

Israel Gohberg

Mark G. Krein was born on April 3, 1907, in Kiev into a Jewish family of modest means. His father was a lumber merchant. From early on he showed a talent for mathematics. At the age of 14 he already attended research seminars. He never got his undergraduate degree. In 1924 he ran away from home to Odessa and in 1926 he was accepted for his doctoral studies by N.G. Chebotarev at Odessa University. He completed his studies in 1929.

An excellent and enthusiastic teacher, he attracted many students. In the thirties he created one of the strongest centers of functional analysis throughout the world at Odessa University. His interests included geometry of Banach spaces, moment problems, integral equations and matrices, spectral theory of linear operators, extension problems and applications. Many of his results from this period, as well as joint results together with his friends and colleagues (N.I. Achiezer, F.R. Gantmacher), and his outstanding students (A.B. Artemenko, M.S. Livsic, D.P. Milman, M.A. Naimark, V.P. Potapov, M.A. Rutman, V.L. Shmuljan) are now characterized as classical and appear in all textbooks on functional analysis.

During World War II, from 1941 to 1944, he held the chair of theoretical mechanics at the Kuibyshev (on the Volga) Industrial Institute. M.G. preferred the chair of theoretical mechanics in a technical institute of higher education to the chair of mathematics. He taught that the work is more interesting and it has more possibilities and responsibilities. In 1944 he returned to Odessa but was soon dismissed from Odessa University. This was the end of the famous center of functional analysis at Odessa University. The administration of Odessa University replaced M.G. Krein by more "reliable" mathematicians, such as N.A. Lednev, and N.I. Gavrilov. The former became famous for his Marxist critique of Einstein's Theory of Relativity, while the latter claimed publicly that he had resolved a number of the outstanding open problems in Mathematics, including the Riemann hypothesis. Each time elementary mistakes were found, and he tried to force the acceptance of his arguments under the pressure of the Communist Party administration. Further comments, I think are superfluous. From 1944 to 1952 M.G. Krein held a part time position as head of the department of functional analysis and algebra at the Mathematical Institute of the Ukrainian Academy of Science in Kiev. He was dismissed from this post in 1952. The official reason given was that he was not a permanent resident of Kiev. The real reason is easy to guess, it happend soon after the tragedy with the Jewish medical doctors. From 1944 to 1954 he held the chair of theoretical mechanics at the Odessa Marine Engineering Institute. For reasons that are still unclear he did not try to extend his contract with this institution, instead he moved to a less prestigious institute - Odessa Civil Engineering Institute. Here he held the Chair of Theoretical Mechanics till his retirement. During the last few years of his life he was a consultant to the Institute of Physical Chemistry of the Ukrainian Academy of Sciences in Odessa.

A list of themes where M.G. Krein's research was fundamental, and in many cases even determined the future of the field, includes: oscillating (totally positive) kernel functions and matrices; problem of moments, orthogonal polynomials, and approximation theory; cones and regular convex sets in Banach spaces; the theory of gaps and spaces with two norms; the extension theory of operators, Hermitian-positive functions and helical arcs; integral operators, string problems and method of directing functionals; stability theories for differential equations; Wiener-Hopf, Toeplitz and singular integral operators, scattering theory and inverse spectral problems; operator theory in spaces with an indefinite metric, indefinite extension problems; non-selfadjoint operators; triangular models; perturbation interpolation and factorization theories; problems in elasticity theory, and ship waves and water resistance.

A profound intrinsic unity and a close interlacing of general abstract and geometric ideas with concrete and analytical results and applications are characteristic of Krein's work.

Krein was a very fine pedagogue and lecturer. He would always share his new ideas and plans with his students and colleagues. He was known for his scientific generosity and enthusiasm, as well as his kindness and attention to young mathematicians. The author of these lines was very privileged to have, during many years, such a teacher, coauthor and friend. He will always remember M.G. Krein with gratitude, affection and admiration.

One of the most eminent mathematicians of our time, Mark Grigorievich Krein, is the author of more than 270 papers and monographs of unsurpassed breadth and quality. His work opened up new areas of mathematics and greatly enriched the more traditional ones. He educated dozens of brilliant students in the USSR and inspired the work of many mathematicians, engineers and physicists all over the world.

In 1982 M.G. Krein was awarded the prestigious international Wolf Prize in Mathematics in Jerusalem. The citation to this prize reads in part as follows: "His work is the culmination of the noble line of research begun by Chebyshev, Stieltjes, S. Bernstein and Markov and continued by F. Riesz, Banach and Szego. Krein brought the full force of mathematical analysis to bear on problems of function theory, operator theory, probability and mathematical physics. His contributions led to important developments in the applications of mathematics to different fields ranging from theoretical mechanics to electrical engineering. His style in mathematics and his personal leadership and integrity have set standards of excellence."

Among his honorary awards, he was elected corresponding member of the Ukrainian Academy, 1939; honorary member of the American Academy of Arts and Sciences, 1968; Foreign Member of the National Academy of Sciences of the United States of America, 1979. He was also awarded the N.M. Krylov Prize of the Ukrainian Academy of Sciences, 1988.

In general, M.G. Krein was a fair, very amiable and kind person. However, all of his life he battled against mediocrity. After the Second World War he had to contend with hostile elements which fought fiercely against him using the officially supported antisemitism which was rife in the Ukraine, and especially so in Odessa. He was accused of Jewish nationalism, presumably for having had too many Jewish students before the War. This accusation was certainly included in his classified file and was held against him all of his life. Presumably, it played a significant role in his two dismissals which were mentioned earlier. He was not allowed to have Jewish students and was deprived of a university base. All attempts on the part of various societies, academies and individuals in the Soviet Union to gain for him some measure of the official recognition which he so richly deserved, were unsuccessful. M.G. Krein was never elected as a full member of Ukrainian Academy of Sciences. Worse than that, there were times when his friends feared that he was in serious danger of arrest. In 1948 in an article in an Odessa local newspaper M.G. Krein was named "rootless cosmopolitan". He was accused of quoting too much foreign mathematicians and following too much their ideas and ignoring the achievements of Russian and Soviet mathematicians. That was considered by the officials as a crime. Usually this was the way that started many campaigns against Jewish scientists, writers and cultural activists that led often to arrests. M.G. was lucky he escaped without an arrest. The campaign run out without harming him essentially.

M.G. Krein responded to his hostile surroundings in the only way open to him, by deep research and hard work. He and many of his students were protected by virtue of his outstanding achievements. In retrospect, it seems clear that he won this very difficult struggle. Firstly, he was able to devote all his life to mathematics (teaching and research), the work he loved so much. Secondly, he was able to spend most of his life in Odessa, a town which he had always regarded with love and affection (some of his friends thought that his life would have been much easier in Moscow or Leningrad). Thirdly, he was always the leader of a strong and dedicated group of colleagues and followers who loved and respected him. (This group existed almost on a private basis, holding many of its meetings in his house, or at the Scientists Club.) Fourthly, he had a great impact on the development of mathematics and its applications throughout the world. Even though he was never allowed to travel abroad, his brilliant work knew no borders.

This fight took a heavy toll on his health, and towards the end of his life he suffered from depression. This condition worsened after the tragic loss within one year of his wife, Rachel, and his only grandson, Aleosha. On October 17, 1989 M.G. Krein died in Odessa (USSR). There he is buried.

M.G. Krein had only one child. His daughter Irma Krein (Kozdoba) has a Ph.D. in Philology and is an expert in Cybernetics. M.G. has also a great-grandson which is also called Mark.

The Raymond and Beverly Sakler
Faculty of Exact Sciences
School of Mathematical Sciences
Tel-Aviv University, Israel

Operator Theory:
Advances and Applications, Vol. 117
© 2000 Birkhäuser Verlag Basel/Switzerland

The Seminar on Ship Hydrodynamics, Organized by Professor M.G. Krein

V.G. Sizov

This paper was intended as an attempt to elucidate the remarkable contribution of M.G. Krein to ship hydrodynamics and study of wave processes in water.

Professor M.G. Krein first turned to the problems on ship hydrodynamics when one of the ship theory teachers at Odessa Institute of Marine Engineers asked him to explain some points of the renowned paper on wave resistance by J.H. Michell [1]. Published back in 1898, it had remained in many aspects interesting, and M.G. Krein was immediately thrilled by it. After the scrutinizing study of [1] he decided to explain its major ideas to a small group of the Institute's staff who were interested in wave problems. As a result, M.G. Krein's seminar on ship hydrodynamics came into existance in 1952. Unfortunately, due to the circumstances existing at that time, this seminar was not long lasting and only 15 sessions were held. Nevertheless, its consequences were remarkable, having stimulated a great interest to the wave propagation in fluids. The subsequent investigations on the posed by it problems were continued by its participants for many years. Their results were reported at numerous conferences and issued in various scientific publications.

In 1980 the annual USSR conference on ship theory (XXIXth Kryloff readings) was held for the first time not in Leningrad, but in Odessa. A summary report "Principal directions and results of investigations of the wave problems in Odessa" was included into its program. M.G. Krein participated both in the conference plenary sessions and sectional meetings.

Unfortunately, the author of this paper is the sole participant of M.G. Krein's seminar alive.

In my report, based on memoirs and notes, I would like to describe the principal problems considered by M.G. Krein during the seminar.

One of such problems dealt with the formulation of the boundary conditions at infinity for the velocity potential induced by a vessel sailing on the surface of fluid.

In paper [1] J.H. Michell pays attention to the fact that the boundary value problem, which he poses, has non-unique solution. The matter is that free waves can be always imposed onto any derived solution. It is stipulated in [1] that the additionally imposed waves are to eliminate any waves at infinite distance in front of the vessel. After this the final expression for the potential is only presented, with no indication where it came from. J.H. Michell pays no attention to the behavior of the potential far astern of the vessel. Thus the problem of removing the existing indefiniteness is not given a clear treatment in [1].

N.Ye. Kochin [2] considering the wave resistance and lifting force on bodies immersed into water as a boundary value for the velocity potential requires that the induced velocities tend to zero at infinity streight ahead of the body while be only limited far away from it in other directions.

As M.G. Krein pointed out that the induced motion occurring in a water basin of any depth and with no side bounds tends to zero both streigt ahead and astern of the vessel, differing only in the order of magnitude of the velocity decrease. This problem is rather thoroughly discussed in the monograph by A.A. Kostyukov [5], who was a permanent participant of the Seminar. The problem is complicated even more by the fact, established by Peters [4], that there exist directions in which the velocities decrease according to another laws.

M.G. Krein suggested that the actual formulation of the conditions at infinity probably consists in the following: *streight ahead of the vessel the velocities tend to zero with the maximum order in magnitude, while astern - as slow as possible.* This statement has not been proved yet.

The Raleigh method for eliminating of free waves based on introducing of fictitious dissipative forces, reaches, of course, its aim since the boundary condition on the free surface for solutions with the additional component cannot be satisfied by any harmonic function limited in the entire half-space. However, this method cannot be deemed as based on some physical principles. Should it be considered merely as a mathematical trick, it is necessary to show that other methods of the same type will lead to the same result, but nobody has shown this up to now.

Thus it becomes necessary to realize the following fact: though the problem considered in [1] is very old, there exist hundreds of works dealing with it and there is no doubt in reliability of the obtained solution - the exact formulation of this problem is still absent.

Another question which was also considered at the Seminar refers to the method of deriving of the Green function for the boundary value problem for the velocity potential of wave motion in fluids.

Green functions obtained by different methods have different expressions, but are, of course, identical. This identity can be proved shown by direct transformations, but these are sometimes not easy to perform. Depending on the nature of the problem under investigation or on the purpose of numerical calculations, one can choose the most adequate representation of this function.

J.H. Michell's method, which was discussed at the Seminar, is based on the use of the generalized Fourier transformation leading to the aim, as compared to other methods, most quickly.

Let us consider one of the boundary value problems which is typical for wave motions of fluids, for example, the problem for the potential $g_1(x, y, z)$ of the source of unit intensity placed in the uniform flow at point (ξ, θ, ζ). Let $g(p, y, z)$

be the Fourier transform of $g_1(x, y, z)$ with respect to the x-coordinate:

$$g(p, y, z) = \int_{-\infty}^{\infty} g_1(x, y, z) e^{ipx} dx \tag{1}$$

Then for g we obtain the following boundary value problem:

$$Ł_g = g_{zz} + g_{yy} - p^2 g = 0 \tag{2}$$

$$\begin{aligned} g_z - kg &= 0 \quad \text{at } z = 0 \\ g_y &= \frac{1}{2} e^{ip\xi} \delta(z - \varsigma) \quad \text{at } y = 0 \end{aligned} \tag{3}$$

The method of solution consists in the presentation of the sought-for function in the form of the Fourier transform by the following system of functions:

$$\psi(z, \lambda) = \cos \lambda z + \frac{k}{\lambda} \sin \lambda z, \tag{4}$$

which satisfy the equations:

$$\psi_{zz} + \lambda^2 \psi = 0 \tag{5}$$

and the same boundary condition 3 as the sought-for function g does. These functions form on the interval $(0, \infty)$ an orthogonal system with the weight

$$\frac{1}{1 + \frac{k^2}{\lambda^2}}$$

That is why it is possible to write:

$$G(p, y, \lambda) = \int_{-\infty}^{\infty} g(p, y, z) \psi(z, \lambda) dz \tag{6}$$

If $k \geq 0$, then the system of functions ψ is complete. If $k < 0$, it is no longer complete, but the addition of the function e^{kz} makes it complete. In the case of wave problems we always have $k < 0$.

That is why the inversion formula takes the following form:

$$g(p, y, z) = \frac{2}{\pi} \int_0^{\infty} G(p, y, \lambda) \psi(z, \lambda) \frac{d\lambda}{1 + \frac{k^2}{\lambda^2}} + \Gamma(p, y) e^{kz} \tag{7}$$

Here functions G and Γ are defined by the substitution of this expression into the equation $L(g) = 0$ with the use of the condition at $y = 0$.

In general this method is applicable to a more general form of equation, namely, having the form:

$$L(g) = g_{zz} + L_1(g) = 0 \tag{8}$$

where L_1- an operator containing partial derivatives only with respect of x and y, but the boundary condition at $z = 0$ should be the same as the one considered above.

Using this method, A.A. Kostyukov found the potential of the source moving under the surface of a deep fluid, and proved its identity with the expression derived by N.Ye. Kochin in another way.

The form of the potential derived by A.A. Kostyukov corresponds to that sequent from J.H. Michell's velocity potential for the ship. It is worth mentioning that J.H. Michell himself did not obtain the explicit expression for the potential of a point source, and only did B.Ya. Levin in 1949 show that J.H. Michell stayed short of the derivation of the potential of a point source from his ship velocity potential.

Using the same method, we were able to obtain the potential of a moving pulsating source and that of a stationary pulsating source and show that the former was identical to the expression containing contour integral and which was derived by L.N. Sretensky and the latter - to the expression obtained by N.Ye. Kochin.

Speaking of the method just described, the following may be of interest.

Analysing an analogous boundary value problem, A.C. Titchmarsh mistakenly considered the above system of functions as a complete one for all values of k. This mistake was repeated by other authors. Later Titchmarsh published a paper correcting the initial results. At the same time, J.H. Michell, long before the work by G. Weil on the generalized Fourier integral, found the correct formula for the generalized Fourier transform by using the limiting transition and obtained the correct expression for the ship velocity potential. For that reason M.G. Krein considered it natural to call the transform used by the J.H. Michell as the Fourier-Michell transform. So did we in our works.

The third question considered at the Seminar dealt with the general formulas for calculating the net force and net torque of wave nature exerted by the fluid on a moving body. Based on the impulse principle, these formulas are derived by considering the vectors-vector relations which are invariant with respect to the closed surface embracing the body surface with singularities, distributed over it, and have the following form:

$$\vec{P} = -\rho \int_S q \cdot grad\varphi_0 dS$$

$$\vec{M} = -\rho \int_S q \cdot (\vec{r} \times grad\varphi_0) dS$$

where q is the source intensity on the surface S, $\vec{r}$-the radius-vector, φ_0 - the part of the velocity potential, which is regular in the entire fluid space including the body volume.

From these formulas, as a particular case, one can obtain the well-known formulas derived by Lagally for the forces acting on isolated singularities.

A.A. Kostyukov gives a detailed derivation of the formulas in [5] Kostyukov. Following the same way, we have obtained analogous formulas for the net forces acting on a body in the non-stationary periodic flow.

Several sessions of the Seminar were dedicated to the problem of the form of a ship of the least wave resistance.

This problem has its own history. Dozens of works have been dedicated to it, including numerous reports at symposiums on the theory of wave resistance, but it still contains a great deal of ambiguous points. Moreover, some of those works contain wrong results and statements.

It looks like this problem was of great interest to M.G. Krein. After the Seminar ceased to exist in 1952 and untill the early 80-ties, he turned to it several times imparting a more complete form to the obtained results.

M.G. Krein formulates the problem as follows. Let the volume displacement of the ship V, the immersed part of the center plane D (projection of the wetted surface of the ship onto its longitudinal vertical plane), and the velocity $\vec{v}$ be known. To find the form of the ship, i.e. a non-negative in the domain D function $y = f(x, z)$ which vanishes in the under-water part of the boundary contour of D, and for which the Michell resistance, i.e. wave resistance, given by J.H. Michell's integral

$$R_w = (Kf, f) = \int_D \int_D k(x - \xi, z + \varsigma) f(x, z) f(\xi, \varsigma) dx dz d\xi d\varsigma \tag{9}$$

with

$$k(x, z) = \frac{4\rho g^4}{\pi v^6} \int_1^\infty e^{-\frac{gz}{v^2}\lambda^2} \cos \frac{gx}{v^2}\lambda \frac{\lambda^4 d\lambda}{\sqrt[2]{\lambda^2 - 1}} \tag{10}$$

is minimal.

M.G. Krein determines the following properties of the functional 9.

1. If the ship is of finite sizes, then its Michell resistance has a positive lower bound which is reachable only in the class of singular distributions.
2. The requirement that the sizes of the ship be finite is essential. If this requirement is omitted, it is possible to point out non-negative functions turning the Michell integral into zero. For instance, if the domain D is an unbounded strip running in the both directions, they are:

$$y_1 = \pm \frac{2}{\pi a x^2} \left(1 - \frac{\sin ax}{ax}\right) \psi(z) \tag{11}$$

$$y_2 = \pm \frac{2}{\pi} \frac{\sin^2 \frac{ax}{2}}{ax^2} \psi(z) \tag{12}$$

where $\psi(z)$ is a function determining the cross-sections form.

The former of these functions corresponds to an infinite ship which is narrowing when getting away from the middle section. The latter corresponds to the caravan of ship following close one to another and also narrowing infinitely.

3. The requirement of non-negativity of the function $f(x, z)$, which follows from the nature of the problem, is also substantial. If this requirement is omitted, it is possible to construct a whole class of functions turning the Michell integral into zero, even in the case when the centreplane area is restricted.

The case of a bi-similar ship whose surface has the form

$$y = \varphi(x)\psi(z) \tag{13}$$

is investigated more rigorously. In this case the domain D represents a rectangle of length $L = 2l$ and height T. All the cross-sections are similar, and this is also valid for the water-planes. Both an above- and under-water ships are considered, thus the domain D being

$$-l \leq x \leq l, \quad H \leq z \leq H + T \tag{14}$$

where H designates the depth of immersion of the upper edge of the domain D. For the above-water ship, $H = 0$.

For the bi-similar ship the Michell resistance is written as

$$R_w = \frac{4\rho g^4}{\pi v^6} \int_{-l}^{l} \int_{-l}^{l} K_\psi(x - \xi)\varphi(x)\varphi(\xi)dxd\xi \tag{15}$$

where the following notations are used:

$$K_\psi(x - \xi) = \int_1^\infty \Psi^2(\lambda) \cos \frac{gx}{v^2}\lambda \frac{\lambda^4 d\lambda}{\sqrt[2]{\lambda^2 - 1}} \tag{16}$$

$$\Psi(\lambda) = \int_H^{H+T} \psi(z) e^{-\frac{gz}{v^2}\lambda^2} dz \tag{17}$$

The cross-sectional forms $\psi(z)$ and displacement V are deemed given, this being equivalent to the setting of the water-plane area. The function $\varphi(x)$ is sought for which minimizes the functional R_w.

An analogous problem was considered by L.N. Sretensky in 1935 [9] Sretensky and by G.Ye. Pavlenko in 1937 [10]. The latter came to the conclusion that for a straight-sided ship of unrestricted draught the form of its water-line can not be represented by functions with squared-integrable derivative. This statement is correct, even though the work contains a mistake.

G.Ye. Pavlenko, using methods of calculus of variations for the function $\varphi(x)$, minimizing the Michell resistance, obtained an integral equation:

$$\int_{-l}^{l} \varphi(\xi) Y_0 \left| \frac{g}{v^2}(x-\xi) \right| d\xi = const \tag{18}$$

where $Y_0 |x|$ is the Neumann function.

Without investigation of this equation, he replaced it with an algebraic system of equations and obtained smooth profiles of the waterlines, which was in contradiction with the results by L.N. Sretensky. It shall be shown that this does not correspond to the actual solution.

Obviously, this contradiction initiated M.G. Krein's interest to the problem. In [5], after reviewing the above results, he noted: "Actual work must be dedicated to finding the reason of the contradiction in investigations of L.N. Sretensky and G.Ye. Pavlenko and further, more deep analysis of the entire problem in general".

When analyzing the problem, M.G. Krein paid attention to the relevance of the requirements of non-negativity of the sought-for function $\varphi(x)$. This requirement does not allow using of the standard methods of calculus of variations, as it restricts possible variations of the desired function. For this reason, usual methods need to be corrected.

This correction, for meeting the requirement of non-negativity, leads to the fact, that sought-for function is uniquely defined from the integral equality - inequality of the form:

$$\int_{-l}^{l} k_{\psi}(x-\xi)\varphi(\xi) d\xi >= \aleph \tag{19}$$

at the same time, value of the constant $\aleph$ is also defined, which with accuracy up to the known multiplier gives searched minimum of R_w.

The written relation must be interpreted in the following way: the equality sign takes place at any point x, not contained within any interval of zeros of the function φ; at other points either the $>$ or $=$ sign takes place.

The use of the mentioned relation represents the specificity of the method of solution of the considered non-standard variational problem.

For the underwater ship ($H > 0$) this relation leads to the solution, representing linear combination of the Dirac delta functions:

$$\varphi_{\min}(x) = \sum_{k=1}^{N} a_k \left[\delta(x-\xi_k) + \delta(x+\xi_k)\right] \tag{20}$$

$$0 \leq \xi_1 \leq \xi_2 \leq \cdots \leq \xi_N \leq l \tag{21}$$

with this, provided H/L is sufficiently large, extremal function consists only of two branches:

$$\varphi_{\min}(x) = \emptyset L \left[\delta(x-\xi_k) + \delta(x+\xi_k)\right] \tag{22}$$

where $\emptyset$ - prismatic coefficient of the ship.

It follows from the above that the lower bound of Michell wave resistance for the deeply immersed underwater ship

$$R_w = \frac{4\rho g^2}{\pi v^6} \emptyset^2 L^2 k_\psi \left(\frac{L}{2}\right) \tag{23}$$

For the above-water ship $H = 0$, $\psi(0) > 0$, and $\varphi_{\min}(x)$ appears to be an ordinary function, which at large Froude numbers is positive, that's why integral equality-inequality for it converts into integral equation:

$$\int_{-l}^{l} k_\psi(x - \xi)\varphi(\xi)d\xi >= \aleph \tag{24}$$

with the kernel of the type:

$$k_\psi(x - \xi) = \frac{v^4}{g^2}\Psi^2(0)\left[\ln\frac{L}{|x|} + H(x)\right] \tag{25}$$

where $H(x)$ is continuously differentiable.

This leads to the following representation of the sought-for function:

$$\varphi_{\min}(x) = \frac{M}{\sqrt[2]{l^2 - x^2}} + \chi(x) \tag{26}$$

where $\chi(x)$ - continuous function, and M - constant, depending on velocity and, as was determine by L.A. Sakhnovich [11], constant M does not turn into zero at any speed.

Now it is clear not only the statement of L.N. Sretensky, that form of the minimizing waterline is not expressed by the function with integrable squared derivative, but also more strong statement is valid, that the squared minimizing function itself is not integrable.

At sufficiently low Froude numbers solution of the integral equation may take negative values. This means, that instead of this equation solution should be searched for from the above-mentioned equality-inequality. Physically this means that the least Michell resistance is reached at realization of the given displacement in the form of two or several ships, going one in a wake of another at some distances. These distances are defined from the equality-inequality simultaneously with the water-plane profile of the separate ships. With the aid of this very work of L.A. Sakhnovich [11] one can show, that irregularities of the above-mentioned nature will take place only onto the external extremities of the extremal ships of the caravan, while middle ships will have water-plane profiles, expressed by continuous functions.

Thus, the problem of defining of the form of the ship of least Michell resistance leads to degenerated forms, expressed either by generalized functions, or by functions will irregularities, at which Michell integral does no longer express wave resistance, as assumptions, put into the base of its derivation, are no longer observed.

Discussing results obtained at the Seminar I told, that in my work, concerned to the successive improvement of the form of the ship in order to reduce its resistance, I considered summarized resistance consisting of Michell's and frictional ones and also the sucking force generated by the propeller. M.G. Krein noted that this changes statement of the problem and leads to elimination of the irregularity in its solution.

Consideration of such problem is reported in our joint works [7] and [8] carried out already upon completion of the Seminar work. However they contain development of investigations started during of the Seminar activity.

When solving the stated problem, it was assumed, that frictional resistance is proportional to the wetted surface of the ship:

$$R_f = 2c\int_D \sqrt{1+f_x^2+f_x^2}dxdz \tag{27}$$

Within the accuracy limits, to which Michell integral represents wave resistance, we can assume:

$$\begin{aligned} R_f &= 2cD + c\int_D (f_x^2+f_x^2)dxdz = 2xD - c \\ &\int_D \Delta f\cdot f dxdz - c\int_{\Gamma_0} f\cdot f_\Pi dxdz \end{aligned} \tag{28}$$

where Γ_0 is a part of boundary of domain D, which lies on the free surface.

As soon as domain D is given, we can minimize the following resistance:

$$\overline{R} = R_f + R_w - 2cD \tag{29}$$

for which the following expression can be obtained:

$$\begin{aligned} \overline{R} &= c\int_D (f_x^2+f_x^2)dxdz + \int_D\int_D k(x-\xi, z+\varsigma) \\ &f(x,z)f(\xi,\varsigma)dxdzd\xi d\varsigma \end{aligned} \tag{30}$$

Should the requirement of non-negativity of the minimizing function be temporarily omitted, then known rules of determining of minimum of squared functional, meeting additional linear condition, which expresses given displacement, defines the sougt-for function:

$$f_{\min} = \frac{V}{V_1} f_1 \tag{31}$$

where f_1 is defined from the following integral-differential boundary value problem:

$$-c\Delta f_1 + K f_1 = \gamma u \tag{32}$$

$$f_1|_{\Gamma_+} = 0 \tag{33}$$

$$f_{1_z}|_{\Gamma_0} = 0 \tag{34}$$

where K is the integral operator generated by the kernel k, $u(x, z) \equiv 1$, γ - specific weight of the fluid, which is introduced for the purpose of dimensionalising of f_1, making it equal to the length dimension; V_1 - displacement, corresponding to the form f_1, i.e.

$$V_1 = 2(f_1, u) \tag{35}$$

The value of minimum resistance itself can be found by formula:

$$\overline{R}_{\min} = \frac{\gamma}{2} \frac{V^2}{V_1} \tag{36}$$

Let us denote as $g(x, y, z, \xi, \zeta)$ the Green function of the operator - Δ, meeting the same boundary conditions on the contour D. With its aid the integral-differential equation transforms into the integral equation:

$$f_1 + \frac{1}{c} G K f_1 = \frac{\gamma}{c} G u \tag{37}$$

where G is the integral operator with the kernel g.

This equation is linear inhomogeneous integral equation of the second kind with regular kernel. One can show, that the corresponding homogeneous equation has unique zero solution. That's why considered equation has the unique solution, which is continuous. Should this solution appear non-negative, this will be the unique solution of the stated problem.

In the case, when this solution takes also negative values, then the situation becomes more complex, than in the case of problem on minimum of Michell resistance only, and an extra investigation is needed for clarification of this question.

¿From the analytic side the problem simplifies, if one consider the ship of bi-similar form and assume that the cross-sectional forms are given with the accuracy up to similarity, and waterplane area is given instead of displacement.

Then provided that condition of non-negativity is temporarily omitted as above the function defining water-plane form is determined from the following integral-differential boundary value problem:

$$-A\frac{d^2\varphi}{dx^2} + B\varphi + \frac{4\rho g^3}{\pi v^6 c} \int_{-l}^{l} k_\psi(x - \xi)\varphi(\xi) d\xi = C \tag{38}$$

$$\varphi(-l) = \varphi(l) = 0 \tag{39}$$

where $A = \int_H^{H+T} \psi^2(z)dz$, $B = \int_H^{H+T} [\frac{d\psi(z)}{dz}]^2 dz$ and constant C is determined from the condition of given water-plane area.

Using the same technique as shown above this boundary value problem can be transformed into the Fredholm linear integral equation of the second kind:

$$\varphi + \lambda G_1 K_\psi \varphi = C G_1 u \tag{40}$$

where

$$\lambda = \frac{4\rho g^3}{\pi v^6 c}$$

and K_ψ and G_1 are integral operators, produced correspondingly by the kernels $k_\psi(x-\xi)$ and $g_1(x,\xi)$, whilst $g_1(x,\xi)$ is the Green function of the differential operator $-A\frac{d^2}{dx^2} + B$ corresponding to the boundary conditions

$$\varphi(-l) = \varphi(l) = 0.$$

Unlike the Green function g, function g_1 is calculated elementary.

Should the solution of this integral equation be non-negative, this will be the solution, defining the water-plane form. If the solution is sign alternating this mean that the circumstance mentioned above comes into effect.

Regretfully, all reported results have not been published in the explicated form. That's why authors of numerous subsequent works published in Germany, Japan and USA, partially repeating initial considerations, failed to obtain sufficiently complete correct results, as they did not notice that they faced the non-standard variational problems.

References

[1] J.H. Michell, *The wave resistance of a ship*, Phil. Mag. **5**, no. 45 (1898), pp. 106–123.

[2] N.Ye. Kochin, On wave resistance and lifting force, acting on the bodies, submerged into the fluid (In Russian). Proceedings of the Conference on the wave-making resistance theory. M., 1937, pp. 65–134.

[3] A.A. Kostyukov, Theory of ship waves and wave-making resistance (In Russian). L., Sudpromgiz Publishers, 1959, p. 311.

[4] A.A. Peters, A new treatment of the ship wave problems. Communications on pure and applied mathematics **2** (1949) pp. 123–148.

[5] M.G. Krein, On definition of the form of the ship of given displacement, which corresponds to the feasible least wave-making resistance of a fluid to propagation of the ship with prescribed velocity (In Russian), unpublished, 1952, typed, p. 52.

[6] M.G. Krein, On the form of a ship with minimum of Michell resistance (In Russian), unpublished, 1960, typed, p. 9 (Paper, submitted to the All-Union Congress on theoretical and applied mechanics. Summaries of the papers of the AS of the USSR, M., 1960).

[7] M.G. Krein and V.G. Sizov, On the form of a ship of minimum total resistance (In Russian), unpublished, 1960, typed, p. 5 (Ibid., M., 1960).

[8] M.G. Krein and V.G. Sizov, On the non-standardness of variational problems of the optimum ship hull form (In Russian). Proceedings of the jubilee session of Bulgarian Institute of Ship Hydrodynamics, Varna, 1981, pp. 77–1, 77–4.

[9] L.N. Sretensky, On the problem of minimum in ship theory (In Russian), PAS of the USSR, 1935, vol. **3**, no. 6, pp. 247–248.

[10] G.Ye. Pavlenko, Ship of least resistance (In Russian). Proceedings of VNITOSS, vol. **11**, Issue 3, 1937, no. 6, pp. 28–62.

[11] L.A. Sakhnovich, Equations with difference kernel on finite interval (In Russian). Successes of mathematical sciences, vol. **35**, issue 4, 1980, pp. 69–129.

Victor Sizov
Odessa State Maritime Academy
ul. Didrikhsona 8
27000 Odessa
Ukraine

Operator Theory:
Advances and Applications, Vol. 117
© 2000 Birkhäuser Verlag Basel/Switzerland

The Works of M.G. Krein on Eigenfunction Expansion for Selfadjoint Operators and their Applications and Development

Yu.M. Berezansky

In this survey, we describe the main idea of belonging to M.G. Krein method of directing functionals and its applications to the spectral theory of ordinary differential selfadjoint operators and to the problem of integral representation of positive definite functions and their generalizations. The second part of the survey is devoted to broad generalizations of this theory.

1 Introduction

In the immense scientific heritage of M.G. Krein there is a series of articles devoted to the construction of the eigenvector expansion for selfadjoint operators. This series is not very large (it consists of the articles [41, 42, 44, 46]) but is very important: for the first time, there was raised the question about eigenvector expansion for general selfadjoint operators with an arbitrary spectrum. M.G. Krein has developed, in these articles, a method of directing functionals, which gives an answer to the question for a large class of such operators, including operators generated by ordinary differential expressions on unbounded intervals, when the spectrum can be fairly arbitrary. Note, that the method of directing functionals was constructed by M.G. Krein under the influence of the article of M.S. Livshic [50] about applications of the operator theory to the moment problem.

In this survey, we describe the main idea of the method of directing functionals and belonging to M.G. Krein applications of the method to the spectral theory of ordinary differential selfadjoint operators and to the problem of integral representation of positive definite functions and their generalizations.

Of course, it is necessary to note that this series of M.G. Krein's articles is connected with many other investigations he has been carrying, but we cannot discuss all of these works here.

The second part of the survey is devoted to broad generalizations of this theory. Note that the idea of differentiation of the resolution of identity of the operator under consideration to get an eigenvector expansion is first found in works of M.G. Krein. Later, this idea was successfully applied in more general situations.

It is impossible for technical reasons to give a detailed reference to a very large number of articles connected with this topic. We often only mention the name of the author and year of the beginning of his work in this direction, the precise reference can be found in the cited books.

2 The General Point of View on the Eigenvector Expansion for a Selfadjoint Operator

Let H be a Hilbert space and A be a selfadjoint operator acting on this space and having discrete simple spectrum $\mathrm{Sp}\,(A) = (\lambda_n)_{n=1}^{\infty}$. Denote by $P(\lambda_n)$ the projector onto the eigensubspace corresponding to the eigenvalue λ_n. Then it follows from the spectral theorem that

$$H = \bigoplus_{n=1}^{\infty} P(\lambda_n)H \quad \text{or} \quad 1 = \sum_{n=1}^{\infty} P(\lambda_n). \tag{2.1}$$

The simplicity of λ_n means that $\dim (P(\lambda_n)H) = 1$. Let $\varphi(\lambda_n) \in P(\lambda_n)H$, $\varphi(\lambda_n) \neq 0$, be fixed. Of course, $A\varphi(\lambda_n) = \lambda_n\varphi(\lambda_n)$. Denote by $\hat{f}(\lambda)$, $\lambda \in \mathbb{C}^1$, the "Fourier transform" (connected with expansion (2.1)) of the vector $f \in H$ defined on the spectrum $\mathrm{Sp}\,(A)$ by the formula

$$H \ni f \longmapsto \hat{f}(\lambda_n) = (f, \varphi(\lambda_n))_H \in \mathbb{C}^1, \quad \lambda_n \in \mathrm{Sp}\,(A), \tag{2.2}$$

and arbitrarily for $\lambda \in \mathbb{R}^1 \backslash \mathrm{Sp}\,(A)$.

Then equalities (2.1) are equivalent to the Parseval equality: $\forall f, g \in H$

$$(f, g)_H = \sum_{n=1}^{\infty} \hat{f}(\lambda_n)\overline{\hat{g}(\lambda_n)}\rho_n = (\hat{f}, \hat{g})_{L^2(\mathbb{R}^1, d\rho(\lambda))}. \tag{2.3}$$

Here $\rho_n = \|\varphi(\lambda_n)\|_H^{-1}$ is the normalization coefficient, and the space L^2 is constructed using the discrete measure $d\rho(\lambda) : \rho(\{\lambda_n\}) = \rho_n > 0$, $\rho(\mathbb{R}^1 \backslash \mathrm{Sp}\,(A)) = 0$.

It was not clear how to generalize these elementary facts for a general selfadjoint operator A with continuous spectrum. But examples of such type expansions (for classical Fourier transform, for Sturm-Liouville problem, etc) showed that the Parseval equality (2.3) must be preserved with some Fourier transform of type (2.2) and measure $d\rho(\lambda)$ connected with the operator A ("spectral measure").

The works [41, 42] of M.G. Krein about directing functionals have given in 1946 the first general answer to this problem.

3 The Method of Directing Functionals

We will give a short exposition of this method due to M.G. Krein for the case of one such functional (in [41, 42, 44] the case of $p < \infty$ functionals was investigated).

Let A be a Hermitian operator acting on a Hilbert space H. We suppose that for some linear dense in H set $D \subset \mathrm{Dom}\,(A)$, $AD \subset D$, there exists a map (directing functional)

$$D \times \mathbb{R}^1 \ni \{f, \lambda\} \longmapsto \Phi(f, \lambda) \in \mathbb{C}^1,$$

linear w.r.t. f and analytic w.r.t. λ, with following properties:

1) The equation $Ax - \lambda x = g$ $(x, g \in D, \lambda \in \mathbb{R}^1)$ has a solution iff $\Phi(g, \lambda) = 0$.
2) For every $\lambda \in \mathbb{R}^1$, there exists a vector $u_\lambda \in D$ such that $\Phi(u_\lambda, \lambda) \neq 0$.

From these assumptions, it is easy to conclude that

$$\forall f \in D, \quad \lambda \in \mathbb{R}^1, \quad \Phi(Af, \lambda) = \lambda\Phi(f, \lambda), \tag{3.1}$$

and, for an arbitrary finite interval (a, b) there exists a vector $u_{(a,b)} \in D$ such that

$$\Phi(u_{(a,b)}, \lambda) \neq 0, \quad \lambda \in (a, b). \tag{3.2}$$

Introduce the "Fourier transform" $\hat{f}(\cdot)$ by setting

$$H \supset D \ni f \longmapsto \hat{f}(\lambda) = \Phi(f, \lambda) \in \mathbb{C}^1. \tag{3.3}$$

It should be noted that in real examples, $\Phi(f, \lambda) = (f, \varphi(\lambda))_H$, where $\varphi(\lambda)$ is some "generalized" eigenvector of the operator A with the eigenvalue λ, therefore the transform (3.3) is similar to (2.2).

The following result takes place: *there exists at least one Borel measure $d\rho(\lambda)$ on the axis $\mathbb{R}^1$ (spectral measure) such that the Parseval equality*

$$(f, g)_H = \int_{\mathbb{R}^1} \hat{f}(\lambda)\overline{\hat{g}(\lambda)}\, d\rho(\lambda), \quad f, g \in D, \tag{3.4}$$

holds.

We will give a sketch of the proof of this theorem assuming, for simplicity, that the operator A is essentially selfadjoint and that there exists $u \in D$ with the property that $\Phi(u, \lambda) = 1$, $\lambda \in \mathbb{R}^1$, instead of assuming the existence of $u_{(a,b)}$ from (3.2).

Let $\mathcal{B}(\mathbb{R}^1) \ni \alpha \mapsto \mathrm{E}(\alpha)$ be the resolution of identity of the operator A, $f \in D$. To prove equality (3.4), it is sufficient to check that

$$\int_0^\lambda dE(s)f - \int_0^\lambda \hat{f}(s)\, dE(s)u = 0, \quad \lambda \in \mathbb{R}^1. \tag{3.5}$$

Indeed, using equality (3.5) and properties of spectral integrals we get $\forall f, g \in D$

$$\begin{aligned}(f, g)_H &= \left(\int_{\mathbb{R}^1} dE(s)f, \int_{\mathbb{R}^1} dE(t)g\right)_H \\ &= \left(\int_{\mathbb{R}^1} \hat{f}(s)\, dE(s)u, \int_{\mathbb{R}^1} \hat{g}(t)\, dE(t)u\right)_H \\ &= \int_{\mathbb{R}^1} \hat{f}(s)\overline{\hat{g}(s)}\, d\rho(s), \qquad d\rho(s) = d(E(s)u, u)_H.\end{aligned}$$

We will prove equality (3.5). Fix $f \in D$, and denote the left-hand side of (3.5) by $r(\lambda)$. It is necessary to prove that the vector-valued function $\mathbb{R}^1 \ni \lambda \mapsto r(\lambda) \in H$ is equal to zero. Because $r(0) = 0$, it is sufficient to check that the derivative $r'(\lambda) = 0$, $\lambda \in \mathbb{R}^1$, i.e. $\forall \lambda \in \mathbb{R}^1$

$$\|r(\lambda + h) - r(\lambda)\|_H = h\varepsilon(h), \tag{3.6}$$

where $\varepsilon(h) \to 0$ if $h \to 0$.

Assume, for example, that $h > 0$. Using expression (3.5) for $r(\lambda)$ we can write:

$$\begin{aligned}
\|r(\lambda + h) - r(\lambda)\|_H &= \left\| \int_{[\lambda,\lambda+h)} dE(s)f - \int_{[\lambda,\lambda+h)} \hat{f}(s)\, dE(s)u \right\|_H \\
&\le \left\| \int_{[\lambda,\lambda+h)} dE(s)f - \int_{[\lambda,\lambda+h)} \hat{f}(\lambda)\, dE(s)u \right\|_H \\
&\quad + \left\| \int_{[\lambda,\lambda+h)} (\hat{f}(\lambda) - \hat{f}(s))\, dE(s)u \right\|_H \\
&= \left\| \int_{[\lambda,\lambda+h)} dE(s)(f - \hat{f}(\lambda)u) \right\|_H \\
&\quad + \left\| \int_{(\lambda,\lambda+h)} (\hat{f}(\lambda) - \hat{f}(s))\, dE(s)u \right\|_H = I_1 + I_2.
\end{aligned} \tag{3.7}$$

We have, using the analyticity of $\hat{f}(s)$,

$$\begin{aligned}
I_2 &= \left\| \int_{(\lambda,\lambda+h)} (\hat{f}(\lambda) - \hat{f}(s))\, dE(s)u \right\|_H \\
&\le \left(\int_{(\lambda,\lambda+h)} |\hat{f}(\lambda) - \hat{f}(s)|^2\, d(E(s)u, u)_H \right)^{1/2} \\
&\le ch(E(\lambda, \lambda + h)u, u)_H =: h\varepsilon_2(h),
\end{aligned} \tag{3.8}$$

where $\varepsilon_2(h) \to 0$ if $h \to 0$.

Consider the first term in (3.7). We have $\Phi(f - \hat{f}(\lambda)u, \lambda) = \Phi(f, \lambda) - \hat{f}(\lambda)\Phi(u, \lambda) = 0$, therefore, by assumption 1), the equation $Ax - \lambda x = f - \hat{f}(\lambda)u$ has a solution $x \in D$. Using this fact, we get, analogously to (3.8), that

$$\begin{aligned}
I_1 &= \left\| \int_{[\lambda,\lambda+h)} dE(s)(f - \hat{f}(\lambda)u) \right\|_H =: \left\| \int_{[\lambda,\lambda+h)} (s - \lambda)\, dE(s)x \right\|_H \\
&= \left(\int_{(\lambda,\lambda+h)} |s - \lambda|^2\, d(E(s)x, x)_H \right)^{1/2} = h\varepsilon_1(h),
\end{aligned}$$

$\varepsilon_1(h) \to 0$ if $h \to 0$. As the result, we have (3.6) with $\varepsilon(h) = \varepsilon_1(h) + \varepsilon_2(h)$.

In the general case, it is necessary to take a certain selfadjoint extension of A (generally speaking, in a larger space). Instead of the condition $\Phi(u, \lambda) = 1$, $\lambda \in \mathbb{R}^1$, it is possible to use (3.2) and linearity of $\Phi(f, \lambda)$ w.r.t. $f \in D$. We stress on some additional properties: 1) the measure $d\rho(\lambda)$ is unique iff one of the defect numbers of the operator A is equal to zero; 2) the set of Fourier transforms $\{\hat{f}(\cdot) \mid f \in D\}$ is dense in $L^2(\mathbb{R}^1, d\rho(\lambda))$ iff A has selfadjoint extensions in H and $d\rho(\lambda)$ is generated by one of such extensions.

In the case of $p < \infty$ directing functionals, the proof sketched above is preserved. In this case, instead of the spectral measure $d\rho(\lambda)$, we have a positive $p \times p$ - matrix measure $d\rho_{j,k}(\lambda)$, $j, k = 1, \ldots, p$. The construction of this measure is, roughly speaking, as before: $d\rho_{j,k}(\lambda) = d(E(\lambda)u_k, u_j)_H$, where $u_1, \ldots, u_p \in D$ are similar to the previous u (or $u_{(a,b)}$).

We recall some works containing a direct generalization and investigation of the method of directing functionals. A.Ya. Povsner [59] has given another approach to the problem, proved the inversion formulas, and investigated the multiplicity of the spectrum of operators having p directing functionals (the multiplicity must be $\leq p$). Yu.M. Berezansky [6] generalized this method to the case of infinitely many directing functionals and has given an application of this generalization to partial difference equations. H. Langer [49] has given a generalization to the case where the values $\Phi(f, \lambda)$ of the functionals lie in some fixed Hilbert space K (instead of $\mathbb{C}^1$ or $\mathbb{C}^p$ as in M.G. Krein's works). A certain variant of the method of directing functionals, more connected with M.S. Livshic's work [50] and its applications to integral representation of Toeplitz and Hankel forms, was proposed by M. Cotlar and C. Sadosky [22].

An exposition of the method of directing functionals is contained in the book of N.I. Akhiezer and I.M. Glazman [3].

4 The Eigenfunction Expansion for General Ordinary Selfadjoint Differential Operators

As it was said in the introduction, the problem of eigenvector expansion was generated by H. Weyl's work [63] on eigenfunction expansion for a singular Sturm-Liouville operator. Therefore, it is natural that M.G. Krein's articles [41, 42, 44, 46] contain applications of this theory to the spectral theory of selfadjoint differential operators. We have already said that the multiplicity of the spectum must be not more than $p < \infty$. Hence it is possible to give applications of the eigenfunction expansion to ordinary differential operators, the corresponding finite systems of such operators, etc.

We consider here briefly the case of a selfadjoint linear differential expression $\mathcal{L}$ of even order $2m$ with smooth coefficients on an interval $(a, b) \subset \mathbb{R}^1$ and satisfying some selfadjoint boundary conditions at a, b, or ∞ ([46], actually in [41, 42, 44]). In this case the directing functionals are of the form

$$\Phi_j(f, \lambda) = \int_a^b f(x)\varphi_j(x, \lambda)\,dx \quad (\lambda \in \mathbb{R}^1), \tag{4.1}$$

where $f \in C^\infty((a,b))$ and are finite in (a,b), and $\varphi_j(x,\lambda)$, $j=1,\dots,p$, $p \le 2m$, are some p solutions of the differential equation $(\mathcal{L}\varphi)(x) = \lambda\varphi(x)$ satisfying some initial conditions connected with the boundary conditions.

For example, this can be seen for Sturm-Liouville expression $(2m=2)$ $p=2$, for the minimal operator on the whole axis $((a,b)=\mathbb{R}^1)$, and $p=1$ for $(a,b)=(0,\infty)$ with the boundary condition given at 0 (the classical H. Weyl's case). In this classical case,

$$(\mathcal{L}u)(x) = -u''(x) + q(x)u(x), \quad \operatorname{Im} q(x) = 0, \quad x \in (0,\infty); \qquad H = L^2((0,\infty),dx). \tag{4.2}$$

Boundary condition could be, for example, given by $u'(0)=0$; $D = \{f \in C^\infty([0,\infty))$, finite near 0 and $\infty\}$.

Denote by $\varphi(x,\lambda)$ a solution of the equation $(\mathcal{L}\varphi)(x) = \lambda\varphi(x)$, $x \in [0,\infty)$; $\varphi(0,\lambda)=1$, $\varphi'(0,\lambda)=0$. We have now one directing functional of the form

$$\Phi(f,\lambda) = \int_0^\infty f(x)\varphi(x,\lambda)\,dx = \hat{f}(\lambda), \quad f \in D, \quad \lambda \in \mathbb{R}^1 \tag{4.3}$$

(it is easy to prove that conditions 1), 2) imposed on the directing functional are fulfilled). Then the general result (3.4) implies the existence of the spectral measure $d\rho(\lambda)$ and that the operator

$$L^2((0,\infty),dx) \supset D \ni f \longmapsto \hat{f}(\lambda) \in L^2(\mathbb{R}^1, d\rho(\lambda)) \tag{4.4}$$

is (after its extension by continuity) a unitary operator between the spaces $L^2((0,\infty),dx)$ and $L^2(\mathbb{R}^1, d\rho(\lambda))$. In particular, we have the Parseval equality

$$(f,g)_{L^2((0,\infty),dx)} = \int_{\mathbb{R}^1} \hat{f}(\lambda)\overline{\hat{g}(\lambda)}\,d\rho(\lambda), \quad f,g \in D. \tag{4.5}$$

For the general case of the expression $\mathcal{L}$, the directing functionals have the form (4.1), and therefore the Fourier transform (4.4)–(4.3) becomes the p-dimensional one:

$$\begin{aligned} D \ni f \longmapsto \hat{f}(\lambda) &= (\hat{f}_1(\lambda),\dots,\hat{f}_p(\lambda)), \quad \hat{f}_j(\lambda) = \Phi_j(f,\lambda) \\ &= \int_a^b f(x)\varphi_j(x,\lambda)\,dx, \end{aligned}$$

where $\Phi_j(f,\lambda)$ are given by (4.1). The Parseval equality (4.4) now becomes the Parseval equality of the form:

$$\begin{aligned} (f,g)_{L^2((a,b),dx)} &= \sum_{j,k=1}^{p} \int_{\mathbb{R}^1} \hat{f}_k(\lambda)\overline{\hat{g}_j(\lambda)}\,d\rho_{j,k}(\lambda) \\ &= \int_{\mathbb{R}^1} d(\rho(\lambda)\hat{f}(\lambda), \hat{g}(\lambda))_{\mathbb{C}^p} \end{aligned} \tag{4.6}$$

with a matrix-valued measure $d\rho(\lambda)$.

Note that the analogous to (4.6) results for ordinary differential equations were obtained in 1950 by K. Kodaira [35], who used a specially developed approach.

In connection with representation (4.6), remark that in H. Langer's article [49] we have, instead of $d(\rho(\lambda)\hat{f}(\lambda), \hat{g}(\lambda))_{\mathbb{C}^p}$, the expression $d(\rho(\lambda)\hat{f}(\lambda), \hat{g}(\lambda))_K$, because now the values of $\hat{f}(\lambda) = \Phi(f, \lambda)$ belong to a Hilbert space K; $d\rho(\lambda)$ is now an operator-valued measure whose values are nonnegative operators in this space K.

The latter type of eigenfunction expansion is typical for Sturm-Liouville expression (4.2) with an operator-valued potential: $\forall x \in [0, \infty)$ $q(x)$ is a selfadjoint operator in the above mentioned space K. In this case, the operator A is acting in the L^2-type space of functions $[0, \infty) \ni x \mapsto u(x) \in K$. So, the Parseval equality in this case has the form of type (4.6) with $d(\rho(\lambda)\hat{f}(\lambda), \hat{g}(\lambda))_K$. Corresponding results belong to V.I. Gorbachuk and M.L. Gorbachuk [29] and F.S. Rofe-Beketov [61] (who used another approach).

5 Positive Definite Functions

The works [41–44, 47] of M.G. Krein contain a very fruitfull idea of applying the eigenfunction expansion to a representation of classical positive definite functions and their various generalizations. We will explain here these results in the simplest situation. More general results will be described in Sections 9, 10, where we will exploit the combination of this idea and a more general construction of the eigenfunction expansion.

Recall that a continuous function $(-2a, 2a) \ni x \mapsto k(x) \in \mathbb{C}^1$, where $0 < a \leq \infty$, is called positive definite if for an arbitrary finite sequence $(\xi)_{j=0}^{\infty}$ of complex numbers ξ_j and points $x^{(j)} \in (-a, a)$,

$$\sum_{j,l=0}^{\infty} k(x^{(j)} - x^{(l)})\xi_j\bar{\xi}_l \geq 0. \tag{5.1}$$

The classical theorem (Bochner for $a = \infty$, and M.G. Krein [37] for $a < \infty$) asserts that $k(\cdot)$ is positive definite iff there is a representation

$$k(x) = \int_{\mathbb{R}^1} e^{i\lambda x}\, d\sigma(\lambda), \quad x \in (-2a, 2a) \tag{5.2}$$

with some positive finite measure $d\sigma(\lambda)$.

A way to prove this result, which was suggested in M.G. Krein's articles [41, 42, 44], is the following. Construct a Hilbert space H_k of functions $(-a, a) \ni x \mapsto \varphi(x) \in \mathbb{C}^1$ with the scalar product

$$(f, g)_{H_k} = \int_{-a}^{a}\int_{-a}^{a} k(y - x) f(y)\overline{g(x)}\, dx\, dy \tag{5.3}$$

(the positivenes of (5.3) follows from (5.1), note that after the completion, the space H_k contains generalized functions f).

The operator

$$C_0^\infty((-a,a)) = D \ni f \longmapsto Af = -i\frac{df}{dx} \tag{5.4}$$

(the index 0 denotes the functions finite w.r.t. (a, b)), as readily seen, is Hermitian in H_k, and

$$\Phi(f,\lambda) = \int_{-a}^{a} f(x)e^{ix\lambda}\,dx = \hat{f}(\lambda), \quad f \in D, \quad \lambda \in \mathbb{R}^1, \tag{5.5}$$

is its directing functional. Using the main result of Section 3 we can write the Parseval equality (3.4): $\forall f, g \in D$

$$(f,g)_{H_k} = \int_{\mathbb{R}^1} \hat{f}(\lambda)\overline{\hat{g}(\lambda)}\,d\rho(\lambda). \tag{5.6}$$

Substituting expressions (5.3) and (5.5) into (5.6) and using the arbitrariness of $f, g \in D$ we conclude that representation (5.2) holds with $d\sigma(\lambda) = d\rho(\lambda)$.

In the Bochner's case ($a = \infty$), the operator A given by (5.4) is essentially selfadjoint and, therefore, the measure $d\sigma(\lambda)$ from (5.2) is defined uniquely. In the case $a < \infty$, the selfajointness of the closure $\tilde{A}$ depends on the behavior of k, every selfadjoint extension of $\tilde{A}$ (possibly, with an exit from H_k) gives some measure $d\sigma(\lambda)$. On the other hand, in each representation (5.2), we can take $x \in \mathbb{R}^1$ and, therefore, extend $k(x)$ to the whole $\mathbb{R}^1$ preserving the positive definiteness. As the result, a number of questions appeared, which were related to the conditions on selfadjointness of $\tilde{A}$, description of its selfadjoint extensions (in other words, description of all extensions of $k(x)$, $x \in (-2a, 2a)$, to the whole axis), etc. Corresponding results are contained in M.G. Krein's articles [37–40, 44] and in the book [9] of Yu.M. Berezansky. In this book for the first time, a theory of rigged spaces was proposed for a study of the problem of extending a positive definite function. Namely, it is convenient to understand the space H_k as a negative space in a certain chain of spaces. This gives a possibility to easily investigate properties of operator (5.4), acting in H_k, for example, to find criteria for its selfadjointness (i.e. the uniqueness of extension of $k(x)$), to get a description of all extensions in the case of nonuniqueness, etc. Such an approach with the use of riggings of spaces in the problem of extension was later repeated by I.V. Kovalishina and V.P. Potapov [36].

R. Bruzual [18, 20] investigated similar problems for generalized Toeplitz kernel, i.e. for a positive definite kernel $K(x, y)$, $x, y \in (-a, a)$, which is represented by four functions $k_{\alpha,\beta}(t)$ $(\alpha, \beta = 1, 2)$: for $x \in [0, a)$, $y \in [0, a)$ $K(x, y) = k_{1,1}(y - x)$, for $x \in [0, a)$, $y \in (-a, 0]$, $K(x, y) = k_{1,2}(y - x)$, etc. (instead of the representation $K(x, y) = k(y - x)$, $x, y \in (-a, a)$, in the classical case).

Note also the recent article of D.Z. Arov and H. Dym [4] devoted to the investigation of the connection between the above mentioned extension problem and two other M.G. Krein extension problems.

These investigations of M.G. Krein are deeply connected with his numerous works on the moment problem. In this survey, it is impossible to give an account of this direction, it needs a special article, and we only mention here his works [44, 45] on the matrix moment problem.

Recall that the classical moment problem is a problem about the possibility of representing (and a study of such a representation) a sequence $s = (s_n)_{n=0}^{\infty}$ of numbers, $s_n \in \mathbb{R}^1$, in the form of the integrals

$$s_n = \int_{\mathbb{R}^1} \lambda^n \, d\sigma(\lambda), \quad n = 0, 1, \dots, \tag{5.7}$$

where $d\sigma(\lambda)$ is some nonnegative measure on the axis. A generalization of this problem is to investigate the possibility of representation of form (5.7) of a sequence of operators s_n acting on a fixed finite-dimensional Hilbert space K; $d\sigma(\lambda)$ is a nonnegative K-operator-valued measure.

In the above mentioned articles, M.G. Krein also investigated the K-operator-valued positive definite functions, for which in the representation of form (5.2), the measure $d\sigma(\lambda)$ is operator-valued. This direction also adjoints to his work [48] on truncated moment problem.

These articles of M.G. Krein have made a great influence on the development of the spectral theory of differential and difference operators with operator-valued coefficients, correponding partial equations, etc. Here, in general, $\dim K = \infty$. See references in the books of Yu.M. Berezansky [9] and V.I. Gorbachuk and M.L. Gorbachuk [30].

6 Generalized Eigenvector Expansions for One Selfadjoint Operator

After the above mentioned works of M.G. Krein, the problem of construction of eigenfunction expansion for an ordinary selfadjoint differential operator was solved. But his theory was not applicable to partial differential operators: in this case it was impossible to construct the directing functionals.

In 1953–1955 there was published a series of important articles devoted to the problem of construction of eigenfunction expansion for selfadjoint partial differential operators (A.Ya. Povzner [60], L. Gårding [23, 24] F.E. Browder [17], L. Hörmander [31]), the first work in this direction belongs to T. Carleman [21]. These articles stimulated the work on the search of a general approach to the problem for a selfadjoint operator A on a Hilbert space H. On the other hand, the above mentioned M.G. Krein's works and the article of F.I. Mautner [56] pointed out to the fruitfullness of the idea of differentiation of the resolution of identity, $E(\lambda)$, of the operator A. Still on the other side, the works [1, 2] of R.A. Alexandryan about

S.L. Sobolev's problem showed that, in these questions, the notion of a generalized function is very useful.

The general approach to generalized eigenfunction expansion was published in 1955 by I.M. Gelfand and A.G. Kostyuchenko [26] and in 1956 by Yu.M. Berezansky [7] in which the above mentioned ideas were exploited. An important development of these works was made by L. Gårding [25], G.I. Kac [32–34], K. Maurin [51–54].

The problem of generalized eigenfunction expansion was later investigated in depth by many mathematicians, a detailed account of the corresponding results is contained in the books of Yu.M. Berezansky [9, 10], Yu.M. Berezansky and Yu.G. Kondratiev [14], I.M. Gelfand and G.E. Shilov [27], I.M. Gelfand and N.Ya. Vilenkin [28], K. Maurin [55]. We present here a list of some of these mathematicians: F.E. Browder (1956), E. Nelson (1958), C. Foias (1959), V.E. Lance (1963), E. Gerlach (1964), Yu.B. Orochko (1965), R.A. Hirschfeld (1965), M. Giertz (1967), M. Cotlar (1968), G.G. Gould (1968), Y. Itagaki (1971), N.A. Derzko (1972), D. Babbit (1972), J. Wloka (1973), K. Napiórkowski (1973), E. Prugovečki (1973), D. Fredricks (1975), V.F. Kovalenko and Yu.A. Semenov (1978), B. Simon (1982), F.A. Berezin and M.A. Shubin (1983). More detailed references can be found in the above mentioned books.

A simple account of the construction of a generalized eigenvector expansion in the latter version can be found in the textbook of Yu.M. Berezansky, G.F. Us, and Z.G. Sheftel [16].

We set forth here the main construction of a generalized eigenvector expansion for one selfadjoint operator A acting on a separable Hilbert space H. Let $\mathcal{B}(\mathbb{R}^1) \ni \alpha \mapsto E(\alpha)$ be its resolution of identity, then

$$1 = \int_{\mathbb{R}^1} dE(\lambda), \quad A = \int_{\mathbb{R}^1} \lambda \, dE(\lambda). \tag{6.1}$$

In the case of discrete spectrum $\operatorname{Sp}(A) = (\lambda_n)_{n=1}^{\infty}$ of the operator A, the resolution of identity is of the form: $E(\alpha) = \sum_{\lambda_n \in \alpha} P(\lambda_n)$, where $P(\lambda_n)$ is the (orthogonal) projector onto the eigensubspace corresponding to λ_n; formulas (6.1) become the following:

$$1 = \sum_{n=1}^{\infty} P(\lambda_n), \quad A = \sum_{n=1}^{\infty} \lambda_n P(\lambda_n). \tag{6.2}$$

Now the question arises as to get formulas of type (6.2) for a general operator A. Here it is necessary to replace the projector $P(\lambda_n)$ with an operator $P(\lambda)$ which, in some sense, "projects the vectors onto generalized eigensubspace of the operator A corresponding to λ". It is natural to introduce the derivative

$$P(\lambda) = \frac{dE(\lambda)}{d\rho(\lambda)}, \quad \lambda \in \mathbb{R}^1, \tag{6.3}$$

w.r.t. some nonnegative measure $\mathcal{B}(\mathbb{R}^1) \ni \alpha \mapsto \rho(\alpha) \in [0, \infty)$. Then formulas (6.1) take the form

$$(6.4) \qquad 1 = \int_{\mathbb{R}^1} P(\lambda)\, d\rho(\lambda), \quad A = \int_{\mathbb{R}^1} \lambda P(\lambda)\, d\rho(\lambda),$$

which generalizes the representation (6.2).

Unfortunately, the derivative (6.3) does not exist, as a rule. But it is possible to prove that if a positive operator-valued measure $\mathcal{B}(\mathbb{R}^1) \ni \alpha \mapsto \Theta(\alpha)$ ($\Theta(\alpha)$ is a nonnegative operator in H) has finite trace, i.e. $\operatorname{Tr} \Theta(\alpha) = \rho(\alpha) < \infty$, $\alpha \in \mathcal{B}(\mathbb{R}^1)$, then the derivative $\frac{d\Theta(\lambda)}{d\rho(\lambda)}$ exists (it is easy to check that $\mathcal{B}(\mathbb{R}^1) \ni \alpha \mapsto \rho(\alpha)$ is a positive measure, the "trace measure").

Using this result the following construction is proposed. Consider the rigging of the zero space H by a positive Hilbert space H_+ and a negative one, H_-:

$$(6.5) \qquad H_- \supset H \supset H_+ \supset D.$$

Here H_+ is a Hilbert space w.r.t. a stronger than in H scalar product, dense in H, H_- is the space conjugate to H_+ (with respect to H); D is an additional linear separable topological space densely and continuously embedded into H_+.

Suppose that the embedding operator $O : H_+ \to H$ is of the Hilbert-Schmidt type (i.e. the embedding is quasinuclear). Then the adjoint (w.r.t. H) operator $O^+ : H \to H_-$ is also of the Hilbert-Schmidt type and embeds H into H_-.

So, we fix such a quasinuclear Hilbert rigging (6.5) (i.e. a chain of spaces). Note, that the classical example of such chain is

$$(6.6) \qquad W_2^{-l}(G) \supset L^2(G, dx) \supset W_2^l(G) \supset C_0^\infty(G),$$

where G is a bounded domain in $\mathbb{R}^d, d < \infty$, $W_2^l(G)(W_2^{-l}(G))$ is the positive (negative) Sobolev space, $l > d/2$.

Consider a selfadjoint operator A in H for which $D \subset \operatorname{Dom}(A)$ and $A \restriction D$ acts continuously from D into H_+ (i.e. A is standardly connected with the chain (6.5)). A vector $\varphi \in H_-$, $\varphi \neq 0$, is, by definition, a generalized eigenvector of the operator A corresponding to an eigenvalue $\lambda \in \mathbb{R}^1$, if $\forall f \in D$

$$(6.7) \qquad (\varphi, Af)_H = \lambda(\varphi, f)_H$$

(i.e. $\hat{A}\varphi = \lambda\varphi$ for some extension $\hat{A}$ of the opereator A on space H_-, this extension is easily constructed). The set $g(A) \subset \mathbb{R}^1$ of all such eigenvalues λ is called the generalized spectrum of A.

Construct the operator-valued measure $\Theta(\alpha) = O^+E(\alpha)O : H_+ \to H_-$, $\alpha \in \mathcal{B}(\mathbb{R}^1)$, whose values are bounded nonnegative operators (i.e. $(\Theta(\alpha)f, f)_H \geq 0$, $f \in H_+$). The quasinuclearity of $O : H_+ \hookrightarrow H$, $O^+ : H \hookrightarrow H_-$ and boundedness of $E(\alpha)$ imply the finiteness of $\operatorname{Tr} \Theta(\alpha)$, and it is possible to apply the above mentioned result on differentiation of the measure Θ w.r.t. its trace measure to the measure under consideration (this result is preserved for operators acting from H_+ into H_-). As the result, we can obtain the following projection spectral theorem.

Let (6.5) *be a quasinuclear Hilbert rigging and A be a selfadjoint operator acting on H and standardly connected with* (6.5)*; $E(\lambda)$ be its resolution of identity; $g(A) \subset \mathbb{R}^1$ be the corresponding generalized spectrum; $O : H_+ \hookrightarrow H$, $O^+ : H \hookrightarrow H_-$.*

Introduce the spectral measure $\rho(\alpha) = \mathrm{Tr}\,(O^+ E(\alpha) O)$, $\alpha \in \mathcal{B}(\mathbb{R}^1)$. Then for ρ-almost all $\lambda \in \mathbb{R}^1$ there exists an operator $P(\lambda) : H_+ \to H_-$, $\|P(\lambda)\|_{HS} \leq 1$, such that

$$(6.8) \qquad O^+ O = \int_{g(A)} P(\lambda)\, d\rho(\lambda), \quad O^+ A O = \int_{g(A)} \lambda P(\lambda)\, d\rho(\lambda).$$

The operator $P(\lambda)$ "projects" H_+ onto the set of generalized eigenvectors with the eigenvalue λ, i.e. $\forall f \in H_+$ $\varphi(\lambda) = P(\lambda) f$ is such a vector. The integrals (6.8) *converge in sense of the Hilbert-Schmidt norm (of operators from H_+ into H_-), $\rho(g(A) \setminus \mathrm{Sp}\,(A)) = 0$.*

Roughly speaking, we can replace the operators in the left-hand side of (6.8) by 1 and A, respectively. It is convenient in many cases to understand the spectral measure as a measure equivalent to the introduced above $d\rho(\lambda)$.

Note also that if the representation of type (6.8) takes place for an arbitrary operator A standardly connected with chain (6.5), then the embedding $O : H_+ \hookrightarrow H$ (and, consequently, $O^+ : H \hookrightarrow H_-$) is quasinuclear.

In the simplest case of simple spectrum, the operator $P(\lambda)$ has the form:

$$(6.9) \qquad P(\lambda) f = (f, \varphi(\lambda))_H\, \varphi(\lambda) \in H_-, \quad f \in H_+,$$

where $\varphi(\lambda) \in H_-$ is a generalized eigenvector of type (6.7). In this case, we can introduce the Fourier transform:

$$(6.10) \qquad \hat{f}(\lambda) = (f, \varphi(\lambda))_H \in \mathbb{C}^1, \quad f \in H_+, \quad \lambda \in \mathbb{R}^1,$$

and the first equality in (6.8) gives the Parseval equality of type (3.4).

It is possible to introduce the Fourier-transform of type (6.10) also in the general case, but now $\hat{f}(\lambda)$ will be a certain vector-valued function of λ.

7 Carleman Operators

The results of Section 6 have natural applications to differential operators. For example, if an operator A is generated by a selfadjoint elliptic expression $\mathcal{L}$ in a bounded domain $G \subset \mathbb{R}^d$, then, as chain (6.5) in this case, it is convenient to take (6.6). In this case, the equality (6.7) means that φ is a generalized solution of the equation $\mathcal{L}u = \lambda u$ and, therefore, it is smooth (up to the boundary if we extend the space $D = C_0^\infty(G))$ a little. As the result, we get theorems on eigenfunction expansion for elliptic operators. Such a situation also holds for ordinary differential equations.

But for these differential operators as well as for some other cases, it is convenient to use the some modification of the constructions of Section 6, namely, when the embedding $H_+ \hookrightarrow H$ is not quasinuclear.

So, let in the space $H = L^2(R, d\mu(x))$, which is constructed from some measure $d\mu(x)$ defined on space R $(\mu(R) \leq \infty)$, we investigate a selfadjoint operator A with the following property: there exists a continuous function $\mathbb{R}^1 \ni \lambda \mapsto a(\lambda) \in \mathbb{C}^1$, $a(\lambda) \neq 0$, $\lambda \in \operatorname{Sp}(A)$, such that

$$(a(A)f)(x) = \int_R K(x,y)f(y)\,d\mu(y), \quad \int_R |K(x,y)|^2\,d\mu(x) < \infty. \tag{7.1}$$

Here $K(x, y)$ is a kernel defined for $\mu \times \mu$-almost all (x, y), the first equality in (7.1) takes place for some dense in $L^2(R, d\mu(x))$ set of functions f, the second inequality in (7.1) holds for μ-almost all $y \in R$. The operator A with these properties is called a Carleman operator.

For Carleman operators, it is possible to use, in the constuction of the expansion, a chain more simple than (6.6). Namely, this chain has form (6.5), but the embedding $H_+ \hookrightarrow H$ is not necessary quasinuclear:

$$L^2(R, p^{-1}(x)\,d\mu(x)) \supset L^2(R,\, d\mu(x)) \supset L^2(R, p(x)\,d\mu(x)) \supset D. \tag{7.2}$$

Here $R \ni x \mapsto p(x) \in [1, \infty)$ is an arbitrary weight with the property

$$\int_R \int_R |K(x,y)|^2 p^{-1}(y)\,d\mu(x)\,d\mu(y) < \infty. \tag{7.3}$$

Carleman operators have many nice properties: their generalized eigenfunctions are ordinary functions from $L^2(R, p^{-1}(x)\,d\mu(x))$ (but not from $L^2(R, d\mu(x))$, as a rule), the projection operator $P(\lambda)$ is an integral operator with the kernel $P(x, y; \lambda) \in L^2(R\times R, p^{-1}(x)p^{-1}(y)\,d\mu(x)\,d\mu(y))$ (spectral kernel of the operator A), etc.

Note that if $\mathbb{R}^1 \ni \lambda \mapsto b(\lambda) \in \mathbb{C}^1$ is a measurable function with the property that $|b(\lambda)| \leq |a(\lambda)|^2$, $\lambda \in \operatorname{Sp}(A)$, then $O^+b(A)O$ is also an integral operator:

$$\begin{aligned}(O^+b(A)Of)(x) &= \int_R B(x,y)f(y)\,d\mu(y), \quad f \in L^2(R, p(x)d\mu(x)); \\ B(x,y) &= \int_{\mathbb{R}^1} b(\lambda)P(x,y;\lambda)\,d\rho(\lambda), \\ &\int_R\int_R |B(x,y)|^2 p^{-1}(x)p^{-1}(y)\,d\mu(x)\,d\mu(y) < \infty\end{aligned} \tag{7.4}$$

($d\rho(\lambda)$ is the spectral measure of A). If the operator $b(A)$ is bounded, then $\forall f \in L^2(R, d\mu(x))$

$$(b(A)f)(x) = \int_R B(x, y) f(y)\, d\mu(y). \tag{7.5}$$

Formulas (7.2)–(7.5) have numerous applications. For example, $\forall u_0 \in L^2(R, d\mu(x))$ $u(x,t) = (e^{itA}u_0(x))$, $t \in [0, \infty)$, is a solution of the evolution equation $du/dt = iAu$, and $u(\cdot, t) \to u_0(\cdot)$ in $L^2(R, d\mu(x))$ if $t \to 0$. The question arises as to for which f this convergence takes place for μ-almost all x. One of positive answers is the following: if $u_0 = c(A)f$, where $|c(\lambda)| \leq |a(\lambda)|^2$, $\lambda \in \mathrm{Sp}\,(A)$, and bounded, then $f \in L^2(R, d\mu(x))$. For the proof, it is necessary to represent $u(x,t)$ in the form (7.4), where $b(\lambda) = e^{it\lambda}c(\lambda)$, and pass to the limit (as $t \to 0$) under the integral signs (relations (7.4), (7.5) permit to do this). Close results are contained in the article of N. Ben-Artzi and A. Devinatz [5].

Ordinary and elliptic selfadjoint operators with smooth coefficients in a domain $G \subset \mathbb{R}^d$, $d \in \mathbb{N}$, are Carleman operators. For elliptic operators of the second order, it is possible to take the function $a(\lambda) = (\lambda - i)^{-m}$, where $\mathbb{N} \ni m > d/4$, for ordinary operators of an arbitrary order, we can put $a(\lambda) = (\lambda - i)^{-1}$.

8 Generalized Eigenvector Expansions for a Family of Commuting Selfadjoint Operators

We will shortly explain this theory, developed in the books of Yu.M. Berezansky [10], Yu.M. Berezansky and Yu.G. Kondratiev [14].

Let $A = (A_x)_{x \in X}$ be a family of (strongly) commuting selfadjoint operators A_x acting on a separable Hilbert space H; the set X of indices x is arbitrary. Denote by $\mathbb{R}^X$ the set of all functions $X \ni x \mapsto \lambda(x) \in \mathbb{R}^1$ and by $C_\sigma(\mathbb{R}^X)$ the σ-algebra spanned by all cylindrical sets from $\mathbb{R}^X$, i.e. the sets of form

$$C(x_1, \ldots, x_n; \alpha) = \{\lambda(\cdot) \in \mathbb{R}^X \mid (\lambda(x_1), \ldots, \lambda(x_n)) \in \alpha\}. \tag{8.1}$$

Here the points $x_1, \ldots, x_n \in X$ are distinct, $\alpha \in \mathcal{B}(\mathbb{R}^n)$, $n \in \mathbb{N}$.

Using the resolutions of identity, E_x, of the operators A_x, it is possible to construct the joint resolution of identity, E, of the family A, defined on $C_\sigma(\mathbb{R}^X)$. It is defined by setting, for cylindrical sets (8.1) with $\alpha = \alpha_1 \times \cdots \times \alpha_n$,

$$E(C(x_1, \ldots, x_n; \alpha_1 \times \cdots \times \alpha_n)) = E_{x_1}(\alpha_1) \ldots E_{x_n}(\alpha_n), \tag{8.2}$$

and further extending this E to the whole $C_\sigma(\mathbb{R}^X)$. So, roughly speaking, E is a "product" of resolutions of identity, E_x, where $x \in X$.

This joint resolution of identity, E, is a projector-valued measure on $\mathbb{R}^X$, defined on $C_\sigma(\mathbb{R}^X)$; we will denote it also by $dE(\lambda(\cdot))$. For every operator A_x, the following representation by spectral integral takes place:

$$A_x = \int_{\mathbb{R}^X} \lambda(x)\, dE(\lambda(\cdot)), \quad x \in X \tag{8.3}$$

(the function $\mathbb{R}^X \ni \lambda(\cdot) \mapsto \lambda(x) \in \mathbb{R}^1$ in (8.3) is integrable).

Suppose now that we have a quasinuclear rigging (6.5), and that every operator A_x is standardly connected with this rigging. Then it is possible to introduce a generalized joint eigenvector $\varphi = \varphi(\lambda(\cdot))$ of the family A, corresponding to the "eigenvalue" $\lambda(\cdot) \in \mathbb{R}^X$. Namely, similar to (6.7), we introduce this notion by putting

$$(8.4) \qquad \begin{aligned} &(\varphi(\lambda(\cdot)), A_x f)_H = \lambda(x)(\varphi(\lambda(\cdot)), f)_H, \\ &0 \neq \varphi(\lambda(\cdot)) \in H_-, \quad f \in D, \quad x \in X. \end{aligned}$$

So, $\varphi(\lambda(\cdot))$ is a joint generalized eigenvector for every operator A_x with the eigenvalue $\lambda(x)$, $x \in X$. As before, the set $g(A) \subset \mathbb{R}^X$ of all such eigenvalues $\lambda(\cdot)$ is called the generalized spectrum of A.

The main result of Section 6 can be generalized the case of the family A in the following manner.

Let (6.5) *be a rigging with the previous properties, $A = (A_x)_{x \in X}$ be a commuting family of selfadjoint operators A_x standardly connected with* (6.5)*; $g(A)$ be its generalized spectrum.*

Introduce the spectral measure $\rho(\alpha) = \mathrm{Tr}\,(O^+ E(\alpha) O)$, $\alpha \in C_\sigma(\mathbb{R}^X)$. Then for ρ-almost all $\lambda(\cdot) \in \mathbb{R}^X$, there exists an operator $P(\lambda(\cdot)) : H_+ \to H_-$, $\|P(\lambda(\cdot))\|_{HS} \leq 1$, such that

$$(8.5) \qquad \begin{aligned} O^+ O &= \int_{g(A)} P(\lambda(\cdot))\, d\rho(\lambda(\cdot)), \\ O^+ A_x O &= \int_{g(A)} \lambda(x) P(\lambda(\cdot))\, d\rho(\lambda(\cdot)), x \in X. \end{aligned}$$

The operator $P(\lambda(\cdot))$ projects H_+ onto the set of joint generalized eigenvectors with the eigenvalue $\lambda(\cdot)$, i.e. $\forall f \in H_+$ $\varphi(\lambda(\cdot)) = P(\lambda(\cdot)) f$ is such a vector. The integrals in (8.5) *converge in the sense of the Hilbert-Schmidt norm $H_+ \to H_-$.*

Note that it is now also possible to introduce the notion of a joint spectrum of the family A if we understand it as the intersection of all closed sets from $C_\sigma(\mathbb{R}^X)$ of total measure E ($\mathbb{R}^X$ is provided with the Tychonoff topology). But this notion is incorrect if the power of the set X is more than countable; there are such families A for which this spectrum is an empty set.

If X is a topological space and $\forall f \in D$ the map $X \ni x \mapsto A_x f \in H_+$ is weakly continuous, then

$$(8.6) \qquad g(A) \subset C(X),$$

where $C(X)$ is the space of real-valued continuous functions on X. The integrals in (8.5) are now "continual integrals".

The results outlined above, of course, are preserved for a family of normal commuting operators A_x. Roughly speaking, in this case it is necessary to replace $\mathbb{R}^X$ with $\mathbb{C}^X$, $C(X)$ in (8.6) will denote the complex-valued functions, etc.

The representations (8.5) take a more special form in the case where the operators A_x of the family are connected by some algebraic relations. Let us clarify this with an example of a family $A = (A_x)_{x \in X}$ of commuting unitary operators.

Let X be a commutative topological group, $A_x D \subset D$, and $\forall f \in D$ the map $X \ni x \mapsto A_x f \in H_+$ is weakly continuous and

$$A_{x+y} f = A_x A_y f, \quad x, y \in X. \tag{8.7}$$

Then from (8.4) it is easy to conclude that $\lambda(x+y) = \lambda(x)\lambda(y),\ x, y \in X$, for $\lambda(\cdot) \in g(A)$. Together with (8.6) and the equality $|\lambda(x)| = 1, x \in X$, this means that every $\lambda(\cdot) \in g(A)$ is a continuous character $\chi(\cdot)$ of the group X. Let $\hat{X}$ be the dual group of such characters, setting $E(\hat{X} \setminus g(A)) = 0$, we can rewrite now the representations (8.5) in the form

$$\begin{aligned} O^+O &= \int_{\hat{X}} P(\chi(\cdot))\, d\rho(\chi(\cdot)), \\ O^+A_xO &= \int_{\hat{X}} \chi(x)P(\chi(\cdot))\, d\rho(\chi(\cdot)), x \in X. \end{aligned}$$

It is clear that we can assume that another type of relations similar to (8.7) holds. In this case, we get representations of type (8.5), where instead of $g(A)$, we can take the set of all scalar-valued continuous solution of the corresponding functional equation.

We demonstrate this approach with an example.

Let X be a real Hilbert space and $A = (A_x)_{x \in X}$ be a family of commuting selfadjoint operators standardly connected with the chain (6.5) *such that $\forall f \in D$ the map $X \ni x \mapsto A_x f \in H_+$ is weakly continuous and*

$$A_{\alpha x + \beta y} f = \alpha A_x f + \beta A_y f, \quad \alpha, \beta \in \mathbb{R}^1, \quad x \in X. \tag{8.8}$$

Then this family admits a representation by the spectral integral

$$A_x = \int_X (\lambda, x)_X\, dE(\lambda), \quad x \in X, \tag{8.9}$$

where E is some unique resolution of the identity defined on the Borel σ-algebra $\mathcal{B}(X)$.

Conversely, the existence of representation of the form (8.9) *for some family $A = (A_x)_{x \in X}$ means that all previous conditions are fulfilled.*

For other results of this type for Jacobi fields, see recent articles of Yu.M. Berezansky [11–13].

Note that the representation (8.5) is also useful for the investigation of some non-commutative relations between selfadjoint or normal operators. Corresponding results are contained in the article of Yu.M. Berezansky, V.L. Ostrovsky, and Yu.S. Samoilenko [15], the book of Yu.M. Berezansky and Yu.G. Kondratiev [14], the article of V.L. Ostrovsky and Yu.S. Samoilenko [58].

9 Expansion of Positive Definite Kernel into Elementary Kernels

In Section 5 we have discussed a nice idea of M.G. Krein which was to apply the eigenfunction expansions to the representation of positive definite functions and their broad generalizations. In the corresponding works, M.G. Krein used the method of directing functionals for the constructions. In article [8] Yu.M. Berezansky proposed to apply, instead of directing functionals, a more general approach explained in Section 6. It resulted in obtaining a broader and more flexible method of integral representation of positive definite functions, kernels etc.

Here we explain in what way it is possible to finish proving the validity of representation (5.2), given in Section 5, by using formulas (6.8). As before, we introduce the space H_k and Hermitian operator (5.4). Then, for some selfadjoint extension B of this operator and its resolution of identity, we use the first formula from (6.8). It follows from this formula that $\forall f, g \in D = C_0^\infty((-a,a))$

$$(9.1)\qquad \begin{aligned}(f,g)_{H_k} &= \int_{g(B)} (P(\lambda)f,g)_{H_k}\, d\rho(\lambda)\\ &= \int_{g(B)} \left(c(\lambda) \int_{-a}^{a}\int_{-a}^{a} e^{i\lambda(y-x)} f(y)\overline{g(x)}\, dx\, dy \right) d\rho(\lambda).\end{aligned}$$

Note that we use here the calculation of $P(\lambda)$ for the operator $B \supset A$. The operator $P(\lambda)$ has now the form (6.9) with $\varphi(\lambda) = \varphi(x,\lambda) = e^{ix\lambda}$, $c(\lambda) \geq 0$. Using the expression (5.3) for $(f,g)_{H_k}$ and arbitrariness of f, g, we conclude that

$$(9.2)\qquad k(y-x) = \int_{\mathbb{R}^1} e^{i\lambda(y-x)}\, d\sigma(\lambda), \quad x, y \in (-a,a),$$

where $d\sigma(\lambda) = c(\lambda) d\rho(\lambda)$ is continued by zero to the whole $\mathbb{R}^1$. The formula (9.2) is the same as (5.2).

It is clear that such a proof can be extended to positive definite kernels and operators more general than $K(x,y) = k(y-x)$ and (5.4).

For example, let $G \subset \mathbb{R}^d$, $d \in \mathbb{N}$, be some domain (bounded or not), and $G \times G \ni (x,y) \mapsto K(x,y) \in \mathbb{C}^1$ be a positive definite kernel. Let $\mathcal{L}$ be some linear differential expression in G, instead of $-i\frac{d}{dx}$ in (5.4). Introduce the Hilbert space H_K by defining the scalar product for functions f, g on G as

$$(9.3)\qquad (f,g)_{H_K} = \int_G \int_G K(x,y) f(y)\overline{g(x)}\, dx\, dy,$$

and suppose that $\forall f, g \in C_0^\infty(G)$ $(\mathcal{L}f, g)_{H_K} = (f, \mathcal{L}g)_{H_K}$. Then the mentioned above method to prove (9.2) with the operator A generated by $\mathcal{L}$ gives the representation

$$(9.4)\qquad K(x,y) = \int_{\mathbb{R}^1} \Omega(x,y;\lambda)\, d\sigma(\lambda), \quad x, y \in G.$$

Here $\Omega(x,y;\lambda)$ are "elementary" positive definite kernels, i.e. in the generalized sense, $\mathcal{L}_x\Omega = \lambda\Omega$, $\bar{\mathcal{L}}_y\Omega = \lambda\Omega$ (it is the kernel of the operator $P(\lambda)$). In many cases (particularly, if $d = 1$, i.e. if $\mathcal{L}$ is an ordinary differential expression), $\Omega(x,y;\lambda)$ can be calculated, and representation (9.4) will become more concrete.

This theory has many aspects. In particular, the following:

1) The generalization of the construction to abstract positive definite kernels which are elements of $H \otimes H$ (instead of functions of two variables), where H is some linear space.
2) The detailed investigation of the problems of essential selfadjointness of an operator A, construction and description of its selfadjoint extensions, etc. In particular, the investigation of the set of all continuations of a positive definite function from a smaller domain to a bigger one, etc. Uniqueness of such extensions, their description.
3) The generalization of investigations 1), 2) to the case of $p \in \mathbb{N}$ commuting operators of type A (now the integral in (9.4) will be extended to $\mathbb{R}^p$).
4) The many-dimensional moment problem, which can be investigated using the above mentioned approach.

Results of this section are contained in the books of Yu.M. Berezansky [10], Yu.M. Berezansky and Yu.G. Kondratiev [14]. Such type of results belong also to many mathematicians. I mention here the following authors (it is possible to find the references to their works in these last two books): M.S. Livshic (1944), A.Ya. Povzner (1944), A.P. Calderon and P. Pepinsky (1952), A. Devinatz (1953), K. Maurin (1958), F.E. Browder (1959), R.B. Zarkhina (1959), G.I. Eskin (1960), I.M. Gelfand and Sia Do-shin (1960), R.S. Ismagilov (1960), A.G. Kostyuchenko and B.S. Mityagin (1960), V.I. Gorbachuk (1962), W. Rudin (1963), B.Ya. Levin and I.E. Ovcharenko (1964), M.L. Gorbachuk (1965), A.E. Nussbaum (1975), L.A. Sakhnovich (1980), J. Friedrich and L. Klotz (1985), and also the article of R. Bruzual [19].

10 Positive Definite Kernels of Infinitely many Variables

The theory outlined in the last section has an essential development to the case of functions of infinitely many variables, i.e. in the situation when, roughly speaking, the domain G from (9.3), (9.4) is a part of the space $\mathbb{R}^\infty$, and an infinite set of operators A comes into consideration. Now in integrals of type (9.4), instead of $\mathbb{R}^1$, we use $\mathbb{R}^\infty$ or a part of it. These results are contained in the book of Yu.M. Berezansky and Yu.G. Kondratiev [14], we present here only two examples of the results.

1) Infinite dimensional moment problem. Let Φ be a nuclear space with involution, $\Phi^{\otimes n}$ be its tensor n-th power ($n \in \mathbb{N}$, $\Phi^{\otimes 0} = \mathbb{C}^1$).

Consider a sequence $s = (s_n)_{n=0}^\infty$, where $s_n \in (\Phi^{\otimes n})'$ (the prime denotes the conjugate space) and is real and symmetric. We assume that

a) s is positive definite. This means that for an arbitrary finite sequence $(\varphi_j)_{j=0}^\infty$, $\varphi_j \in \Phi^{\otimes j}$, the following inequality takes place:

$$\sum_{j,k=0}^{\infty} \langle s_{j+k}, \varphi_j \otimes \bar{\varphi}_k \rangle \geq 0. \tag{10.1}$$

b) Certain estimates on growth of s at ∞ hold. For example, let Φ be a projective limit of Hilbert spaces H_τ, $\tau \in T$. Then there exists $\tau \in T$ such that $\forall \varphi^{(1)}, \ldots, \varphi^{(2n)} \in H_\tau$

$$|\langle s_{2n}, \varphi^{(1)} \otimes \ldots \otimes \varphi^{(2n)} \rangle| \leq m_n^2 \prod_{j=1}^{2n} \|\varphi^{(j)}\|_{H_\tau}, \quad n \in \mathbb{N},$$

and the class $C\{m_n\}$ is quasi-analytic (for example, $m_n = n!$).

Then there exists a measure $\sigma(\lambda) \geq 0$ on the real space $\Phi'_{\mathrm{Re}} \subset \Phi'$ such that, similarly to (5.7),

$$s_n = \int_{\Phi'_{\mathrm{Re}}} \lambda^{\otimes n} \, d\sigma(\lambda), \quad n = 0, 1, \ldots, \tag{10.2}$$

i.e. the sequence s is a moment sequence. Of course, every moment sequence satisfies condition (10.1).

2) Positive definite functions on a band of a Hilbert space. Let H be a real Hilbert space, $e \in H$ some unit vector, $a \in (0, \infty]$. The set $H_a = \{x \in H \mid (x, e)_H \in (-a, a)\}$ is called a band of H. Consider the function $H_{2a} \ni x \mapsto k(x) \in \mathbb{C}^1$ and assume that

a) k is positive definite, i.e. for an arbitrary finite sequence $(\xi_j)_{j=0}^{\infty}$ of complex numbers ξ_j and points $x^{(j)} \in H_a$,

$$\sum_{j,l=1}^{\infty} k(x^{(j)} - x^{(l)}) \xi_j \bar{\xi}_l \geq 0;$$

b) k is continuous in the J-topology induced by neighborhoods of the zero 0, $U(0; A, \varepsilon) = \{x \in H_{2a} \mid (Ax, x)_H < \varepsilon\}$, where $A \geq 0$ is an arbitrary operator on H from the trace class, and $\varepsilon > 0$.

Then we have the following representation for k:

$$k(x) = \int_H e^{i(\lambda, x)_H} \, d\sigma(\lambda), \quad x \in H_{2a}, \tag{10.3}$$

where $d\sigma(\lambda)$ is a nonnegative measure. Conversely, for a function (10.3), conditions a), b) are fulfilled.

So, in the case $a = \infty$, we have the theorem of R.A. Minlos [57] — V.V. Sazonov [62]. In the case $a < 0$, we get for a Hilbert space, a generalization of M.G. Krein's theorem about extension of positive definite functions from an interval to the whole axis, which was outlined in Section 5.

We also mention some authors who have also developed the questions outlined here: Yu.S. Samoilenko (1971), I.M. Gali (1972), S.N. Shifrin (1974), G.F. Us (1974), Yu.G. Kondratiev (1976), S.V. Tishchenko (1979), A.A. Kalyuzhnyi (1982). A detailed reference can be found in the book of Yu.M. Berezansky and Yu.G. Kondratiev [14].

References

[1] R.A. Alexandryan, *On Dirichlet's problem for string equation and on completeness of a system of functions on the circle*, Dokl. AN SSSR **73**, no. 5 (1950), 869–872.

[2] R.A. Alexandryan, *Spectral properties of operators generated by systems of differential equations of S.L. Sobolev's type*, Trudy Moscow Mat. Obshch. **9** (1960), 455–505.

[3] N.I. Akhiezer and I.M. Glazman, *The theory of linear operators in Hilbert space*, 3rd rev. ed., vols. **1, 2**, Vyshcha Shkola, Kharkov, 316+288 (1977, 1978) (English transl. of 1st ed.: vols. 1, 2, Ungar, New York, 1961).

[4] D.Z. Arov and H. Dym, *On three Krein extension problems and some generalizations*, Integr. Equ. Oper. Theory **31**, no. 1 (1998), 1–91.

[5] M. Ben-Artzi and A. Devinatz, *Local smoothing and convergence properties of Schrödinger type equations*, J. Funct. Anal. **101**, no. 2 (1991), 231–254.

[6] Yu.M. Berezansky, *The expansion in eigenfunctions of partial difference equations of second order*, Trudy Moscow Mat. Obshch. **5** (1956), 203–268.

[7] Yu.M. Berezansky, *On eigenfunction expansion of general selfadjoint differential operators*, Dokl. AN SSSR **108**, no. 3 (1956), 379–382.

[8] Yu.M. Berezansky, *A generalization of the Bochner theorem on expansions in eigenfunctions of partial differential equations*, Dokl. AN SSSR **110**, no. 6 (1956), 893–896.

[9] Yu.M. Berezansky, *Expansion in Eigenfunctions of Selfadjoint Operators*, AMS, Providence, R.I. 1968, IX+809 p. (Russian edition: Naukova Dumka, Kiev, 1965).

[10] Yu.M. Berezansky, *Selfadjoint Operators in Spaces of Functions of Infinitely Many Variables*, AMS, Providence, R.I. 1986, XV+383 p. (Russian edition: Naukova Dumka, Kiev, 1978).

[11] Yu.M. Berezansky, *Commutative Jacobi fields in Fock space*, Integr. Equ. Oper. Theory **30**, no. 2 (1998), 163–190.

[12] Yu.M. Berezansky, *On the direct and inverse spectral problems for Jacobi fields*, St. Petersburg Math. J. **9**, no. 6 (1998), 1053–1071.

[13] Yu.M. Berezansky, *On the theory of commutative Jacobi fields*, Methods Funct. Anal. Topology **4**, no. 1 (1998), 1–31.

[14] Yu.M. Berezansky and Yu.G. Kondratiev, *Spectral Methods in Infinite Dimensional Analysis*, vols. 1, 2, Kluwer Acad. Publ., Dordrecht-Boston-London, 1995, XVII+572 p., VIII+427 p. (Russian edition: Naukova Dumka, Kiev, 1988).

[15] Yu.M. Berezansky, V.L. Ostrovsky, and Yu.S. Samoilenko, *Eigenfunction expansion of the families of commuting operators and the representations of commutation relations*, Ukr. Mat. Zh. **40**, no. 1 (1998), 106–109.

[16] Yu.M. Berezansky, Z.G. Sheftel, and G.F. Us, *Functional Analysis*, vols. 1, 2, Birkhäser Verlag, Basel-Boston-Berlin, 1996, XIX+423 p., XVI+293 p. (Russian edition: Vyshcha Shkola, Kiev, 1990).

[17] E. Browder, *The eigenfunction expansion theorem for the general selfadjoint singular elliptic partial differential operator*, Proc. Nat. Acad. Sci. USA **40**, no. 6 (1954), 454–467.

[18] R. Bruzual, *Local semigroups of contractions and some applications to Fourier representation theorems*, Integr. Equ. Oper. Theory **10** (1987), 780–801.

[19] R. Bruzual, *Unitary extensions of two-parameter local semigroups of isometric operators and the Krein extension theorem*, Integr. Equ. Oper. Theory **17** (1993), 301–321.

[20] R. Bruzual, *Representation of measurable positive definite generalized Toeplitz kernels in* **R**, to appear in: Integr. Equ. Oper. Theory.

[21] T. Carleman, *Sur la théorie mathématique de l'équation de Schroedinger*, Ark. Math. Astr. och Fys. **24**, no. 11 (1934), 1–7.

[22] M. Cotlar and C. Sadosky, *Toeplitz and Hankel forms related to unitary representations of the symplectic plane*, Coll. Math. **60/61** (1990), 693–708.

[23] L. Gårding, *Eigenfunction expansion connected with elliptic differential operators*, The Twelfth Congress of Scand. Math. Lund, 1953 (1954), 44–45.

[24] L. Gårding, *Applications of the theory of direct integrals of Hilbert spaces to some integral and differential operators*, Inst. Fluid Dynam. Appl. Math. Lect. Ser., no. 11, Univ. Maryland, College Park, 1954.

[25] L. Gårding, *Eigenfunction expansions*, Supplement I to book of L. Bers, F. John, and M. Schechter, *Partial Differential Equations*, Interscience, New York-London-Sydney, 1964, 301–325.

[26] I.M. Gelfand and A.G. Kostyuchenko, *On eigenfunction expansion of differential and other operators*, Dokl. AN SSSR **103**, no. 3 (1955), 349–352.

[27] I.M. Gelfand and G.E. Shilov, *Generalized Functions. vol. 3: Some Questions in the Theory of Differential Equations*, Academic Press, New York, 1968 (Russian edition: Fizmatgiz, Moscow, 1958).

[28] I.M. Gelfand and N.Ya. Vilenkin, *Generalized Functions. vol. 4: Some Applications of Harmonic Analysis*, Academic Press, New York, 1964 (Russian edition: Fizmatgiz, Moscow, 1961).

[29] V.I. Gorbachuk and M.L. Gorbachuk, *On boundary value problems for differential equation of first order with operator coefficients and on expansion in eigenfunctions of this equation*, Dokl. AN SSSR **208**, no. 6 (1973), 1268–1271.

[30] V.I. Gorbachuk and M.L. Gorbachuk, *Boundary Value Problems for Operator Differential Equations*, Kluwer Acad. Publ., Dordrecht-Boston-London, 1991, XI+347 p. (Russian edition: Naukova Dumka, Kiev, 1984).

[31] L. Hörmander, *On the theory of general partial differential operators*, Acta Math. **94** (1955), 161–248.

[32] G.I. Kac, *On eigenfunction expansion of selfadjoint operators*, Dokl. AN SSSR **119**, no. 1 (1958), 19–22.

[33] G.I. Kac, *Generalized elements of Hilbert space*, Ukr. Mat. Zh. **12**, no. 1 (1960), 13–24.

[34] G.I. Kac, *Spectral expansions of selfadjoint operators on generalized elements of Hilbert space*, Ukr. Mat. Zh. **13**, no. 4 (1961), 13–33.

[35] K. Kodaira, *On ordinaty differential equations of any even order and the corresponding eigenfunction expansions*, Amer. J. Math. **72**, no. 3 (1950), 502–544.

[36] I.V. Kovalishina and V.P. Potapov, *Integral Representation of Hermitian Positive Functions*, Hokkaido University, Sapporo, 1982, p. 129.

[37] M.G. Krein, *On problem of extension of Hermitian positive continuous functions*, Dokl. AN SSSR **26**, no. 1 (1940), 17–21.

[38] M.G. Krein, *On Hermitian operators with deficiency indicies equal to one. I*, Dokl. AN SSSR **43**, no. 8 (1944), 323–326.

[39] M.G. Krein, *On Hermitian operators with deficiency indicies equal to one. II*, Dokl. AN SSSR **44**, no. 4 (1944), 131–134.

[40] M.G. Krein, *On a remarkable class of Hermitian operators*, Dokl. AN SSSR **44**, no. 5 (1944), 191–195.

[41] M.G. Krein, *On a general method of decomposition of positive definite kernels on elementary products*, Dokl. AN SSSR **53**, no. 1 (1946), 3–6.

[42] M.G. Krein, *On Hermitian operators with directing functionals*, Zbirnyk prac' Inst. Mat. AN URSR, no. 10 (1948), 83–106.

[43] M.G. Krein, *Hermitian positive kernels on homogeneous spaces. P. 1*, Ukr. Mat. Zh. **1**, no. 4 (1949), 64–98.

[44] M.G. Krein, *The fundamental propositions of the theory of Hermitian operators with deficiency index (m, m)*, Ukr. Mat. Zh. **1**, no. 2 (1949), 3–66.

[45] M.G. Krein, *Infinite J-matrices and a matrix moment problem*, Dokl. AN SSSR **69**, no. 2 (1949), 125–128.

[46] M.G. Krein, *On an one-dimensional singular boundary problem of the even order in interval* $(0, \infty)$, Dokl. AN SSSR **74**, no. 1 (1950), 9–12.

[47] M.G. Krein, *Hermitian positive kernels on homogeneous spaces. P.* 2, Ukr. Mat. Zh. **2**, no. 1 (1950), 10–59.

[48] M.G. Krein, *The description of all solutions of truncated moment problem and some operator questions*, Mat. issled., Kishinev **2**, no. 2 (1967), 114–132.

[49] H. Langer, *Über die Methode der richtenden Funktionale von M.G. Krein*, Acta Math. Acad. Scie. Hungary **21**, no. 1–2 (1970), 207–224.

[50] M.S. Livshic, *On an application of the theory of Hermitian operators in generalized moment problem*, Dokl. AN SSSR **44**, no. 1 (1944), 3–7.

[51] K. Maurin, *Entwicklung positiv difiniter Kerne nach Eigendistributionen. Differenzierbarkeit der Spektralfunktion eines hypoelliptischen Operators*, Bull. Acad. Polon. Sci. Ser. math. astr. et phys. **6**, no. 3 (1958), 149–155.

[52] K. Maurin, *Allgemeine Eigenfunktionsentwicklungen. Spektraldarstellung abstrakter Kerne. Eine Verallgemeinerung der Distributionen auf Liéschen Gruppen*, Bull. Acad. Polon. Sci. Ser. math. astr. et phys. **7**, no. 8 (1959), 471–479.

[53] K. Maurin, *Eine Bemerkung zur allgemeinen Eigenfunktionsentwicklungen für vertauschbare Operatorensysteme beliebiger Mächtigkeit*, Bull. Acad. Polon. Sci. Ser. math. astr. et phys. **8**, no. 6 (1960), 381–384.

[54] K. Maurin, *Abbildungen vom Hilbert-Schmidtschen Typus und ihre Anwendungen*, Math. Scand. **9** (1961), 359–371.

[55] K. Maurin, *General Eigenfunction Expansions and Unitary Representation of Topological Groups*, PWN, Warsaw, 1968, p. 367.

[56] F.I. Mautner, *On eigenfunction expansions*, Proc. Nat. Acad. Sci. USA **39**, no. 1 (1953), 49–53.

[57] R.A. Minlos, *Generalized random processes and their extensions to measures*, Trudy Moscow Mat. Obshch. **8** (1959), 497–518.

[58] V.L. Ostrovsky and Yu.S. Samoilenko, *Unbounded operators satisfying non-Lie commutation relations*, Repts. Math. Phys. **28**, no. 1 (1989), 91–103.

[59] A.Ya. Povzner, *On method of directing functionals of M.G. Krein*, Zapiski Inst. Mat. Mech. KhGU and Kharkov Mat. Obshch. **20** (1950), 43–52.

[60] A.Ya. Povzner, *On eigenfunction expansion of arbitrary functions in eigenfunctions of operator* $-\Delta u + cu$, Mat. Sbornik **32**, no. 1 (1953), 109–156.

[61] F.S. Rofe-Beketov, *Eigenfunction expansion of infinite systems differential equations in nonselfadjoint and selfadjoint cases*, Mat. Sbornik **51**, no. 3 (1960), 293–342.

[62] V.V. Sazonov, *A remark concerning characteristic functionals*, Teor. Veroyatn. Prim. **3**, no. 2 (1958), 201–205.

[63] H. Weyl, *Über gewöhnliche Differentialgleichungen mit Singularitäten und die zugehörigen Entwicklungen willkürlicher Funktionen*, Math. Annalen **68** (1910), 220–269.

Institute of Mathematics
National Academy of Sciences of Ukraine
3 Tereshchenkivs'ka St.
252601 Kyiv
Ukraine
berezan@mathber.carrier.kiev.ua

1991 Mathematics Subject Classification: Primary 47B15

Operator Theory:
Advances and Applications, Vol. 117
© 2000 Birkhäuser Verlag Basel/Switzerland

M.G. Krein and Extension Theory of Symmetric Operators. Theory of Entire Operators

M.L. Gorbachuk and V.I. Gorbachuk

The aim of this survey is to give a brief exposition of M. Krein's contribution to the extension theory of symmetric operators and the theory of entire operators, to describe the further development of his investigations in these fields and their application to the spectral theory of boundary value problems for differential equations.

The extension theory of symmetric operators is an important branch of analysis that is closely related to various fields of modern mathematics: the moment problem, the theory of analytic functions, boundary value problems for differential equations, probability theory and quantum field theory. The first fundamental results in this theory were obtained by J. von Neumann in 1929 (see [1]), although the origins can probably be traced back to the well-known papers [2, 3] (1909–1910) by H. Weyl, where its principal concepts were presented for a second-order ordinary differential operator. It should be observed that an essential role in the development of extension theory was played by the investigations concerning the classical moment problem. But as an independent part of analysis, this theory could be formed only after M. Krein's investigations. It was a subject that interested him all his life. The problems solved by M. Krein, the methods developed by him led to substantial progress in abstract operator theory as well as in applications. Unlike anybody else he managed to combine many problems, so different at first sight in their statements and methods of solution, by putting in the center of attention, as he used to say "an exciting object hidden behind the scenes: a symmetric operator and its selfadjoint extensions".

M. Krein's investigations in the extension theory of Hermitian operators developed in the following directions:

1) solving concrete problems (the moment problem, the continuation problem for positive definite functions and spiral arcs, the interpolation and extrapolation of functions, boundary value problems for differential equations, etc.);
2) setting up and solving a certain set of problems in the theory of analytic functions (later on, they formed a basis of the analytical apparatus used in the construction of extension theory);
3) constructing an abstract extension theory for symmetric operators.

We shall focus our attention mainly on M. Krein's results in direction 3) and their evolution and application to boundary value problems for differential equations, which are "a real, truly inexhaustible sorce of Hermitian operators". We shall not touch upon other applications or upon direction 2), although we cannot pass by M. Krein's criterion for an entire function to belong to the class (N) in the upper and lower half-planes. We formulate this criterion which is of great importance for constructing the theory of entire Hermitian operators. Recall that an analytic function $f(z)$ is said to be an N-function in a domain G if $\log^+|f(z)|$ has a harmonic majorant there. The criterion is: an entire function $f(z)$ is an N-function in the half-planes $\Im z > 0$ and $\Im z < 0$ if and only if $f(z)$ is of at most exponential type and

$$\int_{-\infty}^{\infty} \frac{\log^+|f(x)|}{1+x^2}\,dx < \infty.$$

Let us turn to extension theory. It is convenient to partition the results into two groups:

1. The extension theory for semibounded operators, published mainly in the fundamental works [4, 5] by M. Krein and [6] by M. Krein and M. Krasnosel'skii (1947). These papers are well-known for specialists and contained in many monographs (see, e.g. [7–9]).
2. The theory of entire operators, published in M. Krein's articles [10–12] (1944), [13] (1946), [14] (1949), Proceedings of ICM-66 [15], and the papers [16, 17] by M. Krein and N. Saakian (1966, 1970).

Note that we consider only the case of a Hilbert space, and do not dwell on Hermitian operators in spaces with indefinite metrics (for the latter we refer to the survey work [18] by T. Azizov, Yu. Ginzburg, and H. Langer).

Krein's publications on the theory of entire operators are largely announcements of results except for [14]. This situation troubled him. To publish the results in detail was M. Krein's dream. He regarded this theory as "one of the most significant of his mathematical achievements". This is why the paper [14] is dedicated to his teacher N. Chebotarev as a sign of the deepest gratitude to him. Probably the same reason incited him to deliver the special course of lectures on entire operators with defect numbers (1, 1) at Odessa Pedagogical Institute (1961). These lectures outlined by his students V. Adamian, D. Arov, A. Nudelman, adapted and extended by authors of this talk, but unpublished during his lifetime, underlie the book [19] "M.G. Krein's lectures on entire operators" published recently on I. Gohberg's initiative. It should also be noted that many of M. Krein's papers on the above topics are collected in the book "Selected works, II" by M. Krein [20] (1996).

To describe the theory of extensions, let A be a closed Hermitian operator with dense domain in a Hilbert space $\mathfrak{H}$: $\overline{\mathcal{D}(A)} = \mathfrak{H}$. For any $\lambda \in \mathbb{C}$ we set

$$\mathfrak{M}_\lambda = (A - \lambda I)\mathcal{D}(A), \quad \mathfrak{N}_\lambda = \mathfrak{H} \ominus \mathfrak{M}_{\bar{\lambda}}, \quad n_\lambda = \dim \mathfrak{N}_\lambda.$$

$\mathfrak{N}_\lambda$ and n_λ are the defect space and the defect number of the operator A corresponding to the given λ. By the well-known Krein-Krasnosel'skii theorem, n_λ remains unchanged for all λ lying in the same connected component from the regularity set of A. If the operator A is semibounded, then its regularity set forms a connected component. Therefore, such an operator admits self-adjoint extensions. Assume that the operator A is lower semibounded, and put

$$m = m(A) = \inf_{f \in \mathcal{D}(A)} \frac{(Af, f)}{(f, f)}.$$

For the sake of simplicity, we consider $m \geq 0$.

As was proved by J. von Neumann, for any $m_1 < m$ there exists a selfadjoint extension $\widetilde{A}$ of A with $m(\widetilde{A}) \geq m_1$. He advanced a hypothesis that there exists an extension $\widetilde{A}$ such that $m(\widetilde{A}) = m$. The extension possessing this property (the so-called hard (Friedrichs) one) was constructed by Stone and Friedrichs. One can find its construction in [21]. M. Krein showed in [4] that the set $\mathfrak{P}(A)$ of all positive selfadjoint extensions $\widetilde{A}$ (i.e. $\widetilde{A} \geq 0$) of the operator A contains two extreme extensions, namely, the hard extension $\widetilde{A}_\mu$ and the soft one $\widetilde{A}_M$ defined as

$$\widetilde{A}_M = A^* \lceil \mathcal{D}(\widetilde{A}_M), \quad \mathcal{D}(\widetilde{A}_M) = \mathcal{D}(A) \dot{+} \ker A^*.$$

All the others lie between $\widetilde{A}_M$ and $\widetilde{A}_\mu$:

$$\widetilde{A}_M \leq \widetilde{A} \leq \widetilde{A}_\mu$$

(the inequalities are understood in the quadratic form sense or, what is the same,

$$\exists a > 0 : (\widetilde{A}_\mu + aI)^{-1} \leq (\widetilde{A} + aI)^{-1} \leq (\widetilde{A}_M + aI)^{-1}).$$

He also indicated the conditions for the equality $\widetilde{A}_M = \widetilde{A}_\mu$ to be fulfilled. In the course of proof, the method of linear-fractional transforms which was proposed by J. von Neumann in his extension theory, and M. Krein's principal theorem on the existence of two extreme extensions $\widetilde{S}_M$ and $\widetilde{S}_\mu$ among all selfadjoint extensions $\widetilde{S}$ preserving the norm ($\|\widetilde{S}\| = \|S\|$) of a nondensely defined bounded symmetric operator S in $\mathfrak{H}$ were used.

The case where the defect numbers of A are finite was studied in greater detail. The generalized resolvent formula was proved. This formula was announced in [13] (1946), although for operators with defect-index (1, 1) it was obtained independently by M. Krein and M. Naimark in 1943. Later on, it was generalized by M. Krein and other mathematicians (M. Vishik, A. Straus, N. Saakian, V. Derkach, M. Malamud) even to nonsymmetric and nondensely defined operators. It plays an important role in various fields of mathematics. Recently there was renewed interest in this formula because of its connection with problems of singular perturbations in quantum field theory. It also underlies the theory of entire operators. In [4], a rule for calculating the number of negative eigenvalues of a selfadjoint extension of a positive operator was given.

The further development of extension theory for positive operators is associated with works by M. Vishik [22] and M. Birman [23], where certain operators acting on $\ker A^*$ are assigned in one-to-one manner to various classes of selfadjoint extensions of a positive definite operator. These classes of extensions are characterized in terms of properties of the operators corresponding to them. In such a way all positive definite extensions and extensions with finite or countable number of negative eigenvalues were described. The classes of solvable, completely and normally solvable selfadjoint extensions were also considered from the same point of view. These investigations were continued by Yu. Arlinskii, A. Kuzhel, V. Mikhailets, G. Nenciu, A. Straus, E. Tsekanovskii and other mathematicians.

The results of M. Krein mentioned above were applied to a number of boundary value problems for differential equations. As a rule, symmetric operators appear there in the following way:

Let l be a formally selfadjoint differential expression in a domain $G \subset \mathbb{R}^n$. Define A as the closure in the space $\mathfrak{H} = L_2(G)$ of the operator generated by l on the set $C_0^\infty(G)$ of all infinitely differentiable functions with compact supports in G. The operator A is called the minimal one generated by l. It turns out that the operator $\tilde{A}y = l[y]$ given on the functions satisfying certain boundary conditions is an extension of the minimal operator A and a restriction of the maximal one A^*. Therefore, the problems arose of

(a) finding the boundary conditions giving all selfadjoint extensions of the minimal operator;

(b) describing the properties of a selfadjoint extension of the minimal operator in terms of boundary conditions corresponding to this extension.

These problems were solved by M. Krein [5] for ordinary differential equations. In this case, the defect numbers of A are finite. But neither Neumann's nor Vishik's-Birman's abstract description enabled one to solve problems (a) and (b) for partial differential equations. The reason is that in this situation the defect-index of the minimal operator is infinite and the structure of the general solution of the equation $l[y] = 0$ is rather complicated. The difficulties were overcome due to important results by F. Rofe-Beketov. In the papers [24, 25], he described in a compact form all selfadjoint extensions of the minimal operator generated by a differential expression whose coefficients are bounded operators on a Hilbert space H. The description is given in the terms of boundary conditions. To do this, for the first time the representation of a selfadjoint linear (binary) relation in a Hilbert space $\mathcal{H}$ was used. By the latter we mean a set $\theta \subset \mathcal{H} \oplus \mathcal{H}$ such that

$$\forall \{f, f'\}, \{g, g'\} \in \theta \quad (f', g)_{\mathcal{H}} - (f, g')_{\mathcal{H}} = 0 \tag{1}$$

and the equality (1) for some $\{f, f'\} \in \mathcal{H} \oplus \mathcal{H}$ and any $\{g, g'\} \in \theta$ implies $\{f, f'\} \in \theta$.

It was shown that a linear relation θ is selfadjoint if and only if there exists a unitary operator U on $\mathcal{H}$ such that for any $\{f, f'\} \in \theta$

$$(U - I)f' - i(U + I)f = 0.$$

Using this representation and the integration by parts formula, F. Rofe-Beketov proved that there exists a one-to-one correspondence between selfadjoint extensions $\widetilde{A}$ of A and unitary operators U on $\mathcal{H} = \oplus_{n\ times} H$ (n is the order of l) such that

$$y \in \mathcal{D}(\widetilde{A}) \Longleftrightarrow (U - I)\widehat{y}' + i(U + I)\widehat{y} = 0,$$

where $\widehat{y}$ and $\widehat{y}'$ are n-dimensional vectors whose components are expressed via the boundary values of $y(t)$ and its derivatives up to $(n - 1)$-order.

As for unbounded operator coefficients (which is the case that is interesting for applications to partial differential equations), the Rofe-Beketov method meets some difficulties as the integration by parts formula cannot be applied immediately to arbitrary elements from the domain of the maximal operator. These difficulties were reduced by M. Gorbachuk (see [26]), who found an analogue of the Green formula for any vector-functions from $\mathcal{D}(A^*)$ in the case of a second-order operator differential equation of elliptic type. In this analogue, certain regularizations stand for the values of $y(t)$ and $y'(t)$ at the ends of the interval in which the differential expression is considered.

If $\dim \mathfrak{H} = \infty$, the interval is finite, and the expression l whose coefficients are bounded operators is regular, then there is no selfadjoint extension of A with discrete spectrum, in contrast to the case of $\dim \mathfrak{H} < \infty$ where the spectrum of any selfadjoint extension of the minimal operator is discrete. If the coefficients of l are unbounded operators, then selfadjoint extensions of A with discrete as well as continuous spectrum may exist. For an operator differential equation of the kind mentioned, the description in terms of the operator U of all the extensions whose spectrum is discrete was given [27]. Moreover, characterized were those of them whose eigenvalues have preassigned asymptotics [28]. A number of other classes of selfadjoint extensions were also described in the same terms. Later on, these results were extended by V. Bruk, L. Vainerman, M. Gorbachuk, V. Gorbachuk, A. Kochubei, V. Mikhailets, and others (see, e.g. [29]) to more general differential expressions. This led to a description in terms of boundary conditions of all selfadjoint extensions of the minimal operators generated by various partial differential equations.

It turned out that the method of description and classification of selfadjoint extensions of symmetric differential operators is of a universal character. V. Bruk [30] and, independently, A. Kochubei [31] showed that these procedures could be carried over to an arbitrary symmetric operator. We recall the principal momemts of this theory.

Let A be a closed Hermitian operator in $\mathfrak{H}$, $\overline{\mathcal{D}(A)} = \mathfrak{H}$. The triple $(\mathcal{H}, \Gamma_1, \Gamma_2)$, where $\mathcal{H}$ is a Hilbert space, Γ_1 and Γ_2 are linear maps from $\mathcal{D}(A^*)$ into $\mathcal{H}$, is called a space of boundary values (SBV) of the operator A if

$$\forall f, g \in \mathcal{D}(A^*) \quad (A^* f, g) - (f, A^* g) = (\Gamma_1 f, \Gamma_2 g)_{\mathcal{H}} - (\Gamma_2 f, \Gamma_1 g)_{\mathcal{H}},$$

and

$$\forall F_1, F_2 \in \mathcal{H} \quad \exists f \in \mathcal{D}(A^*) : \Gamma_1 f = F_1, \ \Gamma_2 f = F_2.$$

Every symmetric operator always has an (SBV). If an (SBV) is given, then the restriction of A^* to the set

$$\{f \in \mathcal{D}(A^*) | (U - I)\Gamma_1 f + i(U + I)\Gamma_2 f = 0\}, \tag{2}$$

where U is a unitary operator on $\mathcal{H}$, is selfadjoint. Conversely, each selfadjoint extension of the operator A is a restriction of A^* to a set of the form (2). The operator U is uniquely determined by an extension. Now various properties of a selfadjoint extension such as the distribution of its spectrum, the fact that the resolvent belongs to a certain ideal, etc. can be characterized in terms of the boundary operator U.

A description of the form (2) was also given for other classes of extensions. Dissipative, solvable, completely solvable, sectorial, lower semibounded extensions, and the extensions preserving gaps were obtained in such a way. For instance, to get all maximal dissipative extensions of A, it is required to take in (2) any contraction on $\mathcal{H}$ as U. Note also that the similar results for nondensely defined symmetric operators were obtained by V. Derkach, M. Malamud, and O. Storozh (for more details about this see in [29, 32]).

Now let us pass to the theory of entire operators. It should be remarked that the concept of an entire operator changed over a period of time, as the theory developed. It is convenient for us to use the definition of an entire operator appearing in M. Krein's one-hour lecture at ICM-66 in Moscow (see [15]). A closed densely defined Hermitian operator in a Hilbert space $\mathfrak{H}$ is called entire (M-entire) if: 1) A is regular; 2) there exists a subspace $M \subset \mathfrak{H}$ (the so-called gauge) such that

$$\forall z \in \mathbb{C} \quad \mathfrak{H} = M \dotplus \mathfrak{M}_z.$$

Let $\mathcal{P}(z)$ denote the projector from $\mathfrak{H}$ onto M parallel to $\mathfrak{M}_z$ (the so-called function of the first kind), and set

$$\mathcal{T}(z) = (A - zI)^{-1}\mathcal{P}(z)$$

(the function of the second kind). If A is entire, then the functions $\mathcal{P}(z)$ and $\mathcal{T}(z)$ are entire. The map

$$U : \mathfrak{H} \ni f \longmapsto f_M(z) = \mathcal{P}(z) f \in M$$

transforms the operator A into the operator of multiplication by z:

$$U(Af) = zUf = zf_M(z).$$

It was shown that there exists a function $\rho(\lambda) : \mathbb{R}^1 \to L[M](L[M]$ is the space of all bounded operators on M), $\rho(0) = 0$, $\rho(\lambda - 0) = \rho(\lambda)$, such that the function $(\rho(\lambda) f, f)$ is bounded, monotonically increasing, and

$$(f, g) = \int_{-\infty}^{\infty} (d\rho(\lambda) f_M(\lambda), g_M(\lambda)). \tag{3}$$

for any $f, g \in \mathfrak{H}$.

The problem arises of finding all the functions $\rho(\lambda)$ giving (3). Let $\Sigma(A, M)$ denote the set of such functions. Then there exists a matrix

$$W(z) = \begin{pmatrix} W_{11}(z) & W_{12}(z) \\ W_{21}(z) & W_{22}(z) \end{pmatrix},$$

whose entries $W_{ik}(z)$ are entire operator-functions with values in $L[M]$, such that

$$\int_{-\infty}^{\infty} \frac{d\rho(\lambda)}{\lambda - z} = (W_{22}(z)\tau(z) + W_{21}(z))(W_{12}(z)\tau(z) + W_{11}(z))^{-1}, \tag{4}$$

where

$$\tau(z) = i(I + \Omega(z))(I - \Omega(z))^{-1},$$

$\Omega(z) \in L[M]$, the function $\Omega(z)$ is analytic in the half-plane $\Im z > 0$, and $\|\Omega(z)\| \leq 1$ in this half-plane. The matrix-function $W(z)$ ($W(0) = I$) is invertible in $L[M \oplus M]$. It is J-expansive if $\Im z \geq 0$ and J-contractive if $\Im z \leq 0$. Here

$$J = i \begin{pmatrix} 0 & -I \\ I & 0 \end{pmatrix}.$$

Formula (4) that characterizes the set $\Sigma(A, M)$ in terms of the resolvent matrix $W(z)$ was obtained by M. Krein [10] (1944) for operators with defect-index (1, 1), and generalized by him later to the case of arbitrary finite defect numbers. But this generalization was not published.

If the defect numbers of A are finite, then the entire functions $W_{ik}(z)$ and $f_M(z)$ are of at most exponential type, and the type does not depend on $f \in \mathfrak{H}$. The function

$$\nabla_M(z) = \sup_{f \in \mathfrak{H}} \frac{\|f_M(z)\|}{\|f\|}$$

is log-subharmonic and of at most exponential type. Its indicator diagram coincides with an intercept of the imaginary axis. So,

$$h_M(\varphi) = \overline{\lim_{r \to \infty}} \frac{\nabla_M(re^{i\varphi})}{r} = \begin{cases} h_M(\frac{\pi}{2}) \sin\varphi & \text{if } \varphi \in [0, \pi] \\ -h_M(-\frac{\pi}{2}) \sin\varphi & \text{if } \varphi \in [-\pi, 0] \end{cases}.$$

The indicator diagram of $\nabla_M(z)$ contains that of any $f_M(z)$ and any $W_{ik}(z)$. In this case, the spectrum of each selfadjoint extension $\widetilde{A}$ of A is discrete, and the eigenvalues of $\widetilde{A}$ coincide with the zeroes of a certain entire function $\Phi_{\widetilde{A}}(z)$ of at most exponential type. Thus, the limit

$$\lim_{r\to\infty} \frac{n(r)}{r}$$

exists, where $n(r)$ is the number of eigenvalues in the interval $(-r, r)$. This limit equals l/π (l is the length of the indicator diagram of $\Phi_{\widetilde{A}}(z)$). If the defect-index of A is infinite, then, as examples show, the functions $f_M(z)$ are not necessarily of exponential type. In this situation, there exists a selfadjoint extension of A whose spectrum contains a preassigned set.

It turns out that if the operator A has finite defect numbers, then on the basis of V. Potapov's results [33], the resolvent matrix $W(z)$ admits the multiplicative representation

$$W(z) = \int_0^{b} \exp(-izJ)H(t)\,dt,$$

where $H(t) \in L[M \oplus M]$ for each $t \in [0, b]$. In particular, it follows from this representation that $W(z)$ is the monodromy matrix of the canonical system

$$\frac{dW(t,z)}{dt} = zJH(t)W(t,z), \quad W(0,z) = I, \quad 0 \le t \le b, \tag{5}$$

that is, $W(z) = W(b, z)$.

If the defect-index of the operator A is $(1, 1)$, the real symmetric matrix $H(t)$ normalized by the condition $SpH(t) = 1$, $t \in [0, b]$, is uniquely determined. This result was formulated in M. Krein's lectures as an undoubted hypothesis. He proved it for positive operators. In the general case, the proof was given by L. de Branges in his paper [34] by using function not operator methods. M. Krein said about this work: "I find this work to be brilliant. In a short period of time, he covered the distance which took me too many years. However, the time I was dealing with those problems, a lot of results needed for the construction of the theory of entire operators had not been discovered yet. I had to prove some of them myself. Today, one can find them in the books by R. Boas or B. Levin. Refering as a rule to R. Boas, L. de Branges repeated many of my statements. But the final result concerning not only positive but any entire operator with defect-index (1, 1) was established by him. This is what I didn't have. I was aspiring to it but had not got there". These words may be compared perhaps only with L. Euler's expression concerning Lagrange's solution of the isoperimetric problem. In his letter to J. Lagrange (1750), L. Euler wrote: "Your analytical solution of the isoperimetric problem contains everything that one can desire in this field. I am very glad that the theory I was interested in almost alone after my first attempts, due to you, came to the greatest perfection".

The set $\Sigma(A, M)$ coincides with the totality of all spectral functions of an incomplete boundary value problem for the system (5), and the functions $f_M(z)$ can be interpreted as the generalized Fourier transform of the corresponding functions from $L_2(M, (0, b))$ if we use the fundamental solution of this equation normalized in an appropriate way.

The theory of entire operators developed by M. Krein in the case of finite defect numbers and by M. Krein and N. Saakian in the case of infinite ones enabled one to consider from a uniform point of view such diverse problems as the classical moment problem (scalar and operator) and the continuation problem for positive definite functions. In the so-called indefinite case, the Hermitian operators associated with these problems are entire. Therefore, their solutions can be obtained from formula (4). However, there are a number of problems in analysis (for example, the continuation problem for spiral arcs) and the theory of differential equations in which the corresponding operators are entire with respect to a generalized not ordinary gauge M. In the case of finite defect numbers, the theory of such operators was constructed by M. Krein, but it was not published. For operators with infinite defect-indices the question remained open for a long time. In his lecture at ICM-66, M. Krein set up several problems. The following two were among them:

1) to extend the theory of entire operators to the case where a gauge consists of generalized elements whose singularity order is 1;
2) to give examples of entire operators representable by partial differential ones.

The first was solved by Yu. Shmulian [35–38] in the beginning of seventies. As for the second, M. Gorbachuk and V. Gorbachuk [39, 40] showed that the operators generated by some hyperbolic partial differential equations are entire with infinite defect numbers. Note that examples of entire operators with infinite defect-indices were considered by Yu. Berezansky [41] back in 1956. Those operators were generated by second-order partial difference expressions. All their spectral functions were described and an analogue of the Weyl circles was constructed. These results were extended by V. Tarnopol'sky [42] to ordinary difference expressions with operator coefficients that is, to the indefinite case of the operator moment problem. Entire operators with infinite defect numbers, arising from the continuation problem for operator-valued positive definite functions, were studied by M. Gorbachuk [43].

In solving problem 1), the theory of spaces with positive and negative norms in the sense of Yu. Berezansky (see [44]) was used. Let us dwell briefly on the principal aspects of Yu. Shmulian's construction. Put

$$\mathfrak{H}_+ = \mathcal{D}(A^*), \quad \|f\|_+ = \|A^* f\| + \|f\|,$$

and let $\mathfrak{H}_-$ denote the dual of $\mathfrak{H}_+$. Then we have the chain of spaces

$$\mathfrak{H}_+ \subseteq \mathfrak{H} \subseteq \mathfrak{H}_-,$$

embedded continuously and densely into each other. Let M be a subspace of $\mathfrak{H}_-$. The operator A is called M-entire if A is regular and

$$\mathfrak{H}_- = M \dot{+} \overline{\mathfrak{M}_z} \quad (\forall z \in \mathbb{C})$$

(the cloasure is taken in $\mathfrak{H}_-$).

Now $\mathcal{P}(z)$ denotes the projector from $\mathfrak{H}_-$ onto M parallel to $\overline{\mathfrak{M}_z}$, the function $\rho(\lambda)$ is such that

$$\int_{-\infty}^{\infty} \frac{d\rho(\lambda)}{1+\lambda^2} < \infty,$$

and

$$C + \int_{-\infty}^{\infty} \left(\frac{1}{\lambda - z} - \frac{\lambda}{1+\lambda^2} \right) d\rho(\lambda)$$

with the constant $C \in L[M]$ replaces the integral in the left hand side of (4).

Problem 2) is solved in such a way. The expression

$$l[y](t) = y''(t) + a(t)y(t) + q(t)y(t), \quad t \in [0, b], \quad b < \infty,$$

is considered, where for any $t \in [0, b]$, $a(t)$ is a nonnegative selfadjoint operator in a Hilbert space H, $\mathcal{D}(a(t)) = \mathcal{D}$ does not depend on t, $q(t) = q^*(t) \in L[H]$, the vector-function $a(t)f$ $(f \in \mathcal{D})$ is continuously differentiable, and the operator-function $q(t)$ is continuous on $[0, b]$. Then there exist $L[H]$-valued solutions $\omega_1(t, \lambda), \omega_2(t, \lambda)$ of the equation

$$l[y](t) = \lambda y(t)$$

satisfying the initial conditions

$$\omega_1(0, \lambda) = I, \quad \omega_1'(0, \lambda) = 0,$$
$$\omega_2(0, \lambda) = 0, \quad \omega_2'(0, \lambda) = I.$$

The map

$$y \longmapsto \widetilde{y}(\lambda) = \int_0^b \omega_1^*(t, \overline{\lambda}) y(t)\, dt$$

transforms the space $\mathfrak{H} = L_2(H, [0, b])$ into H so that the vector-function $\widetilde{y}(\lambda)$ is entire.

Let L_0 be the minimal operator generated by the expression $l[y]$ and the boundary condition

$$y'(0) = 0.$$

Then the operator L_0 is M-entire with

$$M = \{\delta_0 f\}_{f \in H},$$

the functions of the first and the second kind are of the form

$$\mathcal{P}(\lambda)y = \delta_0\widetilde{y}(\lambda),$$

$$\mathcal{T}(\lambda)y = \int_t^b (\omega_1(t,\lambda)\omega_2^*(\xi,\overline{\lambda}) - \omega_2(t,\lambda)\omega_1^*(\xi,\overline{\lambda}))y(\xi)\,d\xi,$$

respectively, and the Parseval equality

$$(y,z) = \int_{-\infty}^{\infty} (d\rho(\lambda)\widetilde{y}(\lambda), \widetilde{z}(\lambda))_H \tag{6}$$

holds. The resolvent matrix $W(z)$ giving the description of all $\rho(\lambda)$ in (6) takes the form

$$W_{11}(\lambda) = 1 + \lambda\int_0^b \omega_1^*(t,\overline{\lambda})\omega_2(t,0)\,dt;$$

$$W_{12}(\lambda) = -\lambda\int_0^b \omega_1^*(t,\overline{\lambda})\omega_1(t,0)\,dt;$$

$$W_{21}(\lambda) = \lambda\int_0^b \omega_2^*(t,\overline{\lambda})\omega_2(t,0)\,dt;$$

$$W_{22}(\lambda) = 1 - \lambda\int_0^b \omega_2^*(t,\overline{\lambda})\omega_1(t,0)\,dt.$$

In particular, if

$$H = \mathbb{C}, \quad a(t) \equiv 0, \quad q : [0,b] \longrightarrow \mathbb{R}^1 \in C[0,b],$$

then formula (4) gives all the distribution functions of the minimal operator generated by the usual Sturm-Liouville expression. If

$$H = L_2(\mathbb{R}^n), \quad a(t) \equiv a = -\Delta, \quad q(t) \equiv 0,$$

then

$$l[y] = \frac{\partial^2 u}{\partial t^2} - \Delta u, \quad y = u(t,\cdot). \tag{7}$$

The minimal operator generated in the space $\mathfrak{H} = L_2([0,b]\times\mathbb{R}^n)$ by the expression (7) and the boundary condition

$$\left.\frac{\partial u}{\partial t}\right|_{t=0} = 0$$

is entire.

References

[1] J. von Neumann, "*Allgemeine Eigenwerttheorie Hermitesher Funktionaloperatoren*", Math. Ann. **102** (1929), 49–131.

[2] H. Weyl, "*Uber gewohnliche lineare Differentialgleichungen mit Singularen Stellen und ihre Eigenfunktionen*", Nachr. Kgl. Ges. Wiss. Göttingen. Math.-phys. Kl. (1909), 37–63.

[3] H. Weyl, "*Uber gewohnliche Differentialgleichungen mit Singularitaten und die Zugehorigen Entwicklungen willkurlicher Funftionen*", Math. Ann. **68** (1910), 220–269.

[4] M.G. Krein, "*The theory of selfadjoint extensions of semibounded Hermitian operators and its applications, part I*" (Russian), Mat. Sbornik **20** (1947), no. 3, 431–495.

[5] M.G. Krein, "*The theory of selfadjoint extensions of semibounded Hermitian operators and its applications, part II*" (Russian), Mat. Sbornik **21** (1947), no. 3, 365–404.

[6] M.G. Krein and M.A. Krasnosel'skii, "*The principal theorems of extension theory of Hermitian operators their applications to the theory of orthogonal polynomials and the moment problem*" (Russian), Uspekhi Mat. Nauk **3** (1947), no. 3, 61–106.

[7] F. Riesz and B. Sz.-Nagy, "*Lectures on functional analysis*" (Russian translation), Moscow, Mir, 1979.

[8] N.I. Akhiezer and I.M. Glazman, "*Theory of linear operators in a Hilbert space*" (Russian), Moscow-Leningrad, Gostekhizdat, 1950.

[9] M.A. Naimark, "*Linear differential operators*" (Russian), Moscow, Gostekhizdat, 1954.

[10] M.G. Krein, "*On Hermitian operators with defect-indexes equal to 1, part I*" (Russian), Dokl. Akad. Nauk SSSR **43** (1944), no. 8, 339–342.

[11] M.G. Krein, "*On Hermitian operators with defect-indexes equal to 1*" (Russian), Dokl. Akad. Nauk SSSR **44** (1944), no. 4, 143–146.

[12] M.G. Krein, "*On a remarkable class of Hermitian operators*" (Russian), Dokl. Akad. Nauk SSSR **44** (1944), no. 5, 191–195.

[13] M.G. Krein, "*On resolvents of a Hermitian operators with defect-index* (m, m)" (Russian), Dokl. Akad. Nauk SSSR **52** (1946), no. 8, 657–660.

[14] M.G. Krein, "*The main principles of representation theory of Hermitian operators with defect-index (m, m)*" (Russian), Ukrain. Mat. Zh. (1949), no. 2, 1–66.

[15] M.G. Krein, "*Analytical problems and results of the theory of linear operators in a Hilbert space*" (Russian), Proceedings of ICM, Moscow, Mir, 1968, 189–216.

[16] M.G. Krein and Sh.N. Saakian, "*Some new results in the theory of resolvents of Hermitian operators*" (Russian), Dokl. Akad. Nauk SSSR **169** (1966), no. 6, 1269–1272.

[17] M.G. Krein and Sh.N. Saakian, "*Resolvent matrix of a Hermitian operator and characteristic functions connected with it*" (Russian), Function. Anal. i Prilozh. **4** (1970), no. 3, 103–104.

[18] T.Ya. Azizov, Yu.P. Ginzburg and H. Langer, "*On M.G. Krein's papers in the theory of spaces with an indefinite metric*", Ukr. Math. J. **46** (1994), no. 1–2, 3–14.

[19] M.L. Gorbachuk and V.I. Gorbachuk, "*M.G. Krein's lectures on entire operators*", Basel/Boston/Berlin, Birkhäuser Verlag, 1997.

[20] M.G. Krein, "*Selected works, II. Banach spaces and operator theory*", Kiev, Insitute of Mathematics, 1996.

[21] N. Dunford and J. Schwartz, "*Linear operators, spectral theory. Selfadjoint operators in Hilbert spaces*", New York/London, Interscience publishers, 1963.

[22] M.I. Vishik, "*On general boundary value problems for elliptic differential equations*" (Russian), Trudy Moscov. Mat. Obšč **1** (1952), 187–246.

[23] M.Sh. Birman, "*On the theory of selfadjoint extensions of positive definite operators*" (Russian), Mat. Sbornik **38** (1956), no. 4, 431–450.

[24] F.S. Rofe-Beketov, "*Selfadjoint extensions of differential operators in a space of vecror-functions*" (Russian), Dokl. Akad. Nauk SSSR **184** (1969), no. 5, 1034–1037.

[25] F.S. Rofe-Beketov, "*On selfadjoint extensions of differential operators in a space of vecror-functions*" (Russian), Teor. Funksii, Funkt. Anal. i Prilozh. **8** (1969), 3–24.

[26] M.L. Gorbachuk, "*Selfadjoint boundary value problems for a second-order differential equation with unbounded operator coefficient*" (Russian), Funkts. Anal. i Prilozh. **5** (1971), no. 1, 10–21.

[27] V.I. Gorbachuk and M.L. Gorbachuk, "*The spectrum of selfadjoint extensions of the minimal operator generated by a Sturm-Liouville equation with operator potential*" (Russian), Ukrain. Mat. Zh. **24** (1972), no. 6, 726–734.

[28] V.I. Gorbachuk and M.L. Gorbachuk, "*Classes of boundary-value problems for the Sturm-Liouville equation with an operator potential*" (Russian), Ukrain. Mat. Zh. **24** (1972), no. 3, 291–304.

[29] V.I. Gorbachuk, M.L. Gorbachuk and A.N. Kochubei, "*The extension theory of symmetric operators and boundary value problems for differential equations*" (Russian), Ukrain. Mat. Zh. **41** (1989), no. 10, 1299–1313.

[30] V.M. Bruk, "*On a class of boundary value problems with spectral parameter in the boundary condition*" (Russian), Mat. Sbornik **100** (1976), no. 2, 210–216.

[31] A.N. Kochubei, "*Extensions of symmetric operators and symmetric binary relations*" (Russian), Mat. Zametki **17** (1975), no. 1, 41–48.

[32] V.A. Derkach and M.M. Malamud, "*Generalized resolvents and boundary value problems for Hermitian operators with gaps*", J. of Functional analysis **95** (1991), no. 1, 1–95.

[33] V.P. Potapov, "*Multiplicative structure of J-contractive matrix-functions*" (Russian), Trudy Moskov. Mat. Obšč **4** (1955), 125–136.

[34] L. Branges, "*Some Hilbert spaces of entire functions IV*", Trans. Amer. Math. Soc. **105** (1962), 43–83.

[35] Yu.L. Shmulian, "*Extended resolvents and extended spectral functions of a Hermitian operator*" (Ukrainian), Dokl. Akad. Nauk Ukrain. SSR, Ser. A (1970), no. 3, 230–234.

[36] Yu.L. Shmulian, "*Representation of a Hermitian operator with a generalized module*" (Ukrainian), Dokl. Akad. Nauk Ukrain. SSR, Ser. A (1970), no. 5, 432–435.

[37] Yu.L. Shmulian, "*Direct and inverse problem of the resolvent matrix theory*" (Ukrainian), Dokl. Akad. Nauk Ukrain. SSR, Ser. A (1970), no. 6, 514–517.

[38] E.P. Tzekanovsky and Yu.L. Shmulian, "*Aspects of the extension theory for unbounded operators in rigged Hilbert spaces*" (Russian), Itogi nauki i tekhniki. Mat. Analiz **14** (1977), 59–100.

[39] V.I. Gorbachuk and M.L. Gorbachuk, "*An example of an entire operator whose defect numbers are infinite*" (Ukrainian), Dokl. Akad. Nauk Ukrain. SSR, Ser. A (1970), no. 7, 579–582.

[40] M.L. Gorbachuk and V.I. Gorbachuk, "*Realization of entire operators by differential ones*", Integr. equ. oper. theory **30** (1998), 200–230.

[41] Yu.M. Berezanskii, "*Expansion in eigenfunctions of second-order partial difference equations*" (Russian), Trudy Moscov. Mat. Obšč **5** (1956), 203–268.
[42] V.G. Tarnopol'skii, "*Absolutely indefinite case for a difference operator with operator coefficients*" (Ukrainian), Dokl. Akad. Nauk Ukrain. SSR, Ser. A (1960), no. 3, 305–308.
[43] M.L. Gorbachuk, "*On description of continuations of a positive definite operator-function*" (Russian), Ukrain. Mat. Zh. **17** (1965), no. 5, 102–110.
[44] Yu.M. Berezansky, "*Expansion in eigenfunctions of self-adjoint operators*" (Russian), Kiev, Naukova dumka, 1965.

Institute of Mathematics
National Academy of Sciences of Ukraine
Kyiv 21.08.1997
47B25, 34G10

Operator Theory:
Advances and Applications, Vol. 117
© 2000 Birkhäuser Verlag Basel/Switzerland

Works by M.G. Krein on Inverse Problems

L.A. Sakhnovich

Among brilliant scientific achievements of M.G. Krein his results on the theory of inverse problems occupy one of the first places. In this paper we are going to describe the main ideas and results of M.G. Krein on the inverse problems as well as their development and continuation in the works of M.G. Krein himself and of other authors.

1. Let the Sturm-Liouville equation

$$-\frac{d^2y}{dy^2} + P(x)y - \lambda y = 0 \tag{1}$$

with a boundary condition be given. Two types of inverse problems arise in connection with this equation. The first type is:

I) To define the potential $P(x)$ by the given spectral characteristics.

The second type is:

II) To define the potential $P(x)$ by the given scattering data.

The first problem was mentioned by Rayleigh [1] who thought about the method of reconstructing the density of the non-homogeneous string by its spectrum. The first result in the theory of inverse problem was obtained by V.A. Ambartsumyan [2], who proved the following theorem.

Theorem: *Let $\lambda_0, \lambda_1, \ldots$ be eigenvalues of the boundary problem*

$$\begin{aligned} y'' + [\lambda - q(x)]y &= 0, \qquad 0 \leqslant x \leqslant \pi \\ y'(0) &= y'(\pi) = 0 \end{aligned}$$

where $q(x)$ is a real continuous on the segment $[0, \pi]$ function. If $\lambda = n^2 (n = 0, 1, \ldots)$ then $q(x) = 0$

The first serious breakthrough in this domain was made by G. Borg [3]. Borg in particular showed that one spectrum generally speaking isn't sufficient for determining $q(x)$ (the result of Ambartsumyan is an exclusion from the general rule). Then A.Ja. Povsner [4] constructed a transformation operator mapping the solution of the simplest equation

$$-\frac{d^2\omega}{dx^2} - \lambda\omega = 0 \tag{2}$$

into the solution $y(x, \lambda)$ of equation (1). V.A. Marchenko [5] proved with the help of the transformation operator that the potential $P(x)$ is defined uniquely by the spectral data. Soon after it there appeared a work by I.M. Gelfand-B.M. Levitan [6] in which the inverse spectral problem was reduced to the solution of a linear integral equation. M.G. Krein in his work [6] deduced the necessary and sufficient conditions of the inverse spectral problem solvability. In his further works [8]–[10] M.G. Krein solved the inverse spectral problem for the Dirac type system and constructed classes of explicit solutions. The second inverse problem (the inverse scattering problem) arose in the quantum scattering theory. This problem was studied both by physicists and mathematicians. Its final solution was obtained in the works by V.A. Marchenko [11] and M.G. Krein [12].

2. Now I shall consider one article by M.G. Krein [8] belonging to his cycle of works on inverse problem. I should like to show what a great number of ideas is contained in this short paper (4 pages). At the beginning of this article M.G. Krein introduces the non-negative operator

$$Kf = \mu\varphi + \int_0^r H(t-s)\varphi(s)\,ds, \qquad \mu > 0. \tag{3}$$

Proceeding from his results on the Hermitian-positive functions [12] M.G. Krein writes the following representation

$$\begin{aligned} \int_0^t (t-s)H(s)\,ds \;=\; & \int_{-\infty}^{\infty}\left(1+\frac{i\lambda t}{1+\lambda^2}-e^{i\lambda t}\right)\frac{d\sigma(\lambda)}{\lambda^2} \\ & + \left(i\gamma - \frac{\mu}{2}\operatorname{sign} t\right)t, \quad (-T<t<T) \end{aligned} \tag{4}$$

where $\gamma = \bar{\gamma}$ and $\sigma(\lambda)$ is a monotonically increasing function such that

$$\int_{-\infty}^{\infty}\frac{d\sigma(\lambda)}{1+\lambda^2} < \infty. \tag{5}$$

I should like to stress that the problem of representing the operator kernel in the form of type (4) is actual even today. A number of works of the interpolation theory is dedicated to constructing representations of type (4) for different classes of operators (see [14], [15], [16]).

Let us come back to the article in question. M.G. Krein introduces the resolvent kernel $\Gamma_r(t, s)$ defined by relation ($\mu = 1$):

$$\Gamma_r(t,s) - \int_0^r H(t-u)\Gamma_r(u,s)\,du = H(t-s). \tag{6}$$

In one of his previous works he proved an important formula

$$\frac{\partial\Gamma_r(t,s)}{\partial r} = -\Gamma_r(t,s)\Gamma_r(t,r). \tag{7}$$

From formulas (6) and (7) M.G. Krein deduces the system of differential equations

$$(8)\qquad \begin{cases} \dfrac{dP(r,\lambda)}{dr} = i\lambda P(r,l) - \overline{A(r)}P_*(r,\lambda), \\ \dfrac{dP_*(r,\lambda)}{dr} = -A(r)P(r,\lambda)\,, \end{cases}$$

where

$$(9)\qquad \begin{cases} P(r,\lambda) = e^{i\lambda r}\left[1 - \displaystyle\int_0^r \Gamma_r(s,0)e^{-i\lambda s}ds\right], \\ P_*(r,\lambda) = 1 - \displaystyle\int_0^r \Gamma_r(0,s)e^{i\lambda s}ds, \end{cases}$$

$$(10)\qquad A(r) = \Gamma_r(0,r).$$

It follows from (8) that

$$(11)\qquad |P_*(r,\lambda)|^2 - |P(r,\lambda)|^2 = 2\,\mathrm{Im}\,\lambda \int_0^r |P(s,\lambda)|^2 ds.$$

Due to relation (11) all the zeroes of the whole function $P(r, \lambda)$ are in the upper half-plane $\mathrm{Im}\,\lambda \geqslant 0$.

This part of the article by M.G. Krein is a kind of an exposition. All the principal characters are introduced. The main connection between them (differential system (8) and relations (4), (10)) are outlined.

In Krein's aproach to the solution of the inverse problem the transformation operator is not used in the explicit form. But it is present behind the scene. The matter is that the operator K (defined by (3)) and the transformation operator V are tied by the relation

$$(12)\qquad K^{-1} = V^*V.$$

The transition from the transformation operator V to the operator K allowed Krein to join the necessary and sufficient conditions of the solvability of the inverse problem.

Formula (12) also gives the solution of the factorisation problem, i.e. the representation of an operator in the form of the product of two triangular operators. This theme was later developed in the famous works by Gohberg and Krein [17] where the problem of factorizing an operator close to the identity operator was solved. I should like to point out that formula (11) also had interesting applications both in the works by M.G. Krein [18] and in the works by some other mathematicians [19], [20], [21]. It turns out that formula (11) makes it possible to generalize the classic Schur-Cohn theorems on the polynomial roots for the entire functions of the form

$$F(z) = 1 + z\int_0^\omega e^{izt}\Phi(t)\,dt.$$

Now we shall proceed to the main part of M.G. Krein's article.

In this part M.G. Krein introduced the generalized Fourier transformation

$$F(\lambda) = (U_p f)(z) = \int_0^\infty f(r)P(r,\lambda)\,dr,$$

for which the Parseval equality

$$\int_0^\infty |f(r)|^2\,dr = \int_{-\infty}^\infty |F(\lambda)|^2\,d\sigma(\lambda)$$

is true. Hence the function $\sigma(\lambda)$ from representation (4) turns to be a spectral function of system (8). M.G. Krein formulates the necessary and sufficient condition under which the operator U_p maps $L^2(0,\infty)$ onto $L^2(\sigma)$, i.e. the condition of the orthogonality of the spectral function $\sigma(\lambda)$. This condition has the form

$$\int_{-\infty}^\infty \frac{\ln \sigma'(\lambda)}{1+\lambda^2} d\lambda = -\infty.$$

M.G. Krein separately considers the important case when

$$\int_{-\infty}^\infty \frac{\ln \sigma'(\lambda)}{1+\lambda^2} d\lambda > -\infty. \tag{13}$$

In this case he formulates four conditions which are equivalent to condition (13). One of these conditions is the existence of the limit

$$\mathcal{P}(\lambda) = \lim_{r\to\infty} P_*(r,\lambda). \tag{14}$$

The function $\mathcal{P}(\lambda)$ and the spectral function $\sigma(\lambda)$ are connected by the relation

$$\mathcal{P}(\lambda) = \frac{1}{\sqrt{2\pi}} \exp\left\{\frac{1}{2\pi}\int_{-\infty}^\infty \frac{1+t\lambda}{t-\lambda}\frac{\ln\sigma'(t)}{1+t^2}\,dt + \alpha i\right\}, \quad \alpha = \bar{\alpha}. \tag{15}$$

As M.G. Krein remarks it follows from formula (15) that $\mathcal{P}(\lambda)$ depends only on the absolutely continuous part of the spectral function $\sigma(\lambda)$.

The enumerated results are continual analogues of the results obtained for the orthogonal polynomials (G. Szegöl [22], A.N. Kolmogorov [23], Geronimus [24], M.G. Krein [25]).

Let us recollect Odessa mathematician P.D. Kalafatty as well, whose work [26] prompted M.G. Krein to investigate the function $P(r,\lambda)$. But where is the inverse problem? Its solution is contained in the above formulas.

1) With the help of formula (4) we construct $H(t)$ by $\sigma(\lambda)$.
2) Then by solving linear equation (3) we find $\Gamma_r(t,s)$.
3) Using formula (10) we find $A(r)$, i.e. we solve the inverse problem.

In conclusion M.G. Krein shows with the help of elementary transformations that the Dirac type system

$$\text{(16)} \qquad \begin{cases} \dfrac{d\Phi}{dr} = -\lambda\Psi - a(r)\Phi + b(r)\Psi \\ \dfrac{d\Psi}{dr} = \lambda\Phi + b(r)\Phi + a(r)\Psi \end{cases}$$

can be reduced to system (8) and here

$$\text{(17)} \qquad a(r) + ib(r) = 2\Gamma_{2r}(0, 2r).$$

The Sturm-Liouville equation can be obtained from system (16) when $b(r) = 0$:

$$\text{(18)} \qquad \Psi'' - V(r)\Psi + \lambda^2\Psi = 0$$

and

$$\text{(19)} \qquad \Phi'' - V_1(r)\Phi + \lambda^2\Phi = 0,$$

where

$$\text{(20)} \qquad V(r) = a^2(r) + a'(r), \qquad V_1(r) = a^2(r) - a'(r).$$

Thus formulas (20) realize the transition from the Dirac type system (16) to the Sturm-Liouville equations (18) and (19). This explains why Miura transformation which is also of form (20) transforms the solution of the MKdV equation in the solution of KdV equation.

The ideas and results of the paper in question led M.G. Krein and other authors to new important results: the factorization of the operators, the operator Bezoutiant, the theory of integral equations with the difference kernels. This combination of the completeness of the obtained results and of the great prospects makes a profound impression. In this connection I want to quote some lines of Russian poet Majkov concerning on artistic creation:

"Let us suddenly come out of darkness
And dispell that darkness
Being beautiful of itself
And of the infinity of the distance lying behind it."

3. A number of M.G. Krein works is dedicated to an effective solution of inverse problems.

In the work [9] M.G. Krein constructed explicit solutions of the inverse spectral problem for the Sturm-Liouville equation

$$\Psi'' - \left[\frac{p(p-1)}{r^2} + q(r)\right]\Psi + \lambda\Psi = 0, \qquad 0 \leqslant r < \infty,$$

where $q(r)$ is a continuous function,

$$\Psi(r, \lambda) \sim r^p, \qquad r \longrightarrow 0, \qquad p = 1, 2, \ldots.$$

M.G. Krein found an effective method of reconstructing $q(r)$ in the case when the spectral function $\sigma(\lambda)$ has the form

$$\sigma(\lambda)=\begin{cases}\dfrac{1}{\pi}\displaystyle\int_0^\lambda R(\mu)\frac{d\mu}{\mu}, & \lambda>0,\\ 0, & \lambda\leqslant 0,\end{cases}$$

where $R(\lambda)$ is a rational non-negative function on the positive half-axis and

$$R(\lambda)=\lambda^p[1+o(1)],\qquad \lambda\longrightarrow\infty.$$

In this work M.G. Krein used the well-known result by M.M. Crumm [27]. Here one of the characteristic features of M.G. Krein was vividly displayed: his ability to pass from general ideas and results to concrete examples and to see in concrete examples roots of new general theories.

We note that papers of V. Bargmann [28], [29] already contain some technique of explicit solution of the inverse problem.

Explicit solutions of inverse spectral problems were unexpectedly applied in the theory of integrable non-linear equations. This gave an impulse to constructing new classes of explicit solutions (R. Hirota [30], M.J. Ablowitz [31], V.A. Marchenko [32], A.L. Sakhnovich [33]).

4. The works on inverse problems were published M.G. Krein mainly in the Reports of the Academy of Sciences, i.e. mostly without any proofs. It is not at all characteristic of M.G. Krein. His results in other domains have as a rule complete proofs. By now the greatest part of the proofs of M.G. Krein's results on inverse problems have been reconstructed (see L.D. Faddeev [34], H. Dym, H.P. McKean [35], L.A. Sakhnovich [36]).

5. Now I shall try to describe in short some new directions in the theory of inverse problems.

I Canonical Differential Systems

The differential system of the form

$$\frac{dY}{dx}=izJ\mathcal{H}(x)Y$$

is called a canonical system. Here J and $\mathcal{H}(x)$ are $(2m)\times(2m)$ matrices and

$$\mathcal{H}(x)\geqslant 0,\qquad J=\begin{bmatrix}0 & E_m\\ E_m & 0\end{bmatrix}.$$

The canonical systems play an important role in many problems of mechanics. The matrix Sturm-Liouville equations, the Dirac type equations, the string matrix equations can be reduced to the canonical form. It has turned out that the inverse spectral problem for the canonical systems can have more then one solution. There exist different classes of canonical systems which have one and the same set of spectral data (see [37]), i.e. they are isospectral classes.

Thus we must divide the set of the canonical systems into classes and together with the spectral data we must give information about the class in which we plan to find the corresponding Hamiltonian $\mathcal{H}(x)$ [37].

II Inverse Problems with a Fixed Energy

Let us consider the equation

$$\frac{d^2\varphi_l}{dr^2} + \left[\lambda - V(r) - \frac{l(l+1)}{r^2}\right]\varphi_l = 0, \qquad r \geqslant 0. \tag{21}$$

The parameter λ is fixed ($\lambda = 1$), $l = 0, 1, 2, \ldots$ and $\varphi(r, \lambda)$ is a function such that

$$\varphi(r, \lambda) \sim r^{l+1}, \qquad r \longrightarrow 0,$$

$$\varphi(r, \lambda) \sim A_l \sin\left(r - \frac{l\pi}{2} + \sigma_l\right), \qquad r \longrightarrow \infty.$$

Problem: The sequence σ_l is given. It is necessary to find the corresponding potential $V(r)$. It has been proved that this problem can have more then one solution (see K. Chadan and P.S. Sabatier [38], B.M. Levitan [39]).

For the formulated inverse problem the method of the classic inverse problem has been used (the method of the transformation operator). This direction of investigations is far from being completed.

I should like to pose another problem concerning this topic. Let us consider the equation

$$\frac{d^2\varphi_l}{dr^2} + \left[\lambda + \frac{2z}{r} - q(r) - \frac{l(l+1)}{r^2}\right]\varphi_l = 0, \qquad z > 0. \tag{22}$$

The discrete spectrum of equation (22) is described by the Ritz formula [40]

$$\lambda_n = \frac{-z^2}{[n + l + \Delta_l + \chi(n, l)]^2},$$

where $\chi(n, l) \to 0, n \to \infty$.
The number Δ_l is called a quantum defect of the spectrum.

Problem: Find a method for recovering the potential $q(r)$ from Δ_l, ($l = 0, 1, 2, \ldots$). Here the formula for constructing Δ_l can be used (see [40]).

III Half-inverse Problems

Let us consider the Sturm-Liouville equation

$$-\frac{d^2y}{dx^2}+P(x)y-\lambda y=0, \qquad -l\leqslant x\leqslant l.$$

There is a number of important and interesting problems of the following type:

1) There is some information concerning the potential $P(x)$.

2) A part of the spectral data is known.

It is necessary to reconstruct $P(x)$. It is natural to call the problems of this type half-inverse ones.

I shall give some examples of half-inverse problems. It is well-known that in the case of a finite segment it is necessary to know two spectra for reconstructing $P(x)$ (see [41]).

Half-inverse problem 1. It is enough to know only one spectrum for reconstructing $P(x)$ if $P(x)$ is an even function

$$P(x)=P(-x).$$

The solution of this problem follows directly from the results of the classic inverse problem.

The following problem is a certain generalization of the above mentioned one.

Half-inverse problem 2. The spectrum of the Sturm-Liouville problem and the uneven part of $P(x)$

$$h_0(x)=P(x)-P(-x)$$

are given. It is necessary to reconstruct $P(x)$.

This problem was considered in my work [42]. M.G. Krein got interested in this direction, but said about the results that it was only a "reconnaissance by a fight".

Lately many works by mathematicians and physicists dedicated to the half-inverse problems on the axis $(-\infty<x<\infty)$ have appeared.

Half-inverse problem 3. The potential $P(x)$ of the Sturm-Liouville equation

$$-\frac{d^2y}{dx^2}+P(x)y-\lambda y=0, \qquad -\infty<x<\infty$$

is known when $x\geqslant 0$. It is necessary to reconstruct $P(x)$ by incomplete spectral data (H. Hochstadt, B. Lieberman [43], F. Gesztezy, B. Simon [44]).

I shall formulate another *unsolved inverse problem.* Let us consider the radial Dirac equation

$$\begin{cases} \left(\frac{d}{dr}+\frac{k}{r}\right)\Psi_1-[m+\lambda-V(r)]\Psi_2=0, & 0\leqslant r\leqslant\infty, \\ \left(\frac{d}{dr}-\frac{k}{r}\right)\Psi_1-[m-\lambda+V(r)]\Psi_2=0, & m>0, \quad k=\pm1,\pm2,\dots. \end{cases}$$

(23)

Setting

$$\Psi=\begin{bmatrix}\Psi_1\\ \Psi_2\end{bmatrix}$$

we shall write system (22) in the matrix form

$$\frac{d\Psi}{dr}+H(r)\Psi=\lambda J\Psi,$$

where

$$H(r)=\begin{bmatrix} k/r & V(r)-m \\ -m-V(r) & -k/r\end{bmatrix}, \qquad J=\begin{bmatrix} 0 & E_m \\ -E_m & 0\end{bmatrix}.$$

The matrix $H(r)$ is usually supposed to be an unknown one and the corresponding spectral problem is considered.

However in the case of the Dirac equation the diagonal elements (k/r) are known and the following half-inverse problem arises.

Half-inverse problem 4. Knowing the diagonal elements of $H(r)$ to find the matrix $H(r)$ by incomplete spectral data belonging to the energy interval $0\leqslant\lambda<\infty$.

References

[1] J.W.S. Rayleigh, *The Theory of Sound,* Dover Publications, N.Y. 1945 (1877).

[2] V.A. Ambartsumian, *Über eine Frage der Eigenwerttheorie*, Zeitschrift für Physik, 1929, Bd.53, 690–695.

[3] G. Borg, *Eine Umkehrung der Sturm-Liouvillishen Eigenvertaufgabe*, Acta Mathematica, (1946), 1–96.

[4] A.Ja. Povzner, *On Differential Equations of Sturm-Liouville Type on Half-Axis*, Matem.sb. **23** (1948), 3–52, (Russian).

[5] V.A. Marchenko, *Some Questions of the Theory of One-Dimensional Differential Linear Operators of the Second Order*, I. Tpudy Moscow. Mat. Obshch. (1952), 327–420 (Russian).

[6] B.M. Gelfand and I.M. Levitan, *On the Determination of a Differential Equation from its Spectral Function*, Izv. Akad. Nauk SSSR, ser.math. **15** (1951), 309–360 (Russian).

[7] M.G. Krein, *Topics in Differential and Integral Equations and Operator Theory*, Operator Theory, Birkhäuser Verlag, 1983.

[8] M.G. Krein, *Continuous Analogues of Propositions for Polynomial Orthogonal on the Unit Circle*, Dokl. Akad. Nauk SSSR, **105** (1955), 637–640.

[9] M.G. Krein, *Continual Analogue of Cristoffel Formula from Theory of Orthogonal Polynomials*, Dokl. Akad. Nauk SSSR, **113**, no. 5 (1957), 970–973 (Russian).

[10] M.G. Krein, *On a Method for the Effective Solution of the Inverse Boundary-Value Problem*, Dokl. Akad. Nauk SSSR, **94**, no. 1 (1954) 13–16 (Russian).

[11] V.A. Marchenko, *Sturm-Liouville Operator and their Applications*, Operator Theory, Advances and Applications, Birkhäuser Verlag 22, 1986 (Translation).

[12] M.G. Krein, *Sur le probleme du prolongement des functions hermitiennes positives et continues*, Dokl. Akad. Nauk SSSR, 26:1 (1940), 17–22 (Russian).

[13] M.G. Krein, *On the Theory of Accelerants and S-Matrices of Canonical Differential Systems*, Dokl. Akad. Nauk SSSR, **111** (1956), 1167–1170.

[14] D. Sarason, *Generalized Interpolation in H^{∞}*, Trans. Amer. Math. Soc. **127** (1967), 179–203.

[15] C. Foias and A.E. Frazho, *The Commutant Lifting Approach to Interpolation Problems*, Birkhäuser, Basel, 1990.

[16] L.A. Sakhnovich, *Interpolation Theory and its Applications*, Kluwer Acad. Publ., 1997.

[17] I. Gohberg and M.G. Krein, *Theory and Applications of Volterra Operators in Hilbert Space*, Amer. Math. Soc. Providence, 1970.

[18] M.G. Krein and H. Langer, *Some Proposition on Analytic Matrix Functions Related to the Theory of Operators in the Space Π_k*, Acta Sci. Math. (Szeged) **43** (1981), 181–205.

[19] I. Gohberg and K. Heinig, *Resultant Matrix and its Generalizations*, Acta Math. Acad. Sci. Hungaricas. **28** (3–4) (1976), 189–209.

[20] I. Gohberg and L. Lezer, *Matrix Generalizations of M.G. Krein Theorems on Orthogonal Polynomials*, Operator Theory, Adv. and Appl. **34** (1988), 137–202.

[21] L.A. Sakhnovich, *The Operator Bezoutiant in the Theory of the Separations of the Roots of Entire Functions*, Functional Anal. Appl. **10** (1976), 45–51, (Russian).

[22] G. Szegö, *Orthogonal Polynomials*, Amer. Math. Soc., N.Y. 1959.

[23] A.N. Kolmogorov, *Interpolation und Extrapolation von Stationären Zuffaligen Folgen*, Izv. Akad. Nauk SSSR, **5** (1941), 3–14.

[24] Ja.I. Geronimus, *Polynomials Orthogonal on a Circle and on an Interval*, New York, Pergamon Press, 1966.

[25] M.G. Krein *On a Problem of Extrapolation of A.N.Kolmogorov*, Dokl. Akad. Nauk SSSR **46** (1945), 306–309 (Russian).

[26] P.D. Kalafatti, *On one new Orthogonal System of Functions*, Dokl. Akad. Nauk SSSR **105** (1955), 631–633.

[27] M.M. Crum, *Associated Sturm-Liouville Systems*, Quart. S.Math. **6** (1955) 121–127.

[28] V. Bargmann, *Remarks on the determination of a central field of force from the elastic scattering phase shifts*, Phys. Rev. v. **75** (1949), 301–303.

[29] V. Bargmann, Rev. Mod. Phys. v. **21** (1949), 488.

[30] R. Hirota, *Direct Methods in Solitons Theory*, Solitons, Springer-Verlag, N.Y., 1980.

[31] M.J. Ablowitz and H. Segun, *Solutions and the Inverse Scattering Transform*, SIAM, Philadelfia, 1981.

[32] V.A. Marchenko, *Nonlinear Equations and Operator Algebras*, Kiev, 1986 (Russian).
[33] A.L. Sakhnovich, *Iterated Backlund-Darboux Transform for Canonical Systems*, Journ. of Functional Analysis, **144**, no. 2 (1997), 359–370.
[34] L.D. Faddeev, *The Inverse Problem of the Quantum Theory of Scattering*, Journ. Math. Phys. **4** (1963), 72–104 (Translation).
[35] H. Dym and H.P. McKean, *Gaussian Processes, Function Theory and Inverse Spectral Problem*, Academic Press, New York, 1976.
[36] L.A. Sakhnovich, *On a Class of Canonical Systems on Half-Axis*, Integral Equations and Operator Theory, **31** (1998), 92–112.
[37] L.A. Sakhnovich, *Spectral Analysis of one Class of Canonical Differential Systems*, Algebra and Anal. **10**, no. 1 (1998), 187–201 (Russian).
[38] K. Chadan and P.C. Sabatier, *Inverse Problems in Quantum Scattering Theory*, Springer-Verlag, N.Y., 1977.
[39] B.M. Levitan, *Inverse Sturm-Liouville Problems*, VNU Science Press BV, Utrecht, 1987.
[40] L.A. Sakhnovich, *On Ritz Formula and Quantum Defects of Spectrum of Radial Schrödinger Equation*, Izv. Akad. Nauk SSSR 30:6 (1966), 1297–1310.
[41] B.M. Levitan and M.G. Gasymov, *Determination of a Differential Equation by Two of its Spectra*, Russ. Math. Surv. 19:2 (1964), 1–63.
[42] L.A. Sakhnovich, *On one Half Inverse Problem*, Uspechi Mat. Nauk **18**, no. 3 (1963), 199–206 (Russian).
[43] H. Hochstadt and B. Lieberman, *An Inverse Sturm-Liouville Problem with Mixed Given Data*, SIAM J. Appl. Math. **34** (1978), 676–680.
[44] F. Gesztezy and B. Simon, *Inverse Spectral Analysis with Partial Information on the Potential*, Helv. Phys. Acta, v. **70** (1997), 60–71.

Operator Theory:
Advances and Applications, Vol. 117
© 2000 Birkhäuser Verlag Basel/Switzerland

The Spectrum of Periodic Point Perturbations and the Krein Resolvent Formula

J. Brüning and V.A. Geyler

We study periodic point perturbations H of a periodic elliptic operator H^0 on a connected complete non-compact Riemannian manifold X, endowed with an isometric, effective, properly discontinuous, and co-compact action of a discrete group Γ. Under some conditions on H^0, we prove that the gaps of the spectrum $\sigma(H)$ are labelled in a natural way by elements of the K_0-group of a certain C^*-algebra. In particular, if the group Γ has the Kadison property then $\sigma(H)$ has band structure. The Krein resolvent formula plays a crucial role in proving the main results.

0 Introduction

The spectral analysis of periodic Schrödinger operators is an interesting problem in physics and mathematics. Among these operators, those with point potential play an important role in view of the fact that the corresponding spectral problem is explicitly solvable [1], [2]. Thus, almost all textbooks on condensed matter physics refer to the well-known Kronig-Penney model [3]. This model was generalized to two and three dimensions in [4]–[6] (see also [2] for details and further references).

On the other hand, investigations of periodic elliptic operators (including Schrödinger operators) on complete Riemannian manifolds have begun in the last decade. Using K-theory for C^*-algebras, J. Brüning and T. Sunada have studied the band structure of the spectrum for such the operators [7], [8], [9]. The results of the cited papers are based on the analysis of the heat kernels. For the case of point perturbations, the heat kernel of the perturbed operator has a complicated form; therefore, we study the resolvent here. The famous Krein resolvent formula [10] provides an adequate tool for obtaining and analyzing the resolvent of a Schrödinger operator perturbed by a point potential. As a result we show in this paper that under certain natural conditions the Krein formula works for the case of point perturbations of elliptic operators on a manifold, too. With the help of this formula we prove that the gaps of a periodic point perturbation of such an operator are labelled by the elements of the K_0-group of an appropriate C^*-algebra. These results may be generalized to the case of gauge-periodic point perturbations of larger classes of elliptic operators [11].

In conclusion, we note that the spectral analysis of periodic Schrödinger operators on manifolds of non-zero curvature is necessary in understanding many physical phenomena like quantum chaos ([12], [13]) and charge transport in non-planar systems [14].

1 Preliminaries

Throughout the paper X denotes a connected complete non-compact Riemannian C^∞-manifold of dimension n; Γ denotes a discrete group which acts on X isometrically, effectively, and properly discontinuously with compact quotient $\Gamma\backslash X$. We shall denote by $d(x, y)$ the Riemannian distance on X and by dx the mesure on X associated with the Riemannian metric; of course, dx is a Γ-invariant measure. It is known (see, e.g., [15]) that there exists a set F (called the *Brillouin zone for* Γ) with the properties:

1) *F is an open and connected set with a negligible boundary;*
2) *$\gamma F \cap F = \emptyset$ if $\gamma \neq e$;*
3) *$\overline{F}$ is compact, the system $(\gamma\overline{F})_{\gamma\in\Gamma}$ is locally finite, and*

$$\bigcup_{\gamma\in\Gamma} \gamma\overline{F} = X .$$

By L we shall denote the standard representation Γ in $L^2(X)$; for $\gamma \in \Gamma$ L_γ is a unitary operator acting by the rule $L_\gamma f(x) = f(\gamma^{-1}x)$.

Let $\tau_0 : C_0^\infty(X) \to C_0^\infty(X)$ be a Γ-invariant formally self-adjoint elliptic operator of order m, $m > n/2$. The closure H^0 of τ_0 in the Hilbert space $\mathcal{H} = L^2(X)$ is a self-adjoint operator with domain $\mathcal{D}(H^0) = W_2^m(X)$ [8], [16]. Note that

$$\mathcal{D}(H^0) \subset C(X) \tag{1}$$

in view of the Sobolev embedding theorem. By $\sigma(A)$ we denote, as usual, the spectrum of a closed operator A and we put $\rho(A) := \mathbf{C}\backslash\sigma(A)$. For $\zeta \in \rho(H^0)$, $R^0(\zeta) := (H^0 - \zeta)^{-1}$ denotes the resolvent of H^0. It follows from (1) that $R^0(\zeta)$ is a bi-Carleman operator for every $\zeta \in \rho(H^0)$ [17]. Recall that a bounded operator A in $L^2(X)$ is called a bi-Carleman operator if there is a measurable function $K_A : X \times X \to \mathbf{C}$ (the integral kernel of A) such that for any $f \in L^2(X)$

$$Af(x) = \int_X K_A(x, y) f(y)\, dy \quad \text{for a.e. } x,$$

and

$$\begin{aligned} &\int_X |K_A(x, y)|^2\, dy < +\infty \quad \text{for a.e. } x, \\ &\int_X |K_A(x, y)|^2\, dx < +\infty \quad \text{for a.e. } y \end{aligned} \tag{2}$$

(see, e.g., [18] for details). We denote by $G^0(x, y;\ \zeta)$ the integral kernel of $R^0(\zeta)$ and by $G^1(x, y;\ \zeta_1, \zeta_2)$ the iterated kernel

$$G^1(x, y;\ \zeta_1, \zeta_2) = \int_X G^0(x, u;\ \zeta_1) G^0(u, y;\ \zeta_2)\, du.$$

In what follows we shall suppose that the principal symbol a_m of τ_0 satisfies the Agmon-Agranovich-Vishik condition:

(AAV) *There exists a constant* $C > 0$ *such that* $|a_m(\nu) + \lambda| \leq C$ *for all* $\lambda > 0$ *and all* $\nu \in T^*X$ *with* $|\nu| = 1$.

The following result is proved in [16] (see Lemmas 4.5, 4.6, and Theorem 4.7); it allows to employ the Krein resolvent formula to point perturbations of H^0:

Theorem A (1) *There exists* $\tilde{E} \in \mathbf{R}$, $\tilde{E} < 0$, *such that for* $E < \tilde{E}$ *the kernel* $G^0(x, y;\ E)$ *is a* C^∞*-function outside the diagonal* $x = y$.

(2) *The operator* H^0 *is semibonded from below. Moreover, for every* $t > 0$ *there exist constants* $E_0(t) < 0$, *and* $k_0(t) > 0$ *such that for any* $x, y \in X$, $x \neq y$, *and for* $E < E_0$

$$|G^0(x, y;\ E)| \leq k_0 d(x, y)^{m-n} \exp(-td(x, y)),$$

if $m < n$, *and*

$$|G^0(x, y;\ E)| \leq k_0(1 + d(x, y)^{m-n} |\log(d(x, y))|) \exp(-td(x, y))$$

otherwise.

To prove Lemma 1 below we need the statement [16]:

Lemma B *Let* $B(x, r) = \{y \in X : d(x, y) < r\}$. *There exists a constant* C_X *such that* $\mathrm{Vol}\,(B(x, r)) \leq \exp(C_X r)$ *for all* $x \in X$ *and* $r > 0$.

Lemma 1 *The following assertions are valid.*

(1) *There is a constant* $\widehat{E} < 0$ *such that for any* $\zeta_1, \zeta_2 \in \rho(H^0)$ *the function* $G^1(x, y;\ \zeta_1, \zeta_2)$ *is at least separately continuous on* $X \times X$ *if* $\zeta_2 < \widehat{E}$ *or* $\zeta_1 < \widehat{E}$.

(2) *For fixed* $\zeta \in \rho(H^0)$ *the function* $G^0(x, y;\ \zeta)$ *is at least separately continuous on* $X \times X$ *outside the diagonal* $x = y$.

(3) *For every* $\varepsilon > 0$ *and* $t > 0$ *there exist constants* $E_1(t, \varepsilon) < 0$, *and* $k_1(t, \varepsilon) > 0$ *such that for* $d(x, y) \geq \varepsilon$

$$|G^0(x, y;\ E)| \leq k_1 \exp(-td(x, y)),$$

whenever $E < E_1$, *and*

$$|G^1(x, y;\ E', E'')| \leq k_1 \exp(-td(x, y)),$$

whenever $E', E'' < E_1$.

(4) *Let* K *be a compact subset of* X *and* x_0 *be a point of* X. *Then for every* $\varepsilon > 0$ *and* $t > 0$ *there exist constants* $E_2(t, \varepsilon) < 0$ *and* $k_2(t, \varepsilon, K, x_0) > 0$ *such that for* $E < E_2$

$$\sup\{|G^0(x, y;\ E)| : y \in K\} \leq k_2 \exp(-td(x, x_0)),$$

whenever $d(x, K) \geq \varepsilon$.

(5) *Let K be a compact subset of X and x_0 be a point of X. Then for every $t > 0$ there exist constants $E_3(t) < 0$, and $k_3(t, K, x_0) > 0$ such that for $E < E_3$*

$$\left[\int_K |G^0(x, y;\ E)|^2\, dy\right]^{1/2} \leq k_3 \exp(-td(x, x_0)),$$

whenever $d(x, K) \geq \varepsilon$.

Proof: Using Theorem A, Lemma B, and the fact that $G^0(x, y;\ \zeta)$ is a bi-Carleman kernel for every $\zeta \in \rho(H^0)$, it is not hard to prove that there is a constant $E^1 < 0$ such that for any $\zeta_1, \zeta_2 \in \rho(H^0)$ we have the following: if $\zeta_2 < E^1$, then the function $G^1(x, y;\ \zeta_1, \zeta_2)$ is continuous with respect to y for fixed x, and if $\zeta_1 < E^1$, then this function is continuous with respect to x for fixed y. To complete the proof of the assertion (1) it is sufficient to apply the identity

$$\overline{G^1(x, y;\ \zeta_1, \zeta_2)} = G^1(y, x;\ \bar{\zeta}_1, \bar{\zeta}_2).$$

The assertion (2) is a consequence of (1) and the Hilbert resolvent identity. The first inequality in (3) immediately follows from Theorem A; the second one is a simple collorary of the first inequality and Lemma B. The proof of the statements (4) and (5) is trivial in virtue of the first inequality in (3). □

In the remainder of this section, we present some necessary facts from M.G. Krein's theory of self-adjoint extensions (see [10] for more details).

Let H^0 be a self-adjoint operator in a Hilbert space $\mathcal{H}$, S a symmetric operator which is a restriction of H^0, and let $\mathcal{N}_\zeta = \mathrm{Ker}(S^* - \zeta)$, where $\zeta \in \rho(H^0)$, be the deficiency subspace of $\bar{S}$. Fix a Hilbert space $\mathcal{G}$ with $\dim \mathcal{G} = \dim \mathcal{N}_\zeta$. A mapping $\zeta \mapsto B(\zeta)$ from $\rho(H^0)$ to the space $\mathcal{L}(\mathcal{G}, \mathcal{H})$ of all bounded operators from $\mathcal{G}$ to $\mathcal{H}$ is called a *Krein Γ-field of the pair* (H^0, S) if the following conditions are satisfied:

(Γ1) *$B(\zeta)$ is a linear topological isomorphism of $\mathcal{G}$ onto $\mathcal{N}_\zeta$;*

(Γ2) *with*

$$U(\zeta, z) := (H^0 - \zeta)(H^0 - z)^{-1} \tag{3}$$

we have

$$B(z) = U(\zeta, z)B(\zeta). \tag{4}$$

If we choose an arbitrary linear topological isomorphism $B(z_0) : \mathcal{G} \to \mathcal{N}_{\zeta_0}$, we can uniquely determine a Γ-field B by $B(z) = U(z_0, z)B(z_0)$. A mapping $Q : \rho(H^0) \to \mathcal{L}(\mathcal{G}, \mathcal{G})$ is said to be a *Krein Q-function* if

$$Q(\zeta) - Q(z)^* = (\zeta - \bar{z})B(z)^* B(\zeta) \tag{5}$$

for each $z, \zeta \in \rho(H^0)$. It follows from Eqs. (3)–(5) that $Q(z)$ is a holomorphic operator-valued function of ζ. This function is uniquely determined by the property (5) up to a self-adjoint summand $C \in \mathcal{L}(\mathcal{G}, \mathcal{G})$. If C in $\mathcal{L}(\mathcal{G}, \mathcal{G})$ is given, we can put

$$Q(z) = C - iy_0 B(z_0)^* B(z_0) + (z - \bar{z}_0) B(z_0)^* B(z), \tag{6}$$

where z_0 is a fixed element of $\rho(H^0)$ and $y_0 = \operatorname{Im} z_0$. Recall that a self-adjoint extension H of S is called *disjoint from* H^0 if $\mathcal{D}(H) \cap \mathcal{D}(H^0) = \mathcal{D}(S)$. The following theorem is the main result of the Krein theory of self-adjoint extensions (see [10] for the proof).

Theorem C *Given an arbitrary self-adjoint (not necessarily bounded) operator A in $\mathcal{G}$, the formula*

$$R_A(z) = R^0(z) - B(z)[Q(z) + A]^{-1} B(\bar{z})^* \tag{7}$$

determines the resolvent of a self-adjoint extension H_A of S that is disjoint from H^0. Moreover, the correspondence $A \mapsto H_A$ estabilishes a bijection between the set of all self-adjoint extensions of S disjoint from H^0 and the set of all self-adjoint operators in $\mathcal{G}$.

Below we need the following property of the operators $U(\zeta, z)$:

Proposition D *The mapping $U(\zeta, z)$ is a linear topological isomorphism of the space $\mathcal{N}_\zeta$ onto $\mathcal{N}_z$ and satisfies the relation $U(\zeta, z) = I + (z - \zeta) R^0(z)$.*

2 Periodic Point Perturbations of H^0

Let us fix a fundamental domain F and some finite subset K of F, and let Λ be the Γ-orbit of K: $\Lambda = \cup_{\gamma \in \Gamma} \gamma \mathrm{K}$. The set Λ may be viewed as the analog of a crystal in Euclidean space. It follows from the properties of the domain F that each point $\lambda \in \Lambda$ has a unique representation of the form $\lambda = \gamma\kappa$ where $\gamma \in \Gamma$, $\kappa \in \mathrm{K}$. Now we define a point perturbation of H^0 supported by Λ: Formally, this is a self-adjoint operator H of the form

$$H = H^0 + \sum_{\substack{\gamma \in \Gamma \\ \kappa \in \mathrm{K}}} \varepsilon_\kappa \delta((\gamma\kappa)^{-1} x), \tag{8}$$

where $\delta(x)$ is the Dirac δ-function and ε_κ are "coupling constants". To assign an operator meaning to the formal expression (8), we use the so-called "restriction-extension procedure" [1], [2]. Thus, we consider the set

$$\mathcal{D}(S) = \{f \in \mathcal{D}(H^0) : f(\lambda) = 0 \quad \forall \lambda \in \Lambda\}, \tag{9}$$

which is well defined since $\mathcal{D}(H^0) \subset C(X)$. Let S be the restriction of H^0 to $\mathcal{D}(S)$; evidently, S is a symmetric operator in $\mathcal{H}$. A self-adjoint extension H of S disjoint from H^0 is then said to be a *point perturbation of H^0 supported by* Λ.

Fix a point perturbation H of H^0. Using the Krein resolvent formula (7) we construct an explicit form of the resolvent $R(\zeta)$ of H for which we need some results from [19]. These results are obtained for the case $\mathcal{H} = L^2(\Omega)$, Ω a domain in $\mathbf{R}^n$, but it is easy to check that they are also valid for the case $\mathcal{H} = L^2(M)$ where M is an arbitrary locally compact space together with a Radon measure.

In what follows we shall denote by I_0 the semi-axis $(-\infty, \tilde{E})$, with $\tilde{E}$ the constant from Theorem A. Let $\zeta \in I_0$; for every $a \in X$ we denote by $g_a(\zeta)$ the function $X \ni x \mapsto G^0(x, a; \zeta)$; if $z \in \rho(H^0)$ is arbitrary, then we put $g_a(z) = U(\zeta, z) g_a(\zeta)$. In view of Proposition D, this definition of $g_a(z)$ does not depend on the choice of ζ in I_0.

Lemma 2 *For some $z \in I_0$ the matrix $(\langle g_\lambda(z) | g_\mu(z)\rangle)_{\lambda,\mu\in\Lambda}$ determines a bounded operator in the standard basis of the space $l^2(\Lambda)$.*

Proof: It follows from Lemma B that there are constants $c_\Lambda > 0$ and $\tilde{c}_\Lambda > 0$ such that for all $\lambda \in \Lambda$ and $r \in \mathbf{R}_+$ we have

$$\#\{\mu \in \Lambda : d(\lambda, \mu) \le r\} \le c_\Lambda \exp(\tilde{c}_\Lambda r),$$

where $\#Y$ is the number of elements in a finite set Y. Denote for simplicity $\langle g_\lambda(z) | g_\mu(z)\rangle$ by $W(\lambda, \mu)$. According to Schur's test [20], the operator W with the matrix $W(\lambda, \mu)$ is bounded on the space $l^2(\Lambda)$ if for some $c' > 0$

$$\sup_{\mu\in\Lambda} \sum_{\lambda\in\Lambda} |W(\lambda, \mu)| \le c' \quad \text{and} \quad \sup_{\lambda\in\Lambda} \sum_{\mu\in\Lambda} |W(\lambda, \mu)| \le c', \tag{10}$$

and in this case we have $\|W\| \le c'$. To find such a constant c' it is sufficient to use Lemma 1(3) and the following assertion which is proved in [21]: □

Lemma E *Let $\varphi : \Lambda \to \mathbf{C}$ be a function such that for some $\mu \in \Lambda$*

$$|\varphi(\lambda)| \le c \exp(-(1 + \delta) \tilde{c}_\Lambda d(\lambda, \mu)),$$

where c and δ are positive constants. Then

$$\sum_{\lambda\in\Lambda} |\varphi(\lambda)| \le c\, \tilde{c}_\Lambda\, \delta^{-1}.$$

For each $\kappa \in \mathrm{K}$ we choose a function $\varphi_\kappa \in C_0^\infty(X)$ such that $\varphi_\kappa(\kappa) = 1$, $\operatorname{supp} \varphi_\kappa \subset F$, and $\operatorname{supp} \varphi_\kappa \cap \operatorname{supp} \varphi_{\kappa'} = \emptyset$ if $\kappa \ne \kappa'$. For every $\lambda \in \Lambda$ we put $\varphi_\lambda = L_\gamma \varphi_\kappa$, if $\lambda = \gamma\kappa$. It is readily seen that the family $\{\varphi_\lambda : \lambda \in \Lambda\}$ lies in $\mathcal{D}(H^0)$ and possesses the properties:

(1) $\varphi_\lambda(\lambda) = 1$ $(\lambda \in \Lambda)$;

(2) $\operatorname{supp} \varphi_\lambda \cap \operatorname{supp} \varphi_{\lambda'} = \emptyset$ if $\lambda \ne \lambda'$;

(3) $\sup\{\|H^0 \varphi_\lambda\| + \|\varphi_\lambda\| : \lambda \in \Lambda\} < \infty$.

Taking into account Lemma 2, we can apply Theorem 3 and Proposition 3 from [19] and get the following result:

Proposition 1 *For every $z \in \rho(H^0)$ the family $\{g_\lambda(z) : \lambda \in \Lambda\}$ is a Riesz basis in $\mathcal{N}_z$. This means that for each family $(\xi_\lambda)_{\lambda\in\Lambda}$ from $l^2(\Lambda)$ the family $(\xi_\lambda\, g_\lambda(z))_{\lambda\in\Lambda}$ is summable in $\mathcal{H}$ and the mapping*

$$B(z) : l^2(\Lambda) \ni (\xi_\lambda) \longmapsto \sum_{\lambda\in\Lambda} \xi_\lambda\, g_\lambda(z) \in \mathcal{H} \tag{11}$$

is a linear topological isomorphism from $l^2(\Lambda)$ onto $\mathcal{N}_z$.

Now we put $\mathcal{G} := l^2(\Lambda)$; using (11) and Proposition D it is easily shown that $B(z)$ is a Krein Γ-field.

Our next purpose is to construct the Krein $\mathcal{Q}$-function for the pair (H^0, S). Fix a point $z_0 \in \mathbf{R}$ such that $z_0 < \widehat{E}$ (cf. Lemma 1(1)), then for all $z \in \rho(H^0)$ and all $a \in X$ the expression $G^0(a, a\,; z) - G^0(a, a\,; z_0)$ is well-defined. Indeed, from the Hilbert resolvent identity we get

$$G^0(a, a\,; z) - G^0(a, a\,; z_0) = (z - z_0)G^1(a, a\,; z, z_0). \tag{12}$$

Now, using Theorem 4 and Proposition 4 from [19] we can determine the Krein $\mathcal{Q}$-function by the infinite matrix $(Q_{\lambda\mu}(z))_{\lambda,\mu\in\Lambda}$ if

$$Q_{\lambda\mu}(z) := \begin{cases} G^0(\lambda, \mu; z), & \text{if } \lambda \neq \mu; \\ G^0(\lambda, \lambda; z) - G^0(\lambda, \lambda; z_0), & \text{if } \lambda = \mu. \end{cases} \tag{13}$$

The results thus obtained are summarized in the following theorem.

Theorem 1 *Let H_A be the point perturbation of H^0 determined by a self-adjoint operator A in the space $l^2(\Lambda)$. Then for every $\zeta \in \rho(H^0)\cap\rho(H_A)$ and $f \in Ł^2(X)$ we have*

$$R_A(\zeta)f = R^0(\zeta)f - \sum_{\lambda\in\Lambda}\left(\sum_{\mu\in\Lambda}[Q(\zeta) + A]^{-1}(\lambda, \mu)\langle g_\mu | f\rangle\right) g_\lambda(\zeta). \tag{14}$$

We are interested in Γ-periodic point perturbations of H^0 only. Proposition 2 below provides a necessary and sufficient condition for H_A to be a Γ-invariant operator. Before stating this proposition we note that there is a natural unitary representation $\tilde{L}$ of the group Γ in $l^2(\Lambda)$: $\tilde{L}_\gamma\varphi(\lambda) = \varphi(\gamma^{-1}\lambda)$, $\varphi \in l^2(\Lambda)$. It is clear that for each $z \in \rho(H^0)$ the operator $Q(z)$ is $\tilde{L}$-invariant, that is, its matrix satisfies the condition $Q_{\lambda+\gamma,\mu+\gamma}(z) = Q_{\lambda\mu}(z)$ for all $\gamma \in \Gamma$, $\lambda, \mu \in \Lambda$. In particular, the diagonal elements $Q_{\lambda\lambda}(z)$ depend only on K since we have $\lambda = \gamma\kappa$ for some $\gamma \in \Gamma$ and $\kappa \in$ K.

Proposition 2 *The operator H_A is Γ-periodic if and only if the operator A is invariant with respect to $\tilde{L}$.*

We omit the easy proof of this proposition.

From now on, we consider only Γ-periodic point perturbations H_A of H^0. From the point of view of physical applications, the most important operators H_A are those where A has a diagonal matrix with respect to the standard basis of $l^2(\Lambda)$ [4], [22]; only these operators appear as limits of Hamiltonians with short-range potentials [4], [23]. On the other hand, even in the case of a bounded $\tilde{L}$-invariant operator A with a non-diagonal matrix, the spectrum of the periodic point perturbation H_A for the Laplacian $H^0 = -\Delta$ may contain a singular component which is a Cantor set [24]. For this reason, we restrict ourselves to the case when the following conditions are fulfilled:

(D) *The operator A has a diagonal matrix* $A_{\lambda\mu} = \alpha_{\lambda\mu}\delta_{\lambda\mu}, \quad \lambda, \mu \in \Lambda;$

(Q) $\lim_{E\to\infty} |Q_{\kappa\kappa}(E)| = \infty$ *for all* $\kappa \in K$.

Yu.G. Shondin has observed (for the case of finite point perturbations) that the conditions (D) and (Q) eliminate some pathological properties of H_A [25]. Namely, under these conditions the operator H_A is "form-local" in the following sense: for any φ and ψ from the form-domain $Q(H_A)$ of H_A the relation $\operatorname{supp}\varphi \cap \operatorname{supp}\psi = \emptyset$ implies $\langle\varphi\,|H_A\psi\rangle = 0$.

From now on, we shall suppose thath the condidtions (D) and (Q) are satisfied. The following theorem is the main result of this section.

Theorem 2 *For every $t > 0$ there are constants $E_4(t) < 0$ and $k_4(t) > 0$ such that for every $E \in \mathbf{R}$, $E < E_4$, the operator $Q(E) + A$ has a bounded inverse with matrix obeying the condition*

$$|[Q(z) + A]^{-1}_{\lambda\mu}| \le k_4 \exp(-td(\lambda, \mu)).$$

Proof: Let $t > 0$ be given. Denote by $D(E)$ the operator in $l^2(\Lambda)$ with matrix $D_{\lambda\mu}(E) = (Q_{\lambda\mu}(E) + A_{\lambda\mu})\delta_{\lambda\mu}$, and set $S(E) = Q(E) + A - D(E)$. Let $C_E = \inf\{|D_{\lambda\lambda}(E)| : \lambda \in \Lambda\}$; according to the condition (Q), $C_E \to \infty$ as $E \to -\infty$. Let $s = \max(t, 2\tilde{c}_\Lambda)$. By Lemma 1 there are $E_4 < 0$ and $c > 0$ such that $|S_{\lambda\mu}(E)| \le c\exp(-2sd(\lambda, \mu))$, whenever $E < E_4$. We can suppose $|E_4|$ is so large that for $E < E_4$ we have $\tilde{c}_\Lambda\, c\, C_E^{-1} \le 1/2$ and $\|D^{-1}(E)S(E)\| < 1$. Then

$$[Q(E) + A]^{-1} = \sum_{j\ge 0}(-D^{-1}(E)S(E))^j.$$

We claim that for all $j \ge 0$

$$|(D^{-1}(E)S(E))^j_{\lambda\mu}| \le (\tilde{c}_\Lambda\, c\, C_E^{-1})^j \exp(-sd(\lambda, \mu)),$$

implying the theorem.

For $j = 0$ we have nothing to prove. If the assertion holds for some $j \geq 0$, we estimate with Lemma E

$$
\begin{aligned}
|(D^{-1}(E)S(E))^{j+1}_{\lambda\mu}| &\leq C_E^{-1} \sum_{\kappa\in\Lambda} |S_{\lambda\kappa}(E)(D^{-1}(E)S(E))^{j}_{\kappa\mu}| \\
&\leq c\, C_E^{-1} (\tilde{c}_\Lambda\, c\, C_E^{-1})^j \sum_{\kappa\in\Lambda} \exp(-2sd(\lambda,\kappa)) \\
\exp(-sd(\kappa,\mu)) &\leq c\, C_E^{-1} (\tilde{c}_\Lambda\, c\, C_E^{-1})^j \exp(-sd(\lambda,\mu)) \\
\sum_{\kappa\in\Lambda} \exp(-2\tilde{c}_\Lambda\, d(\lambda,\kappa)) &\leq (\tilde{c}_\Lambda\, c\, C_E^{-1})^{j+1} \exp(-sd(\lambda,\mu)).
\end{aligned}
$$

□

Corollary: *The operator H_A is semi-bounded from below.*

3 Spectral Structure of H_A

In this section we denote by $\mathcal{K}$ the set of all compact operators in the space $\mathcal{F} = L^2(F)$. We set $C^*_{red}(\Gamma, \mathcal{K}) = C^*_{red}(\Gamma) \otimes \mathcal{K}$, where $C^*_{red}(\Gamma)$ is the reduced group C^*-algebra of Γ [7], [8].

We shall identify $L^2(X)$ with the space $l^2(\Gamma, \mathcal{F})$ by means of the correspondence $\Phi : L^2(X) \ni f \mapsto \varphi \in l^2(\Gamma, \mathcal{F})$, $\varphi(\gamma)(x) = L_\gamma f(x)$. With $\tilde{R}$ the right regular representation of Γ in $l^2(\Gamma)$ we set $R = \tilde{R} \otimes I$, where I is the identity.

Lemma 3 *The mapping Φ is an interwining operator for the representations L and R.*

Proof: This follows by direct calculation. □

This lemma implies that we can identify the space of all Γ-invariant operators in $\mathcal{L}(\mathcal{H}, \mathcal{H})$ with the space $W^*(\Gamma, \mathcal{F})$ of all bounded R-invariant operators $B : l^2(\Gamma, \mathcal{F}) \to l^2(\Gamma, \mathcal{F})$, and we can identify $C^*_{red}(\Gamma, \mathcal{K})$ with a subalgebra of $W^*(\Gamma, \mathcal{F})$. If $B \in W^*(\Gamma, \mathcal{F})$ we define the *Fourier coefficient* $\widehat{B}(\gamma)$ at $\gamma \in \Gamma$ to be the bounded operator on $\mathcal{F}$ given by

$$\widehat{B}(\gamma)v = (B\delta_1^v)(\gamma),$$

where

$$\delta_1^v(\gamma) = \begin{cases} v, & \text{if } \gamma = e; \\ 0 & \text{otherwise.} \end{cases}$$

Recall that the canonical trace Tr_Γ on $C^*_{red}(\Gamma, \mathcal{K})$ is given by

$$\mathrm{Tr}_\Gamma\, B = \mathrm{Tr}\, \widehat{B}(e).$$

We need the following lemma [8]:

Lemma F *If $\widehat{B}(\gamma) \in \mathcal{K}$ for every $\gamma \in \Gamma$ and*

$$\sum_{\gamma \in \Gamma} \|\widehat{B}(\gamma)\| \leq \infty,$$

then $B \in C^*_{red}(\Gamma, \mathcal{K})$.

The main results of the paper are consequences of the following theorem.

Theorem 3 *The resolvent $R_A(\zeta)$ of the operator H_A belongs to $C^*_{red}(\Gamma, \mathcal{K})$ for every $\zeta \in \rho(H_A)$.*

Proof: Since $C^*_{red}(\Gamma, \mathcal{K})$ is closed in $\mathcal{L}(\mathcal{H}, \mathcal{H})$ and $R_A(\zeta)$ is an analytic function on $\rho(H_A)$, it suffices to prove that $R_A(\zeta) \in C^*_{red}(\Gamma, \mathcal{K})$ when ζ runs over some semi-axis $(-\infty, x)$. It is proved in [8] that $\exp(-tH^0) \in C^*_{red}(\Gamma, \mathcal{K})$ for all $t \geq 0$; hence using the Laplace transform we get that $R_A(E) \in C^*_{red}(\Gamma, \mathcal{K})$ for every $E < 0$. Put $V(E) := R^0(E) - R_A(E)$; it remains to show that $V(E) \in C^*_{red}(\Gamma, \mathcal{K})$ for all E in some interval $(-\infty, x)$. We abbreviate

$$M(\lambda, \mu; \zeta) := [Q(\zeta) + A]^{-1}_{\lambda\mu}. \tag{15}$$

According to Theorems 1 and 2 we can find constants $c_E < 0$ and $c_0 > 0$ such that for all $E < c_E$

$$|M(\lambda, \mu; E)| \leq c_0 \exp(-\tilde{c}_0\, d(\lambda, \mu)), \tag{16}$$

and for every $f \in L^2(X)$

$$V(\zeta)f = \sum_{\lambda \in \Lambda} \left(\sum_{\mu \in \Lambda} M(\lambda, \mu; \zeta) \langle g_\mu(\bar{\zeta}) | f \rangle \right) g_\lambda(\zeta). \tag{17}$$

Further, by Lemma 1 we can suppose that the following assertion is true: *For any compact set $C \subset X$, any point $\kappa \in$ K, and any $E < c_E$ there is a constant $k(C, \kappa)$ such that*

$$\left[\int_C |g_\lambda(E)(x)|^2\, dx \right]^{1/2} \leq k \exp(-\tilde{c}_0\, d(\lambda, \kappa)). \tag{18}$$

Then we can choose $\tilde{c}_0$ in such a way that $\tilde{c}_0 > 3\tilde{c}_\Lambda$ where $\tilde{c}_\Lambda$ is the constant from the proof of Lemma 2.

For any $\beta \in \Gamma$ define a matrix $M_\beta(\lambda, \mu; \zeta)$ by the relation

$$M_\beta(\lambda, \mu; \zeta) = \begin{cases} M(\lambda, \mu; \zeta), & \text{if } \lambda = \gamma\kappa,\ \mu = \gamma\beta\kappa' \text{ for some } \gamma \in \Gamma, \kappa,\ \kappa' \in \mathrm{K}; \\ 0 & \text{otherwise.} \end{cases} \tag{19}$$

Since $(g_\lambda(\zeta))_{\lambda\in\Lambda}$ is a Riesz basis, it follows from Lemma E and (16) that for any $f \in L^2(X)$ the series

$$V_\beta(\zeta)f = \sum_{\lambda\in\Lambda}\left(\sum_{\mu\in\Lambda} M_\beta(\lambda,\mu;\zeta)\langle g_\mu(\bar{\zeta})|f\rangle\right) g_\lambda(\zeta) \tag{20}$$

converges and defines a bounded operator in the space $L^2(X)$ (the sum over μ is, in fact, finite). Let us prove that

$$\sum_{\beta\in\Gamma} \|V_\beta(E)\| < +\infty \tag{21}$$

if $E < c_E$. Because $(g_\lambda(\zeta))_{\lambda\in\Lambda}$ is a Riez basis in its own closed linear hull, we have
(1) for each $\varphi \in l^2(\Lambda)$

$$\left\|\sum_{\lambda\in\Lambda}\varphi(\lambda)g_\lambda(E)\right\| \le c_1(E)\|\varphi\|; \tag{22}$$

(2) for any $f \in L^2(X)$

$$\sum_{\lambda\in\Lambda} |\langle g_\lambda(E)|f\rangle|^2 \le c_2^2(E)\|f\|^2. \tag{23}$$

Taking into account (16), (22) and (23) we get

$$\begin{aligned}
\|V_\beta(E)f\|^2 &\le c_1^2\sum_\lambda\left|\sum_\mu M_\beta(\lambda,\mu;E)\langle g_\mu(E)|f\rangle\right|^2\\
&= c_1^2\sum_{\kappa\in\mathrm{K}}\sum_{\gamma\in\Gamma}\left|\sum_{\kappa'\in\mathrm{K}} M(\gamma\kappa,\gamma\beta\kappa';E)\langle g_{\gamma\beta\kappa'}(E)|f\rangle\right|^2\\
&\le c_1^2\sup\{|M(\gamma\kappa,\gamma\beta\kappa';E)|^2 : \gamma\in\Gamma;\kappa,\kappa'\in\mathrm{K}\}\\
&\quad \sum_{\gamma\in\Gamma}\sum_{\kappa,\kappa'\in\mathrm{K}} |\langle g_{\gamma\beta\kappa'}(E)|f\rangle|^2\\
&\le (\#\mathrm{K})\, c_1^2 \sup\{|M(\kappa,\beta\kappa';E)|^2 : \kappa,\kappa'\in\mathrm{K}\}\sum_{\lambda\in\Lambda}|\langle g_\lambda(E)|f\rangle|^2\\
&\le (\#\mathrm{K})\, c_1^2 c_2^2 \max\{\exp(-2\tilde{c}_0\, d(\kappa,\beta\kappa')) : \kappa,\kappa'\in\mathrm{K}\}\,\|f\|^2
\end{aligned} \tag{24}$$

(we have used the identity $|M(\gamma\lambda,\gamma\mu;\zeta)| = |M(\lambda,\mu;\zeta)|$, $\gamma\in\Gamma$, $\lambda,\mu\in\Lambda$ which follows from the fact that $Q(\zeta)$ is $\tilde{L}$-invariant). Thus

$$\|V_\beta(E)\| \le c_2 \sum_{\kappa,\kappa'\in\mathrm{K}} \exp(-\tilde{c}_0\, d(\kappa,\beta\kappa')), \tag{25}$$

and Lemma E and (25) imply (21).

Now we show that

$$\sum_{\beta\in\Gamma} V_\beta(E) = V(E), \tag{26}$$

if $E < c_E$. It is sufficient to prove that for any functions $f_1, f_2 \in C_0^\infty(X)$ we have

$$\sum_{\beta\in\Gamma} \langle f_1 | V_\beta(\zeta) f_2\rangle = \langle f_1 | V(\zeta) f_2\rangle. \tag{27}$$

Let f_1, f_2 be such functions; we prove (27) if we prove that the series

$$\sum_{\lambda,\mu\in\Lambda} M(\lambda, \mu; E)\,\langle g_\mu(E)|f_2\rangle\,\langle f_1|g_\lambda(E)\rangle \tag{28}$$

converges absolutely. Fix a point $\kappa_0 \in \mathrm{K}$; using (18) we get with some $k > 0$:

$$|\langle g_\lambda(E)|f_j\rangle| \le k \exp(-\tilde{c}_0\, d(\lambda, \kappa_0))\|f_j\|. \tag{29}$$

Since $\tilde{c}_0 > \tilde{c}_\Lambda$, (25) follows from Lemma E.

It remains to prove that $V_\beta(E) \in C^*_{red}(\Gamma, \mathcal{K})$ for all $\beta \in \Gamma$ and $E < c_E$. In what follows we fix $\beta \in \Gamma$ and $E < c_E$. First we find the Fourier coefficient $\widehat{V}_\beta(\gamma) = \widehat{V}_\beta(E)(\gamma)$. By direct calculation, we obtain for $u \in L^2(F)$

$$\widehat{V}_\beta(\gamma)(u) = \sum_{\alpha\in\Gamma}\sum_{\kappa,\kappa'\in\mathrm{K}} L_{\alpha\kappa\kappa'}(u), \tag{30}$$

where $L_{\alpha\kappa\kappa'}$ is a one-dimensional continuous linear operator of the form

$$L_{\alpha\kappa\kappa'}(u) = M(\alpha\kappa, \alpha\beta\kappa'; E)\langle g_{\alpha\beta\kappa'}(E)|\tilde{u}\rangle \tilde{g}_{\gamma\alpha\kappa}(E). \tag{31}$$

Here $\tilde{u}$ is the extension of u to the whole manifold X by zero, and $\tilde{g}$ is the restriction of g to F. To prove that $\widehat{V}_\beta(\gamma)$ is a compact operator, we show that

$$\sum_{\alpha,\kappa,\kappa'} \|L_{\alpha\kappa\kappa'}\| < \infty. \tag{32}$$

Fix a point $\kappa_0 \in \mathrm{K}$, then from (18) we deduce that

$$\|\tilde{g}_{\gamma\alpha\kappa}(E)\| \le k(\bar{F}, \kappa_0)\exp(-\tilde{c}_0\, d(\kappa_0, \gamma\alpha\kappa)). \tag{33}$$

Hence

$$\sum_{\alpha\in\Gamma} \|\tilde{g}_{\gamma\alpha\kappa}(E)\| < \infty. \tag{34}$$

On the other hand,

$$\begin{aligned} &\sup\{|M(\alpha\kappa, \alpha\beta\kappa'; E)| : \alpha\in\Gamma,\ \kappa,\kappa'\in\mathrm{K}\}\\ &\quad = \sup\{|M(\kappa, \beta\kappa'; E)| : \kappa,\kappa'\in\mathrm{K}\} =: c_3 < \infty. \end{aligned} \tag{35}$$

Thus

$$\|L_{\alpha\kappa\kappa'}\| \le c_3\, c_4\, \|\tilde{g}_{\gamma\alpha\kappa}(E)\|, \tag{36}$$

where $c_4 := \|g_\lambda(E)\|$ is obviously independent of λ. Thus (32) follows from (34).

Finally we prove that

$$\sum_{\gamma\in\Gamma} \|\widehat{V}_\beta(\gamma)\| < \infty. \tag{37}$$

Let $u \in L^2(F)$, $\|u\| \le 1$, then

$$\|\widehat{V}_\beta(\gamma)(u)\| \le c_3 \sum_{\kappa,\kappa'\in \mathrm{K}} \sum_{\alpha\in\Gamma} |\langle g_{\alpha\beta\kappa'}(E)|\tilde{u}\rangle|\; \|\tilde{g}_{\gamma\alpha\kappa}(E)\|. \tag{38}$$

Fix $\kappa, \kappa' \in \mathrm{K}$ and consider the sum

$$\sum_{\alpha\in\gamma} |\langle g_{\alpha\gamma}(E)|\tilde{u}\rangle|\; \|\tilde{g}_{\gamma\alpha\kappa}\|, \tag{39}$$

where $\gamma = \beta\kappa'$. From (18) we infer

$$\sum_{\alpha\in\Gamma} \|\tilde{g}_{\gamma\alpha\kappa}(E)\| = \sum_{\alpha\in\Gamma} \|\tilde{g}_\alpha(E)\| =: c_5 < \infty, \tag{40}$$

$$\sum_{\alpha\in\Gamma} |\langle g_{\alpha\gamma}(E)|\tilde{u}\rangle| \le c_5, \tag{41}$$

$$\|\tilde{g}_{\gamma\alpha\kappa}(E)\| \le k(\kappa)\exp(-\tilde{c}_0\, d(\gamma\alpha\kappa, \kappa)), \tag{42}$$

$$|\langle g_{\alpha\gamma}(E)|\tilde{u}\rangle| \le k(\kappa)\exp(-\tilde{c}_0\, d(\alpha\nu, \kappa)). \tag{43}$$

Since $d(\alpha\nu,\kappa)) \ge d(\alpha\kappa,\kappa)) - d(\alpha\nu,\alpha\kappa)) = d(\alpha\kappa,\kappa)) - d(\nu,\kappa))$, the inequality (43) may be rewritten as

$$|\langle g_{\alpha\gamma}(E)|\tilde{u}\rangle| \le c'(\kappa,\nu)\exp(-\tilde{c}_0\, d(\alpha\kappa,\kappa)). \tag{44}$$

Write $\Gamma = \Gamma_1 \cup \Gamma_2$, where

$$\Gamma_1 = \{\alpha \in \Gamma : d(\alpha\kappa, \gamma^{-1}\kappa) \le d(\gamma^{-1}\kappa, \kappa)/2\}, \tag{45}$$

$$\Gamma_2 = \{\alpha \in \Gamma : d(\alpha\kappa, \gamma^{-1}\kappa) > d(\gamma^{-1}\kappa, \kappa)/2\}. \tag{46}$$

If $\alpha \in \Gamma_1$, then $d(\alpha\kappa,\kappa) \ge d(\gamma^{-1}\kappa,\kappa) - d(\alpha\kappa,\gamma^{-1}\kappa) \ge d(\gamma^{-1}\kappa,\kappa)/2$. Thus, using (40) and (44), we have

$$\begin{aligned} \sum_{\alpha\in\Gamma_1} |\langle g_{\alpha\nu}(E)|\tilde{u}\rangle|\; \|\tilde{g}_{\gamma\alpha\kappa}(E)\| &\le c_6(\kappa,\nu)\exp\left(-\frac{1}{2}\tilde{c}_0\, d(\gamma^{-1}\kappa,\kappa)\right) \\ &< c_6(\kappa,\nu)\exp\left(-\frac{3}{2}\tilde{c}_\Lambda\, d(\kappa,\gamma\kappa)\right). \end{aligned} \tag{47}$$

Similarly, using (41) and (42) we obtain

$$(48)\qquad \sum_{\alpha\in\Gamma_2} |\langle g_{\alpha\nu}(E)|\tilde{u}\rangle| \; \|\tilde{g}_{\gamma\alpha\kappa}(E)\| \le c_7(\kappa)\exp\left(-\frac{3}{2}\tilde{c}_\Lambda\, d(\kappa,\gamma\kappa)\right).$$

Therefore,

$$(49)\qquad \|\widehat{V}_\beta(\gamma)\| \le c_8(E)\sum_{\kappa\in K}\exp\left(-\frac{3}{2}\tilde{c}_\Lambda\, d(\kappa,\gamma\kappa)\right),$$

and Lemma E implies

$$\sum_\gamma \|V_\beta(\gamma)\| < \infty.$$

The proof follows from Lemma F. □

Corollary 1 *Let $E_1, E_2 \in \mathbf{R}\setminus\sigma(H_A)$, and $E_1 \le E_2$. Then the spectral projector $P_{[E_1,E_2]}$ for the operator H_A belongs to $C^*_{red}(\Gamma,\mathcal{K})$.*

Proof: Indeed, there exists a function φ from $C_0^\infty(\mathbf{R})$ such that $P_{[E_1,E_2]} = \varphi(R_A(E))$ for some $E < E_1$. □

Fix now a number $E' \in \mathbf{R}$ such that $E' < \inf\sigma(H_A)$ and consider the function

$$N(E) = \begin{cases} \mathrm{Tr}_\Gamma\, P_{[E',E]}, & E \ge E'; \\ 0, & E < E'. \end{cases}$$

It is clear that this function is independent of the choice of E'. Moreover, $N(E)$ is constant on each gap of the spectrum of H_A such that the values of $N(E)$ label in a natural way the gaps of H_A [26].

Corollary 2 (Gap Labelling Theorem). *The values of $N(E)$ on gaps of the spectrum of H_A form a countable set of real numbers* $\mathrm{Tr}^*(K_0C^*_{red}(\Gamma))$ *(here $K_0\mathcal{B}$ denotes the K_0-group of a C^*-algebra $\mathcal{B}$).*

Recall that Γ is said to have the *Kadison property* if there exists a constant $c_K > 0$ such that $\mathrm{Tr}_\Gamma\, P \ge c_K$ for every non-zero projector from $C^*_{red}(\Gamma,\mathcal{K})$.

Corollary 3 *If Γ has the Kadison property, then the spectrum of H_A has band structure.*

Acknowledgements

The second named author gratefully acknowledges a grant of Volkswagen-Stiftung. He is also very grateful to the RFFI Foundation (Grant No 96-01-00074) for financial support, as well as to Humboldt University at Berlin for its warm hospitality.

References

1. B.S. Pavlov, *The theory of extensions and explicitly solvable models* (in Russian). Uspekhi Mat. Nauk **42**, no. 6 (1987), 99–131; Engl. transl.: Russ. Math. Surv. **42**, no. 6 (1987), 127–168.
2. S. Albeverio, F. Gesztesy, R. Høegh-Krohn and H. Holden, *Solvable models in quantum mechanics*. Springer-Verlag, Berlin etc., 1988.
3. R. Kronig and W.G. Penney, *Quantum mechanics of electrons in crystal lattices*. Proc. Roy. Soc. (London) **130A** (1931), 499–513.
4. A. Grossmann, R. Høegh-Krohn and M. Mebkhout, *The one-particle theory of periodic point interactions*. Comm. Math. Phys. **77** (1080), 87–100.
5. Yu.E. Karpeshina, *Spectrum and eigenfunctions of Schrödinger operator with zero-range potential of the homogenous lattice type in three dimensional space* (in Russian). Teor. i. Mat. Fiz. **57** (1983), 304–313; Engl. transl.: Theor. and Math. Phys. **57** (1983), 1156–1162.
6. S. Albeverio, F. Gesztesy, R. Høegh-Krohn and H. Holden, *Point interactions in two dimensions: Basic properties, approximations and applications to solid state physics*. J. reine u. angew. Math. **380** (1987), 87–107.
7. T. Sunada, *Group C^*-algebras and the spectrum of a periodic Schrödinger operator on a manifold.* Can. J. Math. **44** (1992), 180–193.
8. J. Brüning and T. Sunada, *On the spectrum of periodic elliptic operators*. Nagoya Math. J. **126** (1992), 159–171.
9. J. Brüning and T. Sunada, *On the spectrum of gauge-periodic elliptic operators*. Astérisque **210** (1992), 65–74.
10. M.G. Krein and H.K. Langer, *Defect subspace and generalized resolvents of an Hermitian operators in the space Π_κ* (in Russian). Funk. Anal. i Prilozhen. **5**, no. 2 (1971), 59–71; Engl. transl.: Funct. Anal. and its Appl. **5** (1971), 217–228.
11. J. Brüning and V. A. Geyler, *Gauge periodic point perturbations on the Lobachevsky plane*. Preprint SFB 288, Berlin, 1998.
12. T. Sunada, *Euclidean versus non-euclidean aspects in spectral geometry*. Progr. Theor. Phys. Suppl. **116** (1994), 235–250.
13. N.E. Hurt, Quantum chaos and mesoscopic systems. Mathematical methods in the quantum signatures of chaos. Kluwer Ac. Publ., Dordrecht etc., 1997.
14. M.L. Leadbeater, C.L. Foden and J. H. Burrougher, e.a. *Magnetotransport in a non-planar two-dimensional electron gas*. Phys. Rev. B. **52** (1995), 8629–8632.
15. D.B. Efremov and M.A. Shubin, *Spectral asymptotics of elliptic operators of the Schrödinger type on the Lobachevsky space* (in Russian). Trudy Sem. I.G. Petrovsky **15** (1991), 3–32.
16. M.A. Shubin, *Spectral theory of elliptic operators on non-compact manifolds*. Astérisque **207** (1992), 35–108.
17. A.V. Bukhvalov, *Applicatons of the method of the theory of order bounded operators in the L^p-spaces* (in Russian). Uspekhi Mat. Nauk **38**, no. 6 (1983), 37–83.
18. V.B. Korotkov, *Integral operators* (in Russian). Nauka, Novosibirsk, 1983.
19. V.A. Geyler, V.A. Margulis and I.I. Chuchaev, *Zero-range potentials and Carleman operators* (in Russian). Sibir. Mat. Zhurn. **36** (1995), 828–841; Engl. transl.: Siberian Math. J. **36** (1995), 714–726.
20. P.R. Halmos and V.S. Sunder, *Bounded linear operators on L^2-spaces*. Springer-Verlag, New York etc., 1978.

21. M.A. Shubin, *Pseudo-difference operators and their Green functions* (in Russian). Izv. AN SSSR. Ser. Mat. **49** (1985), 652–671; Engl. transl.: Math. USSR. Izvestiya. **26** (1986), 605–622.
22. S.A. Gredeskul, M. Zusman, Y. Avishai and M.Ya. Azbel, *Spectral properties and localization of an electron in a two-dimensional system with point scatterers in a magnetic field*. Phys. Reps. **288** (1997), 223–257.
23. V.A. Geyler, *The two-dimensional Schrödinger operators with a uniform magnetic field and its perturbation by periodic zero-range potentials* (in Russian). Algebra i Analiz, **3**, no. 3 (1991), 1–48; Engl. transl.: St.-Petersburg Math. J. **3** (1992), 489–532.
24. S. Albeverio and V.A. Geyler, The band structure of the general periodic Schrödinger operator with point interactions (to be published).
25. Yu.G. Shondin, *Semibounded local Hamiltonians for perturbation of the Laplacian supported by curves with angle points in* $\mathbf{R}^4$ (in Russian). Teoret. i Mat. Fiz. **106** (1996), 179–199.
26. J. Bellissard, Gap labelling theorems for Schrödinger operators. In: *From Number Theory to Physics.* / Eds. Waldschmidt M. e.a. Springer-Verlag, Berlin etc., 1992, 538–630.

J. Brüning
Institute of Mathematics
Humboldt University at Berlin
Unter den Linden 6
10099 Berlin, Germany
bruening@mathematik.hu-berlin.de

V.A. Geyler
Department of Mathematics
Mordovian State University
430000 Saransk
Russia
root@mathan.mordovia.su

AMS Classification: 58625, 81Q10

Operator Theory:
Advances and Applications, Vol. 117
© 2000 Birkhäuser Verlag Basel/Switzerland

The Periodic Choquard Equation

Ya.M. Dymarskii

For a nonlinear periodic eigenvalue problem of the Choquard type we prove the existence of a countable set of normalized eigenfunctions.

1 The Main Results

In this paper, we consider the following nonlinear eigenvalue problem: find $(\lambda, u) \in \mathbf{R} \times \mathbf{C}^2$ solution of

$$(1) \quad -u''(x) + r(x)u(x) + \left(\int_{-\pi}^{\pi} q(u(y), u'(y), y)|x - y|dy \right) u(x) = \lambda u(x),$$

$$(2) \quad u(-\pi) - u(\pi) = u'(-\pi) - u'(\pi) = 0,$$

$$(3) \quad \int_{-\pi}^{\pi} u^2 dx = R^2, \quad R > 0.$$

In this case, λ is called an eigenvalue, and u is called an normalized eigenfunction. The equation (1) is analogous to Ph. Choquard's equation [1] which is an approximation to Hartree-Fock theory for a one component plasma.

Let (λ^*, u^*) be a solution of the nonlinear problem (1)–(3). Then λ^* is an eigenvalue of the linear problem

$$(4) \quad -u''(x) + p(x)u(x) = \lambda u(x), \quad u(-\pi) - u(\pi) = u'(-\pi) - u'(\pi) = 0$$

where

$$(5) \quad p(x) = r(x) + \int_{-\pi}^{\pi} q(u^*(y), u^{*\prime}(y), y)|x - y|dy.$$

It is known that all eigenvalues for the linear problem (4) are interplaced in the following way

$$\lambda_0 < \lambda_1 \leq \lambda_2 < \lambda_3 \leq \lambda_4 < \dots \quad \lim_{n\to\infty} \lambda_n = \infty.$$

It is known also that an eigenfunction corresponding to the n-th eigenvalue has n zeroes in $[-\pi, \pi)$ if n is even and an eigenfunction has $n+1$ zeroes in $[-\pi, \pi)$ if n is odd. The eigenvalue λ^* receives some number and multiplicity as an eigenvalue of the linear problem (4)–(5).

Definition 1 We assign same number and multiplicity to the solution (λ^*, u^*) (and its elements) of the nonlinear problem (1)–(3).

This defenition is equivalent to the one given in [2].

Definition 2 A solution (λ^*, u^*) of the nonlinear problem (1)–(3) is named either simple or multiple if λ^* is either simple or multiple respectively as an eigenvalue of the linear problem (4)–(5).

Comment: If a solution (λ^*, u^*) is multiple then it receives two numbers.

Let us formulate the principal results.

Theorem 1 *Let* $q(u, t, x)$ *is a function of the class* $\mathbf{C}^0(\mathbf{R}^2 \times [-\pi, \pi])$. *Assume that the inequality*

$$0 \le q(u, t, x) < N(|u|^{6-\varepsilon} + |t|^{2-\varepsilon} + 1), \tag{6}$$

where constants $N, \varepsilon > 0$ *are true in the entire domain of defenition. Let* $r(x)$ *is a function of the class* $\mathbf{C}^2[-\pi, \pi]$ *and*

$$r''(x) \ge 0. \tag{7}$$

Then for any $n = 0, 1, 2, \ldots$ *and* $R > 0$ *there exists at least one solution* (λ_n, u_n) *of the problem* (1)–(3) *wich has number* n.

Theorem 2 *Assume that all conditions of the Theorem* 1 *are satisfied and at least one of the lower inequalities* (6) *or* (7) *is strict. Then all solutions of the problem* (1)–(3) *are simple and for any* $n = 0, 1, 2, \ldots$ *and* $R > 0$ *there exists at least one simple solution* (λ_n, u_n) *of the problem* (1)–(3) *wich has number* n.

Corollary 1 *Assume that all conditions of the Theorem* 1 *are satisfied. Then for any* $m = 0, 1, 2, \ldots$ *and* $R > 0$ *there exists an eigenfunction of the problem* (1)–(3) *wich has* $2m$ *zeroes in* $[-\pi, \pi)$.

Corollary 2 *Assume that all conditions of the Theorem* 2 *are satisfied. Then for any* $m = 0, 1, 2, \ldots$ *and* $R > 0$ *there exist at least two eigenfunctions of the problem* (1)–(3) *wich have* $2m$ *zeroes in* $[-\pi, \pi)$.

A pair $(\lambda, 0)$ satisfies the problem (1), (2) for all λ. This pair is called a trivial solution. Let $\mathbf{C}^i(2\pi)$ $(i = 0, 1, \ldots)$ is the Banach space of 2π-periodic $\mathbf{C}^i$-functions. Define the norm in $\mathbf{C}^i(2\pi)$ by the rule: $\|u\|_i = \max(|u(x)| + \cdots + |u^{(i)}(x)|)$.

Definition 3 [3] A number λ^* is called a bifurcation point for the problem (1), (2) if for every $\varepsilon > 0$ the problem (1), (2) has a nontrivial solution (λ, u) satisfying the inequality $|\lambda - \lambda^*| + \|u\|_2 < \varepsilon$.

It is known [3] that all bifurcation points for the problem (1), (2) belong to the set of eigenvalues of the linearized problem

$$-u''(x) + r(x)u(x) + \left(\int_{-\pi}^{\pi} q(0,0,y)|x-y|dy\right) u(x)$$
$$= \lambda u(x),\ u(-\pi) - u(\pi) = u'(-\pi) - u'(\pi) = 0$$

Corollary 3 *Assume that all conditions of the Theorem* 1 *are satisfied. Then all eigenvalues of the linearized problem are bifurcation points.*

Note that the eigenvalues of the linearized problem may be double even for all $n \in \mathbf{N}$. For instance, this happens to be the case for $r(x) = q(0,0,x) \equiv 0$. In the case of double degeneration of the linearized problem, we established the sufficient condition for the absence of a bifurcation and gave the example [4], [5].

Existence therorems of a countable set of normalized eigenfunctions was obtaind by critical points theory [3], [1] for variational symmetric eigenvalue problems and by global bifurcation [6], [7] and global projection [8] methods for nonlinear problems eigenvalues of which are all simple. It is to be noted, that the investigation of fourth order equations [8] was made through the oscillation Gantmaher-Krein theory [9]. The problem (1)–(3) is not variational and may has multiple eigenvalues. It is the speciality of this problem.

First, I would like to describe the idea of the proof. Let

$$S^{\infty}(R) = \{u \in \mathbf{C}^2(2\pi) : (3) \text{ is true}\},$$

$$M = \{(p(x), u(x)) \in \mathbf{C}^0[-\pi,\pi] \times S^{\infty}(R) : u \quad \text{is an eigenfunction of (4)}\},$$

$$Q : S^{\infty}(R) \longrightarrow \mathbf{C}^0[-\pi,\pi] \times S^{\infty}(R);$$
$$Q(u) = \left(r(x) + \int_{-\pi}^{\pi} q(u(y), u'(y), y)|x-y|dy, u\right).$$

The next theorem is obvious.

Theorem 3 *A function* $u^* \in S^{\infty}(R)$ *is an eigenfunction of the problem* (1)–(3) *only in the case* $Q(u^*) \in M$.

Thus, the tool of proof of Theorems 1, 2 may be the "intersection number" [10] of the manifold M and the mapping Q. Unfortunately, I do not know a manner of defenition of a infinite-dimensional intersection number. For this reason I shall apply a finite-dimensional approximation. The same approach was used in the paper [11]. The investigation of an eigenvalue multiplicity runs into severe difficulties too. To overcome these difficulties I shall use the new result which was acquired by V.A. Geyler and M.M. Senatorov [12].

2 The Finite-dimensional Case

Let $\mathbf{R}^k$ be the oriented Euclidean space, $a = (a_1, \dots, a_k) \in \mathbf{R}^k$, the scalar product $\langle a, b\rangle = a_1 b_1 + \dots + a_k b_k$, $S^{k-1}(R) = \{a : \langle a, a\rangle = R^2\}$. Let $L^{(k)}$ be the oriented space of real self-adjoined matrices (dim $L^{(k)} = (k+1)k/2$), $\mathbf{B} \in L^{(k)}$. Consider the k-dimensional nonlinear eigenvalue problem

$$B(a)a = \lambda a, \quad a \in S^{k-1}(R), \tag{8}$$

where

$$B : S^{k-1}(R) \longrightarrow L^{(k)} \tag{9}$$

is a $\mathbf{C}^0$-mapping. We say that a pair (λ, a) satisfying problem (8) is a solution of this problem, λ is called an eigenvalue, and a is called an normalized eigenvector.

Example 1 $B(a) - \mathrm{id}(a) \equiv \mathbf{B}$. In this case the problem (8) is a real self-adjoined eigenvalue problem.

For the problem (8) definitions 1, 2 remain true. Let

$$\begin{aligned} M^{(k)} &= \{(\mathbf{B}, a) \in L^{(k)} \times S^{k-1}(R) : \mathbf{B}a = \lambda a\}, \\ Q_k &: S^{k-1}(R) \longrightarrow L^{(k)} \times S^{k-1}(R); \quad Q_k(a) = (B(a), a). \end{aligned} \tag{10}$$

Lemma 1 *The set $M^{(k)}$ is an oriented $\mathbf{C}^\infty$-manifold,* dim $M^{(k)}$ = dim $L^{(k)}$.

Proof: Let $P_a : \mathbf{R}^k \to \mathbf{R}^{k-1}$ be the orthogonal projection where $\mathbf{R}^{k-1} \perp a$. Then $M^{(k)} = \{(\mathbf{B}, a) : P_a(\mathbf{B}a - R^{-2}\langle \mathbf{B}a, a\rangle a) = 0\}$. At any point $(\mathbf{B}^*, a^*) \in M^{(k)}$ the mapping

$$f : L^{(k)} \times S^{k-1}(R) \longrightarrow \mathbf{R}^{k-1}, \quad f(\mathbf{B}, a) = P_a(\mathbf{B}a - R^{-2}\langle \mathbf{B}a, a\rangle a)$$

is an local $\mathbf{C}^\infty$-epimorphism. □

Definition 4 A point $(\mathbf{B}, a) \in M^{(k)}$ is named either simple or multiple if λ is either simple or multiple respectively as an eigenvalue of the linear problem $\mathbf{B}a = \lambda a$. We assign same number and multiplicity of the eigenvalue λ to the point $(\mathbf{B}, a)$.

The manifold $M^{(k)}$ is stratificated by numbers and multiplicities:

$$\begin{aligned} M^{(k)}(n, l) &= \{(\mathbf{B}, a) : \mathbf{B}a = \lambda a; \ \lambda_{n-1}(\mathbf{B}) < \lambda = \lambda_n(\mathbf{B}) \\ &= \dots = \lambda_{n+l-1}(\mathbf{B}) < \lambda_{n+l}(\mathbf{B})\} \subset M^{(k)}. \end{aligned}$$

Lemma 2 *For any $n \in \mathbf{N}$ the set $M^{(k)}(n, 1)$ is a connected open subset of $M^{(k)}$.*

Proof: Let $(\mathbf{B}_1, a_1)$, $(\mathbf{B}_2, a_2) \in M^{(k)}(n, 1)$. Consider the set $L^{(k)}(n, l) = \{\mathbf{B} \in L^{(k)} : \text{the multiplicity of } \lambda_n(\mathbf{B}) \text{ is equal to } l\}$. The set $L^{(k)}(n, l)$ is a $\mathbf{C}^\infty$-submanifold of $L^{(k)}$; $\operatorname{codim} L^{(k)}(n, l) = (l-1)(l+2)/2$ [13, Addition 10]. In particular, $\operatorname{codim} L^{(k)}(n, 1) = 0$ and $\operatorname{codim} L^{(k)}(n, l) \geq 2$ if $l \geq 2$. Therefore any $\mathbf{B}_1, \mathbf{B}_2 \in L^{(k)}(n, 1)$ are connected by some curve $s_1 \subset L^{(k)}(n, 1)$. The projection

$$\pi : M^{(k)}(n, 1) \longrightarrow L^{(k)}(n, 1); \quad \pi(\mathbf{B}, a) = \mathbf{B}$$

is a 2-leafed covering transformation [14]. Hence, in the first place $M^{(k)}(n, 1)$ is an open subset of $M^{(k)}$. Secondly, there exists the curve $s_1^* \subset M^{(k)}(n, 1)$ which connects $(\mathbf{B}_1, a_1)$ and $(\mathbf{B}_2, a)$ where either $a = -a_2$ or $a = a_2$. In first case there exists a closed loop $s_2 \in L^{(k)}(n, 1)$ such that the point $\mathbf{B}_2$ belong to s_2 and the n-th eigenvector of a given matrix $\mathbf{B}_2$ changes sign when $\mathbf{B}$ transported round s_2 ([13, Addition 10], [15]). Thus $M^{(k)}(n, 1)$ is a connected subset of $L^{(k)}$. □

The next theorem is analogous to the Theorem 3.

Theorem 4 *A vector $a^* \in S^{k-1}(R)$ is an eigenvector of the problem* (8) *only in the case $Q_k(a^*) \in M^{(k)}(n, l)$. The number and the multiplicity of the solution (λ^*, a^*) are defined by indecis (n, l).*

Definition 5 A mapping B is called n-typical if the image of the mapping Q_k doesn't intersect stratums $M^{(k)}(n, l)$ where multiplicites $l \geq 2$.

In other words, a mapping B is n-typical if and only if all solutions of (8) the number of which is equal to n are simple only. The set of n-typical mappings is a set of the second category in the space of $\mathbf{C}^0$-mappings (9). This statement can be proved. But I shall give the obvious sufficient condition for n-typical mapping.

Lemma 3 *If the image $B(S^{k-1}(R)) \subset L^{(k)}(n, 1)$, then B is n-typical.*

Since $\dim M^{(k)}(n, 1) = \dim L^{(k)}$ (see Lemmas 1, 2) for any $n \leq k$ and n-typical mapping B the integer-valued intersection number $\chi(\overline{M}^{(k)}(n, 1), Q_k) = \chi(n, Q_k)$ is determined [10]. If the intersection number is not equal to zero then the problem (8) has a simple solution with number n (see the Theorem 4). The calculation of the intersection number is a difficult problem because the manifold $M^{(k)}(n, 1)$ has the boundary. We know how to calculate the intersection number in the case $B = \text{id}$.

Lemma 4 *Let a matrix $\mathbf{B} \in L^{(k)}$ has a simple eigenvalue λ_n. Let $Q_k(a) \equiv (\mathbf{B}, a)$. Then $|\chi(n, Q_k)| = 2$.*

Proof: Let $\mathbf{B}a_n = \lambda_n a_n$, $a_n \in S^{k-1}(R)$. Since the eigenvalue λ_n is simple, $M^{(k)}(n, 1) \cap \operatorname{Im} Q_k = \{(\mathbf{B}, a_n), (\mathbf{B}, -a_n)\}$. It follows from the proof of the Lemma 2 and the homotopic invariance of the intersection number that points $(\mathbf{B}, \pm a_n)$ have the same sign in the formula of the intersection number. □

In the finish of this section we shall see how a finite-dimensional approximation is carried out.

Denote by u_i $(i = 0, 1, 2, \dots,)$ normalized eigenfunctions of the linear problem (4) where $p(x) \equiv 0$:

$$\begin{aligned} u_0 &= 1/\sqrt{2\pi},\ u_1 = 1/\sqrt{\pi}\sin x,\ u_2 = 1/\sqrt{\pi}\cos x, \\ u_3 &= 1/\sqrt{\pi}\sin 2x, \dots. \end{aligned} \tag{11}$$

Denote by $\mathbf{R}_u^k$ the k-dimentional function space generated by u_i $(i = 0, 1, 2, \dots, k-1)$. Let Pr_k denote the projection

$$Pr_k(u) = \sum_{i=0}^{k-1} \left(\int_{-\pi}^{\pi} u u_i dx \right) u_i.$$

Any element $u \in \mathbf{R}_u^k \cong \mathbf{R}^k$ is a k-dimentional vector. At the same time $u \in \mathbf{R}_u^k$ belong to the space $\mathbf{C}^2(2\pi)$. The k-dimensional nonlinear eigenvalue problem

$$\begin{aligned} -u''(x) + Pr_k \left(\left(r(x) + \int_{-\pi}^{\pi} q(u(y), u'(y), y)|x-y|dy \right) u(x) \right) \\ = \lambda u(x),\ u \in S^{k-1}(R) \end{aligned} \tag{12}$$

will be called the k-dimensional approximation of (1)–(3). In the basis $\{u_0, \dots, u_{k-1}\}$ the problem (12) is of the form (8). An element of the matrix $B(u) = B(\sum_{i=0}^{k-1} a_i u_i) = B(a)$ (9) is

$$\begin{aligned} b_{i,j}(a) = \delta_{i,j}\nu_i^2 + \int_{-\pi}^{\pi} \left(r(x) + \int_{-\pi}^{\pi} q \left(\sum_{i=0}^{k-1} a_i u_i(y), \right.\right. \\ \left.\left. \sum_{i=0}^{k-1} a_i u_i'(y), y \right) |x-y|dy \right) u_i(x) u_j(x) dx. \end{aligned} \tag{13}$$

where $\delta_{i,j} = 1$ when $i = j$, $\delta_{i,j} = 0$ when $i \neq j$, $\nu_i = i$ when i is zero or odd, $\nu_i = i - 1$ when $i \neq 0$ is even.

It will be convenient to denote by Pr_∞ the identity mapping and by $\mathbf{R}_u^\infty$ the space $\mathbf{C}^2(2\pi)$. Thus the problem (1)–(3) is the problem (12) with $k = \infty$.

3 The *a Priori* Estimates

Definition 6 Let (λ, u) is a solution of the problem (12). We shall call a pair (λ, v) an associated solution if

$$\begin{aligned} v \in S^{k-1}(R) \subset \mathbf{R}_u^k,\ \int_{-\pi}^{\pi} uv = 0, -v''(x) \\ + Pr_k \left(\left(r(x) + \int_{-\pi}^{\pi} q(u(y), u'(y), y)|x-y|dy \right) v(x) \right) = \lambda v(x). \end{aligned} \tag{14}$$

We assign same number and multiplicity of (λ, u) to (λ, v).

Clearly, the multiplicity of (λ, v) is equal to two only.

Considering the problem (12) let us denote by $\{(\lambda, u)\}_n^{(k)}$ ($\{(\lambda, v)\}_n^{(k)}$) the set of all solutions (associated solutions respectively), the number of which is equal to n, where $n \leq k \leq \infty$. Let $\{(\lambda, u)\}_n = \cup_{k=n}^{\infty}\{(\lambda, u)\}_n^{(k)}$, $\{(\lambda, v)\}_n = \cup_{k=n}^{\infty}\{(\lambda, v)\}_n^{(k)}$. Let $r_{\min} = \min r(x),\ x \in [-\pi, \pi]$.

Lemma 5 *For all solutions* $(\lambda, u) \in \cup_{n=0}^{\infty}\{(\lambda, u)\}_n$ *and all associated solutions* $(\lambda, v) \in \cup_{n=0}^{\infty}\{(\lambda, v)\}_n$ *the estimates*

$$\int_{-\pi}^{\pi} (u')^2 \leq R^2(\lambda - r_{\min}), \quad \int_{-\pi}^{\pi} (v')^2 \leq R^2(\lambda - r_{\min}) \tag{15}$$

are true.

Proof: Let us multiply the equalities (12) and (14) to u and v respectively and integrate by parts:

$$\begin{aligned}
&\int_{-\pi}^{\pi} (u')^2 + \int_{-\pi}^{\pi} Pr_k\left(\left(r + \int_{-\pi}^{\pi} q(u, u', y)|x - y|dy\right) u(x)\right) u(x)dx = \lambda R^2, \\
&\int_{-\pi}^{\pi} (v')^2 + \int_{-\pi}^{\pi} Pr_k\left(\left(r + \int_{-\pi}^{\pi} q(u, u', y)|x - y|dy\right) v(x)\right) v(x)dx = \lambda R^2.
\end{aligned} \tag{16}$$

Using (6) we obtain

$$\begin{aligned}
&\int_{-\pi}^{\pi} Pr_k\left(\left(r + \int_{-\pi}^{\pi} q(u, u', y)|x - y|dy\right) u(x)\right) u(x)dx \\
&\quad = \int_{-\pi}^{\pi} \left(r + \int_{-\pi}^{\pi} q(u, u', y)|x - y|dy\right) u^2(x)dx \geq R^2 r_{\min}, \\
&\int_{-\pi}^{\pi} Pr_k\left(\left(r + \int_{-\pi}^{\pi} q(u, u', y)|x - y|dy\right) v(x)\right) v(x)dx \\
&\quad = \int_{-\pi}^{\pi} \left(r + \int_{-\pi}^{\pi} q(u, u', y)|x - y|dy\right) v^2(x)dx \geq R^2 r_{\min}.
\end{aligned} \tag{17}$$

It follows from (16) and (17) that (15) is true. □

Lemma 6 *For all solutions* $(\lambda, u) \in \cup_{n=0}^{\infty}\{(\lambda, u)\}_n$ *and any* $w \in S^{k-1}(1) \subset \mathbf{R}_u^k$ *there are positive constants* $C = C(R)$, $\delta = \delta(R)$ *that the estimate*

$$\int_{-\pi}^{\pi} Pr_k\left(\left(r(x) + \int_{-\pi}^{\pi} q(u, u', y)|x - y|dy\right) w(x)\right) w(x)dx < C(\lambda^{1-\delta} + 1)$$

is true.

Proof: Let $min|u(x)| = |u(x^*)|$. Consider the equality

$$u^2(x) = u^2(x^*) + 2\int_{x^*}^{x} u(y)u'(y)dy.$$

Using (3) we obtain

$$u^2(x) < R^2/(2\pi) + 2R\left(\int_{-\pi}^{\pi} u'^2(y)dy\right)^{1/2}. \tag{18}$$

Let $r_{\max} = \max r(x)$. It follows from (18), (3), (6) and (15) that

$$\begin{aligned}
&\int_{-\pi}^{\pi} Pr_k\left(\left(r(x) + \int_{-\pi}^{\pi} q(u,u',y)|x-y|dy\right) w(x)\right) w(x)dx \\
&< \left(r_{\max} + 2\pi\int_{-\pi}^{\pi} q(u,u',y)dy\right)\int_{-\pi}^{\pi} w(x)^2dx \\
&< \left(r_{\max} + 2\pi N\left(\|u\|_0^{4-\varepsilon}\int_{-\pi}^{\pi} u^2dy + \int_{-\pi}^{\pi} |u'|^{2-\varepsilon}dy + 1\right)\right) \\
&< C(\lambda^{1-\delta} + 1).
\end{aligned} \tag{19}$$

□

Theorem 5 *At conditions of the Theorem* 1 *the uniform estimate*

$$|\lambda| + \int_{-\pi}^{\pi} (u'')^2 + \int_{-\pi}^{\pi} (v'')^2 < D = D(R,n)$$

is true for all solutions $(\lambda, u) \in \{(\lambda, u)\}_n$ *and associated solutions* $(\lambda, v) \in \{(\lambda, v)\}_n$.

Proof: At first, for all eigenvalues the estimate $\lambda > r_{\min}$ is true (see (15)).

It follows from Lemma 6 and the variational definition of eigenvalues [16] that $\lambda < \gamma_n$, where γ_n is the n-th eigenvalue of the linear problem

$$-w''(x) + C(\lambda^{1-\delta} + 1)w(x) = \gamma w(x), \quad v \in \mathbf{R}_u^k.$$

Therefore $\lambda < \gamma_n = n^2 + C(\lambda^{1-\delta} + 1)$. Whence it follows that λ is bounded.

Now, from the estimate (15) we find that $\int_{-\pi}^{\pi}(u')^2$ and $\int_{-\pi}^{\pi}(v')^2$ are bounded too.

Finally. Multiply the equality (12) to u'' and integrate

$$-\int_{-\pi}^{\pi} (u'')^2 + \int_{-\pi}^{\pi} Pr_k\left(\left(r(x) + \int_{-\pi}^{\pi} q(u,u',y)|x-y|dy\right) u(x)\right)$$
$$u''(x)dx = \lambda\int_{-\pi}^{\pi} uu''.$$

If $u, v \in \mathbf{R}_u^k$ $(k < \infty)$ then $u'', v'' \in \mathbf{R}_u^k$ too, if $u, v \in \mathbf{C}^2(2\pi)$ then $Pr_\infty = \mathrm{id}$. For this reason and due to (19) we obtain

$$\int_{-\pi}^{\pi}(u'')^2 = \int_{-\pi}^{\pi}\left(r(x) + \int_{-\pi}^{\pi} q(u, u', y)|x-y|dy\right)$$
$$u(x)u''(x)dx + \lambda\int_{-\pi}^{\pi}(u')^2$$
$$< \left(r_{\max} + 2\pi\int_{-\pi}^{\pi} q(u, u', y)dy\right)\int_{-\pi}^{\pi}|u(x)u''(x)|dx$$
$$+ \lambda\int_{-\pi}^{\pi}(u')^2 < C(\lambda^{1-\delta}+1)R\left(\int_{-\pi}^{\pi}(u'')^2\right)^{1/2} + \lambda\int_{-\pi}^{\pi}(u')^2.$$

Since the index of the pover above of $\int_{-\pi}^{\pi}(u'')^2$ on the left of the inequality more then the index on the right, the boundedness of $\int_{-\pi}^{\pi}(u'')^2$ is proved. Reasoning similarly we shall have proved the boundedness of $\int_{-\pi}^{\pi}(v'')^2$. □

We now prove the convergence theorem of finite-dimensional approximations.

Theorem 6 *Let $\{k_i\}_{i=1}^{\infty} \subset \mathbf{N}$ is an increasing sequence such that k-approximation (12), where $k = k_i$ has a solution $(\lambda_n^{(k)}, u_n^{(k)}) \in \{(\lambda, u)\}_n^{(k)}$. Then the problem (1)–(3) has a solution $(\lambda_n, u_n) \in \{(\lambda, u)\}_n^{(\infty)}$. Moreover, if each of a solution $(\lambda_n^{(k)}, u_n^{(k)})$ is multiply, then there exists a multiple solution $(\lambda_n, u_n) \in \{(\lambda, u)\}_n^{(\infty)}$.*

Proof: The problem (12) is equivalent to the operator equation [17]

$$u + \int_{-\pi}^{\pi} G(x,s)Pr_k\left(\left(r(s) + \int_{-\pi}^{\pi} q(u(y), u'(y), y)|s-y|dy\right)u(s)\right)ds$$
$$= (\lambda+1)\int_{-\pi}^{\pi} G(x,s)u(s)ds, \quad u \in S^{k-1}(R), \tag{20}$$

where $G(x, s)$ is the Green function of the linear differential operator

$$d[u] \equiv -u''(x) + u(x), \quad u(-\pi) - u(\pi) = u'(-\pi) - u'(\pi) = 0.$$

It follows from the theorem 5 and the embedding Theorem [18] that the set $\{(\lambda, u)\}_n^{(k)}$ is compact in $\mathbf{R} \times \mathbf{C}^1(2\pi)$. Thus, in $\mathbf{R} \times \mathbf{C}^1(2\pi)$ there exists a limit point (λ_n, u_n) of some subsequence which we denote by $(\lambda_n^{(k)}, u_n^{(k)})$ $(k = k_i)$ for the sake of simplicity. We shall have shown that (λ_n, u_n) is a solution of (1)–(3). Since the system of functions (11) is complete in the 2π-periodic Hilbert space

$L^2(2\pi)$, we have

$$\Big\| \int_{-\pi}^{\pi} G(x,s) \Big\{ \Big(r(s) + \int_{-\pi}^{\pi} q(u_n(y), u_n'(y), y)|s-y|dy \Big) u_n(s)$$

$$- Pr_k \Big(\Big(r(s) + \int_{-\pi}^{\pi} q(u_n^{(k)}(y), (u_n^{(k)})'(y), y)|s-y|dy \Big) u_n^{(k)}(s) \Big) \Big\} ds \Big\|_1$$

$$< \Big\| \int_{-\pi}^{\pi} (\mathrm{id} - Pr_k) G(x,s) \Big(r(s) + \int_{-\pi}^{\pi} q(u_n(y), (u_n)'(y), y)|s-y|dy \Big)$$

$$u_n(s)ds \Big\|_1$$

$$+ \Big\| \int_{-\pi}^{\pi} G(x,s) Pr_k \Big\{ \Big(r(s) + \int_{-\pi}^{\pi} q(u_n(y), (u_n)'(y), y)|s-y|dy \Big) u_n(s)$$

$$- \Big(r(s) + \int_{-\pi}^{\pi} q(u_n^{(k)}(y), (u_n^{(k)})'(y), y)|s-y|dy \Big) u_n^{(k)}(s) \Big\} ds \Big\|_1 \longrightarrow 0$$

with $k = k_i \to \infty$. Therefore for the limit point we obtain

$$\Big\| u_n + \int_{-\pi}^{\pi} G(x,s) \Big(r(s) + \int_{-\pi}^{\pi} q(u_n(y), u_n'(y), y)|s-y|dy \Big) u_n(s)ds$$

$$-(\lambda_n + 1) \int_{-\pi}^{\pi} G(x,s) u_n(s)ds \Big\|_1 < \|u_n - u_n^{(k)}\|_1$$

$$+ \Big\| \int_{-\pi}^{\pi} G(x,s) \Big\{ \Big(r(s) + \int_{-\pi}^{\pi} q(u_n(y), u_n'(y), y)|s-y|dy \Big) u_n(s)$$

$$- Pr_k \Big(\Big(r(s) + \int_{-\pi}^{\pi} q(u_n^{(k)}(y), (u_n^{(k)})'(y), y)|s-y|dy \Big) u_n^{(k)}(s) \Big) \Big\} ds \Big\|_1$$

$$+ \Big\| (\lambda_n + 1) \int_{-\pi}^{\pi} G(x,s) u_n(s)ds - (\lambda_n^{(k)} + 1) \int_{-\pi}^{\pi} G(x,s) u_n^{(k)}(s)ds) \Big\|_1 \longrightarrow 0$$

with $k = k_i \to \infty$. Thus, $(\lambda_n, u_n) \in \mathbf{R} \times \mathbf{C}^1(2\pi)$ is the solution of (20) with $k = \infty$. Since $p(x) = r(x) + \int_{-\pi}^{\pi} q(u_n(y), u_n'(y), y)|x-y|dy \in \mathbf{C}^2$, the eigenfunction $u_n \in \mathbf{C}^2(2\pi)$. The number of the solution (λ_n, u_n) is exactly n [16].

If each of a solution $(\lambda_n^{(k)}, u_n^{(k)})$ is multiply, then reasoning similarly we shall have proved the existence of an associated solution (λ_n, v_n). To put it in another way, (λ_n, u_n) is a multiple solution. □

4 The Simplicity of Solutions

By means of the Geyler-Senatorov theorem [12] a multiplicity of solutions are considered in this section.

Theorem (V.A. Geyler and M.M. Senatorov). *Assume that in the linear problem* (4) *the function* $p(x)$ *satisfies the condition* $p''(x) > 0,\ x \in [-\pi, \pi]$. *Then all eigenvalues of* (4) *are simple only.*

Theorem 7 *If either* $r''(x) > 0$ *or* $q(u, t, x) > 0$ *then for any* $t \in [0, 1]$ *the problem*

$$
\begin{aligned}
&-u''(x) + r(x)u(x) + t\left(\int_{-\pi}^{\pi} q(u(y), u'(y), y)|x - y|dy\right)u(x) = \lambda u(x),\\
&u(-\pi) - u(\pi) = u'(-\pi) - u'(\pi) = 0, \quad \int_{-\pi}^{\pi} u^2 dx = R^2.
\end{aligned} \tag{21}
$$

has simple solutions only.

Proof: For the nonlinear problem (21) the function $p(x)$ is

$$
\begin{aligned}
p(x) &= r(x) + t\left(\int_{-\pi}^{x} q(u^*(y), u^{*\prime}(y), y)(x - y)dy\right.\\
&\quad \left. + \int_{x}^{\pi} q(u^*(y), u^{*\prime}(y), y)(y - x)dy\right)
\end{aligned}
$$

(see (5)). It is easy to see that

$$
\begin{aligned}
p'(x) &= r'(x)\\
&\quad + t\left(\int_{-\pi}^{x} q(u^*(y), u^{*\prime}(y), y)dy - \int_{x}^{\pi} q(u^*(y), u^{*\prime}(y), y)dy\right),\\
p''(x) &= r''(x) + 2tq(u^*(x), u^{*\prime}(x), x) > 0
\end{aligned}
$$

(we are used the inequalities (6) and (7)). The condition of the Geyler-Senatorov theorem is satisfied. □

Corollary 4 *Let either* $r''(x) > 0$ *or* $q(u, t, x) > 0$. *Let* n $(n = 0, 1, 2, \ldots)$ *is fixed. Then for any* $t \in [0, 1]$ *there exists* $k_0 = k_0(n) \in \mathbf{N}$ *such that the problem*

$$
\begin{aligned}
&-u''(x) + Pr_k\left(\left(r(x) + t\int_{-\pi}^{\pi} q(u(y), u'(y), y)|x - y|dy\right)u(x)\right)\\
&\quad = \lambda u(x), \ u \in S^{k-1}(R)
\end{aligned} \tag{22}
$$

with any $k > k_0$ *has simple solutions* $(\lambda, u) \in \{(\lambda, u)\}_n^{(k)}$ *only.*

Proof: The assertion follows immediately from Theorems 6, 7. □

5 The Calculation of the Intersection Number and Proof of the Main Theorems

To calculate the intersection number $\chi(n, Q_k)$ (see (10), (13)) let us make use of the fact that the intersection number is invariant under an admissible homotopy [10].

Theorem 8 *Let $r''(x) > 0$. Let n ($n = 0, 1, 2, \ldots$) is fixed. Then there exists $k_0 = k_0(n) \in \mathbf{N}$ such that the problem* (22) *with any $k > k_0$ has the same intersection number $\chi(n, Q_k) = \pm 2$ for any $t \in [0, 1]$.*

Proof: The assertion follows immediately from the Corollary 4, the Lemmas 3, 4 and the homotopic invarience of the intersection number. □

Proof of the Theorems 1, 2: If $r''(x) > 0$ then the Theorem 2 follows directly from Theorems 7, 8, the solvability of the problem (12) and the convergence Theorem 6.

If $r''(x) \geq 0$ then we can replace the function $r(x)$ by a function $r(x) + \mu r_1(x)$ with $\mu > 0$ and $r_1''(x) > 0$. Approaching the limit as $\mu \to 0$ we complete the proof of the theorems. □

References

[1] P.L. Lions, *The Choquard equation and related questions*, Nonlinear Analysis **4** (1980), 1063–1073.

[2] C. Cosner, *Bifurcation from higher eigenvalues in nonlinear elliptic equations*, Nonlinear Analysis **12** (1988), 271–277.

[3] M.A. Krasnosel'skii, *Topological methods in the theory of nonlinear integral equations*, Pergamon Press, New York, 1964.

[4] Ya.M. Dymarskii, *On tipical bifurcations in a class of operator equations*, Russian Acad. Sci. Dokl. Math. **50** (1995), no. 2, 272–277.

[5] Ya.M. Dymarskii, *On branches of small solutions of certain operator equations*, Ukrainian Mathematical Journal **48** (1996), no. 7, 1017–1027.

[6] L.A. Lusternik, *On some boundary value problem in nonlinear differential equations*, Doklady AN SSSR **33** (1941), 5–8.

[7] P. Rabinowitz, *Some global rezults for nonlinear eigenvalue problems*, J. funct. Analysis **7** (1971), 487–513.

[8] Ya.M. Dymarskii, *On Lusternik theorem for fourth order two-point problem*, in Qualitative and approximate methods of investigation of operator equations, Yaroslavl' (1984), 16–24.

[9] F.R. Gantmaher and M.G. Krein, *Oscillation matrices and kernels*, Moscow, 1950.

[10] M.W. Hirsch, *Differential topology*, Springer-Verlag, New York, 1976.

[11] Yu.G. Borisovich and O.V. Kunakovskaya, *Boundary indices of non-linear operators and the problem of eigenvectors*, in Methods and applications of global analysis, Voronezh University Press (1993), 39–43.

[12] V.A. Geyler and M.M. Senatorov, *The structure of spectrum of the Schrodinger operator*, Matematich. sbornik **188** (1997), no. 5, 21–32.

[13] V.I. Arnold, *Mathematical methods of classic mechanics*, Moscow, 1974.
[14] B.A. Dubrovin, S.P. Novikov and A.T. Fomenko, *Modern geometry*, Moscow, 1979.
[15] H.C. Longuet-Higgins, *The intersection of potential energy surfaces in polyatomic molecules*, Proc. R. Soc. Lond. A **344** (1975), 147–156.
[16] S.H. Gould, *Variational methods for eigenvalue problems*, London, 1966.
[17] R. Courant and D. Hilbert, *Methods of Mathematical Physics*, vol. 1, Interscience Publishers, New York, 1953.
[18] L. Nirenberg, *Topics in nonlinear functional analysis*, New York, 1974.

Department of Mathematics
Lugansk Pedagogical Institute
Oboronnaya str. 2
Lugansk, 348011
Ukraine
kgb@lgpi.lugansk.ua
35 P30

Operator Theory:
Advances and Applications, Vol. 117
© 2000 Birkhäuser Verlag Basel/Switzerland

On the Best Constant in a Poincare-Sobolev Inequality

Yuri V. Egorov

We consider the problem of minimization of the functional

$$F[y] = \frac{(\int_0^1 |y''|^p dx)^{1/p}}{(\int_0^1 |y'|^q dx)^{1/q}},$$

where $p \geq 1, q \geq 1$, in the functional space $W_{p,0}^2(0, 1)$. Close problems were considered by A.P. Buslaev and V.M. Tikhomirov [1] and by B. Dacorogna, C.E. Pfister in [3], and by B. Dacorogna, W. Gangbo, N. Subia in [4]. This problem is important also because of its relation to the classical Lagrange problem and to Wulff's problem.

We consider also another inequality generalizing one studied by W. Walter.

1 The Poincare-Sobolev Inequality

Consider the problem of minimization of the functional

$$F[y] = \frac{(\int_0^1 |y''|^p dx)^{1/p}}{(\int_0^1 |y'|^q dx)^{1/q}}, \tag{1}$$

where $p \geq 1, q \geq 1$, in the functional space $W_{p,0}^2(0, 1)$, the completion of the space $C_0^\infty(0, 1)$ with respect to the norm $\|y\| = (\int_0^1 |y'|^p dx)^{1/p}$ (see [2]).

Application of embedding theorems allows us to prove the existence of an optimal function y_0 such that

$$F[y_0] = \inf_{y \in W_{p,0}^2(0,1)} F[y] = m.$$

It is easy to see that

$$m = \inf_{z \in W_p^1(0,1),\ \int_0^1 z(x)dx=0} G[z],$$

where

$$G[z] = \frac{(\int_0^1 |z'|^p dx)^{1/p}}{(\int_0^1 |z|^q dx)^{1/q}},$$

and that $G[y_0'] = m$.

The Euler-Lagrange equation has the form

$$(|z'|^{p-2}z')' + m^p|z|^{q-2}z = C_1, \tag{2}$$

where C is a constant and the function $z_0 = y_0'$ is a solution of (2), if

$$\int_0^1 |z_0|^q dx = 1.$$

The equation (2) is integrable: from (2) it follows that

$$(p-1)|z'|^{p-2}z'z'' + m^p|z|^{q-2}zz' = C_1 z'$$

and therefore

$$\frac{p-1}{p}|z'|^p + \frac{m^p}{q}|z|^q = C_1 z + C_2.$$

Since

$$\int_0^1 |z|^q dx = 1, \int_0^1 |z'|^p dx = m^p, \int_0^1 z(x)dx = 0,$$

we have

$$C_2 = m^p\left(1 - \frac{1}{p} + \frac{1}{q}\right)$$

and

$$\begin{aligned} |z'|^p &= \frac{p}{p-1}\left[m^p\left(1 - \frac{1}{p} + \frac{1}{q}\right) + C_1 z - \frac{m^p}{q}|z|^q\right] \\ &= \frac{m^p}{q(p-1)}(pq - q + p + Cz - p|z|^q), \end{aligned} \tag{3}$$

where $C = C_1\, p\, q\, m^{-p}$. If $z(x) > 0$ for all small enough positive x, we have $C > 0$.

Theorem 1 *If $2p \geq q \geq 1$, $p \geq 1$, then the function z_0, minimizing the functional G in the space of functions $z \in W_p^1(0,1)$ satisfying $\int_0^1 z(x)dx = 0$, is odd with respect to the point $x = 1/2$, i.e. $z_0(x) = -z_0(1-x)$ and the constant C in* (3) *vanishes. Moreover, $z_0(x) = z_0(1/2 - x)$ for $0 < x < 1/2$ and for $0 < x < 1/4$ we have*

$$x = \frac{[q(p-1)]^{1/p}}{m}\int_0^{z_0}\frac{dz}{(pq + p - q - p|z|^q)^{1/p}}$$

and

$$m = 4q^{1/p-1}(1-1/p)^{1/p}(q+1-q/p)^{1/q-1/p}B(1/q, 1-1/p), \tag{4}$$

if $p > 1$, $m = 4$ if $p = 1$.

Proof: We can suppose that $z_0(x) > 0$ for $0 < x < b$ and $z_0(x) < 0$ for $b < x < 1$, since otherwise we can reconstruct z_0 by displacing to left the intervals, where $z_0(x) \geq 0$ without changing the values $G[z_0]$ and $\int_0^1 z_0 dx$. On the other hand, if z_0 vanishes in $]0, b[$, then $C_1 = 0$ and by the uniqueness theorem $z_0(x) \equiv 0$.

It is clear that z_0 is symmetrical on the interval $[0, b]$ with respect to the point $b/2$ and on the interval $[b, 1]$ with respect to the point $(b+1)/2$.

We have

$$\max_{0\leq x\leq b} z_0(x) = z_0(b/2) = M_1 > 0, \quad \min_{b\leq x\leq 1} z_0(x) = z_0((b+1)/2) = M_2 < 0$$

and moreover,

$$pq + p - q + CM_j - p|M_j|^q = 0, \; j = 1, 2.$$

Put

$$I(M) = \int_0^M \frac{zdz}{(pq + p - q + Cz - p|z|^q)^{1/p}}.$$

We have

$$\begin{aligned} I(M_1) &= \frac{m}{[q(p-1)]^{1/p}} \int_0^{b/2} z(x)dx, \\ I(M_2) &= -\frac{m}{[q(p-1)]^{1/p}} \int_b^{(b+1)/2} z(x)dx \end{aligned}$$

and therefore, $I(M_1) = I(M_2)$. If $C \neq 0$, then obviously $M_1 \neq -M_2$ and the function z_0 is not symmetrical. Inversely, if $C = 0$, then $M_1 = -M_2$ and $b = 1/2$.

We have

$$\begin{aligned} I(M) &= \int_0^M \frac{zdz}{(pq + p - q + Cz - p|z|^q)^{1/p}} \\ &= \int_0^1 \frac{M^2 tdt}{(pq + p - q + CMt - p|Mt|^q)^{1/p}} \\ &= M^2 \int_0^1 \frac{tdt}{[(pq + p - q)(1-t) + p|M|^q(t - t^q)]^{1/p}}. \end{aligned}$$

The function $I(M)$ is even. We shall prove that the equality $I(M_1) = I(M_2)$ implies that $M_1 = -M_2$ and therefore, $b = 1/2$ and $C = 0$.

We have for $M > 0$

$$\begin{aligned} I'(M) &= \int_0^1 \frac{tdt}{[(pq + p - q)(1-t) + p|M|^q(t - t^q)]^{1/p}} \\ &\quad \times \left[2M - \frac{qM^{q+1}(t - t^q)}{[(pq + p - q)(1-t) + p|M|^q(t - t^q)]}\right] \\ &= \int_0^1 \frac{t[2M(pq + p - q)(1-t) + (2p - q)|M|^{q+1})(t - t^q)]dt}{[(pq + p - q)(1-t) + p|M|^q(t - t^q)]^{1/p+1}}. \end{aligned}$$

The function under the sign of integral is positive if

$$2p \geq q, \ pq + p - q \geq 0.$$

To obtain (4) it suffices to remark that

$$\int_0^M \frac{dz}{(pq + p - q - p|z|^q)^{1/p}} = \frac{1}{4},$$

if $M^q = q + 1 - q/p$. □

Remark: The proof shows that the resultat is true also if $q > 2p \geq q - \varepsilon$ for $\varepsilon > 0$ small enough.

The previous theorem allows to study the problem of Lagrange in the particular case when $\alpha > 1$. Namely, we can solve the following problem of minimization:

$$M_\alpha = \inf_{y \in H_0^2(0,1)} L[Q, y], \text{ where } L[Q, y] = \frac{\int_0^1 Q y''^2 dx}{\int_0^1 y'^2 dx}$$

in the class of positive potentials Q such that $\int_0^1 Q^\alpha dx = 1, \alpha > 1$.

In [2] we have proved that $M_\alpha < \infty$ and that the extremal functions $Q_0(x)$, $y_0(x)$ do exist.

Theorem 2 *If* $\alpha > 1$, *then*

$$M_\alpha = \frac{4(2\alpha + 1)}{\alpha} \left(\frac{\alpha + 1}{2\alpha + 1} \right)^{1 - 1/\alpha} B \left(\frac{1}{2}, \frac{1}{2} + \frac{1}{2\alpha} \right)^2,$$

the optimal functions Q_0 *and* y_0 *are even with respect to the point* $x = 1/2$ *and are definied by* (5) *and* (6).

Proof: Note that by the Hölder inequality

$$\int_0^1 Q y''^2 dx \leq \left(\int_0^1 Q^\alpha dx \right)^{1/\alpha} \left(\int_0^1 |y''|^p dx \right)^{2/p},$$

where $p = 2\alpha/(\alpha - 1) > 2$. Therefore

$$L[q, y] \leq G[y'],$$

where

$$G[z] = \frac{(\int_0^1 |z'|^p dx)^{2/p}}{\int_0^1 z^2 dx}.$$

Theorem 1 implies that

$$m_\alpha = \inf_{z \in H_0^1(0,1), \int_0^1 z dx = 0} G[z]$$

is positive, that the optimal function z_0 satisfies the equation

$$|z_0'|^p = \frac{pm^p}{p-1}\left(\frac{3}{2} - \frac{1}{p} - \frac{1}{2}|z_0|^2\right),$$

and $z_0(x) = z_0(1-x)$. Therefore, for $0 < x < 1/4$ we have

$$x = \left(\frac{p-1}{p}\right)^{1/p} \frac{1}{m} \int_0^{z_0} \frac{dy}{(\frac{3}{2} - \frac{1}{p} - \frac{1}{2}y^2)^{1/p}}, \tag{5}$$

$z_0(x) = z_0(1/2 - x)$ for $0 < x < 1/2$, $z_0(x) = z_0(3/2 - x)$ for $1/2 < x < 1$. The optimal function Q_0 is defined as

$$Q_0(x) = |z_0'(x)|^{2/(\alpha-1)} \tag{6}$$

and vanishes at the points $x = 1/4$ and $x = 3/4$, but the order of zero is $\gamma = 1/\alpha$, is less than 1, i.e.

$$Q_0(x) = a(1/4 - x)^\gamma [1 + o(1)]$$

if $x \to 1/4-$.

Let

$$\mu = \inf_{z \in H_0^1(0,1), \int_0^1 z dx = 0} \frac{\int_0^1 Q_0 z'^2 dx}{\int_0^1 z^2 dx}.$$

Then the optimal function y satisfies the equation

$$(Q_0 y')' + \mu y = C,$$

it is bounded in $[0, 1]$ and its derivative y' is also bounded. On the other hand,

$$(Q_0 z_0')' + m z_0 = 0.$$

Let $y_1(x) = y(x) - y(1-x)$. Then

$$(Q_0 y_1')' + \mu y_1 = 0.$$

If $y_1 \not\equiv 0$, then $L[Q_0, y_1] = \mu$, i.e. y_1 is the extremal function. This function vanishes at $x = 0$ and at $x = 1/2$. If y_1 vanishes somewhere in $]0, 1/2[$, then by the Sturm theorem $\mu > m$, what is impossible. But if $y_1(x) > 0$ in $]0, 1/2[$, then the Sturm theorem implies that $\mu = m$ and $y_1(x) = C z_0(x)$.

If $y_1 \equiv 0$, then $y(x) = y(1-x)$. Therefore $\int_0^{1/2} y(x) dx = 0$ and there exists a point $b \in]0, 1/2[$ such that $y(b) = 0$. The function $Q_0 y' = z$ satisfies the equation

$$z'' + \mu z / Q_0 = 0,$$

Moreover, $z(1/2) = 0$ and $z(b') = 0$, where $0 < b' < 1/2$. On the other hand, the function $Q_0 z_0' = u$ satisfies the equation

$$u'' + mu/Q_0 = 0,$$

$u(1/4) = 0, u(x) > 0$ for $0 < x < b'$, $u(x) < 0$ for $b' < x < 1/2$.

By the Sturm theorem we have $\mu > m$, what is absurd. □

2 Nonsymmetric Solutions

We have shown above that if $2p \geq q$ then there are no other solutions except the symmetric one. If $2p < q$ then there are always two values $M_- < 0$ and $M_+ > 0$ such that

$$pq + p - q + CM_\pm - p|M_\pm|^q = 0, \quad I(M_+) = I(-M_-).$$

So along with a symmetric solution there can be another one, which is non-symmetric. We will show that for big values of q this non-symmetric solution gives a smaller value of m, so that the optimal solution is really asymmetric.

The following theorem is not proved yet.

Theorem 3 (Conjecture). *For any $p \geq 1$ there exists a $Q(p)$ such that for $q > Q(p)$ the function z_0, minimizing the functional G in the space of functions $z \in W^1_{p,0}(0, 1)$, satisfying the condition $\int_0^1 z(x)dx = 0$, is asymmetric, i.e. $z_0(x) \neq -z_0(1-x)$. This function is symmetric, i.e. coincides with the function described in Theorem 1, for $q \leq Q(p)$.*

We have no proof, except of a numerical one. So this is really only a conjecture.[1]

We will state the following theorem, which gives a weaker result but can be proved.

Theorem 4 *If*

$$1 \leq p < \frac{q+1}{4}, \tag{7}$$

then the function z_0, minimizing the functional G in the space of functions $z \in W^1_{p,0}(0, 1)$, satisfying the condition $\int_0^1 z(x)dx = 0$, is asymmetric, i.e. $z_0(x) \neq -z_0(1-x)$.

Proof: Let us construct a function $y(x)$ such that $y(x) > 0$ for $0 < x < b$, $y(x) < 0$ for $b < x < 1$, $\int_0^1 y(x)dx = 0$, and the function y gives minimal values to the functionals

$$G_1[z] = \frac{(\int_0^b |z'|^p dx)^{1/p}}{(\int_0^b |z|^q dx)^{1/q}}, \quad G_2[z] = \frac{(\int_b^1 |z'|^p dx)^{1/p}}{(\int_b^1 |z|^q dx)^{1/q}},$$

---

[1] *Added in the proof.* Now (in February 1999) the author has proved it with $Q(p) = 3p$.

Let m_j for $j = 1, 2$ be the minimal value of G_j in the classes $W^1_{p,0}(0, b)$ and $W^1_{p,0}(b, 1)$, respectively. Then $m_1^p = m_0 b^{1-p-p/q}$ and $m_2^p = m_0(1-b)^{1-p-p/q}$. We can assume that $\int_0^b y(x)^q dx = 1$. Then $\int_b^1 y(x)^q dx = (b/(1-b))^{q-1}$ and

$$G[y] = c\frac{b^{1-p-p/q} + (b/(1-b))^{p-p/q}(1-b)^{1-p-p/q}}{[1+(b/(1-b))^{q-1}]^{p/q}} \equiv cf(b),$$

where c is independent of b. The function f is even, $f'(1/2) = 0$ and $f''(1/2) < 0$, if inequality (7) holds. It means that f has a locally maximal value at $b = 1/2$ and two points of minimum at $b = 1/2 \pm \varepsilon_{p,q}$. Remark that $\inf G \leq G[y]$ and y is symmetric if and only if $b = 1/2$. Therefore, the optimal solution is asymmetric if condition (7) holds. □

Remark: In fact, $f''(1/2) < 0$ and the optimal function is asymmetric if $(2q+1)^2 p - q(q^2+3q+1) < 0$. In particular, this is true if $q \geq 4p - 123/64$.

3 A Walter Problem

The purpose of this section is to prove the inequality

$$\int_0^x (x-s)^{\alpha-1} f(s)^\alpha ds \leq \left(\int_0^x f(s)ds\right)^\alpha, \quad \alpha \geq 1,$$

if $g(s)$ is a positive nondecreasing function. This problem was raised by W. Walter in [6] in relation with the work [7] by P.J. Bushell and W. Okrasinski.

Changing $x = ts$ we obtain the inequality

$$\alpha\int_0^1 (1-s)^{\alpha-1} g(s)^\alpha ds \leq \left(\int_0^1 g(s)ds\right)^\alpha \tag{8}$$

with $\alpha \geq 1$ and $g(s)$ being a positive nondecreasing function.

We will prove a more general inequality.

Theorem 5 *Let f and h be continuous positive functions, f be increasing, h be decreasing and*

$$h(x) \leq C(1-x)^{\alpha-1}. \tag{9}$$

Then

$$\int_0^1 h(s)f(s)^\alpha ds \leq \int_0^1 h(s)ds\left(\int_0^1 f(s)ds\right)^\alpha, \quad \alpha \geq 1.$$

Lemma 1 *Let g be increasing on $(0, 1)$ function, $\int_0^1 g(t)dt = 0$, h be decreasing positive. Then*

$$\int_0^1 g(t)h(t)dt \leq 0.$$

Proof: There exists a point $t_0 \in (0,1)$ such that $g(t_0) = 0$, $g(t) \leq 0$ for $0 \leq t \leq t_0$ and $g(t) \geq 0$ for $t_0 \leq t \leq 1$. Therefore,

$$\int_0^1 g(t)h(t)dt \leq h(t_0)\int_0^{t_0} g(t)dt + h(t_0)\int_{t_0}^1 g(t)dt = 0.$$

□

Lemma 2 *Let $g_1, g_2, g_2 - g_1$ be increasing on $(0, 1)$ functions,*

$$\int_0^1 g_1(t)dt = \int_0^1 g_2(t)dt,$$

and h be decreasing positive. Then

$$\int_0^1 g_2(t)h(t)dt \leq \int_0^1 g_1(t)h(t)dt.$$

Proof of Theorem: Put

$$M \equiv \sup_{\int_0^1 f(t)dt=1, f'(t)\geq 0} \int_0^1 h(t)f(t)^\alpha dt,$$

where h is decreasing positive, satisfying (9), $\alpha \geq 1$.

Since f is increasing and $\int_0^1 f(t)dt = 1$, we have $f(t) \leq C_1(1-x)^{-1}$. Therefore,

$$\int_0^1 h(t)f(t)^\alpha dt \leq C\int_0^1 (1-t)^{\alpha-1}f(t)^\alpha dt \leq CC_1^{\alpha-1}\int_0^1 f(t)dt = CC_1^{\alpha-1}$$

and M is finite.

By Lemma 2 the functional increases if f is substituted for f_1, where

$$\int_0^1 f(t)dt = \int_0^1 f_1(t)dt = 1$$

and $f - f_1$ is increasing. For any increasing function f the function $f - 1$ is increasing, too and therefore, $\int_0^1 h(t)f(t)^\alpha dt \leq \int_0^1 h(t)dt$, i.e. $f_0 \equiv 1$, $M = \int_0^1 h(t)dt$. □

Another proof of the Walter conjecture was proposed recently (after the Odessa Conference) by S.M. Malamud and by A.A. Jagers.

Another **Walter conjecture:** the inequality

$$\alpha\int_0^x (x-s)^{\alpha-1}f(s)^\alpha ds \leq \left(\int_0^x f(s)ds\right)^\alpha,$$

is true if $0 < \alpha < 1$, and f is decreasing. This can be proved in the same way.

References

[1] A.P. Buslaev and V.M. Tikhomirov, *Spectra of nonlinear differential equations and widths of Sobolev classes*, Math. USSR Sbornik **71** (1992), no. 2, pp. 427–446.

[2] Yu. Egorov and V.A. Kondratiev, *Spectral Theory of Elliptic Operators*, Birkhauser, 1996.

[3] B. Dacorogna and C.E. Pfister, *Wulff theorem and best constant in Sobolev inequality*, J. Math. Pures Appl. **71** (1992), pp. 97–118.

[4] B. Dacorogna, W. Gangbo and N. Subia, *Sur une généralisation de l'inégalité de Wirtinger*, Ann. Inst. Henri Poincaré **9**, 1 (1992), pp. 29–50.

[5] Yu. Egorov, *On a Kondratiev problem*, C.R. Acad. Sci. Paris, t.324, Série I, 1997, pp. 503–507.

[6] W. Walter, *Problem: An integral inequality by Bushell and Okrasinski*, in: General Inequalities 6, Proceedings of the 6th International Conference on General Inequalities, Oberwolfach, Dec. 9–15, 1990, International Series on Numerical Mathematics, vol. **103** (1992), pp. 495–496, Birkhäuser, Basel.

[7] P.J. Bushell and W. Okrasinski, *Nonlinear Volterra equations with convolutional kernel*, J. London Math. Soc. (2) **41** (1990), pp. 503–510.

L'université Paul Sabatier
UFR MIG, MIP, 118, route de Narbonne
31062 Toulouse
France
egorov@mip.ups-tlse.fr

AMS Classification 47A20, 46A15

Operator Theory:
Advances and Applications, Vol. 117
© 2000 Birkhäuser Verlag Basel/Switzerland

On Solutions of Parabolic Equations from Families of Banach Spaces Dependent on Time

S.D. Eidelman and S.D. Ivasyshen

We present the definitions of families of Banach spaces dependent on time and, with their help, we trace the evolution of solutions of broad classes of parabolic equations. In this case, we obtain exact results on a correct solvability of the Cauchy problem and an integral representation of solutions.

The present paper is dedicated to the memory of Mark Krein, the Person and Mathematician, with the feeling of admiration and deep gratitude.

Fourty years ago, at the Conference on functional analysis (Odessa, 1958) the first of the authors told about his results on the theory of linear parabolic by Petrovskii systems. He made proposal to characterize the evolution in t of solutions of the Cauchy problem by their affiliation to a family of Banach spaces (to a certain space for every t). Mark Krein did like this structure, and he discussed it with interest. It turned out that the structure allows one to receive exact results on a correct solvability of the Cauchy problem and on a representation of solutions, defined in the open layer $(0, T] \times \mathbb{R}^n$, through their limit values on the hyperplane $\{t = 0\}$ for parabolic equations of various structures.

Below we present the initial definitions of families of Banach spaces, the essential semigroup property of functions involved to these definitions, various generalizations of the initial structure, and their applications to the derivation of the above-mentioned results.

1 Statement of the Problem. General Results

Let T be a given positive number, and $\Pi^n_{(0,T]} \equiv \{(t,x) | t \in (0,T], x \in \mathbb{R}^n\}$be a layer in the space $\mathbb{R}^{n+1}$, $n \geq 1$. Let's consider a parabolic (for now we do not define more exactly what sense it has) equation of the form

$$(1) \qquad (\partial_t - A(t,x,\partial_x))u(t,x) = 0, \quad (t,x) \in \Pi^n_{(0,T]},$$

with the (initial) condition

$$(2) \qquad u|_{t=0} = \varphi$$

on the lower boundary of the layer $\Pi^n_{(0,T]}$.

By Φ we denote the space of initial data such that there exists a classical solution u of the Cauchy problem (1), (2) for any $\varphi \in \Phi$. By U we denote the space of all such solutions u.

As early as in the 50s, the first of the authors [1] found for the parabolic by Petrovskii equations (1) that if Φ is a space of functions exponentially growing as $|x| \to \infty$, then the corresponding space U consists of functions, which exponentially grow with respect to x for every fixed $t \in (0, T]$ as $|x| \to \infty$.

In this case, the type of growth depends on t, generally saying. This means that a solution $u(t, x)$, $(t, x) \in L_{(0,T]}$, with respect to x belongs to a certain space U_t for every fixed $t \in (0, T]$. Hence, the evolution in time of solutions of problem (1), (2) can be described by using their affiliation to the corresponding spaces U_t. The second of the authors [2] proved that the space U can be defined as the space of classical solutions u of equation (1) possessing the following properties:

$$\begin{aligned} &\forall t \in (0, T] \ : \ u(t, \cdot) \in U_t; \\ \exists C > 0 \quad &\forall t \in (0, T] \ : \ \|u(t, \cdot)\|_{U_t} \leq C. \end{aligned}$$

The space U defined in such a manner can be characterized as the range of the Poisson operator P, generated by the fundamental solution Z of the Cauchy problem for equation (1) and defined on the space Φ. Moreover, the operator $P : \Phi \to U$ is an isomorphism.

Thus, on the proper choice of the space Φ and determination of a family of the appropriate spaces U_t, $t \in (0, T]$, the following assertions are true.

Theorem 1 *If $\varphi \in \Phi$, then there exists the unique solution u of the Cauchy problem* (1), (2), *which belongs to the space U and is defined by the formula*

$$u = P\varphi. \tag{3}$$

In this case, the way to satisfy the initial condition (2) depends on a choice of the space Φ and requires a refinement.

Theorem 2 *If u is a solution of equation* (1) *from the space U, then there exists a unique element $\varphi \in \Phi$ such that representation* (3) *is valid.*

Thus, the question of validity of Theorems 1, 2 is reduced to a choice of proper spaces Φ and determination of appropriate families of spaces U_t, $t \in (0, T]$.

Sufficiently broad families of spaces Φ and U_t, $t \in (0, T]$, were described by one of the authors for parabolic by Petrovskii systems of equations [1]. He also proved Theorem 1 for such systems. The second author [2] supplemented the considered families of spaces, proved.

Theorem 2, and extended results to a more broad class of systems of equations, namely, $\overrightarrow{2b}$-parabolic systems [3]. In [4, 5], analogous results are obtained for degenerate parabolic equations of arbitrary orders, which are generalizations of the classical diffusion equation with inertia of A.N. Kolmogorov. Recently, the authors have defined a new class of degenerate parabolic equations, namely, degenerate equations of the Kolmogorov type with $\overrightarrow{2b}$-parabolic part in the main group of variables, and have begun its study [6, 7]. The results mentioned above are also applicable to equations of this new class.

In the subsequent sections, we present the definitions of spaces Φ and $U_t, t \in (0, T]$, for parabolic by Petrovskii equations of arbitrary orders, $\overrightarrow{2b}$-parabolic equations, and some degenerate equations of the Kolmogorov type with $\overrightarrow{2b}$-parabolic part in the main group of variables. For the last class of equations, we give a detailed proof of Theorem 2. The last section contains some remarks and generalizations.

We use the following notations: $m, n, b, b_1, \ldots, b_n$ are given natural numbers, and $m \leq n$; $N \equiv m+n$; $\overrightarrow{2b} \equiv (2b_1, \ldots, 2b_n)$; $q \equiv 2b/(2b-1)$, $q_j \equiv 2b_j/(2b_j-1)$, $1 \leq j \leq n$, Z_+^r is a set of all r-dimensional multiindices; $|k| \equiv \sum_{j=1}^n k_j$, $\|k\| \equiv \sum_{j=1}^n (k_j/(2b_j))$ if $k \equiv (k_1, \ldots, k_n) \in Z_+^n$; i is the imaginary unit; $\{X \equiv (x, y), \Xi \equiv (\xi, \eta)\} \subset \mathbb{R}^n$ if $\{x \equiv (x_1, \ldots, x_n)\xi \equiv (\xi_1, \ldots, \xi_n)\} \subset \mathbb{R}^n$, $\{y \equiv (y_1, \ldots, y_m), \eta \equiv (\eta_1, \ldots, \eta_m)\} \subset \mathbb{R}^m$; $\Pi_H^r \equiv H \times \mathbb{R}^r$ if $H \subset R$, $r = n$ or $r = N$ and T is a given positive number.

2 Parabolic by Petrovskii Equations

Consider equation (1) of order $2b$, i.e., the equation

$$\left(\partial_t - \sum_{|k|\leq 2b} a_k(t,x)\partial_x^k\right) u(t,x) = 0, \ (t,x) \in \Pi_{(0,T]}^n. \tag{4}$$

Assume that the following conditions are valid:

α_1) equation (4) is uniformly parabolic by Petrovskii in the layer $\Pi_{[0,T]}^n$, i.e.,

$$\exists \delta > 0 \ \forall \sigma \in \mathbb{R}^n \ \forall \Pi_{[0,T]}^n : Re \sum_{|k|=2b} a_k(t,x)(i\sigma)^k \leq -\delta|\sigma|^{2b};$$

α_2) coefficients of equation (4) are bounded in $\Pi_{[0,T]}^n$, continuous in t uniformly in $x \in \mathbb{R}^n$, and satisfy the Hölder condition in x in $\mathbb{R}^n$ uniformly in t;

α_3) there exists an equation, conjugated by Lagrange to equation (4), and its coefficients satisfy condition α_2.

As is known [1], if conditions α_1 and α_2 are satisfied, there exists the fundamental solution (f.s.) of the Cauchy problem for equation (4), i.e., such a function $Z(t, x; \tau, \xi)$, $0 \leq \tau < t \leq T$, $\{x, \xi\} \subset \mathbb{R}^n$ that the formula

$$u(t,x) = \int_{\mathbb{R}^n} Z(t,x;\tau,\xi)\varphi(\xi)d\xi, \ (t,x) \in \Pi_{(\tau,T]}^n, \tag{5}$$

defines the solution of equation (4) in $\Pi^n_{(\tau,T]}$, which satisfies the condition

$$u(t,x)|_{t=\tau} = \varphi(x), \quad x \in \mathbb{R}^n,$$

for any $\tau \in [0,T)$ and an arbitrary continuous bounded function φ. For the f.s. Z, the following estimates are true:

$$\begin{aligned} &\left|\partial_t^{k_0}\partial_x^k Z(t,x;\tau,\xi)\right| \le C(t-\tau)^{-k_0-(n+|k|)/(2b)} \exp\{-c\rho(t-\tau,x,\xi)\}, \\ &0 \le \tau < t \le T,\ \{x,\xi\} \subset \mathbb{R}^n,\ 2bk_0 + |k| \le 2b, \end{aligned} \tag{6}$$

where $C > 0$, $c > 0$, and $\rho(t,x,\xi) \equiv t^{1-q}|x-\xi|^q$.

But if condition α_3 is also satisfied, then the f.s. Z possesses the property of normality, and the formula of convolution is valid for it.

In order to define the spaces Φ and U_t, $t \in (0,T]$, for equation (4), we present the following reasoning. If we assume that the initial function φ satisfies the inequality

$$|\varphi(x)| \le C\exp\{a|x|^q\}, \quad x \in \mathbb{R}^n, \quad a \ge 0, \tag{7}$$

then, as we have indicated above, the solution u of equation (4), constructed by using the function φ and defined by formula (5), generally saying, possesses the estimate

$$|u(t,x)| \le C\exp\{k(t,a)|x|^q\}.$$

By using (5) for $\tau = 0$ and estimates (6), (7), we obtain that, for determination of the function k, we should estimate from above the function

$$f(\xi) \equiv -c_0 t^{1-q}|x-\xi|^q + a|\xi|^q, \quad \xi \in BbbR^n,$$

for fixed $(t,x) \in \Pi^n_{(0,T]}$, $a \ge 0$ and $c_0 \in (0,c)$, where c is the constant from estimates (6). Since $f(\xi) \le -c_0 t^{1-q}||x|-|\xi||^q$ it is sufficient to find a maximum of the function $f_0(r) \equiv -c_0 t^{1-q}||x|-|r||^q + ar^q$, $r \ge 0$. This maximum equals $k(t,a)|x|^q$, where

$$k(t,a) \equiv c_0 a(c_0^{2b-1} - a^{2b-1}t)^{1-q}, \quad t \in (0,T], \tag{8}$$

if we take $a \ge 0$ to be such that the inequality $T < (c_0/a)^{2b-1}$ is satisfied. Thus, the estimate

$$-c_0 t^{1-q}|x-\xi|^q + a\,|\xi|^q \le k(t,a)|x|^q,\ t \in [0,T],\ \{x,\xi\} \subset \mathbb{R}^n,$$

is valid.

Note that $k(0,a) = a$, and function (8) has the following semigroup property:

$$k(t-\tau, k(\tau,a)) = k(t,a),\ 0 \le \tau \le t \le T. \tag{9}$$

This equality is verified by the immediate count.

For $\nu = \pm 1$, we set

$$V_\nu(t,x) \equiv \exp\{-\nu k(t,a)|x|^q\}, \quad (t,x) \in \Pi^n_{[0,T]}.$$

As spaces Φ, we take the spaces Φ^a_p, $1 \le p \le \infty$.

For $1 < p \le \infty$, the space Φ^a_p consists of measurable by Lebesgue functions $\varphi : \mathbb{R}^n \to \mathbb{C}$, for which the norm

$$\|\varphi\|^a_p \equiv \|\varphi(\cdot)V_1(0,\cdot)\|_{L_p(\mathbb{R}^n)} \tag{10}$$

is finite.

Let $\mathcal{B}_n$ be a σ-algebra of Borel sets of the space $\mathbb{R}^n$. We define the space Φ^a_1 as a space of all denumerably additive functions $\varphi : \mathcal{B}_n \to \mathbb{C}$ (generalized Borel measures in $\mathbb{R}^n$), which satisfy the condition

$$\|\varphi\|^a_1 \equiv \int_{\mathbb{R}^n} V_1(0,x)d\,|\varphi|\,(x) < \infty, \tag{11}$$

where $|\varphi|$ is a full variation of φ.

Spaces $L^{k(t,a)}_p$ of measurable by Lebesgue functions $v : \mathbb{R}^n \to \mathbb{C}$, for which the norm

$$\|v\|^{k(t,a)}_p \equiv \|v(\cdot)V_1(t,\cdot)\|_{L_p(\mathbb{R}^n)} \tag{12}$$

is finite, are the spaces U_t, $t \in (0,T]$ corresponding to spaces Φ^a_p.

By P, we denote the Poisson operator, which corresponds to equation (4) and is defined by the formulae

$$(P\varphi)(t,x) \equiv \int_{\mathbb{R}^n} Z(t,x;0,\xi)\left\{\begin{matrix} \varphi(\xi)d\xi, \ 1 < p \le \infty, \\ d\varphi(\xi), \ p = 1, \end{matrix}\right\}, \ (t,x) \in \Pi^n_{(0,T]}, \tag{13}$$

on elements $\varphi \in \Phi^a_p$.

For the formulation of the main results, we still introduce spaces W^a_1 and W^a_0 of all respectively measurable by Lebesgue and continuous functions $\psi : \mathbb{R}^n \to \mathbb{C}$, which possess the appropriate properties

$$\|\psi\|_{W^a_1} \equiv \|\psi(\cdot)V_{-1}(T,\cdot)\|_{L_1(\mathbb{R}^n)} < \infty \tag{14}$$

and

$$N^a_0(x) \equiv |\psi(x)|\,V_{-1}(T,x) \longrightarrow 0, \ |x| \longrightarrow \infty. \tag{15}$$

For equation (4) satisfying conditions α_1, α_2, and α_3, Theorems 1, 2 have the following concrete form.

Theorem 3 1. *Let* $\varphi \in \Phi^a_p$, $1 \le p \le \infty$. *Then function* (3), *where* P *is defined by formulae* (13), *is the unique solution of equation* (4) *in the layer* $\Pi^n_{(0,T]}$, *possessing the following properties:*

$$\forall t \in (0,T] : u(t,\cdot) \in L^{k(t,a)}_p;$$

$$\exists C > 0 \forall t \in (0,T] : \|u(t,\cdot)\|^{k(t,a)}_p \le C\,\|\varphi\|^a_p;$$

for $1 < p < \infty$ $\lim_{t\to 0} \|u(t,\cdot)-\varphi(\cdot)\|_p^{k(t,a)}$, *and, for* $p=1$ *and* $p=\infty$, $u(t,\cdot)\to\varphi$, $t\to 0$, *weakly, i.e., for any* ψ, *respectively, from the spaces* W_0^a *and* W_1^a, *the relations*

$$\lim_{t\to 0}\int_{\mathbb{R}^n}\psi(x)u(t,x)dx = \begin{cases} \int_{\mathbb{R}^n}\psi(x)d\varphi(x), & p=1,\\ \int_{\mathbb{R}^n}\psi(x)\varphi(x)dx, & p=\infty,\end{cases}$$

are true.

2. *Let* u *be a solution of equation* (4), *satisfying the condition*

$$\forall t\in(0,T]: \|u(t,\cdot)\|_p^{k(t,a)} \le C$$

with some $C>0$ *and* $p\in[1,\infty]$. *Then there exists the unique element* $\varphi\in\Phi_p^a$ *such that the solution* u *is representable in the form* (3), (13).

3 $\overrightarrow{2b}$-parabolic Equations

Consider the equation

$$\left(\partial_t - \sum_{\|k\|\le 1} a_k(t,x)\partial_x^k\right)u(t,x)=0,\ (t,x)\in\Pi_{(0,T]}^n, \tag{16}$$

by assuming that it is uniformly $\overrightarrow{2b}$-parabolic in $\Pi_{[0,T]}^n$, i.e., the condition

$$\exists\delta>0\,\forall\sigma\in\mathbb{R}^n\,\forall(t,x)\in\Pi_{[0,T]}^n : Re\sum_{\|k\|=1} a_k(t,x)(i\sigma)^k \le -\delta\sum_{j=1}^n \sigma_j^{2b_j}$$

is satisfied. We suppose also that conditions α_2 and α_3 are valid for coefficients of equation (16). Under these assumptions, there exists the f.s. Z of the Cauchy problem for equation (16) possessing the required properties. In particular, estimates (6) are true, in which

$$\rho(t,x,\xi)\equiv\sum_{j=1}^n t^{1-q_j}\left|x_j-\xi_j\right|^{q_j}.$$

As proved in [2], for equation (16), Theorem 3 is valid, in which Φ_p^a, $L_p^{k(t,a)}$, W_0^a and W_1^a are substituted by, respectively, the spaces $\Phi_p^{\vec a}$, $L_p^{\vec k(t,a)}$, $W_0^{\vec a}$ and $W_1^{\vec a}$. Define them.

In the case of equation (16), types of a possible growth of solutions in various spatial variables, generally saying, are different. Hence, here we deal with the type of growth which is a vector function $\vec k(t,\vec a)\equiv(k_1(t,a_1),\dots,k_n(t,a_n))$, where $\vec a\equiv(a_1,\dots,a_n)$,

$$k_j(t,a_j)\equiv c_0a_j(c_0^{2b_j-1}-a^{2b_j-1}t)^{1-q_j},\ 1\le j\le n. \tag{17}$$

Here, $c_0, a_1, \ldots, a_n$ are given numbers such that $0 < c_0 < c$, $a_j \geq 0$, $1 \leq j \leq n$, $T < \min_{1\leq j\leq n}(c_0/a_j)^{2b_j-1}$, where c is the constant from estimates (6) for equation (16). Note that, for each function (17), equality (9) is true.

Norms $\|\cdot\|_p^{\vec{a}}$, $\|\cdot\|_p^{\vec{k}(t,\vec{a})}$, $\|\cdot\|_{W_1^{\vec{a}}}$ and the function $N_0^{\vec{a}}$, which define, respectively, the spaces $\Phi_p^{\vec{a}}$, $L_p^{\vec{k}(t,\vec{a})}$, $W_0^{\vec{a}}$ and $W_1^{\vec{a}}$ are assigned by formulas (10)–(12), (14) and (15), where

$$V_\nu(t,x) \equiv \exp\left\{-\nu\sum_{j=1}^{n} k_j(t,a_j)|x_j|^{q_j}\right\}, \quad (t,x) \in \Pi^n_{[0,T]}.$$

4 Degenerated Equations of the Kolmogorov type with $\overrightarrow{2b}$-parabolic Part in the Main Group of Variables

Here, we consider an equation of the form

$$\left(\partial_t - \sum_{j=1}^{m} x_j\partial_{y_j} - \sum_{\|k\|\leq 1} a_k(t)\,\partial_x^k\right) u(t,X) = 0, \quad (t,X) \in \Pi^N_{(0,T]} \tag{18}$$

by assuming that coefficients $a_k : [0,T] \to \mathbb{C}$, $\|k\| \leq 1$, are continuous and the condition of $\overrightarrow{2b}$-parabolicity in the group of variables t, x is satisfied [7]. The last means that the inequality

$$\exists\delta > 0\ \forall\sigma \in \mathbb{R}^n\ \forall t \in [0,T] : Re\sum_{\|k\|=1} a_k(t)\,(i\sigma)^k \leq -\delta\sum_{j=1}^{n}\sigma_j^{2b_j}$$

is satisfied.

In [7], we constructed the f.s. $Z(t,X;\tau,\Xi)$, $0 \leq \tau < t \leq T$, $\{X,\Xi\} \subset \mathbb{R}^n$, of the Cauchy problem for equation (18) and studied its properties. In particular, we obtained the estimates

$$\begin{aligned}
&|\partial_x^r\partial_y^s Z(t,X;\tau,\Xi)| + |\partial_\xi^r\partial_\eta^s Z(t,X;\tau,\Xi)|\\
&\quad\leq C_{rs}(t-\tau)^{-M_{rs}}\exp\{-c\rho(t-\tau,X,\Xi)\},\\
&0 \leq \tau < t \leq T,\ \{X,\Xi\} \subset \mathbb{R}^N,\ r \in Z_+^n,\ s \in Z^m,
\end{aligned} \tag{19}$$

where

$$\begin{aligned}
M_{rs} &\equiv \sum_{j=1}^{n}(1/(2b_j))(r+1) + \sum_{j=1}^{m}(1+1/(2b_j))(s_j+1),\\
\rho(t,X,\Xi) &\equiv \sum_{j=1}^{n} t^{1-q_j}|x_j-\xi_j|^{q_j} + \sum_{j=1}^{m} t^{1-2q_j}|y_j+tx_j-\eta_j|^{q_j}.
\end{aligned}$$

Results of the previous sections are also applicable to equation (18). Define the spaces required for this equation. Introduce the following collections of functions:

$$\begin{aligned}
\vec{k}(t,\vec{a}) &\equiv (k_{11}t, a_{11}), \dots, k_{1n}(t, a_{1n}), k_{21}(t, a_{21}), \dots, k_{2m}(t, a_{2m})),\\
\vec{s}(t) &\equiv (s_{11}(t), \dots, s_{1n}(t), s_{21}(t), \dots, s_{2m}(t));\\
k_{lj}(t, a_{lj}) &\equiv c_0 a_{lj}(c_0^{2b_j-1} - a_{lj}^{2b_j-1} t^{2b_j(l-1)+1})^{1-q_j},\\
s_{1j}(t) &\equiv k_{1j}(t, a_{1j}) + 2^{q_j-1}\theta(m-j)t^{q_j}k_{2j}(t, a_{2j}),\\
s_{2j}(t) &\equiv 2^{q_j-1}k_{2j}(t, a_{2j}),\ 0 \le t \le T,
\end{aligned}$$

where $c_0 \in (0, c)$, c is the constant from estimates (19), $\vec{a} \equiv (a_{11}, \dots, a_{1n}, a_{21}, \dots, a_{2m})$ is a collection of nonnegative numbers such that

$$T < \min_{l,j}(c_0/a_{lj})^{(2b_j-1)/(2b_j(l-1)+1)},$$

$\theta(\tau) = 1$ for $\tau \ge 0$ and $\theta(\tau) = 0$ for $\tau \le 0$.

Note that $\vec{k}(0, \vec{a}) = \vec{a}$ and $k_{lj}(t, a_{lj}) \ge a_{lj}$, $t \in [0, T]$, and the inequality

$$(20) \qquad -c_0\rho(t, X, \Xi) + [\vec{a}, \Xi] \le [\vec{k}(t, \vec{a}), X(t)],\ t \in (0, T],\ \{X, \Xi\} \subset \mathbb{R}^N,$$

is also true, where $X(t) \equiv (x_1, \dots, x_n, y_1 + tx_1, \dots, y_m + tx_m)$ and the notation

$$(21) \qquad [\vec{a}, \Xi] \equiv \sum_{j=1}^{n} a_{1j}|\xi_j|^{q_j} + \sum_{j=1}^{m} a_{2j}|\eta_j|^{q_j}$$

is used. In addition, we note that the relations

$$\begin{aligned}
k_{1j}(t-\tau, k_{1j}(\tau, a_{1j})) &= k_{1j}(t, a_{1j}),\ 1 \le j \le n;\\
k_{2j}(t-\tau, k_{2j}(\tau, a_{2j})) &\le k_{2j}(t, a_{2j}),\ 1 \le j \le m,\ 0 \le \tau \le t \le T,
\end{aligned}$$

are valid.

Let $1 \le p \le \infty$ and $v : \mathbb{R}^n \to \mathbb{C}$ be a given measurable by Lebesgue function. For any $t \in [0, T]$, we define norms

$$\begin{aligned}
\|v\|_p^{\vec{k}(t,\vec{a})} &\equiv \|v(X)\exp\{-[\vec{k}(t,\vec{a}), X(t)]\}\|_{L_p(\mathbb{R}^N)},\\
\|v\|_p^{\vec{s}(t)} &\equiv \|v(X)\exp\{-[\vec{s}(t), X]\}\|_{L_p(\mathbb{R}^N)}
\end{aligned}$$

and introduce the following spaces: $L_p^{\vec{k}(t,\vec{a})}$, $t \in [0, T]$, $1 \le p \le \infty$, are spaces of measurable by Lebesgue functions $v : \mathbb{R}^N \to \mathbb{C}$, for which norms $\|v\|_p^{\vec{k}(t,\vec{a})}$ are finite; $\Phi_p^{\vec{a}} \equiv L_p^{\vec{k}(0,\vec{a})}$ if $1 < p \le \infty$; $\Phi_1^{\vec{a}}$ is the space of generalized Borel measures φ in $\mathbb{R}^N$, which satisfy the condition

$$\|\varphi\|_1^{\vec{a}} \equiv \int_{\mathbb{R}^N} \exp\{-[\vec{a}, X]\} d\,|\varphi|\,(X) < \infty;$$

$W_1^{\vec{a}}$ and $W_0^{\vec{a}}$ are spaces of all respectively measurable by Lebesgue and continuous functions $\psi : \mathbb{R}^N \to \mathbb{C}$, which satisfy, respectively, the conditions

$$\|\psi(X)\exp\{[\vec{s}(T), X]\}\|_{L_1(\mathbb{R}^N)} < \infty$$

and

$$\|\psi(X)\exp\{[\vec{s}(T), X]\}\| \longrightarrow 0,\ |X| \longrightarrow \infty.$$

Since

$$|y_j + tx_j|^{q_j} \leq 2^{q_j-1}(|y_j|^{q_j} + t^{q_j}|x_j|^{q_j}),\ 1 \leq j \leq m,$$

we have

$$\exp\{-[\vec{s}(t), X]\} \leq \exp\{-[\vec{k}(t,\vec{a}), X(t)]\}$$

and, therefore,

$$\|v\|_p^{\vec{s}(t)} \leq \|v\|_p^{\vec{k}(t,\vec{a})},\ 0 \leq t \leq T,\ 1 \leq p \leq \infty.$$

Note that $s_{1j}(t) \geq a_{1j}$, $s_{2j}(t) > a_{2j}$, therefore, for $\varphi \in \Phi_p^{\vec{a}}$, we have

$$\|\varphi\|_p^{\vec{s}(t)} \leq \|\varphi\|_p^{\vec{a}},\ 0 \leq t \leq T,\ 1 < p \leq \infty.$$

The following theorems are true.

Theorem 4 *For any element $\varphi \in \Phi_p^{\vec{a}}$, $1 \leq p \leq \infty$, the formula*

$$(22)\quad u(t,X) \equiv \int_{\mathbb{R}^n} Z(t,X;0,\Xi)\begin{cases} \varphi(\xi)\,d\xi, & 1 < p \leq \infty, \\ d\varphi(\xi), & p = 1, \end{cases} \quad (t,x) \in \Pi^n_{(0,T]},$$

defines the unique solution of equation (18) *in $\Pi^N_{(0,T]}$, which possesses the following properties:*

$$\forall t \in (0,T] : u(t,\cdot) \in L_p^{\vec{k}(t,\vec{a})};$$
$$\exists C > 0 \forall t \in (0,T] : \|u(t,\cdot)\|_p^{\vec{k}(t,\vec{a})} \leq C\,\|\varphi\|_p^{\vec{a}};$$

for $1 < p < \infty$, $\lim_{t\to 0} \|u(t,\cdot) - \varphi(\cdot)\|_p^{\vec{s}(t)} = 0$, and, for $p = 1$ and $p = \infty$, $u(t,\cdot) \to \varphi$, $t \to 0$ weakly as, i.e., for any ψ, respectively, from the spaces $W_0^{\vec{a}}$ and $W_1^{\vec{a}}$, the relations

$$\lim_{t\to 0}\int_{\mathbb{R}^n} \psi(X)\,u(t,X)\,dX = \begin{cases} \int_{\mathbb{R}^n} \psi(X)\,d\varphi(X), & p = 1, \\ \int_{\mathbb{R}^n} \psi(X)\,\varphi(X)\,dX, & p = \infty \end{cases}$$

are valid.

Theorem 5 *Let u be a solution of equation* (18) *satisfying the condition*

$$\forall t \in (0, T] : \|u(t, \cdot)\|_p^{\vec{k}(t,\vec{a})} \leq C \tag{23}$$

with some $C > 0$ and $p \in [1, \infty]$. Then there exists the unique element $\varphi \in \Phi_p^{\vec{a}}$ such that the solution u is representable in the form (22).

The proof of Theorem 4 is very cumbersome. To perform it, it is necessary in the appropriate way to modify the procedure of proofs for analogous theorems related to equation (4) and (16). In this case, we substantially used properties of the f.s.

Let as prove Theorem 5. We use the notation

$$V_\nu(t, X) \equiv \exp\{-\nu[\vec{k}(t, \vec{a}), X(t)]\}, \ \nu = \pm 1.$$

Let, at first, $1 < p \leq \infty$. Condition (22) implies that the sequence

$$\left\{u\left(\frac{1}{\nu}, X\right) V_1\left(\frac{1}{\nu}, X\right), \ X \in \mathbb{R}^N, \ \nu \geq 1\right\} \tag{24}$$

is bounded in the space $L_p(\mathbb{R}^N)$. This space is isometric to a space dual to $L_{p'}(\mathbb{R}^N)$. By the theorem on weak compactness of a bounded set in the dual space, sequence (23) is weakly compact in $L_p(\mathbb{R}^N)$. Therefore, there exists its subsequence

$$\left\{u\left(\frac{1}{\nu(r)}, X\right) V_1\left(\frac{1}{\nu(r)}, X\right), \ X \in \mathbb{R}^N, \ r \geq 1\right\} \tag{25}$$

which weakly converges to some function $\psi \in L_p(\mathbb{R}^N)$, i.e.,

$$\begin{aligned} \forall v \in L_{p'}(\mathbb{R}^N) : \ &\lim_{r\to\infty} \int_{\mathbb{R}^N} \overline{v(\Xi)} V_1\left(\frac{1}{\nu(r)}, \Xi\right) u\left(\frac{1}{\nu(r)}, \Xi\right) d\Xi \\ &= \int_{\mathbb{R}^N} \overline{v(\Xi)} \psi(\Xi)\, d\Xi. \end{aligned} \tag{26}$$

We set $\varphi(\Xi) \equiv \psi(\Xi) V_{-1}(0, \Xi)$, $\Xi \in \mathbb{R}^N$. Then $\varphi \in \Phi_p^{\vec{a}}$, and relation (26) reads

$$\begin{aligned} \forall v \in L_{p'}(\mathbb{R}^N) : \ &\lim_{r\to\infty} \int_{\mathbb{R}^N} \overline{v(\Xi)} V_1\left(\frac{1}{\nu(r)}, \Xi\right) u\left(\frac{1}{\nu(r)}, \Xi\right) d\Xi \\ &= \int_{\mathbb{R}^N} \overline{v(\Xi)} V_1(0, \Xi) \varphi(\Xi) u\left(\frac{1}{\nu(r)}, \Xi\right) d\Xi. \end{aligned} \tag{27}$$

We take a fixed point $(t, X) \in \Pi_{(0,T]}^N$ and consider a function

$$v(\Xi) \equiv \overline{Z(t, X; 0, \Xi)} V_{-1}(0, \Xi), \ \Xi \in \mathbb{R}^N. \tag{28}$$

By using inequalities (19) and (20), we get

$$|v(\Xi)| \leq Ct^{-M_{00}} \exp\{-(c-c_0)\rho(t,X,\Xi)\} V_{-1}(t,X), \ \Xi \in \mathbb{R}^N, \tag{29}$$

which implies that $v \in L_{p'}(\mathbb{R}^N)$. Therefore, by virtue of equality (27), we have

$$\begin{aligned} &\lim_{r\to\infty} \int_{\mathbb{R}^N} Z(t,X;0,\Xi)\, V_1\left(\frac{1}{\nu(r)},\Xi\right) V_{-1}(0,\Xi)\, u\left(\frac{1}{\nu(r)},\Xi\right) d\Xi \\ &= \int_{\mathbb{R}^N} Z(t,X;0,\Xi)\,\varphi(\Xi)\, d\Xi. \end{aligned} \tag{30}$$

We may assume that $1/\nu(r) \leq t/2$, $r \geq 1$. By using the formula of Green-Ostrogradskii and properties of the f.s., we prove by analogy with the case of parabolic by Petrovskii equations [1] that

$$u(t,X) = \int_{\mathbb{R}^N} Z\left(t,X;\frac{1}{\nu(r)},\Xi\right) u\left(\frac{1}{\nu(r)},\Xi\right) d\Xi. \tag{31}$$

By virtue of this equality, we have

$$\begin{aligned} &u(t,X) - \int_{\mathbb{R}^N} Z(t,X;0,\Xi)\,\varphi(\Xi)\, d\Xi \\ &= \int_{\mathbb{R}^N} \left(Z\left(t,X;\frac{1}{\nu(r)},\Xi\right) - Z(t,X;0,\Xi)\right) u\left(\frac{1}{\nu(r)},\Xi\right) d\Xi \\ &+ \int_{\mathbb{R}^N} Z(t,X;0,\Xi)\left(1 - V_1\left(\frac{1}{\nu(r)},\Xi\right) V_{-1}(0,\Xi)\right) u\left(\frac{1}{\nu(r)},\Xi\right) d\Xi \\ &+ \left(\int_{\mathbb{R}^N} Z(t,X;0,\Xi) V_1\left(\frac{1}{\nu(r)},\Xi\right) V_{-1}(0,\Xi) u\left(\frac{1}{\nu(r)},\Xi\right) d\Xi \right. \\ &\left. - \int_{\mathbb{R}^N} Z(t,X;0,\Xi)\varphi(\Xi) d\Xi \right) \equiv \sum_{j=1}^{3} I_j^{(r)}, \quad r \geq 1. \end{aligned} \tag{32}$$

In order to obtain representation (22), it is sufficient to prove that, for $j = 1, 2, 3$,

$$\lim_{r\to\infty} I_j^{(r)} = 0. \tag{33}$$

It follows from (30) that (33) is valid for $j = 3$. Prove (33) for $j = 2$. By using the Hölder inequality and estimates (23), we have

$$\begin{aligned} |I_2^{(r)}| &\leq \left\| u\left(\frac{1}{\nu(r)},\cdot\right)\right\|_p^{\vec{k}(\frac{1}{\nu(r)},\vec{a})} \left(\int_{\mathbb{R}^N} F_r(\Xi)\, d\Xi\right)^{1/p'} \\ &\leq C\left(\int_{\mathbb{R}^N} F_r(\Xi)\, d\Xi\right)^{1/p'}, \end{aligned} \tag{34}$$

where $F_r(\Xi) \equiv |Z(t,X;0,\Xi)|^{p'}|V_{-1}(\frac{1}{\nu(r)},\Xi) - V_{-1}(0,\Xi)|^{p'}$, $\Xi \in \mathbb{R}^N$, $r \geq 1$.

Let us study properties of the functions F_r, $r \geq 1$. Inequalities (19), (20) imply the estimate

$$\begin{aligned}(F_r(\Xi))^{1/p'} &\leq Ct^{-M_{00}} \exp\{-(c-c_0)\rho(t,X,\Xi)\} \\ &\times \left(\exp\left\{-c_0\rho(t,X,\Xi) + \left[\vec{s}\left(\frac{1}{\nu(r)}\right),\Xi\right]\right\}\right. \\ &\left.+\exp\{-c_0\rho(t,X,\Xi)+[\vec{a},\Xi]\}\right) \\ &\leq Ct^{-M_{00}} \exp\{-(c-c_0)\rho(t,X,\Xi)\} \\ &\times \left(\exp\left\{\left[\vec{k}\left(t,\vec{s}\left(\frac{1}{\nu(r)}\right)\right),X(t)\right]\right\} + V_{-1}(t,X)\right).\end{aligned}$$

Here, $r \geq r_0$, where r_0 is taken so that

$$t < \min_{l,j}\left(c_0/s_{lj}\left(\frac{1}{\nu(r_0)}\right)\right)^{(2b_j-1)/(2b_j(l-1)+1)}.$$

Since, for any $r \geq r_0$,

$$k_{lj}\left(t,s_{lj}\left(\frac{1}{\nu(r)}\right)\right) \leq k_{lj}\left(t,s_{lj}\left(\frac{1}{\nu(r_0)}\right)\right), \tag{35}$$

we have

$$\begin{aligned}(F_r(\Xi))^{1/p'} &\leq Ct^{-M_{00}} \exp\{-(c-c_0)\rho(t,X,\Xi)\} \\ &\times \left(\exp\left\{\left[\vec{k}\left(t,\vec{s}\left(\frac{1}{\nu(r)}\right)\right),X(t)\right]\right\} + V_{-1}(t,X)\right), \\ &\Xi \in \mathbb{R}^N,\ r \geq r_0.\end{aligned}$$

This implies the existence of an integrable majorant for the sequence $\{F_r, r \geq r_0\}$. Further, since $\lim_{r\to\infty} F_r(\Xi) = 0$ for any $\Xi \in \mathbb{R}^N$, by virtue of the Lebesgue theorem on majorant convergence, we have

$$\lim_{r\to\infty}\int_{\mathbb{R}^N} F_r(\Xi)\,d\Xi = 0.$$

In view of (34), we get (33) for $j=2$.

To prove (33) for $j=1$, we utilize the inequality

$$\begin{aligned}|Z(t,X;h,\Xi) - Z(t,X;0,\Xi)| &\leq Cht^{-M_{00}-1}\left(1+\sum_{j=1}^{m} t^{-1/(2b_j)}|x_j|\right) \\ &\exp\{-c\rho(t,X,\Xi)\},\end{aligned} \tag{36}$$

which is valid for any $t \in (0, T]$, $\{X, \Xi\} \subset \mathbb{R}^N$, and h sufficiently small. This inequality is proved by taking into account the property of normality of the f.s. and estimates (19).

By using inequalities (20), (23), (35), (36) and the Hölder inequality, we have, for $r \geq r_0$,

$$
\begin{aligned}
|I_1^{(r)}| \leq\ & C\frac{1}{\nu(r)} t^{-M_{00}-1}\left(1+\sum_{j=1}^{m} t^{-1/(2b_j)}|x_j|\right) \\
& \exp\left\{\left[\vec{k}\left(t, \vec{s}\left(\frac{1}{\nu(r)}\right)\right), X(t)\right]\right\} \times t^{M_0/p'} \left\| u\left(\frac{1}{\nu(r)}, \cdot\right)\right\|_p^{\vec{k}(\frac{1}{\nu(r)}, \vec{a})} \\
\leq\ & C\frac{1}{\nu(r)} t^{-1-M_0/p}\left(1+\sum_{j=1}^{m} t^{-1/(2b_j)}|x_j|\right) \\
& \exp\left\{\left[\vec{k}\left(t, \vec{s}\left(\frac{1}{\nu(r_0)}\right)\right), X(t)\right]\right\} \longrightarrow 0,
\end{aligned}
$$

for $r \to \infty$.

Let now $p = 1$. Condition (23) implies that sequence (24) is bounded in the space $L_1(\mathbb{R}^N)$. This space is dual to no other Banach space, but it is embedded to the space $M(\mathbb{R}^N)$ of all finite generalized measures, defined on the σ-algebra $\mathcal{B}_N$ of Borel sets of the space $\mathbb{R}^N$. The space $M(\mathbb{R}^N)$ is isometric to a space, dual to the space $C_0(\mathbb{R}^N)$ of functions continuous in $\mathbb{R}^N$, which tend to zero at infinity. The boundedness in $L_1(\mathbb{R}^N)$ of sequence (24) implies the boundedness of the corresponding sequence of generalized measures in $M(\mathbb{R}^N)$, and, hence, a weak compactness of the latter. Therefore, there exists subsequence (25) and a generalized measure $\mu \in M(\mathbb{R}^N)$ such that

$$
\begin{aligned}
& \forall v \in C_0(\mathbb{R}^N) : \lim_{r\to\infty} \int_{\mathbb{R}^N} \overline{v(\Xi)} V_1\left(\frac{1}{\nu(r)}, \Xi\right) u\left(\frac{1}{\nu(r)}, \Xi\right) d\Xi \\
& = \int_{\mathbb{R}^N} \overline{v(\Xi)} d\mu(\Xi).
\end{aligned} \tag{37}
$$

For bounded sets $A \in \mathcal{B}_N$, we take

$$\varphi(A) \equiv \int_A V_{-1}(0, X) d\mu(X)$$

and then

$$\int_A V_1(0, X)\, d\varphi(X) = \mu(A).$$

If A is an unbounded set from $\mathcal{B}_N$, we consider a monotonically nondecreasing sequence of bounded sets $A_r \in \mathcal{B}_N$, $r \geq 1$, such that $\cup_{r=1}^{\infty} A_r = A$, and take

$$\int_A V_1(0, X)\, d\varphi(X) \equiv \lim_{r\to\infty} \int_{A_r} V_1(0, X)\, d\varphi(X) = \lim_{r\to\infty} \mu(A_r) = \mu(A).$$

The function φ, defined in such a way, belongs to $\Phi_1^{\vec{a}}$ and equality (37) is written in the form

$$\forall v \in C_0(\mathbb{R}^N): \lim_{r\to\infty} \int_{\mathbb{R}^N} \overline{v(\Xi)} V_1\left(\frac{1}{\nu(r)}, \Xi\right) u\left(\frac{1}{\nu(r)}, \Xi\right) d\Xi = \int_{\mathbb{R}^N} \overline{v(\Xi)} V_1(0, \Xi)\, d\varphi(\Xi). \tag{38}$$

It follows from estimate (29) that function (28) belongs to the space $C_0(\mathbb{R}^N)$ for any fixed point $(t, X) \in \Pi^N_{(0,T]}$. Therefore, by virtue of (38), we get that

$$\lim_{r\to\infty} \int_{\mathbb{R}^N} Z(t, X; 0, \Xi)\, V_1\left(\frac{1}{\nu(r)}, \Xi\right) V_{-1}(0, \Xi)\, u\left(\frac{1}{\nu(r)}, \Xi\right) d\Xi = \int_{\mathbb{R}^N} Z(t, X; 0, \Xi)\, d\varphi(\Xi). \tag{39}$$

The further reasoning is the same as in the case of $p > 1$: by using (31), we write the equality

$$u(t, X) - \int_{\mathbb{R}^N} Z(t, X; 0, \Xi)\, d\varphi(\Xi) = I_1^{(r)} + I_2^{(r)} + \tilde{I}_3^{(r)}, \quad r \geq 1, \tag{40}$$

where $I_j^{(r)}$, $j = 1, 2$, are the same as in (32), and

$$\tilde{I}_3^{(r)} \equiv \int_{\mathbb{R}^N} Z(t, X; 0, \Xi)\, V_1\left(\frac{1}{\nu(r)}, \Xi\right) V_{-1}(0, \Xi) u\left(\frac{1}{\nu(r)}, \Xi\right) d\Xi - \int_{\mathbb{R}^N} Z(t, X; 0, \Xi) d\varphi(\Xi)$$

then we prove relation (33) for $j = 1, 2$. As far as $\tilde{I}_3^{(r)} \to 0$, $r \to \infty$, by virtue of (39), it follows from (40) that representation (22) is true.

Thus, we have proved the existence of the element $\Phi_p^{\vec{a}}$ such that the given solution of equation (18), which satisfies condition (23), is the Poisson integral of an element φ. Uniqueness of φ directly follows from Theorem 4.

5 Remarks and Generalizations

The results presented above are valid for solutions of parabolic by Petrovskii systems of equations and their generalizations, namely, $\overrightarrow{2b}$-parabolic systems [2, 8] and degenerated equations of type (18) with 3 groups of spatial variables [7]. They are generalized on parabolic by Petrovskii systems and confluent parabolic by Petrovskii equations of the Kolmogorov type, which contain weak degenerations on the initial hyperplane [9].

References

1. S.D. Eidelman, "*Parabolic systems*" (Russian), Moscow, Nauka, 1964.
2. S.D. Ivasyshen, "*Integral representation and initial values of solutions of* $\overrightarrow{2b}$*-parabolic systems*" (Russian), Ukrain. Mat. Zh. **42** (1990), no. 4, 500–506.
3. S.D. Eidelman, "*On a class of parabolic systems*" (Russian), Dokl. Akad. Nauk SSSR **133** (1960), no. 1, 40–43.
4. S.D. Ivasyshen and L.N. Androsova, "*On an integral representation and initial values of solutions of certain degenerating parabolic equations*" (Russian), Dokl. AN UkrSSR (1991), no. 1, 16–19.
5. S.D. Ivasyshen and L.N. Androsova, "*On an integral representation of solutions of a class of degenerate parabolic equations of the Kolmogorov type*" (Russian), Differents. Uravn. **27** (1991), no. 3, 479–487.
6. S.D. Eidelman and S.D. Ivasyshen, "*On a new class degenerating parabolic equations*" (Russian), Usp. Mat. Nauk **51** (1996), no. 5, 227.
7. S.D. Ivasyshen and S.D. Eidelman, "$\overrightarrow{2b}$*-parabolic equations with degeneration in a part of variables*" (Russian), Dokl. Akad. Nauk **360** (1998), no. 3, 303–305.
8. S.D. Ivasyshen and O.S. Kondur, "*Properties of a class of solutions of a uniform parabolic by Petrovskii system of arbitrary order*" (Ukrainian), Dop. NAN Ukr. (1996), no. 11, 12–15.
9. O.G. Voznyak and S.D. Ivasyshen, "*Fundamental solutions of the Cauchy problem for a class of degenerate parabolic equations and their application*" (Ukrainian), Dop. NAN Ukr. (1996), no. 10, 11–16.

International Solomon University
Kyiv Chernivtsi State University
Chernivtsi
35K30, 35K65.

Operator Theory:
Advances and Applications, Vol. 117
© 2000 Birkhäuser Verlag Basel/Switzerland

Canonical Systems on the Line with Rational Spectral Densities: Explicit Formulas

I. Gohberg, M.A. Kaashoek and A.L. Sakhnovich

To the memory of M.G. Krein

Explicit formulas for the direct and inverse spectral problems for a canonical system on the full line with rational spectral density are obtained via a reduction to the half line case.

0 Introduction

¿This paper deals with the full line canonical system of differential equations

$$\begin{aligned} \frac{d}{dx}u(x,\lambda) &= i(\lambda j + jV(x))u(x,\lambda), j \\ &= \begin{bmatrix} I_m & 0 \\ 0 & -I_m \end{bmatrix}, V = \begin{bmatrix} 0 & v \\ v^* & 0 \end{bmatrix}, \quad x \in \mathbb{R}, \end{aligned} \tag{0.1}$$

where I_m is $m \times m$ identity matrix, the potential V is a $2m \times 2m$ matrix function with the $m \times m$ block entry v being locally summable on $(-\infty, \infty)$, and λ is the spectral parameter. Here and in the sequel we use the notations $\mathbb{R}$, $\mathbb{C}$ and $\mathbb{C}_+$ for the real line, the complex plane and the open upper half plane, respectively. As usual we associate with the system (0.1) the differential operator

$$(H_0 f)(x) = -ij\frac{d}{dx}f(x) - V(x)f(x). \tag{0.2}$$

By definition the domain of H_0 is the subspace $D(H_0) \subset L^2_{2m}(-\infty,\infty)$ of absolutely continuous $\mathbb{C}^{2m}$-valued functions f with compact support such that $-ij\frac{df}{dx} - Vf \in L^2_{2m}(-\infty,\infty)$. The operator H_0 is symmetric and its closure is self-adjoint (cf., [6] and the remark at the end of Section 1). Following H. Weyl, we consider the spectral properties of the canonical system (0.1) in terms of a diagonalization of the operator H_0, i.e., we introduce a diagonalizing operator U,

$$(Uf)(z) = \frac{1}{\sqrt{2\pi}}\int_{-\infty}^{\infty} u(x,z)^* f(x)dx, \quad z \in \mathbb{R}, \tag{0.3}$$

where $u(x,z)$ is the fundamental solution of (0.1) normalized by $u(0,z) = I_{2m}$. It is easy to see that U diagonalizes H_0, i.e., for $f \in D(H_0)$ we have

$$\begin{aligned} (UH_0 f)(z) &= \frac{1}{\sqrt{2\pi}}\int_{-\infty}^{\infty}\left(-ij\frac{du(x,z)}{dx} - V(x)u(x,z)\right)^* \\ f(x)dx &= z(Uf)(z). \end{aligned} \tag{0.4}$$

A nondecreasing $2m \times 2m$ matrix function τ is called a *spectral function* of the system (0.1) if U is an isometric mapping from $L^2_{2m}(-\infty, \infty)$ into the space $L^2_{2m}(\mathbb{R}, d\tau)$ with inner product

$$\langle f_1, f_2 \rangle_{d\tau} = \int_{-\infty}^{\infty} f_2(z)^*(d\tau(z)) f_1(z).$$

The Weyl functions (see [4] or [5] for a definition) of the systems on the full line and on both half lines are connected, and the spectral problems on the full line may be reduced to those on the half line (see Chapter 7 in [4], Chapter 5 in [8], and the references therein). Recently explicit formulas for solutions of various problems for canonical systems on the half line with rational scattering matrix were derived in [1], and in a self-contained way and in a somewhat different and more general setting in [2]. In this paper we shall apply the results of [2] to solve spectral problems on the full line for the case when the spectral function has a rational spectral density and a finite number of jumps.

The paper consists of three sections (not counting this introduction). In the first we review the results from [2] that are needed in this paper and add some auxiliary results. Section 2 describes the reduction of the full line problems to the half line and the solution of the direct problem. The final section describes the spectral function and solves the inverse spectral problem.

1 Spectral Results on the Half Line

In this section we review some results from [2] concerning the half line equation. Consider a canonical system on the half line:

$$\begin{aligned} \frac{d}{dx}\tilde{u}(x,\lambda) &= i(\lambda\tilde{j} + \tilde{j}\tilde{V}(x))\tilde{u}(x,\lambda), \; x \geq 0, \\ \tilde{j} &= \begin{bmatrix} I_p & 0 \\ 0 & -I_p \end{bmatrix}, \; \tilde{V} = \begin{bmatrix} 0 & \tilde{v} \\ \tilde{v}^* & 0 \end{bmatrix}. \end{aligned} \tag{1.1}$$

Here the potential $\tilde{V}$ is a $2p \times 2p$ matrix function with the $p \times p$ block entry $\tilde{v}$ being locally summable on $[0, \infty)$. The differential operator

$$(\tilde{H}_0 f)(x) = -i\tilde{j}\frac{d}{dx}f(x) - \tilde{V}(x)f(x) \tag{1.2}$$

is defined on the subspace $D(\tilde{H}_0) \subset L^2_{2p}(0, \infty)$ of absolutely continuous $\mathbb{C}^{2p}$-valued functions f with compact support such that $-i\tilde{j}\frac{df}{dx} - \tilde{V}f \in L^2_{2p}(0, \infty)$, and

$$\begin{bmatrix} I_p & -I_p \end{bmatrix} f(0) = 0. \tag{1.3}$$

Then the operator

$$(\tilde{U}f)(z) = \frac{1}{\sqrt{2\pi}} \int_0^{\infty} [I_p \; I_p]\tilde{u}(x,z)^* f(x)dx, \; z \in \mathbb{R}, \tag{1.4}$$

where $\tilde{u}(x, z)$ is the fundamental solution of (1.1), diagonalizes $\tilde{H}_0$, i.e.,

$$(\tilde{U}\tilde{H}_0 f)(z) = z(\tilde{U} f)(z), \quad f \in D(\tilde{H}_0). \tag{1.5}$$

A nondecreasing $p \times p$ matrix function τ is called a *spectral function* of the system (1.1) if $\tilde{U}$ is an isometric mapping from $L^2_{2p}(0, \infty)$ into $L^2_p(\mathbb{R}, d\tau)$. In [2] explicit formulas are obtained for the case when the potentials $\tilde{v}$ have the form

$$\tilde{v}(x) = 2i\gamma_1^* e^{ix\alpha^*} S(x)^{-1} e^{ix\alpha}(\gamma_1 + i\gamma_2), \tag{1.6}$$

where

$$S(x) = I_n + \int_0^x \Lambda(t)\Lambda(t)^* dt, \; x \geq 0, \; \Lambda(x) = [e^{-ix\alpha}\gamma_1 \quad - e^{ix\alpha}(\gamma_1 + i\gamma_2)], \tag{1.7}$$

and the $n \times p$ matrices γ_1, γ_2 and the $n \times n$ matrix α satisfy the identity

$$\alpha - \alpha^* = i\gamma_1\gamma_1^* - i(\gamma_1 + i\gamma_2)(\gamma_1 + i\gamma_2)^*. \tag{1.8}$$

The matrices γ_1, γ_2 and α satisfying (1.8) form an admissible triple. Potentials of the form (1.6) are said to be *determined by the admissible triple* γ_1, γ_2 and α.

Theorem 1.1 *Assume that the potential $\tilde{v}$ of the half line canonical system* (1.1) *is determined by the admissible triple γ_1, γ_2 and α, and put $\beta = \alpha + \gamma_1\gamma_2^*$. Let z_k be the real points of $\sigma(\beta)$,*

$$z_k \in (\sigma(\beta) \cap \mathbb{R}), \; 1 \leq k \leq s, \tag{1.9}$$

and let the matrices ν_k and the matrix function W be given by the equalities

$$\nu_k = 2\pi \operatorname{res}_{z=z_k} \gamma_1^*(zI_n - \beta)^{-1}\gamma_1 \quad (1 \leq k \leq s), \tag{1.10}$$

$$W(z) = I_m + C(zI_{2n} - A)^{-1}B, \tag{1.11}$$

where the matrices A, B and C are defined by

$$C = [\,\gamma_1^* \quad \gamma_2^*\,], \; B = \begin{bmatrix} \gamma_2 \\ \gamma_1 \end{bmatrix}, \; A = \begin{bmatrix} \beta^* & \gamma_2\gamma_2^* \\ 0 & \beta \end{bmatrix}. \tag{1.12}$$

Then the matrix function

$$\tau(z) = \int_0^z W(r)dr + \sum_{z_k < z} \nu_k \tag{1.13}$$

is a nondecreasing piecewise absolutely continuous spectral function of the canonical system (1.1). *Moreover the function*

$$\phi(z) = 2i\left\{\frac{1}{2}I_p + (i\gamma_1^* + \gamma_2^*)(zI_n - \beta)^{-1}\gamma_1\right\} \tag{1.14}$$

is a Weyl function of the canonical system (1.1).

The last statement in the above theorem means (by one of the characteristic properties of a Weyl function) that ϕ admits a Nevanlinna representation

$$\phi(\lambda) = \frac{1}{\pi}\left(\nu + \int_{-\infty}^{\infty}\left(\frac{1}{z-\lambda} - \frac{z}{1+z^2}\right)d\tau(z)\right). \tag{1.15}$$

Functions ϕ such that $i(\phi(\lambda)^* - \phi(\lambda)) \geq 0$ ($\lambda \in \mathbb{C}_+$) are called Nevanlinna functions. For the rational Nevanlinna function ϕ in (1.14) the representation (1.15) may be easily proven directly. Indeed, notice that from (1.10), (1.11) and (1.14) it follows (see [2], formula (4.21)) that

$$W(z) = \frac{1}{2i}\{\phi(z) - \phi(\overline{z})^*\}, \ \nu_k = -\pi \operatorname{res}_{z=z_k}\phi(z) \ (1 \leq k \leq s). \tag{1.16}$$

Therefore $\phi(z)$ may be split into the sum $\phi(z) = \phi_c(z) + \phi_s(z)$, where $\phi(z) - \phi(\overline{z})^* = \phi_c(z) - \phi_c(\overline{z})^*$ and $\phi_c(z)$ is holomorphic in $\mathbb{C}_+ \cup \mathbb{R}$ and $\phi_s(z) = \frac{1}{\pi}\sum_{k=1}^{s}\frac{\nu_k}{z_k - z}$. In view of (1.13) and (1.16), to prove (1.15) it remains to show that

$$\phi_c(\lambda) = \frac{1}{\pi}\left(\tilde{\nu} + \int_{-\infty}^{\infty}\frac{\phi_c(z) - \phi_c(\overline{z})^*}{2i}\left(\frac{1}{z-\lambda} - \frac{z}{1+z^2}\right)dz\right). \tag{1.17}$$

To do this one uses the residue calculus to obtain

$$\frac{1}{2i\pi}\int_{-\infty}^{\infty}\frac{(\phi_c(z) - iI_p)dz}{z-\lambda} = \phi_c(\lambda) - iI_p,$$

$$\frac{1}{2i\pi}\int_{-\infty}^{\infty}\frac{(\phi_c(\overline{z})^* + iI_p)dz}{z-\lambda} = 0,$$

$$\frac{1}{\pi}\int_{-\infty}^{\infty}\left(\frac{1}{z-\lambda} - \frac{z}{1+z^2}\right)dz = i,$$

and hence (1.17) holds true.

The rational Nevanlinna function ϕ given by (1.14) also appears in another way. The transfer matrix function

$$w_{\alpha,\Lambda}(x,\lambda) = I_{2p} + i\tilde{j}\Lambda(x)^* S(x)^{-1}(\lambda I_n - \alpha)^{-1}\Lambda(x) \tag{1.18}$$

was essential in the considerations of [2]. Writing $w_{\alpha,\Lambda}(0,z)$ as

$$w_{\alpha,\Lambda}(0,z) = \begin{bmatrix} a(z) & b(z) \\ c(z) & d(z) \end{bmatrix}, \tag{1.19}$$

a direct calculation yields

$$\phi(z) = i(a(z) + c(z))(a(z) - c(z))^{-1}. \tag{1.20}$$

We add to the above results the following proposition.

Proposition 1.2 *Let ϕ be a proper rational Nevanlinna function such that*

$$\lim_{\lambda\to\infty} \phi(\lambda) = iI_p. \tag{1.21}$$

Then ϕ is a Weyl function of the canonical system (1.1) *with a potential determined by an admissible triple γ_1, γ_2 and α. In particular ϕ admits the representation* (1.14).

Proof: The proper rational Nevanlinna function ϕ admits the decomposition $\phi(z) = \phi_c(z) + \phi_s(z)$, where $\phi_c(z)$ is holomorphic in $\mathbb{C}_+ \cup \mathbb{R}$ and $\phi_s(z) = \frac{1}{\pi}\sum_{k=1}^{s} \frac{\nu_k}{z_k - z}$ $(\nu_k \geq 0)$. Taking into account (1.21) one shows that ϕ_c admits the representation (1.17) by using the arguments appearing after (1.17). Define $W(z)$ by the first part of (1.16), and notice that $\phi(z) - \phi(\overline{z})^* = \phi_c(z) - \phi_c(\overline{z})^*$ and that ν_k $(1 \leq k \leq s)$ satisfy the second part of (1.16). Then (1.15) follows from (1.16) and (1.17). In other words our ϕ admits the representation (1.15) with τ of the form

$$\tau(z) = \int_0^z W(r)dr + \sum_{z_k<z} \nu_k \quad (\nu_k \geq 0, 1 \leq k \leq s), \tag{1.22}$$

where W and ν_k are given by (1.16).

In particular, from (1.16) and the conditions of the proposition it follows that W is nonnegative rational matrix function, which has no poles on the real line and has the value I_p at infinity. From the general theory it easily follows that W admits the factorization

$$W(z) = g(\overline{z})^* g(z), \tag{1.23}$$

where the factor g has a minimal realization

$$g(z) = I_m + \tilde{\gamma}_2^*(zI_r - \tilde{\beta})^{-1}\tilde{\gamma}_1, \sigma(\tilde{\beta}) \subset \mathbb{C}_-. \tag{1.24}$$

(The fact is deduced, for instance, in [7].) Notice that we can choose our realization so that

$$\tilde{\beta}^* - \tilde{\beta} = i\tilde{\gamma}_2\tilde{\gamma}_2^*. \tag{1.25}$$

The nonnegative matrices ν_k can be presented in the form $\nu_k = 2\pi\Phi_k^*\Phi_k$ with $p_k \times p$ matrices Φ_k. Put now

$$\beta = \begin{bmatrix} \tilde{\beta} & 0 \\ 0 & Z \end{bmatrix}, \quad \gamma_1 = \begin{bmatrix} \tilde{\gamma}_1 \\ \Phi \end{bmatrix}, \quad \gamma_2 = \begin{bmatrix} \tilde{\gamma}_2 \\ 0 \end{bmatrix}, \tag{1.26}$$

where

$$Z = \operatorname{diag}\{z_1 I_{p_1}, z_2 I_{p_2}, \ldots\}, \quad \Phi = \begin{bmatrix} \Phi_1 \\ \Phi_2 \\ \vdots \end{bmatrix}.$$

By virtue of (1.24)–(1.26) we obtain

$$g(z) = I_m + \gamma_2^*(zI_n - \beta)^{-1}\gamma_1, \; \nu_k = 2\pi \operatorname{res}_{z=z_k} \gamma_1^*(zI_n - \beta)^{-1}\gamma_1, \tag{1.27}$$

and

$$\beta^* - \beta = i\gamma_2\gamma_2^*. \tag{1.28}$$

The last equation is equivalent to (1.8) and hence γ_1, γ_2 and α form an admissible triple and determine a potential $\tilde{v}$.

One can see that the functions W and matrices ν_k given by (1.10)–(1.12) and (1.23) and (1.27) coincide, i.e., the representation (1.10)–(1.13) for τ follows from (1.22).

By Theorem 1.1 we have now that ϕ given in Proposition 1.2 and the right hand side of (1.14) admit a Nevanlinna representation with the same τ and may differ only by a constant. According to (1.21) they have the same limit at infinity and therefore coincide. □

The matrix function τ given by (1.10)–(1.13) will be called the *τ-function determined by the admissible triple* γ_1, γ_2 and α. From the proof of the Proposition 1.2 it follows that each function τ of the form (1.22) with nonnegative rational matrix function W, which has no poles on the real line and has the value I_p at infinity, is a τ-function determined by an admissible triple.

Remark: By Propositions 1.4 and 1.3 of [6] the closure $\operatorname{clos}\tilde{H}_0$ of the operator $\tilde{H}_0$ is a self-adjoint operator. From the considerations of Theorem 2.1 below it follows that the operator H_0 can be represented as $H_0 = Q^*\tilde{H}_0 Q$ for some operator $\tilde{H}_0$ of the form (1.2) and a unitary mapping Q from $L^2_{2m}(-\infty,\infty)$ onto $L^2_{4m}(0,\infty)$. Thus we have $\operatorname{clos}H_0 = (\operatorname{clos}H_0)^*$ too. The uniqueness (up to normalization) of the spectral functions of systems (0.1) and (1.1) for a given potential may be obtained now in the usual way (see references in [6] for the case of the continuous potential and [8] for a more general situation of a locally summable potential).

2 Reduction of the Full Line Problem to the Half Line; Structured Admissible Triples and Weyl Functions

In this section we reduce the spectral problem for the full line equation (0.1) to a half line problem.

Theorem 2.1 *The spectral function of the full line canonical system* (0.1) *coincides with the spectral function of the half line canonical system* (1.1) *with*

$$p = 2m, \; \tilde{v}(x) = \begin{bmatrix} 0 & v(x) \\ v(-x)^* & 0 \end{bmatrix}. \tag{2.1}$$

Proof: Let $u(x, \lambda)$ be the fundamental solution of (0.1) normalized by $u(0, \lambda) = I_{2m}$. Put

$$u_+(x,\lambda) = u(x,\lambda), u_-(x,\lambda) = u(-x,\lambda), J = \begin{bmatrix} I_m & 0 & 0 & 0 \\ 0 & 0 & 0 & I_m \\ 0 & 0 & I_m & 0 \\ 0 & I_m & 0 & 0 \end{bmatrix}, \quad x \geq 0.$$

(2.2)

The representation

$$\tilde{u}(x,\lambda) = J \begin{bmatrix} u_+(x,\lambda) & 0 \\ 0 & u_-(x,\lambda) \end{bmatrix} J, \quad x \geq 0 \tag{2.3}$$

holds. Indeed, $J = J^{-1}$ and hence both sides of (2.3) coincide at $x = 0$. It also easily follows from (0.1) and (2.2) that the right hand side of (2.3) satisfies

$$\frac{d}{dx} J \begin{bmatrix} u_+(x,\lambda) & 0 \\ 0 & u_-(x,\lambda) \end{bmatrix}$$

$$J = i\left(\lambda J \begin{bmatrix} j & 0 \\ 0 & -j \end{bmatrix} J + J \begin{bmatrix} j & 0 \\ 0 & -j \end{bmatrix} \begin{bmatrix} V(x) & 0 \\ 0 & V(-x) \end{bmatrix} J\right) \tag{2.4}$$

$$J \begin{bmatrix} u_+(x,\lambda) & 0 \\ 0 & u_-(x,\lambda) \end{bmatrix} J, \quad x \geq 0.$$

Now we have $J \begin{bmatrix} j & 0 \\ 0 & -j \end{bmatrix} J = \tilde{j}$ and $J \begin{bmatrix} V(x) & 0 \\ 0 & V(-x) \end{bmatrix} J = \tilde{V}(x)$ with $\tilde{V}$ as in (1.1) and $\tilde{v}$ given by (2.1). Therefore according to (2.4) the right hand side of (2.3) satisfies (1.1) with $\tilde{v}$ given by (2.1). So (2.3) is valid. Putting $\tilde{f}(x) = J \begin{bmatrix} f(x) \\ f(-x) \end{bmatrix}$ for $x \geq 0$ and using (0.3), (1.4) and (2.3) we get

$$(\tilde{U}\tilde{f})(z) = \frac{1}{\sqrt{2\pi}} \int_0^\infty [u_+(x,z) \; u_-(x,z)] J \tilde{f}(x) dx = (Uf)(z). \tag{2.5}$$

Hence the spectral functions coincide. □

Since the potentials $\tilde{v}$ in (2.1) have a special structure, the spectral functions of the systems on the full line are determined by specially structured admissible triples.

Theorem 2.2 *Let τ be the $2m \times 2m$ τ-function determined by the admissible triple*

$$\alpha = \begin{bmatrix} \alpha_- & 0 \\ 0 & \alpha_+ \end{bmatrix}, \gamma_1 = \begin{bmatrix} 0 & \gamma_1^- \\ \gamma_1^+ & 0 \end{bmatrix},$$

(2.6)

$$\gamma_2 = i\left(\gamma_1 - \begin{bmatrix} \gamma_1^- + i\gamma_2^- & 0 \\ 0 & \gamma_1^+ + i\gamma_2^+ \end{bmatrix}\right),$$

with $r \times r$ *matrix* α_-, $(n-r) \times (n-r)$ *matrix* α_+, $r \times m$ *matrices* γ_1^- *and* γ_2^- *and* $(n-r) \times m$ *matrices* γ_1^+ *and* γ_2^+. *Then* τ *is the spectral function of system* (0.1) *with the potential* v *given by*

$$(2.7) \qquad \begin{aligned} v(x) &= -2i\gamma_+^* e^{ix\alpha_+^*} S_+(x)^{-1} e^{ix\alpha_+} \hat{\gamma}_+, \\ v(-x) &= 2i\hat{\gamma}_-^* e^{-ix\alpha_-^*} S_-(x)^{-1} e^{-ix\alpha_-} \gamma_-(x > 0), \end{aligned}$$

where $\gamma_\pm = \gamma_1^\pm$, $\hat{\gamma}_\pm = -(\gamma_1^\pm + i\gamma_2^\pm)$, *and*

$$(2.8) \qquad S_\pm(x) = I + \int_0^x (e^{-it\alpha_\pm} \gamma_\pm \gamma_\pm^* e^{it\alpha_\pm^*} + e^{it\alpha_\pm} \hat{\gamma}_\pm \hat{\gamma}_\pm^* e^{-it\alpha_\pm^*}) dt.$$

Proof: Substitute (2.6) into (1.7). One obtains

$$(2.9) \qquad S(x) = \begin{bmatrix} S_-(x) & 0 \\ 0 & S_+(x) \end{bmatrix}.$$

Substitute now (2.6) and (2.9) into (1.6). Then one gets $\tilde{v}$ of the form (2.1) with $v(x)$ and $v(-x)$ given by (2.7). By virtue of Theorem 1.1 the function τ is the spectral function of (1.1) with $\tilde{v}$ defined by (1.6). Applying Theorem 2.1 we see that τ is the spectral function of the system (0.1) on the full line with v defined by (2.7). □

Remark: Notice that the admissibility of (2.6) is equivalent to the fact that $\gamma_1^\pm$, $\gamma_2^\pm$ and $\alpha_\pm$ form admissible triples. Moreover by (2.7) $v(x)$ is determined by the admissible triple γ_1^+, γ_2^+ and α_+ and $v(-x)^*$ is determined by the admissible triple γ_1^-, γ_2^- and α_-. So Theorem 2.2 may be reformulated as a solution of the direct spectral problem.

Theorem 2.3 *Let* v *be the potential of the full line canonical system* (0.1). *Assume* $v(x)$ *is determined by the admissible triple* γ_1^+, γ_2^+ *and* α_+ *for* $x > 0$ *and* $v(-x)^*$ *is determined by the admissible triple* γ_1^-, γ_2^- *and* α_- *for* $x < 0$. *Then the* τ*-function determined by the admissible triple* (2.6) *is the spectral function of the system* (0.1).

The set of such τ has the following characteristic property.

Proposition 2.4 *Let* τ *be the* $2m \times 2m$ τ*-function determined by the admissible triple* (2.6). *Put*

$$(2.10) \qquad \tilde{\phi}(\lambda) = (I_{2m} + i\phi(\lambda))(I_{2m} - i\phi(\lambda))^{-1},$$

where

$$(2.11) \qquad \phi(\lambda) = \frac{1}{\pi} \lim_{l \to \infty} \int_{-l}^{l} \frac{d\tau(z)}{z - \lambda} (\lambda \in \mathbb{C}_+).$$

Then $\tilde{\phi}$ has a special structure, namely

$$\tilde{\phi}(\lambda) = \begin{bmatrix} 0 & \phi_-(\lambda) \\ \phi_+(\lambda) & 0 \end{bmatrix}, \tag{2.12}$$

with $m \times m$ strictly proper rational matrix functions ϕ_+ and ϕ_- such that

$$\phi_\pm(\lambda)^*\phi_\pm(\lambda) \leq I_m (\lambda \in \mathbb{C}_+). \tag{2.13}$$

Proof: We shall show first that $\phi(\lambda)$ given by (1.14) admits the representation (2.11). Indeed, from (1.14) it follows that

$$\lim_{\eta\to\infty} (\phi(i\eta) + \phi(i\eta)^*) = 0. \tag{2.14}$$

On the other hand we know that ϕ admits representation (1.15). From (1.15) it follows

$$\lim_{\eta\to\infty} (\phi(i\eta) + \phi(i\eta)^*) = \frac{2}{\pi}\left(\nu - \lim_{\eta\to\infty} \int_{-\infty}^{\infty} \frac{z(\eta^2 - 1)}{(z^2 + \eta^2)(1 + z^2)} d\tau(z)\right). \tag{2.15}$$

By (2.14) and (2.15) we have

$$\nu = \lim_{\eta\to\infty} \int_{-\infty}^{\infty} \frac{z(\eta^2 - 1)}{(z^2 + \eta^2)(1 + z^2)} d\tau(z). \tag{2.16}$$

In view of (1.13) the right hand side of (2.16) for $l > \max_{k\leq s}|z_k|$ may be rewritten as

$$\nu = \int_{-l}^{l} \frac{z}{1 + z^2} d\tau(z) + \lim_{\eta\to\infty} \int_{\Gamma} \frac{z(\eta^2 - 1)}{(z^2 + \eta^2)(1 + z^2)} W(z)dz,$$

where $\Gamma = (-\infty, l) \cup (l, \infty)$. Hence

$$\begin{aligned} \nu &= \int_{-l}^{l} \frac{z}{1 + z^2} d\tau(z) + \int_{\Gamma} \frac{z}{1 + z^2}(W(z) - I_{2m})dz \\ &= \lim_{l\to\infty} \int_{-l}^{l} \frac{z}{1 + z^2} d\tau(z). \end{aligned} \tag{2.17}$$

Therefore the right hand sides in (1.15) and (2.11) coincide.

Recall that (1.14) yields (1.20). By (1.20) and (2.11) we obtain

$$\tilde{\phi}(\lambda) = c(\lambda)a(\lambda)^{-1}. \tag{2.18}$$

By (1.7), (1.18) and (1.19) we see

$$\begin{aligned} w_{\alpha,\Lambda}(0, \lambda) &= \begin{bmatrix} a(\lambda) & b(\lambda) \\ c(\lambda) & d(\lambda) \end{bmatrix} = I_{4m} + i\tilde{j}\begin{bmatrix} \gamma_1^* \\ -(\gamma_1 + i\gamma_2)^* \end{bmatrix} \\ & (\lambda I_n - \alpha)^{-1}[\gamma_1 \quad -(\gamma_1 + i\gamma_2)]. \end{aligned} \tag{2.19}$$

Taking into account (2.6) we may express a and c via $\gamma_\pm$, $\hat{\gamma}_\pm$ and $\alpha_\pm$, where

$$\gamma_\pm = \gamma_1^\pm, \quad \hat{\gamma}_\pm = -(\gamma_1^\pm + i\gamma_2^\pm). \tag{2.20}$$

From (2.19) and (2.20) it follows

$$\begin{aligned} a(\lambda) &= I_m + i\begin{bmatrix} \gamma_+^*(\lambda I-\alpha_+)^{-1}\gamma_+ & 0 \\ 0 & \gamma_-^*(\lambda I-\alpha_-)^{-1}\gamma_- \end{bmatrix}, \\ c(\lambda) &= -i\begin{bmatrix} 0 & \hat{\gamma}_-^*(\lambda I-\alpha_-)^{-1}\gamma_- \\ \hat{\gamma}_+^*(\lambda I-\alpha_+)^{-1}\gamma_+ & \end{bmatrix}. \end{aligned} \tag{2.21}$$

Hence in view of (2.18) and (2.21) we obtain (2.12), where

$$\phi_\pm(\lambda) = -i\hat{\gamma}_\pm^*(\lambda I-\alpha_\pm)^{-1}\gamma_\pm(I_m - i\gamma_\pm^*(\lambda I-\alpha_\pm)^{-1}\gamma_\pm)^{-1}. \tag{2.22}$$

We can rewrite (2.22) as

$$\begin{aligned} \phi_\pm(\lambda) &= -i\hat{\gamma}_\pm^*(\lambda I-\alpha_\pm)^{-1}\gamma_\pm(I_m - i\gamma_\pm^*(\lambda I-\beta_\pm)^{-1}\gamma_\pm)^{-1} \\ &= -i\hat{\gamma}_\pm^*(\lambda I-\beta_\pm)^{-1}\gamma_\pm \end{aligned} \tag{2.23}$$

with

$$\beta_\pm = \alpha_\pm - i\gamma_\pm\gamma_\pm^*. \tag{2.24}$$

By (2.10) we have

$$\begin{bmatrix} \tilde{\phi}(\lambda) \\ I_{2m} \end{bmatrix} = \begin{bmatrix} iI_{2m} & I_{2m} \\ -iI_{2m} & I_{2m} \end{bmatrix}\begin{bmatrix} \phi(\lambda) \\ I_{2m} \end{bmatrix}(I_{2m} - i\phi(\lambda))^{-1},$$

i.e.,

$$\begin{bmatrix} \phi(\lambda) \\ I_{2m} \end{bmatrix} = \frac{1}{2}\begin{bmatrix} -iI_{2m} & iI_{2m} \\ I_{2m} & I_{2m} \end{bmatrix}\begin{bmatrix} \tilde{\phi}(\lambda) \\ I_{2m} \end{bmatrix}(I_{2m} - i\phi(\lambda))^{-1}.$$

Hence we have

$$\phi(\lambda) = i(I_{2m} - \tilde{\phi}(\lambda))(I_{2m} + \tilde{\phi}(\lambda)). \tag{2.25}$$

From (2.11) and (2.25) it follows

$$\begin{aligned} (\phi(\lambda) - \phi(\lambda)^*)/i &= (I_{2m} + \tilde{\phi}(\lambda)^*)^{-1}(I_{2m} - \tilde{\phi}(\lambda)^*\tilde{\phi}(\lambda)) \\ &\quad (I_{2m} + \tilde{\phi}(\lambda))^{-1} \geq 0. \end{aligned} \tag{2.26}$$

Taking into account (2.12) and (2.26) one gets (2.13). □

3 Spectral Functions and Inverse Spectral Problems

Let τ be the $2m \times 2m$ τ-function determined by an admissible triple, and let $\tilde{\phi}$ be defined by (2.10) and (2.11). In the previous section we showed that if the triple has the form (2.6), then for $\tilde{\phi}$ the formulas (2.12) and (2.13) hold. In this section we shall prove the converse statement, namely: if (2.12) and (2.13) hold, then τ is determined by an admissible triple of the form (2.6). To prove this we shall need a result on the realization of the rational contractive matrix function.

Proposition 3.1 *Let ψ be a strictly proper rational $p \times p$ matrix function which is contractive in $\mathbb{C}_+$. Then it admits a realization*

$$\psi(\lambda) = -i\hat{\gamma}^*(\lambda I_n - \theta)^{-1}\gamma, \tag{3.1}$$

where

$$\theta - \theta^* = -i(\gamma\gamma^* + \hat{\gamma}\hat{\gamma}^*). \tag{3.2}$$

Proof: Consider the matrix function

$$\phi(\lambda) = i(I_p + \psi(\lambda))(I_p - \psi(\lambda))^{-1}. \tag{3.3}$$

Analogously to (2.26) we have

$$\begin{aligned}(\phi(\lambda) - \phi(\lambda)^*)/i &= (I_p - \psi(\lambda)^*)^{-1}(I_p - \psi(\lambda)^*\psi(\lambda)) \\ &\quad (I_p - \psi(\lambda))^{-1} \geq 0, \lambda \in \mathbb{C}_+.\end{aligned} \tag{3.4}$$

So we see that ϕ is a Nevanlinna function, i.e. $i(\phi^* - \phi) \geq 0$ $(\lambda \in \mathbb{C}_+)$. From (3.3) we get also that

$$\lim_{\lambda \to \infty} \phi(\lambda) = iI_p. \tag{3.5}$$

Therefore by Proposition 1.2 the function ϕ admits the representation (1.14) with γ_1, γ_2 and β given by (1.26):

$$\phi(\lambda) = 2i\left\{\frac{1}{2}I_p + (i\gamma_1^* + \gamma_2^*)(\lambda I_n - \beta)^{-1}\gamma_1\right\}. \tag{3.6}$$

Then taking into account that by (3.3) we have $\psi(\lambda) = -(I_p + i\phi(\lambda))(I_p - i\phi(\lambda))^{-1}$, and substituting (3.6) into the last formula we get

$$\begin{aligned}\psi(\lambda) &= I_p - 2\left\{I_p + 2\left(\frac{1}{2}I_p + (i\gamma_1^* + \gamma_2^*)(\lambda I_n - \beta)^{-1}\gamma_1\right)\right\}^{-1} \\ &= I_p - (I_p - (i\gamma_1^* + \gamma_2^*)(\lambda I_n - \beta + \gamma_1(i\gamma_1^* + \gamma_2^*))^{-1}\gamma_1).\end{aligned} \tag{3.7}$$

Putting

$$\hat{\gamma} = -(\gamma_1 + i\gamma_2), \quad \gamma = \gamma_1, \quad \theta = \beta - \gamma_1(i\gamma_1^* + \gamma_2^*), \tag{3.8}$$

we obtain (3.1) from (3.7). Moreover (3.2) follows from (1.28) and (3.8). □

Proposition 3.1 shows in particular that the rational, strictly proper and contractive ψ admits a realization (3.1) such that the corresponding Riccati equation

$$T\theta - \theta^* T + i(T\gamma\gamma^* T + \hat{\gamma}\hat{\gamma}^*) = 0$$

has a positive definite solution. (See also [3] and references therein.) Now we can improve Proposition 2.4.

Theorem 3.2 *A $2m \times 2m$ matrix function τ of the form (1.22) with a finite number of jumps and with a nonnegative rational matrix function $W(z)$, which has no poles on the real line and has the value I_{2m} at infinity, is determined by an admissible triple of the form (2.6) iff the function*

$$\tilde{\phi}(\lambda) = (I_{2m} + i\phi(\lambda))(I_{2m} - i\phi(\lambda))^{-1}, \tag{3.9}$$

where

$$\phi(\lambda) = \frac{1}{\pi} \lim_{l\to\infty} \int_{-l}^{l} \frac{d\tau(z)}{z-\lambda} (\lambda \in \mathbb{C}_+), \tag{3.10}$$

has a special structure, namely

$$\tilde{\phi}(\lambda) = \begin{bmatrix} 0 & \phi_-(\lambda) \\ \phi_+(\lambda) & 0 \end{bmatrix} \tag{3.11}$$

with $m \times m$ strictly proper rational matrix functions ϕ_+ and ϕ_- such that

$$\phi_\pm(\lambda)^* \phi_\pm(\lambda) \le I_m (\lambda \in \mathbb{C}_+). \tag{3.12}$$

Proof: By Theorem 1.1 and the proof of Proposition 1.2 the set of τ-functions determined by admissible triples coincides with the functions τ of the form (1.22) with a finite number of jumps and with $W(z)$ a nonnegative rational matrix function, which has no poles on the real line and has the value I_{2m} at infinity. Then the necessity of (3.11) and (3.12) follows from Proposition 2.4.

Suppose (3.11) and (3.12) hold. Then according to Proposition 3.1 we have

$$\phi_\pm(\lambda) = -i\hat{\gamma}_\pm^*(\lambda I - \beta_\pm)^{-1}\gamma_\pm, \quad \beta_\pm - \beta_\pm^* = -i(\gamma_\pm\gamma_\pm^* + \hat{\gamma}_\pm\hat{\gamma}_\pm^*). \tag{3.13}$$

Put

$$\alpha_\pm = \beta_\pm + i\gamma_\pm\gamma_\pm^*, \quad \gamma_1^\pm = \gamma_\pm, \quad \gamma_2^\pm = i(\gamma_\pm + \hat{\gamma}_\pm). \tag{3.14}$$

According to (3.13) and (3.14) the matrices $\gamma_1^\pm$, $\gamma_2^\pm$ and $\alpha_\pm$ form admissible triples. Substitute $\gamma_1^\pm$, $\gamma_2^\pm$ and $\alpha_\pm$ into (2.6). Then γ_1, γ_2 and α form an admissible triple also. Compare now (2.10), (2.12) and (2.23) with (3.9), (3.11) and (3.13). We see that the τ-function determined by γ_1, γ_2 and α defines via (2.11) the same ϕ as the initial τ, i.e. it coincides with τ and τ is determined by γ_1, γ_2 and α of the form (2.6). □

The results of the Theorems 2.2 and 3.2 may be reformulated as the solution of the inverse spectral problem.

Theorem 3.3 *Let a* $2m \times 2m$ *matrix function* τ *satisfy the conditions of Theorem* 3.2. *Then* τ *is the spectral function of the system* (0.1) *with the potential* v *given by* (2.7) *and* (2.8), *where* $\gamma_\pm$, $\hat{\gamma}_\pm$ *and* $\alpha_\pm = \beta_\pm + i\gamma_\pm\gamma_\pm^*$ *are obtained from the representation* (3.13) *of* $\phi_\pm$ *given by* (3.9)–(3.11).

The potential v may be recovered by ϕ_+ on one half line or by ϕ_- on the other one. For this purpose use Theorem 3.3 and formulas (2.7) and (2.8).

References

[1] D. Alpay and I. Gohberg, *Inverse spectral problem for differential operators with rational scattering matrix functions*, J. Diff. Eq. **118** (1995), 1–19.

[2] I. Gohberg, M.A. Kaashoek and A.L. Sakhnovich, *Canonical systems with rational spectral densities*: explicit formulas and applications, Math. Nachr. **194** (1998), 93–125.

[3] P. Lancaster and L. Rodman, *Algebraic Riccati equations*, Clarendon Press, Oxford, 1995.

[4] B.M. Levitan, *Inverse Sturm-Liouville problems*, VSP, Zeist, 1987.

[5] B.M. Levitan and I.S. Sargsjan, *Sturm-Liouville systems and Dirac operators*, Mathematics and its Applications, vol. **69**, Kluwer Academic, Dordrecht, 1991.

[6] F.E. Melik-Adamjan, *Canonical differential operators in a Hilbert space*, Izv. Akad. Nauk Arm. SSR Math. **12** (1977), 10–31 (Russian).

[7] L. Roozemond, *Canonical pseudo-spectral factorization and Wiener-Hopf integral equations*, in: Constructive methods of Wiener-Hopf factorization (eds. I. Gohberg and M.A. Kaashoek), OT **21**, Birkhäuser Verlag, Basel, 1986, pp. 127–156.

[8] L.A. Sakhnovich, *Spectral theory of canonical differential systems, method of operator identities*, OT, Birkhäuser Verlag, to appear.

I. Gohberg
School of Mathematical Sciences
Raymond and Beverly Sackler
Faculty of Exact Sciences
Tel Aviv University
Ramat Aviv 69978,
Israel

M.A. Kaashock
Faculteit der Exacte Wetenschappen
Vrije Universiteit
De Boelelaan 1081a
1081 HV Amsterdam,
The Netherlands.

A.L. Sakhnovich
Branch of Hydroacoustics
Marine Institute of Hydrophysics,
NASU Preobradzenskaya 3
270100 Odessa,
Ukraine

AMS Classification Primary: 34L05, 34A55; Secondary: 34A05, 34B20, 47N20

Operator Theory:
Advances and Applications, Vol. 117
© 2000 Birkhäuser Verlag Basel/Switzerland

Oscillations in Systems with Periodic Coefficients and Sector-restricted Nonlinearities

*A. Halanay and VL. Răsvan**

One of the fields of Applied Mathematics that grew up from the pioneering papers of A.M. Liapunov and M.G. Krein is the theory of dynamical systems with periodic coefficients. The results of M.G. Krein and V.A. Yakubovich concerning stability of linear periodic Hamiltonian systems turned out to have applications in the so-called linear-quadratic theory of the controlled systems. The present paper deals with a "subset" of this theory: existence of nonlinear oscillations (periodic and almost periodic solutions) in systems with sector-restricted nonlinearities (the so-called absolutely stable systems). Since almost all results obtained for differential equations have their discrete-time (more or less) counterpart, both continuous time and discrete time periodic cases are presented here. The existence conditions are expressed in terms of an associated periodic Hamiltonian system that is required to be dichotomic and strongly disconjugate. This property may be checked in terms of the properties of an associated matrix Riccati equation or of some Linear Matrix Inequalities.

1 Introduction. Problem Statement

This paper has two motivations and, for this reason, two sources, in the sense that it originates from the interaction of two lines of research. It is worth to mention that both of them go "back in history" to A.M. Liapunov and they followed for a long time independently. A. The first line of research originates in the papers of Liapunov and Zhukovskii and was concerned with stability results for periodic Hill equation. In the 40-ies and 50-ies (in our century) there were many attempts to generalize these classical results and also to improve them. It was M.G. Krein who understood that these classical results might be improved within the framework of the theory of Hamiltonian systems of positive type. His monumental paper [11] is in fact a new starting point in this field and pioneering in many directions that made him famous as mathematician (to mention only the theory of spaces with indefinite metrics). From all results and directions suggested by this by now classical memoir we would like to mention the theory of the multipliers of first and second kind for periodic Hamiltonian systems of positive type. With respect to this, the paper [11] generated a quite long line of research whose results are incorporated in papers written by M.G. Krein himself [12] or in papers written jointly with co-workers [13], [14]. It is important to mention here the results concerning instability of Hamiltonian systems which are in fact results on exponential dichtomy [18], [19]. Many of the above mentioned results are incorporated in the monograph [20]. B.

*Realized during author's stage at Weizmann Institute of Science, Dept. of Theoretical Mathematics, as Meyerhoff Visiting Professor.

The second line of research is more recent and starts with a paper belonging to Lurie and Postnikov [16]; apparently it deals with a very special problem, global asymptotic stability of the equilibrium at the origin for the system

$$\dot{x} = Ax - b\phi(c^*x) \tag{1}$$

where $x \in \mathbb{R}^n$, A is a $n \times n$ matrix, b and c are n-dimensional vectors and $\phi(\cdot)$ is a nonlinear function satisfying the sector restriction

$$0 \leq \phi(\sigma) \leq \bar{\phi}\sigma^2 \tag{2}$$

Worth to mention that this stability is robust with respect to the nonlinear function, i.e. it is valid for any nonlinear function satisfying (2). We shall not insist here on all development of this problem in the last half-century, development that occupies thousands of papers and dozens of monographs. For the economy of the present paper we just note that the absolute stability result (the above described property of the zero equilibrium - global asymptotic stability for all nonlinear functions of the class defined by (2) - is called absolute stability) may be established by constructing a suitably chosen Liapunov function of the form

$$V(x) = x^*Hx + \beta \int_0^{c^*x} \phi(\sigma)d\sigma \tag{3}$$

where H is a Hermitian matrix and β a real parameter (suitable choice means finding H and β in order that $V(x)$ should have the required properties (positivity, decrease along the solutions etc.) In the 60-ies it was discovered by Yakubovich [21], [22] that the Liapunov function (3) may be used (provided $\beta = 0$) to establish existence and exponential stability of forced oscillations for the system

$$\dot{x} = Ax - b\phi(c^*x) + f(t) \tag{4}$$

It has been proved in those papers that (4) has a unique solution which is bounded on the entire axis, it is exponentially stable and is periodic or almost periodic if f is periodic or almost periodic respectively. Several years later Yakubovich [23] was able to prove that the same Liapunov function may give existence of a certain type of self-sustained oscillations for (1) and, generally speaking, various properties of (3) may give for (1) all kinds of qualitative behaviour [24]. C. The main crossing point of these two lines of research seems to be (according to our opinion) that of the applied method. It is now a well established fact that a Liapunov function of the form (3) with the required properties may be constructed provided a certain frequency domain inequality holds - the Yakubovich-Kalman-Popov lemma [17] (worth to mention that one of the proofs of this lemma strongly rely on factorization results that go back to Gohberg and Krein [7]). In some sense this result is connected to the so-called optimal stabilization problem i.e. of finding the control function $u(t)$ in order that the corresponding solution of the controlled system

$$\dot{x} = Ax + bu(t) \tag{5}$$

should asymptotically (even exponentially) tend to 0 and the quadratic performance index

$$J(x, u) = \int_0^{\infty} (\kappa u^2(t) + u^*(t)l^*x(t) + x^*(t)lu(t) + x^*(t)Mx(t))dt \tag{6}$$

should be minimized on the set of admissible pairs $(x(t), u(t))$ (for which (6) is finite). This connection is expressed by the fact that the minimal value of (6) is given by a quadratic form whose matrix H is determined from Yakubovich-Kalman-Popov lemma provided a frequency domain inequality holds. It was discovered also by Yakubovich [25] that this frequency domain inequality (mentioned above) is equivalent to exponential dichotomy and strong disconjugacy of an associated $2n$-dimensional Hamiltonian system. There are two implications of this discovery that has become the turning point of the entire linear-quadratic theory (generic name unifying qualitative theory of systems (1) and optimal stabilization problem): first of them is concerned with numerical methods and the second one leads straightforwardly to earlier Krein results on Hamiltonian systems. The main step on this way has been performed by Yakubovich [25] who was able to construct a rather general linear quadratic theory for (5)–(6) under the assumption that the coefficients $A(t), b(t), \kappa(t), l(t), M(t)$ are T-periodic. D. We are now in position to state the problem that will be tackled in what follows. We shall consider the nonlinear system

$$\dot{x} = A(t)x - b\phi(t, c^*(t)x) + f(t) \tag{7}$$

under the following basic assumptions: i) the matrix $A(t)$ and the vectors $b(t)$, $c(t)$ have continuous and T-periodic elements; ii) the time-varying nonlinearity $\phi(t, \sigma)$ is piecewise continuous and T-periodic with respect to variable t, uniformly with respect to σ and continuous in σ; also it satisfies the condition

$$0 \leq \frac{\phi(t, \sigma_1) - \phi(t, \sigma_2)}{\sigma_1 - \sigma_2} \leq \overline{\Phi} \tag{8}$$

for any $\sigma_1 \neq \sigma_2$ and all $t \in [0, T)$; iii) $f(t)$ has bounded on the whole real axis components, possibly periodic or almost periodic. In the periodic case the period of $f(t)$ may equal T, the period of system's coefficients, but this is not compulsory. Under these assumptions existence and uniqueness is ensured for the Cauchy problem (i.e. for any t_0 and x_0 there exists $x(t; t_0, x_0)$, a unique solution of (7) for $t > t_0$). We are interested however in existence and stability of a certain "limit" solution of (7) - a solution that is defined on the whole real axis and is "of the same type" as the forcing term (periodic or almost periodic). Since almost all results obtained for differential equations have their discrete-time (more or less) counterpart, we shall consider here also the discrete-time version of the above problem. Both continuous-time and discrete-time results strongly rely on [25] (including its discrete-time counterpart) and on the properties of invariant manifolds for flows obtained by Halanay [8], [9] for continuous-time as well as for discrete-time flows.

2 Main Result for the Continuous-time Case

We shall consider here the nonlinear system (7) under the basic assumptions i)–iii) stated in Section 1 of the present paper. Our aim is to give an answer to the problem formulated also in Section 1 namely existence and stability of the unique bounded solution that is defined on the entire axis. The sufficient conditions given below are expressed in the language of system's coefficients A, b, c, more precisely in the terms of exponential dichotomy and strong disconjugacy of an associated periodic linear Hamiltonian system; this Hamiltonian system is associated to the equations that define a periodic quadratic Liapunov function. In order to state the main result we shall define the following Hamiltonian system

$$\begin{aligned} \dot{x} &= \left(A(t) - \frac{1}{2\Phi}b(t)c^*(t)\right)x + \frac{1}{\Phi}b(t)b^*(t)p \\ \dot{p} &= -\frac{4}{\Phi}c(t)c^*(t)x - \left(A(t) - \frac{1}{2\Phi}b(t)c^*(t)\right)^* p \end{aligned} \tag{9}$$

Let $Z(t)$ denote the $2n \times 2n$ transition matrix of (9) and $Z(T)$ be the monodromy matrix. If the eigenvalues of $Z(T)$ i.e. the multipliers of (9) are not on the unit circle i.e. the following frequency domain condition holds

$$det[Z(T) - e^{i\omega}I_{2n}] \neq 0, \ \forall \omega \in [-\pi, \pi) \tag{10}$$

then the Hamiltonian system is exponentially dichotomic (in fact unstable); since the multipliers of a periodic Hamiltonian system are located symmetrically with respect to the unit circle system (9) has n linearly independent real solutions $(x_j(t), p_j(t))$, $j = \overline{1, n}$ that tend exponentially to zero. Introducing the matrix

$$X(t) = (x_1(t) \dots x_n(t)) \tag{11}$$

the fulfillment of the condition

$$det X(t) \neq 0, \ \forall t \in [0, T] \tag{12}$$

is called strongly disconjugacy (non-oscillatory behaviour) of the Hamiltonian. We are now in position to start

Theorem 1 *Let the nonlinear system* (7) *satisfy the basic assumptions i)–iii) of Section* 1 *and also the following: iv) the multipliers of* $A(t)$ *are inside the unit disk i.e.* $A(t)$ *defines an exponentially stable evolution; v) the Hamiltonian system* (9) *is exponentially dichotomic and nonoscillatory (strongly disconjugate) i.e. it satisfies* (10) *and* (12). *Then system* (7) *has a bounded on the entire real axis solution which is exponentially stable. If* $f(t)$ *is periodic and its period a rational multiple of* T *then this solution is periodic. If the period of* $f(t)$ *is an irrational multiple of* T *or if* $f(t)$ *is only almost periodic then the solution is almost periodic.*

The proof of this Theorem, which is sketched in the Appendix, relies on obtaining some estimates of the solutions and of the difference of two solutions of (7) in order to make use of the result concerning invariant manifolds for flows in Banach spaces [8].

3 Significance of the Main Result. Computational Remarks

The existence result of Theorem 1 displays an "almost linear behaviour" of (7); indeed it is well known that the linear system

$$\dot{x} = Ax + f(t) \tag{13}$$

where A is a Hurwitz matrix and $f(t)$ is bounded on the whole real axis has a unique bounded on the whole real axis solution which is exponentially stable; moreover this solution is periodic if f is periodic and almost periodic if f is almost periodic. There exist analogous results for timevarying systems: if the system

$$\dot{x} = A(t)x + f(t) \tag{14}$$

is such that $A(t)$ defines an exponentially stable evolution for the free system $\dot{x} = A(t)x$ and $f(t)$ is bounded on the whole real axis, then (14) has a unique bounded on the whole real axis solution which is exponentially stable; this solution is periodic if $A(t)$ and $f(t)$ are periodic with a rational ratio of their periods and almost periodic in the cases when one of them or both are almost periodic or periodic with uncommensurable period. Obviously Theorem 1 states that such properties are valid also for the nonlinear system (7).

Application of Theorem 1 requires the solution of two kinds of problems: checking of the sufficient conditions for the existence of forced oscillations and computation of the forced oscillations themselves. The numerical check of Theorem's conditions requires the following operations: computation of the multipliers of $A(t)$ and check of (10) - which again requires computation of some multipliers - and of (12) which requires finding of some subspaces of solutions that tend asymptotically to zero. This is connected with the stabilizing solution of the associated (to the Hamiltonian system) periodic Riccati equation; a long list of references exists for this problem and we send the reader to some of them [2], [3], [4].

4 Forced Nonlinear Oscillations in Discrete-time Systems

There are several motivations for the topics of this section. The straightforward one is that almost all results obtained for differential equations have their discrete-time counterpart, obtained for difference equations. On the other side the study of forced oscillations in discrete-time affine systems is motivated by such applications as digital signal processing by nonlinear signal processors [26]. Here also the "almost linear behaviour" i.e. existence of a unique bounded on the whole real axis solution that is exponentially stable and of the same type as the forcing term is of interest. We would like to insist on almost periodic signals since they correspond to modulated signals; in the discrete-time case almost periodic sequences (discrete signals) are obtained in a natural way by sampling periodic signals when the sampling period and the period of the continuous time signal are in an irrational ratio [9]. It has to be mentioned here that periodic oscillations in pulse-modulated systems have been studied in a slightly different way than that of the present paper [15], [5], [6]. Our framework will be that of the difference equations with periodic coefficients. We shall consider here the system

$$x_{k+1} = A_k x_k - b_k \phi_k(c_k^* x_k) + f_k \tag{15}$$

under the following basic assumptions: i) the matrix sequences $\{A_k\}_k$, the vector sequences $\{b_k\}_k$, $\{c_k\}_k$ and the sequence $\{\phi_k(\cdot)\}_k$ are N-periodic sequences; ii) $\phi_k(\sigma)$ are continuous with respect to σ and satisfy

$$0 \leq \frac{\phi_k(\sigma_1) - \phi_k(\sigma_2)}{\sigma_1 - \sigma_2} \leq \overline{\Phi} \tag{16}$$

for any $\sigma_1 \neq \sigma_2$ and $k = \overline{0, N-1}$; iii) f_k has bounded components for all integers k, possibly periodic or almost periodic. Also in the periodic case the period of f_k may equal N, the period of system's coefficients, but this is not compulsory. In order to state the main result on discrete-time systems, we need introduction of the following linear discrete-time Hamiltonian system:

$$\begin{aligned} x_{k+1} &= \left(A_k - \frac{1}{2\overline{\Phi}} b_k c_k^*\right) x_k - \frac{1}{\overline{\Phi}} b_k b_k^* p_{k+1} \\ p_k &= -\frac{1}{4\overline{\Phi}} c_k c_k^* x_k + \left(A_k - \frac{1}{2\overline{\Phi}} b_k b_k^*\right)^* p_{k+1} \end{aligned} \tag{17}$$

We may now state:

Theorem 2 *Consider system* (15) *under the basic assumptions i)–iii) and assume additionally the following: iv) the multipliers of A_k are inside the unit disk D_1 of the complex plane i.e. A_k defines an exponentially stable evolution; v) the triple (A_k, b_k, c_k) and $\overline{\Phi}$ are such that*

$$det\left(A_k - \frac{1}{2\overline{\Phi}} b_k c_k^*\right) \neq 0,\ 0 \leq k \leq N-1 \tag{18}$$

vi) the Hamiltonian system (17) *is exponentially dichotomic and strongly disconjugate (non-oscillatory). Then there exists a bounded sequence satisfying* (15) *for all $k \in \mathbf{Z}$, which is periodic if f_k is periodic and almost periodic if f_k is almost periodic. Moreover this solution of* (15) *is exponentially stable.*

5 Comments and Remarks. Computational Aspects

We associate to the Hamiltonian system (17) the discrete-time Riccati matrix equation

$$\begin{aligned} &H_k - A_k^* H_{k+1} A_k - \left(\frac{1}{\overline{\Phi}} - b_k^* H_{k+1} b_k\right)^{-1} \left(\frac{1}{2} c_k - A_k^* H_{k+1} b_k\right) \\ &\left(\frac{1}{2} c_k - A_k^* H_{k+1} b_k\right)^* = 0 \end{aligned} \tag{19}$$

If the Hamiltonian system (17) is exponentially dichotomic and strongly disconjugate and (18) holds then it may be shown [10] that (19) has a N-periodic global solution such that

$$\frac{1}{\overline{\Phi}} - b_k^* H_{k+1} b_k > 0 \tag{20}$$

and this periodic solution is stabilizable in the following sense: if the controlled system

$$x_{k+1} = A_k x_k + b_k \mu_k \tag{21}$$

is considered, by choosing the control (input) sequence as follows

$$\mu_k = -\left(\frac{1}{\overline{\Phi}} - b_k^* H_{k+1} b_k\right)^{-1} \left(\frac{1}{2} c_k - A_k^* H_{k+1} b_k\right)^* x_k \tag{22}$$

the "closed loop" linear system

$$x_{k+1} = \left[A_k - b_k \left(\frac{1}{\overline{\Phi}} - b_k^* H_{k+1} b_k\right)^{-1} \left(\frac{1}{2} c_k - A_k^* H_{k+1} b_k\right)^*\right] x_k \tag{23}$$

is exponentially stable. Now, if the Hamiltonian system (17) is exponentially stable and strongly disconjugate, the same properties are valid for the perturbed Hamiltonian

$$\begin{aligned} x_{k+1} &= \left(A_k - \frac{1}{2\overline{\Phi}} b_k c_k^*\right) x_k - \frac{1}{\overline{\Phi}} b_k b_k^* p_{k+1} \\ p_k &= \left(-\frac{1}{4\overline{\Phi}} c_k c_k^* + \delta I\right) x_k + \left(A_k - \frac{1}{2\overline{\Phi}} b_k b_k^*\right)^* p_{k+1} \end{aligned} \tag{24}$$

with $\delta > 0$ sufficiently small. *This follows from the discrete-time version of the results of M.G. Krein on perturbed Hamiltonians of positive type* [11], discrete-time version that may be obtained from the original results of [11]. Exponential dichotomy and strong disconjugacy of (24) ensure existence of a N-periodic sequence of Hamiltonian matrices $\{H_k\}$ satisfying the discrete-time Riccati inequality

$$\begin{aligned} & H_k - A_k^* H_{k+1} A_k - \left(\frac{1}{\overline{\Phi}} - b_k^* H_{k+1} b_k\right)^{-1} \left(\frac{1}{2} c_k - A_k^* H_{k+1} b_k\right) \\ & \left(\frac{1}{2} c_k - A_k^* H_{k+1} b_k\right)^* \geq \delta I \end{aligned} \tag{25}$$

and also (20). At this point it is more convenient, from computational point of view, to replace (25) by a *Linear Matrix Inequality* (LMI). Indeed (25) looks like an inequality satisfied by a Schur complement and since (20) holds for N steps (due to periodicity) inequalities (20) and (25) may be replaced by a Linear Matrix Inequality (LMI)

$$\begin{pmatrix} H_k - A_k^* H_{k+1} A_k & \frac{1}{2}c_k - A_k^* H_{k+1} b_k \\ (\frac{1}{2}c_k - A_k^* H_{k+1} b_k)^* & \frac{1}{\phi} - b_k^* H_{k+1} b_k \end{pmatrix} \geq \delta I_{n+1} \tag{26}$$

together with the condition $H_N = H_0$. In fact this is a Dynamic Linear Matrix Inequality but since we assumed that H_k is N-periodic, a simple dimension augmentation reduces (26) to a $N(n+1) \times N(n+1)$ LMI that may be solved using the existing software. For a suitable δ the Riccati inequality may be replaced by a Riccati equation whose solution is again stabilizable in the above defined sense.

6 Concluding Remarks

The results presented here are only a part of the possible extensions of qualitative properties of systems with sector restricted nonlinearities to the case of periodic coefficients both in continuous and discrete-time. Work on such extensions as self-sustained oscillations is in progress. In all these cases the main tool is replacement of V.M. Popov type frequency domain inequalities of the constant coefficients case by exponential dichotomy and strong disconjugacy of some associated linear-periodic Hamiltonian systems. And this sends again and again to the classical results of M.G. Krein.

Acknowledgement

The programme of studying systems with sector nonlinearities and periodic coefficients using the results of M.G. Krein and V.A. Yakubovitch has been worked up essentially by Professor Aristide Halanay during 1997. These are the first achievements within this programme. Unfortunately, on December 6, 1997, Professor Halanay left this world after a heart attack while he was still full of creative energy and ideas. The fulfillment of the above mentioned programme would be the best homage to his memory.

References

[1] I. Barbălat and A. Halanay, *Conditions de comportement "presque lineaire" dans la théorie des oscillations*, Rev. Roum. Sci. Techn-Electrotech. et Energ. vol. **29** (1974), pp. 321–341.

[2] S. Bittanti, P. Colaneri and G. Guadabassi, *Analysis of periodic Liapunov and Riccati equations via canonical decomposition, SIAM J. Control and Optimization*, vol. **24** (1986), pp. 1138–1149.

[3] S. Bittanti, P. Colaneri and G. DeNicolao, *A note on the Maximal Solution of the Periodic Riccati equation*, IEEE Trans. Aut. Contr. vol. **AC-34** (1989), pp. 1316–1319.

[4] P. Van Dooren, *A generalized eigenvalue approach for solving Riccati equations*, SIAM J. Sci. and Stat. Comp. vol. **2** (1981), pp. 121–135.

[5] A.Kh. Gelig, *Dynamics of pulse systems and neural networks*, Leningrad Univ. Publ. House, 1982 (in Russian).

[6] A.Kh. Gelig and A.N. Churilov, *Oscillations and stability of nonlinear pulse systems*, St. Petersburg Univ. Publ. House, 1993 (in Russian).

[7] I.Ts. Gohberg and M.G. Krein, *Systems of integral equations on semi-axis, with kernel depending on argument difference*, Usp. Mat. Nauk, 1958, no. 2, (in Russian).

[8] A. Halanay, *Invariant manifolds for systems with time lags, in Differential Equations and dynamical systems* (Hale and La Salle eds.) pp. 199–213, Academic Press, 1967.

[9] A. Halanay and D. Wexler, *Qualitative Theory of pulse systems*, Editura Academiei, Bucharest, 1968 (in Romanian, Russian Edition by Nauka, Moscow, 1971).

[10] A. Halanay and V. Ionescu, Time-varying discrete linear systems, Birkhäuser, 1994.

[11] M.G. Krein, *Foundations of the theory of λ-zones of stability of a canonical system of linear differential equations with periodic coefficients*, In memoriam: A.A. Andronov, Izd. Akad. Nauk SSSR, Moscow, 1955, pp. 413–98 (English version in AMS Transl. (2), vol. **120** (1983), pp. 1–70).

[12] M.G. Krein, *On Tests for Stable Boundness of Solutions of Periodic Canonical Systems*, Prikl. Mat. Mekh. vol. **19** (1955), pp. 641–680 (English version in AMS Transl. (2), vol. **120** (1983), pp. 71–110).

[13] M.G. Krein and G.Ia. Lyubarskii, *About analytic properties of multipliers of periodic canonical systems of positive types*, Izv. Akad. Nauk SSSR Ser. Mat. vol. **26** (1962, pp. 549–572) (English version in AMS Transl.(2) vol. **89** (1970), pp. 1–28).

[14] M.G. Krein and V.A. Yakubovich, *Hamiltonian Systems of Linear Differential Equations with periodic Coefficients*, Proc. Int'l Symp. on Nonlinear Vibrations vol. **1**, pp. 277–305, Izd. Akad. Nauk Ukrain. SSR, Kiev, 1963 (English version in AMS Trans. (2) vol. **120** (1983), pp. 139–168).

[15] V.M. Kuntsevich and Ju.N. Chekhovoy, *Nonlinear Control Systems with Frequency and Pulse-Width Modulation*, Naukova Dumka, Kiev, 1970 (in Russian).

[16] A.I. Lurie and V.N. Postnikov, *About the theory of stability for controlled systems*, Prikl. Mat. Mekh. vol. **8**, no. 3 (1944) (in Russian).

[17] V.M. Popov, *Hyperstability of Control Systems*, Springer Verlag, 1973.

[18] V.A. Yakubovich, *Structure of the group of symplectic matrices and of the set of unstable canonical systems with periodic coefficients*, Mat. Sbornik, vol. **44**(86) (1958), pp. 313–352 (in Russian).

[19] V.A. Yakubovich, *Oscillatory properties of solutions of canonical systems*, Mat. Sbornik, vol. **56**(98) (1962), pp. 3–42 (in Russian).

[20] V.A. Yakubovich and V.M. Starzhinskii, *Linear differential equations with periodic coefficients*, Nauka, Moscow, 1972 (English version, J. Wiley, 1975).

[21] V.A. Yakubovich, *Method of matrix inequalities in stability theory for nonlinear controlled systems*. I. Absolute stability of forced oscillations, Avtom. i telemekh. vol. **XXV** (1964) pp. 1017–1029.

[22] V.A. Yakubovich, *Periodic and almost periodic limit regimes of controlled systems with several, generally speaking, discontinuous nonlinearities*, Dokl. Akad. Nauk SSSR vol. **171**, no. 3, pp. 533–536 (in Russian).

[23] V.A. Yakubovich, *Frequency domain conditions for self-sustained oscillations in nonlinear systems with a single time-invariant nonlinearity*, Sib. Mat. Journal vol. **12** (1973), no. 5 (in Russian).

[24] V.A. Yakubovich, *Frequency domain methods for qualitative study of nonlinear controlled systems*, VII Int. Konferenz ü. nichtlin. Schwingungen, Bd. I, 1, Akademie Verlag, Berlin, 1977 (in Russian).

[25] V.A. Yakubovich, *Linear-quadratic optimization problem and frequency domain theorem for periodic systems I*, Sib. Mat. Journal vol. **27** (1986), no. 4, pp. 181–200.

[26] G. Wade, *Signal coding and processing*, Cambridge Univ. Press, 1994.

Appendix

Proof of the Theorems: The proof of the main results of the paper relies on some estimates of the solutions and of the difference of two solutions of the systems (continuous or discrete-time) in order to make use of the result concerning invariant manifolds for flows in Banach spaces [8], [9]. For the sake of completness we reproduce below these results. □

Theorem A. 1 *Let $c(t;\tilde{t},\tilde{c})$ be a general flow in the Banach space C and suppose that there exist real positive constants $l, \theta, \alpha < 1, \beta, \gamma$ such that:*

a) $\| \tilde{c} \| \leq l$ *implies* $\| c(t;\tilde{t},\tilde{c}) \| \leq l$ *for* $\tilde{t}+\theta \leq t \leq \tilde{t}+2\theta$;

b) $\| \tilde{c}_1 \| \leq l, \| \tilde{c}_2 \| \leq l$ *imply* $\| c(t;\tilde{t},\tilde{c}_1) - c(t;\tilde{t},\tilde{c}_2) \| < \alpha \| \tilde{c}_1 - \tilde{c}_2 \|$ *for* $\tilde{t}+\theta \leq t \leq \tilde{t}+2\theta$;

c) $\| c(t;\tilde{t},\tilde{c}_1) - c(t;\tilde{t},\tilde{c}_2) \| \leq \beta e^{\gamma(t-\tilde{t})} \| \tilde{c}_1 - \tilde{c}_2 \|$ *for* $t > \tilde{t}$.

Then there exists $p : \mathbb{R} \to C$ and $\nu > 0$ such that

1) $\| p(t) \| \leq l$;

2) $p(t) = c(t; t_1, p(t_1)),\ t \geq t_1,\ t_1 \in \mathbb{R}$;

3) $\| c(t;\tilde{t},\tilde{c}) - p(t) \| \leq \beta e^{-\nu(t-\tilde{t})} \| \tilde{c} - p(\tilde{t}) \|$ *for* $\| \tilde{c} \| \leq l,\ t \geq \tilde{t}$;

4) *if* $c(t+T;\tilde{t}+T,\tilde{c}) \equiv c(t;\tilde{t},\tilde{c})$ *then* $p(t+T) \equiv p(t)$;

5) *if every sequence $\{h_n\}_n$, $h_n \to \infty$, contains as subsequence h_{nk} such that for every bounded solution $c(t;\tilde{t},\tilde{c})$ the sequence $c(t+h_{nk};\tilde{t}+h_{nk},\tilde{c})$ converges uniformly on each compact subset of the half axis $t \geq \tilde{t}$, uniformly with respect to $\tilde{t}$, $\tilde{c}$ for $\tilde{t} \in \mathbb{R}$, $\| \tilde{c} \| \leq l$ then p is almost periodic.*

Theorem A. 2 ([9], *Chapter* 2) *Consider the discrete time system $c_{k+1} = f(c_k)$ under the following assumptions: there exist the positive constants $l > 0$, $\alpha < 1$, β, ρ and the integer $K > 0$ such that:*

a) $\| \tilde{c} \| \leq l$ *implies* $\| c_k(\tilde{k},\tilde{c}) \| \leq l$ *for* $\tilde{k}+K \leq k \leq \tilde{k}+2K$;

b) $\| \tilde{c}_1 \| \leq l,\ \| \tilde{c}_2 \| \leq l$ *imply* $\| c_k(\tilde{k},\tilde{c}_1) - c_k(\tilde{k},\tilde{c}_2) \| \leq \alpha \| \tilde{c}_1 - \tilde{c}_2 \|$ *for* $\tilde{k}+K \leq k \leq \tilde{k}+2K$;

c) $\| c_k(\tilde{k},\tilde{c}_1) - c_k(\tilde{k},\tilde{c}_2) \| \leq \beta\rho^{k-\tilde{k}} \| \tilde{c}_1 - \tilde{c}_2 \|$ *for* $k > \tilde{k}$.

Then there exists a sequence $\{p_k\}_k$ *and* $\nu \in (0, 1)$ *such that:*

1) $\| p_k \| \leq l$;
2) $p_k = c_k(\tilde{k}, p_{\tilde{k}})$ *or* $p_{k+1} = f_k(p_k)$, $k \in \mathbf{Z}$ *i.e.* p_k *is a solution of the system;*
3) $\| c_k(\tilde{k}, \tilde{c}) - p_k \| \leq \beta \nu^{k-\tilde{k}} \| \tilde{c} - p_{\tilde{k}} \|$ *for* $\| \tilde{c} \| \leq l$, $k > \tilde{k}$ *i.e.* p_k *is exponentially stable;*
4) *if* $f_{k+N} = f_k$ *then* $p_{k+N} = p_k$;
5) *if* $\{f_k(\cdot)\}_k$ *is almost periodic then* $\{p_k\}_k$ *is almost periodic.*

Proof of Theorem 1: Consider the quadratic form $V(t, x) = x^* H(t)x$ where $H(t) = H^*(t) = H(t + T)$ will be determined in the sequel. We differentiate $V^*(t) = V(t, x(t))$ along the solutions of (7) and find

$$\frac{dV^*}{dt} = x^*(t)H(t)[A(t)x(t) - b(t)\phi(t, c^*(t)x(t))] \tag{27}$$
$$+[A(t)x(t) - b(t)\phi(t, c^*(t)x(t))]^* H(t)x(t)$$
$$+x^*(t)\dot{H}(t)x(t) + x^*(t)H(t)f(t) + f^*(t)H(t)x(t)$$

Taking into account the sector condition we shall apply the so-called S-procedure by adding and subtracting the nonnegative quantity $\phi(t, \sigma)(\sigma - \phi(t, \sigma)/\overline{\Phi})$; denoting $u(t) = -\phi(t, c^*(t)x(t))$ we shall have

$$\frac{dV^*}{dt} = -\{x(t)(-H(t))(A(t)x(t) + b(t)u(t)) + (A(t)x(t) + b(t)u(t))^* \tag{28}$$
$$(-H(t))x(t) + x^*(t)(-\dot{H}(t))x(t) + u(t)(c^*(t)x(t) + u(t)/\overline{\Phi}\}$$
$$+u(t)(c^*(t)x(t) + u(t)/\overline{\Phi}) + x^*(t)H(t)f(t) + f^*(t)H(t)x(t)$$

We shall use in the following the main theorem of [25]. According to the notations of [25] we have

$$\mathcal{G}(t, x, u) = u^2/\overline{\Phi} + \frac{1}{2}(uc^*(t)x + x^*c(t)u) \tag{29}$$

and the assumptions of Theorem 1 imply the assumptions of the main theorem of [25] for the linear-quadratic pair:

$$l\dot{x} = A(t)x + b(t)u(t) \tag{30}$$
$$J = \int_0^\infty \mathcal{G}(t, x(t), u(t))dt$$

the linear Hamiltonian system (9) being the one associated to the above pair according to [25]. The condition of Theorem 1 on the Hamiltonian (dichotomy and disconjugacy) is in fact condition (C) of [25]; but (C) $\rightarrow$ (B) in [25] i.e. there exists a

Hermitian periodic matrix $R_\delta(t) = R_\delta^*(t) = R_\delta(t+T)$ with absolutely continuous elements such that

$$(31)\quad \begin{aligned} &x^* R_\delta(t)(A(t)x + b(t)u) + (A(t)x + b(t)u)^* R_\delta(t)x + x^* \dot{R}_\delta(t)x \\ &+ \mathcal{G}(t, x, u) \geq \delta(|x|^2 + |u|^2) \end{aligned}$$

for $\delta > 0$ sufficiently small and a.e. on $[0, T)$. If we take $H(t) = -R_\delta(t)$ then we shall have

$$(32)\quad \begin{aligned} \frac{dV^*}{dt} &\leq -\delta(|\phi(t, c^*(t)x(t))|^2 + |x(t)|^2) - x^*(t)R_\delta(t)f(t) \\ &- f^*(t)R_\delta(t)x(t) \leq -\delta|x(t)|^2 + c_1|x(t)| \end{aligned}$$

since f and R_δ are bounded on $[0, T)$. On the other hand the inequality (31) holds for any x and u, in particular for $u = 0$. Therefore

$$(33)\quad H(t)A(t) + A^*(t)H(t) + \dot{H}(t) \leq -\delta I$$

and since $A(t)$ defines an exponentially stable evolution it follows that $H(t) \geq \frac{\delta}{4}\alpha I$ with $\alpha > 0$ sufficiently small. The two inequalities for V and its derivative allow to make use of the arguments [1] and obtain the estimate

$$(34)\quad |x(t)| \leq \beta_1 + \beta_2 e^{-\mu(t-\tilde{t})}|x(\tilde{t})|,\ t \geq \tilde{t}$$

where β_1, β_2, μ are strictly positive constants. Consider now two different solutions $x_1(t)$ and $x_2(t)$ of (7) and let $z(t) = x_1(t) - x_2(t)$. Consider now $\tilde{V}(t) = V(t, z(t))$. We shall have

$$(35)\quad \begin{aligned} \frac{dV^*}{dt} &= z^*(t)H(t)[A(t)z(t) - b(t)(\phi(t, c^*(t)x_1(t)) - \phi(t, c^*(t), x_2(t)))] \\ &+[A(t)z(t) - b(t)(\phi(t, c^*(t)x_1(t)) - \phi(t, c^*(t)x_2(t)))]^* H(t)z(t) \\ &+z^*(t)H(t)z(t) \end{aligned}$$

where $H(t) = -R_\delta(t)$ as above. We denote $u(t) = -\phi(t, c^*(t)x_1(t)) + \phi(t, c^*(t)$ $x_2(t))$, take into account the assumption on $\phi(\sigma)$ i.e. (8) and apply the S-procedure by adding and subtracting in (35) the nonnegative quantity $u(t)(c^*z(t) + u(t)/\overline{\Phi})$. Using (31) we shall have

$$(36)\quad \frac{d\tilde{V}}{dt} \leq -\delta|z(t)|^2$$

and denoting $\mu = \frac{1}{2}min\{\delta, \alpha\frac{\delta}{4}\}$ we obtain

$$(37)\quad |x_1(t) - x_2(t)| \leq \beta e^{-\mu(t-\tilde{t})}|x_1(\tilde{t}) - x_2(\tilde{t})|$$

The estimates (35) and (37) obtained for a solution of (7) and for the difference of two solutions allow to obtain the fulfillment of the assumptions of Theorem A.1

on invariant manifolds. Indeed the fulfillment of (a) follows from (35) by taking $l = 2\beta_1$ and θ such that $\beta_2 e^{-\mu\theta} < \beta_1$. Further (37) is nothing else but (c)with $\gamma = -\mu < 0$. We may then choose θ such that $\beta e^{-\mu\theta} < 1$ and (b) holds. In fact by choosing θ as the largest among the two numbers determined above, all assumptions of Theorem A.1 hold and the proof ends. □

Proof of Theorem 2: As in the case of Theorem 1, we shall apply Theorem A.2 on invariant manifolds for discrete-time flows on Banach spaces; its assumptions are fulfilled for the flow defined by the solutions of (15) if the following estimates for the solution of (15) hold.

$$|x_k| \leq \beta_1 + \beta_2 \nu^{k-\tilde{k}} |x_{\tilde{k}}|,\ k > \tilde{k} \tag{38}$$

$$|x_k^1 - x_k^2| \leq \beta \nu^{k-\tilde{k}} |x_{\tilde{k}}|,\ k > \tilde{k} \tag{39}$$

with $\beta > 0$, $\beta_1 > 0$, $\beta_2 > 0$, $0 < \nu < 1$. Indeed, if (38) holds then we may choose $l = 2\beta_1$ and $K > 0$ such that $\beta_2 \nu^K < \beta_1$ and (a) of Theorem A.2 holds. Remark further that (39) is nothing else but (c) in Theorem A.2 with $\rho = \nu < 1$. We may then choose K such that $\beta \nu^K < 1$ and (b) holds. In fact by choosing K as the largest among the two positive integers determined above, all assumptions of Theorem A.2 hold and the proof of Theorem 2 ends provided (38) and (39) are true. At this end we shall consider the quadratic Liapunov function

$$V_k(x) = x^* H_k x \tag{40}$$

where H_k, $k = \overline{0, N-1}$ are the matrices whose existence follow from the assumption (vi) of Theorem 2 on the Hamiltonian system(17); indeed, as pointed out in Section 5, the same assumption holds for the perturbed Hamiltonian (24) hence there exists a N-periodic sequence of Hermitian matrices H_k satisfying (25) and (20). It is exactly this periodic sequence $\{H_k\}_k$ that we introduce in (40). If x_k^1 and x_k^2 are two solutions of (15) we may define $u_k = -\phi_k(c_k^* x_k^1) + \phi_k(c_k^* x_k^2)$, $z_k = x_k^1 - x_k^2$ and have

$$V_{k+1}(z_{k+1}) - V_k(z_k) = (A_k z_k + b_k u_k)^* H_k (A_k z_k + b_k u_k) - z_k^* H_k z_k \tag{41}$$

Taking into account (16) we deduce that $u_k(c_k^* z_k + u_k/\overline{\Phi}) \leq 0$ for u_k and z_k defined above. We may apply the S-procedure by adding and subtracting in (41) this quantity. This allow deducing from (25) that

$$V_{k+1}(z_{k+1}) - V_k(z_k) \leq -\delta |z_k|^2 \tag{42}$$

On the other hand (25) and (20) are equivalent to (26) and (26) gives

$$A_k^* H_{k+1} A_k - H_k \leq -\delta I$$

But A_k defines an exponentially stable evolution and from the converse Liapunov theorem (in the discrete case, see e.g. [9]) it follows that $H_k \geq \delta I$. Since H_k is bounded we have $H_k \leq \Lambda I$ hence $V_k(x) \leq \Lambda|x|^2$. We deduce that

$$V_{k+1}(z_{k+1}) \leq \left(1 - \frac{\delta}{\Lambda}\right) V_k(z_k)$$

and this will give (39)

$$|z_k| \leq \sqrt{\frac{\Lambda}{\delta}} \left(1 - \frac{\delta}{\Lambda}\right)^{\frac{k-\tilde{k}}{2}} |z_{\tilde{k}}|$$

For (38) we consider a solution of (15) and define $u_k = -\phi_k(c_k^* x_k)$ hence $u_k(c_k^* x_k + u_k/\overline{\Phi}) \leq 0$ from (16). With the same Liapunov function (40) we find, after the S-procedure and using also (25) and (20), that

$$V_{k+1}(x_{k+1}) - V_k(x_k) \leq -\delta|x_k|^2 + f_k^* H_{k+1}(A_k x_k + b_k u_k) + (A_k x_k + b_k u_k)^* H_{k+1} f_k + f_k^* H_{k+1} f_k$$

Since f_k is bounded we deduce from the above inequality that

$$V_{k+1}(x_{k+1}) - V_k(x_k) \leq -\delta|x_k|^2 + \delta_1|x_k| + \delta_2$$

and, denoting $w_k = V_k(x_k)$

$$w_{k+1} \leq \left(1 - \frac{\delta}{\Lambda}\right) w_k + \frac{\delta_1}{\sqrt{\Lambda}} \sqrt{w_k} + \delta_2$$

which gives, after some manipulation

$$|x_k| \leq \sqrt{\frac{\Lambda}{\delta}} \rho_0^{\frac{k-\tilde{k}}{2}} |x_{\tilde{k}}| + \frac{\sqrt{\Lambda}}{\delta}(\delta_1 + \sqrt{\delta_1^2 + 4\delta_1\delta_2})$$

that is (38) since

$$\rho_0 = 1 - \frac{\delta}{\Lambda}\left(1 - \frac{2}{1 + \sqrt{1 + 4\delta\delta_2/\delta_1}}\right) < 1$$

□

Dept. of Mathematics
Bucharest University
Romania

Dept. of Automatic Control
Craiova University
A.I. Cuza Str., no.13
RO-1100, Craiova
Romania
Tel/Fax: +40.51.143198
vrasvan@automation.ucv.ro

Operator Theory:
Advances and Applications, Vol. 117
© 2000 Birkhäuser Verlag Basel/Switzerland

Differential Operator Matrices of Mixed Order with Periodic Coefficients

*R. Hryniv, A. Shkalikov and A. Vladimirov**

We study spectral properties of 2×2 operator matrices H defined in the Hilbert space $\mathbb{H} = L_2(\mathbb{R}) \times L_2(\mathbb{R})$ by linear differential systems of mixed order with periodic coefficients. We prove that the spectrum $\sigma(H)$ of H has a band and gap structure and consists of two band sequences one of which, when infinite, has a finite accumulation point, and give sufficient conditions for this accumulation to take place.

1 Introduction

In this note we consider an operator matrix of the form

$$H = \begin{pmatrix} -\frac{d^2}{dx^2} + p(x) & q(x) \\ q(x) & u(x) \end{pmatrix} \tag{1}$$

acting in the Hilbert space $\mathbb{H} = L_2(\mathbb{R}) \times L_2(\mathbb{R})$, the functions $p(x), q(x)$, and $u(x)$ being periodic with the same period. The operator H represents the simplest operator model for mixed order differential systems which frequently occur in hydrodynamics and magnetohydrodynamics (see, e.g., [LL] and [L]). Elliptic systems of mixed order were studied by Agmon, Douglis and Nirenberg [ADN], Grubb and Geymonat [GG] in 50–70-ies. An abstract approach to the study of operators generated by mixed order differential systems was recently developed in the papers of Atkinson, Langer, Mennicken, Shkalikov, Adamyan, Motovilov, Saurer [ALMS], [S], [M], [AL], [MS], [ALMSa]. Important developments of this method for applications to concrete problems were carried out by Hardt, Mennicken, Naboko, Faierman, and Möller [HMN], [FMM].

The operator of the concrete form (1) on the finite interval was studied in detail by Langer, Mennicken and Möller [LMM]. Shkalikov [S] and Konstantinov [K] considered this operator (and a more general one) on the whole axis $\mathbb{R}$ for functions $q(x)$ that vanish at $\pm\infty$.

The aim of our work is to investigate the spectrum of operator H in the periodic case. Some new phenomena which will be observed here are also inherited by more general operators of mixed order with periodic coefficients. However, in this note we shall concentrate our attention only to the model operator (1) and shall always

*This work was supported by Russian Foundation for Fundamental Research (RFFI), grant No. 98-01-01000 and grant No. 96-15-96091.

assume that *the functions $p(x), q(x)$, and $u(x)$ are continuous, real-valued, and 1-periodic.*

For $q(x) \equiv 0$ the spectrum $\sigma(H)$ of the operator H is the union of the spectra $\sigma(u)$ of the multiplication operator by $u(x)$ and $\sigma(S)$ of the periodic Schrödinger operator $S := -\frac{d^2}{dx^2} + p(x)$. It is well known that $\sigma(u)$ coincides with the interval $I_0 := [u_-, u_+]$, which is the range of the function $u(x)$, while $\sigma(S)$ is absolutely continuous and has a *band and gap* structure. The latter means that $\sigma(S)$ consists of a sequence of closed intervals (bands) I_n^0, $n = 1, 2, \dots$, separated by gaps (which can degenerate into the empty set), and the endpoints of I_n^0 are eigenvalues of the restriction of the operator S to the period subject to periodic or antiperiodic boundary conditions.

We shall see that non-zero $q(x)$ dramatically changes the spectrum $\sigma(H)$ giving rise to a new band sequence.

Theorem 1 *Suppose that $q(x) \not\equiv 0$. Then the spectrum $\sigma(H)$ of the operator H has a band and gap structure and is the union of the following sets (see Fig. 1):*

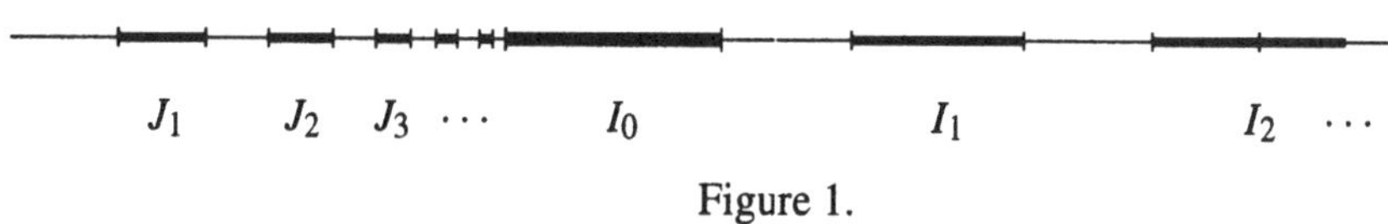

Figure 1.

a) *the interval $I_0 := [u_-, u_+]$;*

b) *infinitely many intervals $I_n := [\mu_{2n-1}, \mu_{2n}]$, $n \in \mathbb{N}$, to the right of I_0 which tend to $+\infty$;*

c) *a finite or infinite number of intervals $J_k := [\nu_{2k-1}, \nu_{2k}]$, $k = 1, \dots, \omega$, $\omega \leq \infty$, to the left of I_0.*

It also turns out that the number of spectral bands J_k (more exactly, finiteness or infiniteness of ω) is very sensitive to the behaviour of $q(x)$ on the set

$$\tau := \{x \in [0, 1] \mid u(x) = u_-\}$$

of critical points of $u(x)$. Namely, the following theorem holds true.

Theorem 2 *Suppose that the function u is twice continuously differentiable in some neighbourhood of the set τ.*

a) *If the set τ is finite and for every $x \in \tau$ the inequality $8q^2(x) < u''(x)$ holds, then there are finitely many spectral bands J_k, i.e., $\omega < \infty$.*

b) *If for some critical point $x_0 \in \tau$ we have $8q^2(x_0) > u''(x_0)$, then $\omega = \infty$ and the spectral bands J_k accumulate to the point u_- as $k \to \infty$.*

Note that if $\omega = \infty$, then it is possible to find asymptotics of J_k at u_- as $k \to \infty$. This and some other questions are studied in detail in the forthcoming paper [HSV].

2 The Spectral Decomposition

Since the operator H commutes with the shifts $x \mapsto x + n$, $n \in \mathbb{Z}$, the abstract harmonic analysis reduces the spectral study of H to that for the family of the operators H_θ, $\theta \in [0, 2\pi)$, defined on

$$\begin{aligned}\mathfrak{D}(H_\theta) &= \{(y_1, y_2) \in W_2^2[0,1] \times L_2[0,1] \mid y_1(1) = e^{i\theta} y_1(0),\\ y_1'(1) &= e^{i\theta} y_1'(0)\}\end{aligned}$$

by expression (1). More exactly, we have the following statement.

Theorem 3 ([RS, Theorem XIII.85d]). *Let $d\mu$ denote the Lebesgue measure on $[0, 1]$. Then the point $\lambda_0 \in \mathbb{R}$ belongs to $\sigma(H)$ if and only if for any $\varepsilon > 0$ we have*

$$d\mu\{\theta \in [0, 2\pi) \mid \sigma(H_\theta) \cap (\lambda_0 - \varepsilon, \lambda_0 + \varepsilon) \neq \emptyset\} > 0. \tag{2}$$

The operators H_θ, $\theta \in [0, 2\pi)$, being defined by differential expression (1) on the unit interval, meet all the assumptions of [ALMS, Theorem 2.2]. Therefore the spectrum $\sigma(H_\theta)$ has the following structure.

Theorem 4 ([ALMS, Theorem 2.2]). *For every $\theta \in [0, 2\pi)$, the spectrum $\sigma(H_\theta)$ of the operator H_θ consists of the essential part $\sigma_{ess}(H_\theta)$, which is the range $I_0 = [u_-, u_+]$ of the function $u(x)$, and the discrete part $\sigma_d(H_\theta)$, which coincides with the spectrum of the transfer-function $\mathcal{M}_\theta(\lambda)$, defined for $\lambda \notin I_0$ on the domain*

$$\mathfrak{D}(\mathcal{M}_\theta) := \{y(x) \in W_2^2[0,1] \mid y(1) = e^{i\theta} y(0),\ y'(1) = e^{i\theta} y'(0)\}$$

by the differential expression

$$\mathcal{M}_\theta(\lambda) y(x) = -y''(x) + \left(p(x) - \lambda + \frac{q^2(x)}{\lambda - u(x)}\right) y(x).$$

Recall that $\lambda_0 \in \mathbb{C}$ belongs to the *spectrum* $\sigma(\mathcal{M}_\theta)$ of the transfer-function $\mathcal{M}_\theta(\lambda)$ if the operator $\mathcal{M}_\theta(\lambda_0)$ is not boundedly invertible. $\lambda_0 \in \sigma(\mathcal{M}_\theta)$ is an *eigenvalue* of $\mathcal{M}_\theta(\lambda)$ if $\mathrm{Ker}\mathcal{M}_\theta(\lambda_0) \neq \{0\}$, and then the number $\dim \mathrm{Ker}\mathcal{M}_\theta(\lambda)$ is its *multiplicity*.

Since for any $\lambda_0 \notin I_0$ the operator $\mathcal{M}_\theta(\lambda_0)$ has a compact resolvent, the spectrum of $\mathcal{M}_\theta(\lambda)$ consists only of eigenvalues. These eigenvalues are real since, by Theorem 4, they belong to the spectrum $\sigma(H_\theta)$ of the selfadjoint operator H_θ. Moreover, the eigenvalues of $\mathcal{M}_\theta(\lambda)$ are semisimple as $\mathcal{M}_\theta'(\lambda) \ll 0$ for all real λ and have multiplicity not greater than 2 since $\mathcal{M}_\theta(\lambda)$ is defined by the differential expression of order 2.

3 The Study of the Transfer-function $\mathcal{M}_\theta(\lambda)$

We saw in the previous section that $\sigma(H_\theta)$ is determined mainly by the transfer-function $\mathcal{M}_\theta(\lambda)$ and so we shall study in this section the spectral properties of $\mathcal{M}_\theta(\lambda)$. In the sequel $\Lambda = (\lambda_-, \lambda_+)$ will denote any of the two intervals $(-\infty, u_-)$ and (u_+, ∞) constituting the set $\mathbb{R} \setminus I_0$.

Let $u_1(x,\lambda)$ and $u_2(x,\lambda)$ be solutions of the equation

$$-y''(x) + p(x,\lambda)y(x) = 0, \quad p(x,\lambda) := p(x) - \lambda + \frac{q^2(x)}{\lambda - u(x)} \tag{3}$$

satisfying the initial conditions

$$\begin{aligned} u_1(0,\lambda) &= 1, & u_1'(0,\lambda) &= 0; \\ u_2(0,\lambda) &= 0, & u_2'(0,\lambda) &= 1. \end{aligned}$$

Then the *fundamental matrix*

$$M(x,\lambda) := \begin{pmatrix} u_1(x,\lambda) & u_2(x,\lambda) \\ u_1'(x,\lambda) & u_2'(x,\lambda) \end{pmatrix}$$

depends continuously on $x \in [0,1]$ and analytically on $\lambda \in \Lambda$, $\det M(x,\lambda) \equiv 1$, and for any solution $y(x,\lambda)$ of equation (3) the equality

$$\begin{pmatrix} y(x,\lambda) \\ y'(x,\lambda) \end{pmatrix} = M(x,\lambda) \begin{pmatrix} y(0,\lambda) \\ y'(0,\lambda) \end{pmatrix} \tag{4}$$

holds. Therefore a point $\lambda_0 \in \Lambda$ is an eigenvalue of the transfer-function $\mathcal{M}_\theta(\lambda)$ if and only if the number $e^{i\theta}$ is an eigenvalue of the matrix $M(1,\lambda_0)$, or, which is equivalent due to $M(1,\lambda) \in \mathrm{SL}(2,\mathbb{R})$, if and only if λ_0 is a solution to the equation

$$2\cos\theta = \mathrm{Tr} M(1,\lambda). \tag{5}$$

We shall use this equality to study dependence of the spectrum of $\mathcal{M}_\theta(\lambda)$ on θ.

Lemma 5 *The spectrum of the transfer-function $\mathcal{M}_\theta(\lambda)$, $\theta \in [0,2\pi)$, forms a discrete set in Λ. Moreover, the eigenvalues of $\mathcal{M}_\theta(\lambda)$ are locally analytic and monotonic functions of $\theta \in (0,\pi) \cup (\pi, 2\pi)$.*

Proof: For fixed $\theta \in [0,2\pi)$ and $\lambda \in \Lambda$, the operator $\mathcal{M}_\theta(\lambda)$ is bounded below and has a discrete spectrum. Since the potential $p(x,\lambda)$ in (3) strictly decreases in $\lambda \in \Lambda$, the transfer-function $\mathcal{M}_\theta(\lambda)$ strictly decreases in $\lambda \in \Lambda$ as well, and by [HM] the spectrum of $\mathcal{M}_\theta(\lambda)$ is discrete in Λ.

Suppose now that $\theta_0 \in (0,\pi) \cup (\pi, 2\pi)$ and that $\lambda_0 \in \Lambda$ is an eigenvalue of $\mathcal{M}_{\theta_0}(\lambda)$. Then, according to (5), $|\mathrm{Tr} M(1,\lambda_0)| < 2$ and there exists a neighbourhood $\mathcal{O}$ of λ_0 such that the value of θ for which $\lambda \in \sigma(\mathcal{M}_\theta) \cap \mathcal{O}$ is given by an analytic function $\theta(\lambda) = \arccos \frac{1}{2}\mathrm{Tr}M(1,\lambda)$. We shall prove that this function is locally invertible.

Denote by $\mu_1(\lambda,\theta) \leq \mu_2(\lambda,\theta) \leq \ldots$ the eigenvalues of $\mathcal{M}_\theta(\lambda)$ repeated according to multiplicities. Note that these eigenvalues are in fact simple if $\theta \neq 0$ and $\theta \neq \pi$. Indeed, otherwise there would exist a real number μ and two linearly independent solutions $y_1(x)$ and $y_2(x)$ of the equation

$$-y''(x) + p(x,\lambda)y(x) = \mu y(x) \tag{6}$$

satisfying the boundary conditions

$$y(1) = e^{i\theta}y(0), \qquad y'(1) = e^{i\theta}y'(0). \tag{7}$$

But then every solution of equation (6) would satisfy boundary conditions (7), which is not true for $\overline{y_1(x)}$.

Since the operator $\mathcal{M}_{\theta_0}(\lambda_0)$ is not invertible, there exists a unique index k such that $\mu_k(\lambda_0,\theta_0) = 0$. We shall show that $\mu_k(\lambda,\theta)$ is locally jointly analytic in λ and θ, that with the function $\theta(\lambda)$ defined above we have $\mu_k(\lambda,\theta(\lambda)) \equiv 0$ for $\lambda \in \mathcal{O}$ and then shall use the implicit function theorem to find the inverse function $\theta^{-1}(\lambda)$.

Note that, for θ fixed, $\mathcal{M}_\theta(\lambda)$ is a holomorphic operator family in $\lambda \in \Lambda$ of type (A) (see [Ka, §VII.2]). Therefore the simple eigenvalue $\mu_k(\lambda,\theta)$ is locally analytic in λ. Simple arguments show that in fact $\mu_k(\lambda,\theta)$ is globally analytic in $\lambda \in \Lambda$. Denote by $y_k(\lambda,\theta)$ the corresponding normalized eigenfunction that is analytic in λ. Then, differentiating the identity

$$(\mathcal{M}_\theta(\lambda)y_k(\lambda,\theta), y_k(\lambda,\theta)) = \mu_k(\lambda,\theta)$$

with respect to λ and using selfadjointness of $\mathcal{M}_\theta(\lambda)$, we get

$$\frac{\partial \mu_k}{\partial \lambda} < 0. \tag{8}$$

Similarly, for fixed $\lambda \in \Lambda$ the operator family $\mathcal{M}_\theta(\lambda)$ is analytic in $\theta \in [0,2\pi)$ (see [RS, §XIII.16]) and by the same arguments $\mu_k(\lambda,\theta)$ is analytic in $\theta \in (0,\pi) \cup (\pi,2\pi)$. Due to Osgood's lemma the function $\mu_k(\lambda,\theta)$ is jointly analytic in some neighbourhood of the point (λ_0,θ_0). Therefore, by virtue of (8) and the implicit function theorem, in some neighbourhood of θ_0 the equation $\mu_k(\lambda,\theta) = 0$ defines a unique function $\lambda(\theta)$, which is evidently the inverse of $\theta(\lambda)$. This means that, in that neighbourhood of θ_0, the function $\lambda(\theta)$ is analytic and strictly monotonic. Since $\theta_0 \in (0,\pi) \cup (\pi,2\pi)$ was taken arbitrarily, the lemma is proved. □

There exists a strong connection between the spectrum of $\mathcal{M}_\theta(\lambda)$ and oscillation properties of solutions of equation (3) (see Theorem 10 below). To establish it, we need some auxiliary results.

Fix a nonzero vector $\vec{X} \in \mathbb{R}^2$ and choose a branch $\phi_{\vec{X}}(\lambda)$ of the multivalued function $\arg(M(x,\lambda)\vec{X})$ which is continuous in $x \in [0,1]$ and $\lambda \in \Lambda$ and satisfies the condition $\arg(M(0,\lambda)\vec{X}) \in [0, 2\pi)$. Since the potential $p(x,\lambda)$ strictly decreases in λ, the matrix $M(1,\lambda)$ rotates vectors on the plane $\mathbb{R}^2$ in the positive direction, i.e., the function $\phi_{\vec{X}}(\lambda)$ strictly increases along $\lambda \in \Lambda$ (see [H, Ch. 11]).

Lemma 6 *Fix a nonzero vector* $\vec{X} = (x_1, x_2) \in \mathbb{R}^2$ *and denote by* Λ_n, $n \in \mathbb{Z}$, *closed intervals constituting the complement* $\Lambda \setminus \Lambda'$, *where* $\Lambda' := \{\lambda \in \Lambda \mid |TrM(1,\lambda)| < 2\}$ *(see Fig. 2). Then each segment* Λ_n *contains a unique point* λ^* *such that* $\vec{X}$ *is an eigenvector of the matrix* $M(1,\lambda^*)$ *(or, equivalently, such that* $\phi_{\vec{X}}(1,\lambda^*) - \phi_{\vec{X}}(0,\lambda^*)$ *is a multiple of* π*).*

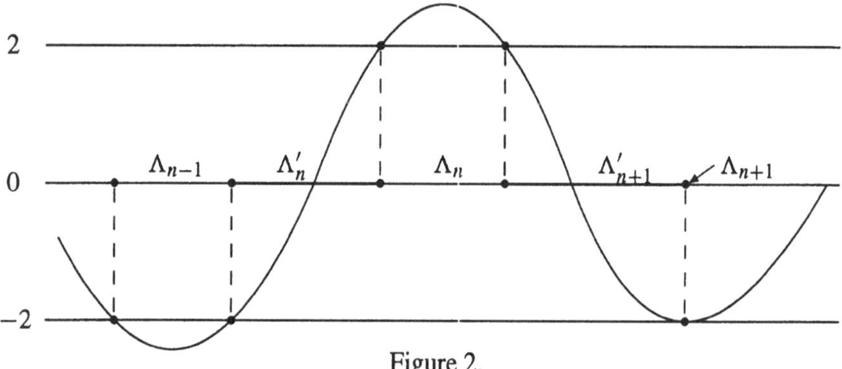

Figure 2.

Proof: First we prove uniqueness. Let λ_1 and λ_2 be two subsequent points at which $M(1,\lambda)$ has an eigenvector X. Denote by μ_1 and μ_2 the corresponding real eigenvalues; then $\mathrm{Tr}M(1,\lambda_k) = \mu_k + 1/\mu_k$, $k = 1, 2$. Since $\phi_{\vec{X}}(\lambda)$ strictly increases, we have $\phi_{\vec{X}}(1,\lambda_2) = \phi_{\vec{X}}(1,\lambda_1) + \pi$. Therefore μ_1 and μ_2 (and hence $\mathrm{Tr}M(1,\lambda_1)$ and $\mathrm{Tr}M(1,\lambda_2)$) are of the opposite signs. This shows that the points λ_1 and λ_2 cannot belong to the same interval Λ_n.

The points λ^* such that $\vec{X}$ is an eigenvector of $M(1,\lambda)$ are zeros of the function

$$f_{\vec{X}}(\lambda) := \det(\vec{X}; M(1,\lambda)\vec{X}).$$

Note that $f_{\vec{X}}(\lambda)$ differs by a positive multiplier from $\sin(\phi_{\vec{X}}(1,\lambda) - \phi_{\vec{X}}(0,\lambda))$ and therefore changes sign at each its zero.

Denote by Λ'_n and Λ'_{n+1} two intervals adjacent to Λ_n. To prove that $f_{\vec{X}}(\lambda)$ has a zero in Λ_n it suffices to show that $f_{\vec{X}}(\lambda)$ assumes values of opposite signs on Λ'_n and Λ'_{n+1}. Since the function $f_{\vec{Y}}(\lambda)$ is continuous in $\vec{Y} \in \mathbb{R}^2 \setminus \{0\}$ and $\lambda \in \Lambda$ and does not vanish on Λ'_n, the values of $f_{\vec{Y}}(\lambda)$ for any $\vec{Y} \in \mathbb{R}^2 \setminus \{0\}$ and any $\lambda \in \Lambda'_n$ are of the same sign. The same is true if the interval Λ'_n is replaced by Λ'_{n+1}.

Take now any $\lambda' \in \Lambda_n$. Since $|\mathrm{Tr}M(1,\lambda')| \geq 2$, the matrix $M(1,\lambda')$ has a real eigenvalue with a corresponding real eigenvector $\vec{Y}$. Then $f_{\vec{Y}}(\lambda') = 0$; moreover,

λ' is a unique zero of the function $f_{\vec{Y}}(\lambda)$ in Λ_n and hence $f_{\vec{Y}}(\lambda)$ takes values of the opposite signs on Λ'_n and Λ'_{n+1}. According to the arguments above the same is true for the function $f_{\vec{X}}(\lambda)$, and the lemma is proved. □

Corollary 7 *Denote by $\mathcal{N}_\theta(\alpha, \beta)$ the number of eigenvalues of the transfer-function $\mathcal{M}_\theta(\lambda)$ in the interval (α, β) counted with multiplicities. Then for an arbitrary vector $\vec{X} \in \mathbb{R}^2 \setminus \{0\}$ and any interval $(\alpha, \beta) \Subset \Lambda$ we have*

$$-2 \leq \mathcal{N}_\theta(\alpha, \beta) + \frac{\phi_{\vec{X}}(1, \alpha) - \phi_{\vec{X}}(1, \beta)}{\pi} \leq 2. \tag{9}$$

Proof: Let $\lambda_1 < \lambda_2 < \cdots < \lambda_n$ be all the points from (α, β) such that $\vec{X}$ is an eigenvector of the matrix $M(1, \lambda)$. If $\theta \neq 0$ and $\theta \neq \pi$, then every interval $(\lambda_k, \lambda_{k+1})$, $k = 1, \ldots, n-1$, contains exactly one eigenvalue of $\mathcal{M}_\theta(\lambda)$, while each of the intervals (α, λ_1) and (λ_n, β) contains at most one of them. Next, we have the relations $\phi_{\vec{X}}(1, \lambda_{k+1}) - \phi_{\vec{X}}(1, \lambda_k) = \pi$, $k = 1, \ldots, n-1$, and $\phi_{\vec{X}}(1, \beta) - \phi_{\vec{X}}(1, \lambda_n) \leq \pi$, $\phi_{\vec{X}}(1, \lambda_1) - \phi_{\vec{X}}(1, \alpha) \leq \pi$. Combining all these facts, we arrive at inequality (9).

For $\theta = 0$ or $\theta = \pi$ eigenvalues of $\mathcal{M}_\theta(\lambda)$ of multiplicity 2 are possible. If, e.g., λ_k is such an eigenvalue from (α, β), then $M(1, \lambda_k) = \pm I$ and the corresponding interval Λ_k degenerates into the single point λ_k. In this situation we shall regard each of the intervals $[\lambda_{k-1}, \lambda_k]$ and $[\lambda_k, \lambda_{k+1}]$ (if they exist) as containing exactly one eigenvalue of $\mathcal{M}_\theta(\lambda)$. Under such an interpretation all the above-mentioned arguments remain true and so inequality (9) holds for $\theta = 0$ or $\theta = \pi$ as well. □

Fix a nonzero vector $\vec{X} = (x_1, x_2) \in \mathbb{R}^2 \setminus \{0\}$. By virtue of equality (4) those $x^* \in [0, 1]$ for which $\phi_{\vec{X}}(x^*, \lambda) = \pi(n+1/2)$, $n \in \mathbb{Z}$, are just zeros of the solution $y(x, \lambda)$ of equation (3) subject to the initial conditions $y'(0, \lambda) = x_1$, $y(0, \lambda) = x_2$. According to the Sturm theory, at such points x^* the function $\phi_{\vec{X}}(x, \lambda)$ is strictly increasing in x, so the total number of zeros of the solution $y(x, \lambda)$ on the segment $[0, 1]$ differs at most by 2 from the quantity $[\phi_{\vec{X}}(1, \lambda)/\pi]$.

Definition 8 We call equation (3) *oscillatory at a point* $\lambda^* \in \overline{\Lambda}$ if there exists a sequence λ_n, $\lambda_n \to \lambda^*$, such that some solution $y(x, \lambda)$ of equation (3) for $\lambda = \lambda_n$ has at least n zeros on the segment $[0, 1]$.

Remark 9 If equation (3) is oscillatory at a point λ^*, then $\limsup_{\lambda \to \lambda^*} \phi_{\vec{X}}(1, \lambda) = \infty$. Since the function $\phi_{\vec{X}}(1, \lambda)$ is continuous on Λ, this is possible only for $\lambda^* = \lambda_-$ or $\lambda^* = \lambda_+$.

Theorem 10 *The eigenvalues of $\mathcal{M}_\theta(\lambda)$, $\theta \in [0, 2\pi)$, may accumulate only to the point λ_+. This takes place (simultaneously for all $\theta \in [0, 2\pi)$) if and only if differential equation* (3) *is oscillatory at the point λ_+.*

Proof: Since $\phi_{\vec{X}}(1,\lambda) > 0$ for all $\vec{X} \in \mathbb{R}^2 \setminus \{0\}$ and all $\lambda \in \Lambda$, we have $\phi_{\vec{X}}(1,\beta) - \phi_{\vec{X}}(1,\alpha) \leq \phi_{\vec{X}}(1,\beta)$ for any $\alpha < \beta$, and so by Corollary 7 no point $\alpha \in [\lambda_-, \lambda_+)$ can be an accumulation point of eigenvalues of the transfer-function $\mathcal{M}_\theta(\lambda)$.

By the same Corollary 7 $\lambda = \lambda_+$ is an eigenvalue accumulation point if and only if $\phi_{\vec{X}}(1,\lambda) \to \infty$ as $\lambda \to \lambda_+$ for some (and then for any) nonzero vector $\vec{X} \in \mathbb{R}$. This holds if and only if the number of zeros of the solution $y(x,\lambda)$ of equation (3) grows unboundedly as $\lambda \to \lambda_+$, i.e. if and only if equation (3) is oscillatory at λ_+. The proof is complete. □

Remark 11 Note that for the potential $p(x,\lambda)$ defined by (3) we have $p(x,\lambda) \to -\infty$ uniformly in $x \in [0,1]$ as $\lambda \to +\infty$. Therefore, according to the Sturm theory, equation (3) is oscillatory at $+\infty$, and so the eigenvalues of $\mathcal{M}_\theta(\lambda)$ do accumulate to $+\infty$.

4 Proof of the Main Theorems

Proof of Theorem 1: Denote by $I_n := [\mu_{2n-1}, \mu_{2n}]$ the closure of the n-th interval constituting the set

$$\{\lambda \in (u_+, \infty) \mid |\mathrm{Tr} M(1,\lambda)| < 2\}.$$

By Remark 11 there are infinitely many such intervals; moreover, such an ordering is possible since the points for which $\mathrm{Tr} M(1,\lambda) = 2\cos\theta_0$, $\theta_0 \in [0, 2\pi)$, constitute a discrete subset of $[u_+, \infty)$. At the endpoints μ_k of I_n we have $|\mathrm{Tr} M(1,\lambda)| = 2$, and so they are eigenvalues of $\mathcal{M}_\theta(\lambda)$ for $\theta = 0$ or $\theta = \pi$.

In the same way we introduce the subsets J_k, $k = 1, \dots, \omega$, $\omega \leq \infty$, of the interval $(-\infty, u_-)$ and denote by σ the union of the sets I_0, I_n, $n \in \mathbb{N}$, and J_k, $k = 1, \dots, \omega$. This is exactly the set described by the theorem. Note that σ is closed since any two neighbouring segments from this union have at most one common point.

Since for each $\theta \in [0, 2\pi)$ the spectrum $\sigma(H_\theta)$ of the operator H_θ lies in σ, relation (2) fails for $\lambda \notin \sigma$ and so $\sigma(H) \subset \sigma$.

To establish the reverse inclusion we notice first that $I_0 \subset \sigma(H)$ because $I_0 \subset \sigma(H_\theta)$ for all $\theta \in [0, 2\pi)$. Next, take an arbitrary point λ_0 from any of the intervals I_n, $n \in \mathbb{N}$. It is an eigenvalue of the transfer-function $\mathcal{M}_{\theta_0}(\lambda)$ for $\theta_0 = \arccos \frac{1}{2} \mathrm{Tr} M(1,\lambda_0)$ and $\theta_0 = 2\pi - \arccos \frac{1}{2} \mathrm{Tr} M(1,\lambda_0)$. Since by Lemma 5 the eigenvalues of $\mathcal{M}_\theta(\lambda)$ depend continuously on θ, for any $\varepsilon > 0$ there is an eigenvalue of $\mathcal{M}_\theta(\lambda)$ in $(\lambda_0 - \varepsilon, \lambda_0 + \varepsilon)$ for all $\theta \in [0, 2\pi)$ sufficiently close to θ_0. Therefore condition (2) is satisfied for $\lambda = \lambda_0$ and so by Theorem 3 $\lambda_0 \in \sigma(H)$.

The same arguments apply to the intervals J_k, $k = 1, \dots, \omega$, and so $\sigma \subset \sigma(H)$. The theorem is proved. □

Proof of Theorem 2: In view of the construction of the spectral bands J_k (see the proof of the previous theorem) $\omega = \infty$ if and only if the transfer-function

$\mathcal{M}_\theta(\lambda)$ for every $\theta \in [0, 2\pi)$ has infinitely many eigenvalues in $(-\infty, u_-)$. By Theorem 10 this is the case if and only if differential equation (3) is oscillatory at the point u_-. Now all the statements follow from [MSS, Theorems 1.5 and 2.4].

□

References

[AL] V.M. Adamyan and H. Langer, *Spectral properties of a class of rational operator valued functions*, J. Operator Theory **33** (1995), 259–277.

[ALMSa] V. Adamyan, H. Langer, R. Mennicken and J. Saurer, *Spectral components of selfadjoint block operator matrices with unbounded entries*, Math. Nachr. **178** (1996), 43–80.

[ALMS] F.V. Atkinson, H. Langer, R. Mennicken and A.A. Shkalikov, *The essential spectrum of some matrix operators*, Math. Nachr. **167** (1994), 5–20.

[ADN] S. Agmon, A. Douglis and L. Nirenberg, *Estimates near the boundary for solutions of elliptic partial differential equations satisfying general boundary conditions. I*, Comm. Pure Appl. Math. **12** (1959), 623–727 and II, **17** (1964), 35–92.

[FMM] M. Faierman, R. Mennicken and M. Möller, *The essential spectrum of singular ordinary differential operators of mixed order. I, II*, Math. Nachr. (to appear).

[GG] G. Grubb and G. Geymonat, *The essential spectrum of elliptic systems of mixed order*, Math. Ann. **227** (1977), 247–276.

[HMN] V. Hardt, R. Mennicken and S.N. Naboko, *Systems of singular differential operators of mixed order and applications to one-dimensional MHD problem*, Math. Nachr. (to appear).

[H] P. Hartman, *Ordinary Differential Equations*. John Wiley & Sons: New York, 1964.

[HM] R.O. Hryniv and T.A. Melnyk, *On a singular Rayleigh functional*, Mat. Zametki **60** (1996), no. 1, 130–134 (Russian); translation in Math. Notes **60** (1996), no. 1–2, 97–100.

[HSV] R.O. Hryniv, A.A. Shkalikov and A.A. Vladimirov, *Spectral analysis of periodic differential operator matrices of mixed order* (to appear).

[Ka] T. Kato, *Perturbation Theory of Linear Operators*. Springer-Verlag, Berlin, 1976.

[K] A.Yu. Konstantinov, *Spectral theory of some matrix differential operators of mixed order*, Universität Bielefeld, Preprint no. 97-060.

[LL] L.D. Landau and E.M. Lifshitz, *Fluid Mechanics*. London: Pergamon Press, 1959.

[LMM] H. Langer, R. Mennicken and M. Möller, *A second order differential operator depending non-linearly on the eigenvalue parameter*. Oper. Theory Adv. Appl. **48** (1990), 319–332.

[L] A.E. Lifschitz, *Magnetohydrodynamics and Spectral Theory*. Kluwer Acad. Publishers: Dodrecht, 1989.

[MSS] R. Mennicken, H. Schmid and A.A. Shkalikov, *On the eigenvalue accumulation of Sturm-Liouville problem depending nonlinearly on the spectral parameter*, Math. Nachr. **189** (1998), 157–170.

[MS] R. Mennicken and A.A. Shkalikov, *Spectral decomposition of symmetric operator matrices*, Math. Nachr. **179** (1996), 259–273.

[M] A.K. Motovilov, *The removal of an energy dependence from the interaction in two body systems*, J. Math. Phys. **32** (1991), 3509–3518.

[RS] M. Reed and B. Simon, *Methods of Modern Mathematical Physics, vol. IV.* Academic Press: London, 1978.

[S] A.A. Shkalikov, *On essential spectrum of matrix operators*, Math. Notes **58** (1995), no. 6, 945–949.

Institute for Applied Problems
of Mechanics and Mathematics
3b Naukova str.
Lviv 290601
Ukraine
hryniv@mebm.lviv.ua

Department of Mechanics
and Mathematics
Moscow State University
Moscow 119899
Russia
shkal@nw.math.msu.su

Current address:
Department of Mathematics and Statistics
University of Calgary
Calgary, AB
Canada T2N 1N4
hryniv@math.ucalgary.ca

AMS Subject Classification: 47A10 (47E05, 34B05, 34L05)

Operator Theory:
Advances and Applications, Vol. 117
© 2000 Birkhäuser Verlag Basel/Switzerland

Asymptotics of Generalized Eigenvectors for Unbounded Jacobi Matrices with Power-like Weights, Pauli Matrices Commutation Relations and Cesaro Averaging

Jan Janas and Serguei Naboko

We consider unbounded, selfadjoint Jacobi matrices with weights $\lambda_n = n^{\alpha}(1+\Delta_n)$, $\lim \Delta_n = 0$ and $\alpha \in (\frac{1}{2}, 1)$. The asymptotics for generalized eigenvectors of fixed energy is obtained. This allows to carry out, by using so called grouping in block method, analysis of absolutely continuous spectra of our class of operators. The main role play here algebraic properties of Pauli matrices arising in natural way in the analysis of corresponding transfer matrices. It happenes that the commutation relations between Pauli matrices lead to the appearance of Cesaro-like averages in our study.

1 Introduction

The paper is devoted to spectral analysis of selfadjoint Jacobi matrices. It is in a way a continuation of our previous study on Jacobi matrices with power-like weights [9], however it can be read separately. The main idea of the present paper is based again on grouping in blocks approach, applied for the first time in [9], but not only on this. Concerning the ideas of grouping in block method see the end of Preliminaries. It turns out that in our analysis of products of large number of transfer matrices, algebra of Pauli matrices plays essential role. In order to explain this more percisely let us recall that essentially selfadjoint (under suitable conditions, see the line below (2.1)) Discrete String Operator J associated to Jacobi matrix with weights λ_n acts in l^2 by the formula

$$u_n \longrightarrow \lambda_{n-1}u_{n-1} + \lambda_n u_{n+1},\ n \in \mathbb{N}\backslash\{0\}.$$

Consider the relevant spectral equation

$$\lambda_{n-1}u_{n-1} + \lambda_n u_{n+1} = \lambda u_n,\ n > 1,\ \lambda \in \mathbb{R}, \tag{1.1}$$

where $u = \{u_n\}_{n=1}^{\infty}$ is a complex sequence (so called generalized eigenvectors). Since $n = 1$ is omitted in (1.1) the dimension of the space of generalized eigenvectors (not necessary in l^2) is equal to 2, for any λ. Asymptotic behaviour of generalized eigenvectors as $n \to \infty$ is a proper tool for detailed analysis of spectral properties of J. The theory due to Gilbert-Pearson [17], [10] provides precise description of the absolutely continuous spectrum of J as the set of $\lambda's$ for which there are no subordinated solutions to (1.1). Recall that u is a subordinated solution of (1.1) if for any linearly independent solution v of (1.1) $\lim_{n\to\infty}(\sum_{k=1}^{n}|u_k|^2)(\sum_{k=1}^{n}|v_k|^2)^{-1} = 0$. This remarkable notion reduces spectral analysis of J to finding asymptotics of solutions of (1.1) almost like in physical textbooks on quantum mechanics. As usual we consider instead of scalar equation (1.1) the vector equation in $\mathbb{C}^2$, $\vec{u}_{n+1} = B_n\vec{u}_n$, where $\vec{u}_n = \begin{pmatrix} u_{n-1} \\ u_n \end{pmatrix}$ and B_n is the transfer matrix given by (2.3). Therefore we have to study the asymptotics of $B_n \dots B_2$, as $n \to \infty$. In our particular case of power like weights $\lambda_n = n^\alpha(1+\Delta_n)$, $\lim_n \Delta_n = 0$, the structure of the transfer matrix leads to appearance of Pauli matrices in the principal part of asymptotics of B_n, as $n \to \infty$ (see (2.5)). Our approach here is based on commutations of exponential terms in the asymptotics of $B_n B_{n-1}$ for sufficiently long period of $n's$ defined by grouping in block methods. The length of these periods (intervals) is determined by the "quantization condition" (2.11). In this way the commutation relations between Pauli matrices begin to play essential role in our study. We repeat analogous commutation procedure a few times but with different collection of Pauli matrices. Therefore our study strongly depends on the interplay between commutation relations of Pauli matrices and the asymptotic behaviour of large products of transfer matrices. It turns out that commutation procedure leads to appearance of Cesaro-like averages in our formulas. This explains less restrictive character of conditions on weights in this paper in comparison with [9]. Moreover, although the asymptotics of solutions is of WKB type (due to later applications of Harris-Lutz like transform [7] and Kiselev's ideas [11] for "3/4" case) we could obtain (by using more recent Kiselev's results for "1/2" case or even "2/3" one [3]) a different asymptotics including new terms of Cesaro-like averages over the mentioned intervals perturbations $\{\Delta_n\}$. Actually there are even not one Cesaro averaging but its iterations which appeared in condition (b) of Theorem 5.1. Finally observe that the possibility of using Harris-Lutz transform is also based on some simple properties of Pauli matrices algebra.

2 Preliminaries

Given a sequence $\{\lambda_k\}$ of positive numbers, denote by Λ the diagonal operator given by $\Lambda e_n = \lambda_n e_n$. Let U be the unilateral shift $Ue_n = e_{n+1}$. The operator

(2.1) $$J = (U\Lambda)^* + U\Lambda$$

is essentially selfadjoint in l^2 provided the Carleman condition holds $\Sigma_k \lambda_k^{-1} = +\infty$, see [1]. In what follows the weights are given by

(2.2) $$\lambda_n = n^\alpha(1+\Delta_n),$$

where $\alpha \in (\frac{1}{2}, 1)$ and $\lim_n \Delta_n = 0$. As we mentioned in the Introduction analysis of the system (1.1) is strongly related to the products $B_n \cdot B_{n-1} \dots B_2$ of the transfer matrices

$$B_n := \begin{pmatrix} 0 & 1 \\ -\frac{\lambda_{n-1}}{\lambda_n} & \frac{\lambda}{\lambda_n} \end{pmatrix}. \tag{2.3}$$

Let

$$\sigma_1 = \begin{pmatrix} 0 & 1 \\ 1 & 0 \end{pmatrix}, \sigma_2 = \begin{pmatrix} 0 & i \\ -i & 0 \end{pmatrix}, \sigma_3 = \begin{pmatrix} 1 & 0 \\ 0 & -1 \end{pmatrix}$$

be the Pauli matrices. Due to complicated formulae in this paper we prefer to use another notations for a similar collection of matrices. We denote

$$S := \sigma_1, P := \frac{1}{i}\sigma_2, T := -\sigma_3.$$

The following standard commutation algebra plays essential role below,

$$[P, T] = 2S, \ [P, S] = -2T, \ [S, T] = 2P \tag{2.4}$$

Let us define a collection of new parameters which will be used below,

$$\begin{aligned} \varepsilon_n &:= \Delta_n - \Delta_{n-1}, \ \delta_n := \frac{1}{2}(\varepsilon_n - \varepsilon_{n-1}), \ \varphi_n := \frac{\lambda}{2}\left(\frac{1}{\lambda_n} + \frac{1}{\lambda_{n-1}}\right) \\ \beta_n &:= \frac{\alpha}{n} + \varepsilon_n, \ \psi_n := \frac{1}{2}(\beta_n + \beta_{n-1}). \end{aligned}$$

In what follows we always assume that $\{\varepsilon_n \Delta_n\} \in l^1$. Since

$$\begin{aligned} \frac{\lambda_{n-1}}{\lambda_n} &= 1 - \beta_n + O(\varepsilon_n \Delta_n) + O\left(\frac{\varepsilon_n}{n}\right), \\ &(\lambda_{n-1}\lambda_n)^{-1} = O(n^{-2\alpha}) \end{aligned}$$

and (using (2.2))

$$\lambda\left(\frac{1}{\lambda_n} - \frac{1}{\lambda_{n-1}}\right) = O(n^{-1-\alpha}) + O(\varepsilon_n n^{-\alpha})$$

we have

$$B_n B_{n-1} = \begin{pmatrix} -1 + \beta_{n-1} & \lambda/\lambda_{n-1} \\ -\lambda/\lambda_n & -1 + \beta_n \end{pmatrix} + s_n,$$

where $\{\|s_n\|\} \in l^1$ and $\|\cdot\|$ denotes the matrix norm. Note that φ_n and β_n are square summable, the first because $\alpha > 1/2$ and the second due to our assumption $\{\varepsilon_n \Delta_n\} \in l^1$. For the proof see [9]. Straightforward reasoning shows that the product $B_n B_{n-1}$ can be written as

$$-e^{-\psi_n} \exp(-\varphi_n P) \exp(-\delta_n T)(I + r_n), \tag{2.5}$$

where r_n are again some 2×2 matrices such that $\{\|r_n\|\} \in l^1$ due to the preceding formulae and estimations and also $\alpha > \frac{1}{2}$.

Let α, β be two small real numbers, for any bounded operators A_1, A_2 in a Banach space

$$\exp(\alpha A_1)\exp(\beta A_2) = \exp(\beta A_2)\exp(\alpha A_1)(I + R), \tag{2.6}$$

where $R = \alpha\beta[A_1, A_2] + O(\alpha^2\beta) + O(\alpha\beta^2)$ and $O(\alpha\beta^2)$, $O(\alpha^2\beta)$ are operators with norms estimated by corresponding $O(\cdot)$. This general, folklore type result will be used below. For its proof see for example [9].

Using relations (2.4) we have immediately the following elementary commutation relations with errors for exponents.

Remark 2.1 Let $\{\alpha_n\}, \{\beta_n\}$ be two sequences in l^2. Then

$$\exp(\alpha_n P)\exp(\beta_n T) = \exp(\beta_n T)\exp(\alpha_n P)(I + R_n^1), \tag{2.7}$$

$$\exp(\alpha_n P)\exp(\beta_n S) = \exp(\beta_n S)\exp(\alpha_n P)(I + R_n^2), \tag{2.8}$$

$$\exp(\alpha_n S)\exp(\beta_n T) = \exp(\beta_n T)\exp(\alpha_n S)(I + s_n), \tag{2.9}$$

$$\exp(\alpha_n S)\exp(\beta_n T) = \exp(\alpha_n S + \beta_n T)(I + t_n), \tag{2.10}$$

where $\|R_n^1\|, \|R_n^2\|, \|s_n\|, \|t_n\|$ are in l^1.

Relations (2.7)–(2.10) will be used below in analysis of some products of the transfer matrices. In this analysis also the more precise formula (2.6) will be useful.

As it was mentioned in the Introduction one of the main ideas of the present paper is a continuation of the grouping in blocks approach to the analysis of solutions of (1.1). We always take $\lambda \in \mathbb{R}$ and $\lambda \neq 0$. Let us recall briefly the main idea of grouping in blocks procedure (see for details [9]).

Let $\mathbb{N}\backslash\{0, 1\} = \cup_s[a_{s-1}, a_s) = \cup_s\Omega_s$, where a_s are choosen (independently of λ) such that

$$\sum_{t\in\Omega_s} \frac{1}{\lambda_t} = 2 + O(s^{-\frac{\alpha}{1-\alpha}}). \tag{2.11}$$

Since $\sum_{t\in\Omega_s} \lambda_t^{-1}$ is not equal precisely to 2 and $\lambda_t \asymp t^\alpha$ (we use the standard notation $x_n \asymp y_n$ for equivalent sequence x_n, y_n) we can obtain in (2.11) the $O(s^{-\frac{\alpha}{1-\alpha}})$ term. Moreover, (2.11) implies that $a_s \asymp s^{(1-\alpha)^{-1}}$ and we may always take a_s to be even (because changing a_s by 1 can be absorbed in $O(s^{-\frac{\alpha}{1-\alpha}})$ term for any s). Next we replace the study of complicated dynamical system $\vec{u}_{n+1} = B_n\vec{u}_n$ by a simpler, averaged one determined by the above block decomposition: $\mathbb{N}\backslash\{0, 1\} = \cup_s\Omega_s$. This is the main idea of the grouping in block method. It turns out that this new dynamical system can be studied by standard Fourier series. It should be also emphasized that the Harris-Lutz transform, (or rather its slightly changed version) is not applicable to the orginal system, can be applied to the new one. This will become clear below (for more details see again [9]).

The following notation will be useful below. For a sequence A_k of matrices $\prod_{k=1}^N A_k := A_N \ldots A_1$. For a given subset $\Omega \subset \mathbb{N}$ the sum $\sum'_{k\in\Omega} A_k$ (resp. the ordered product $\prod'_{k\in\Omega} A_k$) denotes summation (resp. multiplication) over all odd indices $k \in \Omega$. The symbol $|\Omega|$ denotes the number elements of Ω. In what follows we always assume that all the brackets to the right from the sign $\prod$ are included in that ordered product.

3 Analysis of Products of Transfer Matrices Over One Block

This section explains how the grouping in blocks idea combined with algebric properties of Pauli matrices gives asymptotics of solutions to (1.1) at infinity. Let Ω_s be the decomposion of $\mathbb{N}\backslash\{0,1\}$ given by (2.11). We can assume without any loss of generality that a_s are even. Consider the matrix product $\prod_{j=a_{k-1}+1}^{a_k-1\prime} B_j B_{j-1}$ over one block Ω_k. We reorder the above product in the following way. All terms $\exp(-\delta_j T)$ are brought together to the left and we keep all $\exp(-\varphi_j P)$ at fixed places. This can be done by repeating use of formula (2.6) with $A_1 = P$ and $A_2 = T$. Other reorderings do not give any reasonable results. The reason for our choice of ordering is the following. First, the order of $\sum_{j\in\Omega_k}^{n\prime} \varphi_j$ is not small due to the "quantization condition" (2.11) but $\sum'_{j\in\Omega_k} |\delta_j|$ can be (and really will be, see (§6)) decreasing. Therefore "gathering" of errors during the moving procedure will be not small for another reodering. Second (less important) being **unitary** exponent with P is a bit more convenient for the estimation of errors. Using formula (2.5) the above product can be written as

$$\prod_{j\in\Omega_k}{}' [-e^{-\psi_j} \exp(-\varphi_j P) \exp(-\delta_j T)(I + r_j)]. \tag{3.1}$$

By repeated use of formula (2.6) we express (3.1) in the form

$$\begin{gathered}(-1)^{\alpha(k)} \exp\left(-\sum_{\Omega_k}{}' \psi_j\right) \exp(-\eta_{a_k-1}T) \prod_{j\in\Omega_k}{}' \exp(-\varphi_j P)\\ [I + 2\varphi_j\eta_j S + O(\varphi_j\eta_j^2) + O(\varphi_j^2\eta_j)](I + r_j^{(1)}),\end{gathered} \tag{3.2}$$

where $\eta_s := \sum_{j\in\Omega_k}^{s}{}' \delta_j$, $\alpha(k) := \frac{a_k - a_{k-1}}{2}$ and $\{\|r_j^{(1)}\|\} \in l^1$. Here and below we always consider this sort of formulae with sums over one block piecewisely. More exactly we use them only for indices running over our fixed block Ω_k (determined by the position of the index within Ω_k). Therefore the summation starts from the beginning of Ω_k to the given index. We hope this convention will not lead to misunderstanding. "Old" errors r_j are transformed into "new" errors $r^{(1)}$ after the above commutation procedure. However their norms are also summable provided we impose the additional (not restrictive) assumption $\eta_n \in l^\infty$. Note that a few lines below our condition on $\{\eta_n\}$ will be replaced by the more restrictive (3.3). Note that $O(\varphi_j\eta_j^2)$ and $O(\varphi_j^2\eta_j)$ are uniform with respect to the block number k. It means that constants in estimations of both terms don't depend on k. The same convention will be assumed for all $O(\cdot)$ terms below. In what follows $r_j^{(s)}$ will always mean some 2×2 matrices such that $\{\|r_j^{(s)}\|\} \in l^1$ (in j). We want to write the error matrix given in the bracket $[I + 2\varphi_j\eta_j S + O(\varphi_j\eta_j^2) + O(\varphi_j^2\eta_j)]$ of (3.2) in the form $\exp(2\eta_j\varphi_j S)(I + r_j^{(2)})$. This can be done provided both $O(\cdot)$ terms belong to l^1. For this reason below we will assume that

$$\eta_n = O(n^{-y}),\ y > \frac{1-\alpha}{2},\ n \in \mathbb{N}. \tag{3.3}$$

It follows that $O(\varphi_j^2\eta_j)$ (resp. $O(\varphi_j\eta_j^2)$) are of order $O(j^{-y-2\alpha})$ (resp. $O(j^{-\alpha-2y})$) and so both $O(\cdot)$ terms are summable. However, $\eta_j\varphi_j$ is only of order $O(j^{-\alpha-y})$ and so it may not belong to l^1 in general. Now, let us to consider the product (see (3.2))

$$\prod_{j\in\Omega_k}{}' \exp(-\varphi_j P)\exp(2\eta_j\varphi_j S)(I + r_j^{(3)}). \tag{3.4}$$

Again we reorder (3.4) by moving all $\exp(2\varphi_j\eta_j S)$ to the left and keeping all $\exp(-\varphi_j P)$ at fixed places in the block Ω_k. We do this by applying $|\Omega_k|/2$ times relations (2.6) with $A_1 = P$ and $A_2 = S$ and also (2.4). Hence the product (3.4) can be expressed as

$$\begin{aligned} &\exp\left(2\sum_{\Omega_k}{}'\varphi_j\eta_j S\right)\prod_{j\in\Omega_k}{}'\exp(-\varphi_j P)\\ &\left\{I + 4\varphi_j\sum_{\Omega_k}^{j}{}'\varphi_s\eta_s T + O\left[\varphi_j\left(\sum_{\Omega_k}^{j}\varphi_s\eta_s\right)^2\right]\right.\\ &\left.+O\left(\varphi_j^2\sum_{\Omega_k}^{j}\varphi_s\eta_s\right)\right\}(I + r_j^{(4)}). \end{aligned} \tag{3.5}$$

The terms $O[\varphi_j(\sum_{\Omega_k}^j{}'\varphi_s\eta_s)^2]$ and $O[\varphi_j^2(\sum_{\Omega_k}^j{}'\varphi_s\eta_s)]$ appear by applying repeatedly (2.7) and (2.4). Note that $\varphi_j\sum_{\Omega_k}^j{}'\varphi_s\eta_s$ is of order $O(j^{-\alpha-y})$ due to (3.3) and (2.11) so it again does not belong to l^1 in general. On the other hand both $O(\cdot)$ terms from (3.5) belong to l^1 due to our assumption (3.3). Indeed, they have orders $O(j^{-\alpha-2y})$ similarly as the above $O(\cdot)$ terms from (3.2). Denote (using again piecewise definition)

$$\phi_n := \sum_{\Omega_k}^{n}{}'\varphi_s\eta_s,\ n \in \Omega_k. \tag{3.6}$$

Then the above arguments allow to write formula (3.5) in the following form

$$\exp(2\phi_{a_k-1}S)\prod_{j\in\Omega_k}{}'\exp(-\varphi_j P)\exp(4\varphi_j\phi_j T)(I+r_j^{(5)}). \tag{3.7}$$

In this way we deal again with the products of matrices exp $(-\varphi_j P)$ and exp $(4\varphi_j\phi_{jk}T)$ similarly to the starting product given in (3.1). We will repeat right the same procedure as we did for formula (3.1).This is essential consequence of the commutation relations (2.7). Now, applying again (2.7) to the product (3.7) we move (for the same reason as before) all the factors $\exp(4\varphi_j\phi_{jk}T)$ to the left and this leads to

$$\exp(2\phi_{a_k-1}S)\exp\left(4\sum_{j\in\Omega_k}{}'\varphi_j\phi_j T\right)$$

$$\prod_{j\in\Omega_k}{}'\exp(-\varphi_j P)\left[I-2\varphi_j\sum_{s\in\Omega_k}^{j}{}'\varphi_s\phi_s S\right. \tag{3.8}$$

$$\left.+O\left(\varphi_j^2\sum_{s\in\Omega_k}^{j}{}'\varphi_s\phi_s\right)+O\left(\varphi_j\left(\sum_{s\in\Omega_k}^{j}{}'\varphi_s\phi_s\right)^2\right)\right](I+r_j^{(6)}).$$

The last but one bracket looks like the one in (3.5) but with ϕ_s instead of η_s and coefficient 2 instead of 4 and T replaced by S. We see that Cesaro averages appear here. Let us briefly explain the reason of their appearance. During the first commutation procedure, when all exponents $\exp(\ldots T)$ where moved to the left new error terms with arised S (see (3.2)). They have a **slightly changing** within one block coefficients φ_j. Since $\sum_{j\in\Omega_k}{}'\varphi_j\approx 2\lambda$ is almost fixed, after the second commutation procedure when all exponents with S were moved to the left again, "the sum of errors" is given by (3.6). Here the summation over one block and multiplication by **a slightly changing** in Ω_k coefficients φ_s for n close to the right end of Ω_k gives us the Cesaro average (multiplied by λ) modulo l^1 (in the index block number k) term. To be more precise let us write the following estimations (they prove rigorously the above considerations).

$$\phi_n=\sum_{s\in\Omega_k}^{n}{}'\varphi_s\eta_s=\sum_{\Omega_k}^{n}{}'\varphi_{a_k-1}\eta_s+\sum_{\Omega_k}^{n}{}'\eta_s(\varphi_s-\varphi_{a_k-1})$$

and

$$\left|\sum_{\Omega_k}^{n}{}'\eta_s(\varphi_s-\varphi_{a_k-1})\right|\le\max_{s\in\Omega_k}|\varphi_s-\varphi_{a_k-1}|\max_{t\le n}\sum_{\Omega_k}^{t}|\eta_r|$$

$$=O(k^{-1/1-\alpha})O(k^{\frac{\alpha-y}{1-\alpha}})+O\left(\max_{n\in\Omega_k}|\Delta_n|/k^{\alpha/1-\alpha}\right)O(k^{\frac{\alpha-y}{1-\alpha}}).$$

Since $|\Omega_k| \asymp a_k^\alpha$, $y > \frac{1-\alpha}{2}$ and $\{\Delta_n\} \in l^2$ these estimations explain comments made above. Note that in general in the procedure $\sum_{s\in\Omega_k} \varphi_s(''')$ the number $|\Omega_k|$ is of the same order as φ_s^{-1} uniformly in $s \in \Omega_k$. Therefore this two operators $\sum_{\Omega_k}$ and multiplication by φ_s "neutralize" each other. In what follows we will use this idea several times.

We want that the expression in the above bracket $[\cdots]$ of (3.8) to be of the form $I + R_j$, where $\{\|R_j\|\} \in l^1$. This forces the next assumption

$$\sum_k \sum_{n\in\Omega_k} \left| {}'\varphi_n \sum_{s\in\Omega_k}^{n} {}'\varphi_s \phi_s \right| < \infty. \tag{3.9}$$

Hence the product in (3.8) can be rewritten as

$$\begin{aligned} &\exp(2\phi_{a_k-1} S) \exp\left(4 \sum_{j\in\Omega_k} {}'\varphi_j \phi_j T\right) \\ &\prod_{j\in\Omega_k} {}' \exp(-\varphi_j P)(I + r_j^{(7)}). \end{aligned} \tag{3.10}$$

Finally, due to unitarity of $\exp(-\varphi_j P)$ formula (3.8) can be replaced by

$$\begin{aligned} &\exp(2\phi_{a_k-1} S) \exp\left(4 \sum_{j\in\Omega_k} {}'\varphi_j \phi_j T\right) \\ &\exp\left(-\sum_{j\in\Omega_k} {}'\varphi_j P\right) \prod_{j\in\Omega_k} {}'(I + r_j^{(8)}). \end{aligned} \tag{3.11}$$

Since $|\phi_j| \le M(k^{-y/1-\alpha})$, $j \in \Omega_k$, by definition of ϕ_j, we have

$$\left|\sum_{\Omega_k} {}'\varphi_j \phi_j\right| \le c\,|\Omega_k|\, a_k^{-\alpha} a_k^{-y} \le c_1 k^{-\frac{y}{1-\alpha}},$$

and $|\phi_{a_k-1}| \le M_2 k^{-\frac{y}{1-\alpha}}$, for some positive constans c, c_1, M_2. Thus by using Remark 2.1 and (3.3) we write (3.11) in the form

$$\begin{aligned} &\exp\left(4 \sum_{j\in\Omega_k} {}'\varphi_j \phi_j T\right) \exp(2\phi_{a_k-1} S) \\ &\exp\left(-\sum_{\Omega_k} {}'\varphi_j P\right) \prod_{j\in\Omega_k} {}'(I + r_j^{(9)}). \end{aligned} \tag{3.12}$$

Here we simply joined the error of commutation of $\exp(\cdots S)$ and $\exp(\cdots T)$ to the product $\prod_{j\in\Omega_k}{}'(I+r_j^{(8)})$. Combining (3.2) and (3.12) we can rewrite the product $\prod_{j\in\Omega_k}{}' B_j B_{j-1}$ as

$$(-1)^{\alpha(k)}\exp\left(-\sum_{\Omega_k}{}'\psi_j\right)\exp(-\omega_k T)$$

(3.13)

$$\exp(2\phi_{a_k-1}S)\exp\left(-\sum_{\Omega_k}{}'\varphi_j P\right)\prod_{j\in\Omega_k}{}'(I+r_j^{(9)}),$$

where $\omega_k := \eta_{a_k-1} - 4\sum_{j\in\Omega_k}{}'\varphi_j\phi_j$.

Since ω_k and ϕ_{a_k-1} are square summable (in k), due to (3.3) so

$$\exp(-\omega_k T)\exp(2\phi_{a_k-1}S) = (I-\omega_k T + 2\phi_{a_k-1}S)(I+r_k^{(10)}).$$

Therefore, (3.13) is equivalent to

$$(-1)^{\alpha(k)}\exp\left(\sum_{\Omega_k}{}'\psi_j\right)(I-\omega_k T+2\phi_{a_k-1}S)$$

(3.14)

$$\exp\left(-\sum_{\Omega_k}{}'\varphi_j P\right)(I+r_k^{(11)}).$$

We also used here that

$$\prod_{j\in\Omega_k}(I+r_j^{(9)}) = I + r_k,\ \{\|r_k\|\}\in l^1.$$

Hence (3.14) can be written as

$$(-1)^{\alpha(k)}\exp\left(-\sum_{\Omega_k}{}'\psi_j\right)(I-\omega_k T+2\phi_{a_k-1}S)\exp(-\lambda P)(I+r_k^{(12)}).$$

(3.15)

By definition of ψ_j we have

$$\sum_{k=1}^{m}\sum_{j\in\Omega_k}{}'\psi_j = \frac{\alpha}{2}\sum_{s=2}^{a_m-1}\frac{1}{s} + \Delta_{a_m-1} - \Delta_1,$$

and so the product of $B_n B_{n-1}$ over m blocks Ω_s, $s = 1, \ldots, m$ can be written as

$$(-1)^{(a_m-2)/2} \exp\left(-\frac{\alpha}{2}\sum_{s=2}^{a_m-1}\frac{1}{s} - \Delta_{a_m-1} + \Delta_1\right) \tag{3.16}$$

$$\prod_{k=1}^{m}(I + -\omega_k T + 2\phi_{a_k-1}S)\exp(-\lambda P)(I + r_k^{(12)}).$$

Remark 3.1 We could repeat the commutation procedure as many times as we wish. However, it has not give us any serious advantages except the next iteration of Cesaro averaging for perturbations. The reason for this is the cyclicity property of commutators of Pauli matrices and so we simply come back in a way to the original situation but with averaged coefficients. Therefore there is no reason (in our opinion) to repeat our procedure many times.

4 Asymptotics of Transfer Matrices Over Blocks

Now, we shall apply Harris-Lutz transform to the averaged system (in block index k). This is a happy coincidence that both matrices S and T satisfy the relations

$$S_k := \exp(k\lambda P) S \exp(-k\lambda P) = \begin{pmatrix} \sin 2k\lambda & \cos 2k\lambda \\ \cos 2k\lambda & -\sin 2k\lambda \end{pmatrix} \tag{4.1}$$

$$T_k := \exp(k\lambda P) T \exp(-k\lambda P) = \begin{pmatrix} -\cos 2k\lambda & -\sin 2k\lambda \\ -\sin 2k\lambda & \cos 2k\lambda \end{pmatrix}, \tag{4.2}$$

and so anti-diagonal elements coincide in both matrices S_k and T_k. This property gives us an opportunity to apply the Harris-Lutz approach, see the form of U_k and V_k below. It happens that it is a consequence of the same property of matrices S and T (but not P). Really the following trivial proposition holds true:

Proposition: *Consider the product $e^{\Theta P} X e^{-\Theta P}$ of 2×2 matrices with an arbitrary parameter Θ and $P = \begin{pmatrix} 0 & 1 \\ -1 & 0 \end{pmatrix}$. Then the products have equal antidiagonal elements if and only if the same holds true for X.*

Proof: A given 2×2 matrix Y has equal antidiagonal elements if only if $tr\,PY = 0$. Put $Y := e^{\Theta P} X e^{-\Theta P}$. Then $tr\,YP = tr(Xe^{-\Theta P} P e^{\Theta P}) = tr\,XP$. □

According to an analog of Kiselev's idea [11] (see also [9] for details) not only the structure of matrices S_k and T_k is essential for possibility of applying Harris-Lutz transform but also the rate of growth of coefficients ω_k and ϕ_{a_k-1}. Namely we assume that

$$\omega_k = O(k^{-3/4-\omega}) \text{ and } \phi_{a_k-1} = O(k^{-3/4-\omega}),\ \omega > 0. \tag{4.3}$$

The following slight extension of Kisielev result will be used later: If $\varepsilon_k = O(k^{-\alpha})$, $\alpha > \frac{1}{2}$, then

$$(\alpha) \qquad \sum_{k=n}^{\infty} \varepsilon_k e^{ik\Theta} = O(n^{-\alpha+\frac{1}{2}+\omega}), \text{ for all } \omega > 0$$

and almost all $\Theta \in (-\pi, \pi)$. This statement with its proof can be found for example in [8]. Therefore we intend to use here so-called '3/4' Kiselev's approach. Recently much stronger '1/2' results for continous Schrödinger operators were proved by Kiselev [3] and Remling [12]. There is a good chance now to obtain similar results in our case too but this is not our goal at all. Our interests in this work were concentrated on other topics, like interplay between Pauli algebra properties and spectral analysis of unbounded Jacobi matrices.

Define the sequence Z_k of 2×2 matrices by

$$Z_k = \exp(-(k-1)\lambda P)(I + V_k)(I + U_k),$$

where U_k, V_k are certain 2×2 matrices such that $\lim_k \|U_k\| = \lim_k \|V_k\| = 0$. The role of these multipliers will be explained later. Consider the following product (see (3.16))

$$\prod_{k=k_o}^{m} Z_{k+1}[Z_{k+1}^{-1}(I - \omega_k T + 2\phi_{a_k-1} S)\exp(-\lambda P)Z_k]Z_k^{-1},$$

where k_o is so large that $\|U_k\| < 1$ and $\|V_k\| < 1$ for $k \geq k_o$ and so Z_k are inertible.

Since $\exp(-k\lambda P)$ are unitary and $\lim_k U_k = \lim_k V_k = 0$ we have

$$Z_k^{-1}(I + r_k^{(12)})Z_k = I + r_k^{(13)}.$$

Suppose that

$$V_k = \begin{pmatrix} 0 & b_k \\ b_k & 0 \end{pmatrix}, \quad b_k \in \mathbb{R}.$$

Then $(I + V_{k+1})^{-1} = (1 - b_{k+1}^2)^{-1}(I - V_{k+1})$ and we compute (here we used the first multiplier in Z_k for deleting $\exp(-\lambda P)$ in (3.16))

$$\begin{aligned}(4.4) \qquad &(1 - b_{k+1}^2)^{-1}(I - V_{k+1})[I - \omega_k T_k + 2\phi_{a_k-1} S_k](I + V_k) \\ &= (1 - b_{k+1}^2)^{-1}[I - V_{k+1} + V_k - V_{k+1}V_k \\ &\quad - \omega_k T_k + 2\phi_{a_k-1} S_k + \omega_k V_{k+1} T_k - 2\phi_{a_k-1} V_{k+1} S_k \\ &\quad - \omega_k T_k V_k + 2\phi_{a_k-1} S_k V_k + \omega_k V_{k+1} T_k V_k - 2\phi_{a_k-1} V_{k+1} S_k V_k].\end{aligned}$$

Now, choose V_k satisfying

$$(4.5) \qquad V_k - V_{k+1} = -2\phi_{a_k-1} S_k^{ant},$$

where

$$S_k^{ant} = \begin{pmatrix} 0 & \cos 2k\lambda \\ \cos 2k\lambda & 0 \end{pmatrix}.$$

It follows that

$$b_k = -2\sum_{s=k}^{\infty} \phi_{a_s-1} \cos 2s\lambda. \tag{4.6}$$

Since ϕ_{a_k-1} satisfies (4.3) so applying (α) we have

$$(*)\, b_k = O(k^{-\frac{1}{4}-w+\varepsilon}),$$

for almost all $\lambda \in \mathbb{R}$ and $\varepsilon > 0$ is arbitrary small.

Observe that

$$V_{k+1}V_k = V_{k+1}^2 - 2\phi_{a_k-1}V_{k+1}S_k^{ant} = b_{k+1}^2 I + O(k^{-1-2w+\varepsilon}),$$

for almost all $\lambda \in \mathbb{R}$. Using the above equalities we can rewrite (4.4) for $\varepsilon < 2\omega$

$$(1 - b_{k+1}^2)^{-1}\left[I - b_{k+1}^2 I + 2\phi_{a_k-1}\begin{pmatrix} \sin 2k\lambda & 0 \\ 0 & -\sin 2k\lambda \end{pmatrix} - \omega_k T_k + r_k^{(14)}\right]. \tag{4.7}$$

This is easy to check because all the other terms have norms which imply that they are summable (use (*) to estimate these terms). But $\phi_{a_k-1}b_{k+1}^2$ and $\omega_k b_{k+1}^2$ are also summable and we write (4.7) as:

$$I + 2\phi_{a_k-1}\begin{pmatrix} \sin 2k\lambda & 0 \\ 0 & -\sin 2k\lambda \end{pmatrix} - \omega_k T_k + r_k^{(15)}. \tag{4.8}$$

In this way we removed the anti-diagonal oscilating terms in the matrix S_k by using the second multiplier $I + V_k$ from Z_k. Now we are going to do the same with the ani-diagonal oscilating parts of T_k by applying the last multiplier from Z_k. Repeating again the Harris-Lutz idea we look for U_k in the form

$$U_k = \begin{pmatrix} 0 & a_k \\ a_k & 0 \end{pmatrix}, \; a_k \in \mathbb{R}.$$

Then (similarly as above) we have

$$(I + U_{k+1})^{-1}\left[I + 2\phi_{a_k-1}\begin{pmatrix} \sin 2k\lambda & 0 \\ 0 & -\sin 2k\lambda \end{pmatrix} - \omega_k T_k\right]$$

$$(I + U_k) = (1 - a_{k+1}^2)^{-1}\left[I + 2\phi_{a_k-1}\begin{pmatrix} \sin 2k\lambda & 0 \\ 0 & -\sin 2k\lambda \end{pmatrix}\right.$$

$$\left. - \omega_k T_k - U_{k+1} - 2\phi_{a_k-1}U_{k+1}\begin{pmatrix} \sin 2k\lambda & 0 \\ 0 & -\sin 2k\lambda \end{pmatrix}\right. \tag{4.9}$$

$$- \omega_k U_{k+1} T_k + U_k + 2\phi_k \begin{pmatrix} \sin 2k\lambda & 0 \\ 0 & -\sin 2k\lambda \end{pmatrix} U_k$$

$$- \omega_k T_k U_k - U_{k+1} U_k - 2\phi_k U_{k+1} \begin{pmatrix} \sin 2k\lambda & 0 \\ 0 & -\sin 2k\lambda \end{pmatrix}$$

$$\left. U_k + \omega_k U_{k+1} T_k U_k \right].$$

Choose U_k satisfying

$$U_k - U_{k+1} - \omega_k T_k^{ant} = 0, \tag{4.10}$$

here

$$T_k^{ant} = \begin{pmatrix} 0 & -\sin 2k\lambda \\ -sin 2k\lambda & 0 \end{pmatrix}.$$

Then using again (4.3) (now for ω_k) and (α) we obtain

$$(**)\ a_k = O(k^{-\frac{1}{4}-w+\varepsilon}),$$

for arbitrary small $\varepsilon > 0$ and almost all $\lambda \in \mathbb{R}$.

Similarly as above (4.10) and (**) imply (for $\varepsilon < 2\omega$) that (4.9) looks like

$$\begin{aligned} &(1 - a_{k+1}^2)^{-1} \left[I - a_{k+1}^2 I + 2\phi_{a_k - 1} \begin{pmatrix} \sin 2k\lambda & 0 \\ 0 & -\sin 2k\lambda \end{pmatrix} \right. \\ &\left. - \omega_k \begin{pmatrix} -\cos 2k\lambda & 0 \\ 0 & \cos 2k\lambda \end{pmatrix} + r_k^{(16)} \right]. \end{aligned} \tag{4.11}$$

All left terms are included in $r_k^{(16)}$. Note that $a_{k+1}^2 \omega_k$ and $\omega_k \phi_{a_k - 1}$ are also summable and so (4.11) can be written as

$$I + 2\phi_{a_k - 1} \begin{pmatrix} \sin 2k\lambda & 0 \\ 0 & -\sin 2k\lambda \end{pmatrix} + \omega_k \begin{pmatrix} \cos 2k\lambda & 0 \\ 0 & -\cos 2k\lambda \end{pmatrix} + r_k^{(17)}. \tag{4.12}$$

Observe that by Paley theorem [15] the series $\sum_k \phi_{a_k - 1} \cos 2k\lambda$ and $\sum_k \omega_k \sin 2k\lambda$ are almost everywhere convergent. Now, combining (4.12), the relations $Z_k^{-1}(I + r_k^{(12)}) Z_k = I + r_k^{(13)}$ and the a.e. convergence of the both just mentioned series we can write the product $\prod_{j\in\Omega_k}{}' B_j B_{j-1}$ (for $k \geq k_o$) in the following way (see (3.13), (3.14) etc.)

$$
\begin{aligned}
&(-1)^{\alpha(k)} \exp\left(-\sum_{j\in\Omega_k}{}' \psi_j\right) Z_{k+1} \left[I + 2\phi_{a_k-1} \begin{pmatrix} \sin 2k\lambda & 0 \\ 0 & -\sin 2k\lambda \end{pmatrix} \right. \\
&\qquad \left. + \omega_k \begin{pmatrix} \cos 2k\lambda & 0 \\ 0 & -\cos 2k\lambda \end{pmatrix} \right] (I + r_k^{(18)}) Z_k^{-1} \\
&= (-1)^{\alpha(k)} \exp\left(-\sum_{j\in\Omega_k}{}' \psi_j\right) Z_{k+1} \exp\left[2\phi_{a_k-1} \begin{pmatrix} \sin 2k\lambda & 0 \\ 0 & -\sin 2k\lambda \end{pmatrix} \right. \\
&\qquad \left. + \omega_k \begin{pmatrix} \cos 2k\lambda & 0 \\ 0 & -\cos 2k\lambda \end{pmatrix} \right] (I + r_k^{(19)}) Z_k^{-1},
\end{aligned} \tag{4.13}
$$

here we also used that $\{\omega_k\}$, $\{\phi_{a_k-1}\}$ are from l^2. Therefore the product of $B_n B_{n-1}$ over $m - k_o + 1$ blocks Ω_k, $k = k_o, \ldots, m$ can be written as follows (see (3.16))

$$
\begin{aligned}
&(-1)^{\frac{a_m - a_{k_o-1}}{2}} \exp\left(-\frac{\alpha}{2} \sum_{s=k_o}^{a_m-1} \frac{1}{s} - \Delta_{a_m-1} + \Delta_{a_{k_o}-1} \right) \\
&Z_{m+1} \prod_{k=k_o}^{m} \exp\left[2\phi_{a_k-1} \begin{pmatrix} \sin 2k\lambda & 0 \\ 0 & -\sin 2k\lambda \end{pmatrix} \right. \\
&\qquad \left. + \omega_k \begin{pmatrix} \cos 2k\lambda & 0 \\ 0 & -\cos 2k\lambda \end{pmatrix} \right] (I + r_k^{(19)}) Z_{k_o}^{-1}.
\end{aligned} \tag{4.14}
$$

5 Absolute Continuity of Spectra and Asymptotics of Solutions

Formula (4.14) allows to obtain asymptotics of $\vec{u}_{a_m}$. Indeed, denoting $\Theta_m := \frac{\alpha}{2} \sum_{s=k_o}^{a_m-1} \frac{1}{s} + \Delta_{a_m-1} - \Delta_{a_{k_o}-1}$ and

$$
D_k(\lambda) := 2\phi_{a_k-1} \begin{pmatrix} \sin 2k\lambda & 0 \\ 0 & -\sin 2k\lambda \end{pmatrix} + \omega_k \begin{pmatrix} \cos 2k\lambda & 0 \\ 0 & -\cos 2k\lambda \end{pmatrix}
$$

we have (for $n = a_m$)

$$
\begin{aligned}
\vec{u}_{n+1} &= (-1)^{\frac{a_m - a_{k_o}}{2}} \exp(-\Theta_m) Z_{m+1} \\
&\left[\prod_{k=k_o}^{m} \exp(D_k(\lambda))(I + r_k^{(20)}) \right] Z_{k_o}^{-1} B_{a_{k_o}-1} \cdots B_2 \begin{pmatrix} u_1 \\ u_2 \end{pmatrix}.
\end{aligned}
$$

But the diagonal matrix series $\sum_k D_k(\lambda)$ converges almost everywhere [15] and so

$$\left\| \left[\prod_{k=k_o}^{l} \exp\left(D_k(\lambda)\right) \right]^{\pm 1} \right\| = \| e^{\pm \sum_{k=k_o}^{l} D_k(\lambda)} \|$$

is uniformly bounded in l also a.e. It follows that replacing $\sum_{s=k_o}^{a_m-1} \frac{1}{s}$ approximately by $ln a_m$

$$\begin{aligned} \vec{u}_{n+1} &= (-1)^{\frac{a_m - a_{k_o}}{2}} n^{-\alpha/2} Z_{m+1} \left[\prod_{k=k_o}^{m} \exp(D_k(\lambda)) \right] \\ & \left[\prod_{k=k_o}^{m} (I + r_k^{(21)}) \right] Z_{k_o}^{-1} B_{a_{k_o}-1} \cdots B_2 \begin{pmatrix} u_1' \\ u_2' \end{pmatrix}, \end{aligned}$$

for some nonzero $\begin{pmatrix} u_1' \\ u_2' \end{pmatrix}$. Using commutativity of $\exp(D_k(\lambda))$ and $P^2 = -I$ we obtain

$$\begin{aligned} \text{(5.1)} \qquad \vec{u}_{n+1} &= P^{a_m - a_{k_o}} n^{-\alpha/2} \exp(-\lambda m P) \\ & (I + V_{m+1})(I + U_{m+1}) \exp\left(\sum_{k=k_o}^{m} D_k(\lambda) \right) \\ & \left[\prod_{k=k_o}^{m} (I + r_k^{(21)}) \right] Z_{k_o}^{-1} B_{a_{k_o}-1} \cdots B_2 \begin{pmatrix} u_1' \\ u_2' \end{pmatrix}. \end{aligned}$$

Since $\lim_m V_m = \lim_m U_m = 0$, all matrices B_s are invertible and $\|r_k^{(21)}\| \in l^1$ we have

$$\text{(5.2)} \qquad \vec{u}_{n+1} = P^{a_m - a_{k_o}} n^{-\frac{\alpha}{2}} \exp(-\lambda m P) \exp\left(\sum_{k=k_o}^{m} D_k(\lambda) \right) (\vec{c} + o(1))$$

with a nonzero vector $\vec{c} \in \mathbb{C}^2$. Again by almost everywhere convergence of the series $\sum_k D_k(\lambda)$ we simply obtain the asymptotic

$$\text{(5.3)} \qquad \vec{u}_{n+1} = P^n n^{-\alpha/2} \exp(-\lambda m P)(\vec{d} + o(1)), \quad n = a_m,$$

for some nonzero vector $\vec{d}$. Therefore the asymptotic formula for $\vec{u}_n$ is known for special sequence of indices $n = a_m$. However (3.3) implies that the similar asymptotic formula holds for arbitrary $n \in \Omega_k$. This is clear. We claim that the matrix $\prod_{j=a_{k-1}+1}^{n}{}' B_j B_{j-1}$ is close to $\exp(-\sum_{j=a_{k-1}}^{n}{}' \varphi_j P)$ (see (3.12)).

Indeed, recall that $\sum_{j\in\Omega_k}^{n}{}'\varphi_j\phi_j$ and $\sum_{j\in\Omega_k}^{n}{}'\varphi_j\eta_j$ tend to zero as $k\to\infty$, uniformly in n within Ω_k. $\|r_n^{(9)}\|\in l^1$, all the matrices $\exp(-\sum_1^l\varphi_jP)$ are unitary and

$$\sum_{s=a_{k-1}}^{n}\psi_s=\frac{1}{2}(\Delta_{a_n-1}-\Delta_{a_{k-1}})+|\Omega_k|O\left(\frac{1}{n}\right)$$

by definition of ψ_s. On the other hand $\lim_k(a_k/a_{k-1})=1$. Therefore the variation of the coefficient $n^{-\alpha/2}$ within Ω_k is negligible. All these facts imply (see (3.13) or even (2.5)) the above mentioned claim.

Hence for $n\in\Omega_{m+1}$ we have using the "quantization condition (2.11)"

$$\begin{aligned}\vec{u}_{n+1}&=n^{-\alpha/2}P^n\exp\left(-\sum_{j=a_m}^{n}\varphi_jP\right)\exp(-\lambda mP)(\vec{d}+o(1))\\&=n^{-\alpha/2}P^n\exp\left(-\frac{\lambda}{2}\sum_{s=1}^{n}\frac{1}{\lambda_s}P\right)(\vec{f}+o(1))\end{aligned}\tag{5.4}$$

with $\vec{f}\neq 0$. Diagonalizing P and using (5.4) we obtain immediately.

Theorem 5.1 *Assume that* $\{\Delta_n\}$ *satisfies the following conditions*

(a) $\{\varepsilon_n\Delta_n\}\in l^1$,

(b) $\sum_k\sum_{n\in\Omega_k}|{}'\varphi_n\sum_{s\in\Omega_k}^{n}{}'\varphi_s\phi_s|<+\infty$, (*a Cesaro like averages*)

(c) $\phi_{a_k-1}=O(k^{-3/4-\omega})$, $\omega_k=O(k^{-3/4-\omega})$, $\omega>0$

(d) $\eta_n=O(n^{-y})$, $y>\frac{1-\alpha}{2}$, $n\in\mathbb{N}$,

here $\phi_n=\sum_{\Omega_k}^{n}{}'\varphi_s\eta_s$, $n\in\Omega_k=[a_{k-1},a_k)$, $\omega_k=\eta_{a_k-1}-4\sum_{\Omega_k}{}'\varphi_j\phi_j$.
Then for two solutions $u_n,\overline{u}_n$ *of* (1.1) *we have for almost every* $\lambda\in\mathbb{R}$

$$u_n=i^nn^{-\alpha/2}\exp\left(\frac{i\lambda}{2}\sum_{s=1}^{n}\frac{1}{\lambda_s}\right)(1+o(1)).\tag{5.5}$$

Asymptotic formula (5.5) allows to apply Gilbert-Pearson subordination theory, see [10].

For this reason we need the following, proved in [9],

Generalized Stolz-Behncke Lemma:

Let J be an arbitrary selfadjoint Jacobi matrix with nonzero non-diagonal elements $\{\lambda_k\}$ satisfying the Carleman condition $\sum_k|\lambda_k|^{-1}=+\infty$. If every solution $\{v_n\}$ of the spectral reccurence equation (1.1) satisfies the estimate

$$\sum_{k=1}^{N}|v_k|^2=O\left(\sum_{k=1}^{N}|\lambda_k|^{-1}\right),\ N\longrightarrow\infty,\tag{5.6}$$

then there are no subordinated solutions of (1.1).

In case $\lambda_k \equiv 1$ very close result was obtaimed by Stolz, see [13], [14], [2]. The asymptotic formula (5.5) immediately implies that (5.6) is satisfied for both our solutions $u_n, \overline{u}_n$. Hence we can apply the results of [10] and so the absence of subordinated solutions for a.e. λ leads to the presence of absolutely continuos spectrum. This is one of the main points of Gilbert-Pearson theory [10].

Theorem 5.2 *Suppose that weights* $\{\lambda_k\}$ *satisfy conditions of Theorem* 5.1. *Then the absolutely continuos spectrum of* J *covers the whole real line.*

Note that the above result does not state that J has purely absolutely continuous spectrum. For the construction of eigenvalues lying on the abssolutely continuous spectrum see [9]. The problem of pure absolute continuity was studied in detail by J. Dombrowski for the case of monotonic weights λ_n [4], [5], [6]. She considered even general case of tridiagonal matrices T i.e. $T = J + Diag(q_n)$. Her methods are completely different and are based on commutator inequalities relatad to the matri T.

6 Comments and Examples on Cesaro Averaging

Let us consider a simplifed from of condition (b) of Theorem 5.1

$$\sum_k \left| {\sum_{\Omega_k}}' \varphi_n \phi_n \right| < \infty. \tag{6.1}$$

In this condition we only omitted one averaging over blocks. Although conditions (b) and (6.1) are formally speaking independent, it is clear that for $\phi_n \geq 0$, (6.1) implies (b). In the paper [9] we introduced another condition

$$\sum_{s=1}^{\infty} \varphi_s |\eta_s| < \infty, \tag{6.2}$$

which also allowed to reduce the study of the whole dynamical system to the averged one (see §2). It's very easy to check that both conditions (b) and (6.1) can be derived from (6.2). The inverse implications are not true in general, and therefore the Cesaro-like averaging procedure gives us less restrictive conditions for the spectral analysis of Jacobi matrices with power like weights. This will be illustrated below by two examples.

Non-oscillating examples of Jacobi matrices. Let us we consider construction of examples which are based on the locatization of perturbations Δ_n in weights near the end points of intervals Ω_k. This geometrical behaviour of Δ_n allows to explore the new possibilities of Cesaro averaging. In order to illustrate such possibility consider weights with Δ_n given by the piecewise formula

$$\Delta_n = \begin{cases} 0, & n \in \Omega_k \backslash \rho_k \text{ or } n \text{ is odd} \\ k^{-x/1-\alpha}, & n \in \rho_k \text{ and } n \text{ is even } n \neq a_k - 2 \\ 0, & n = a_k - 2 \end{cases} \tag{6.3}$$

here $\Omega_k \supset \rho_k = [d_k, a_{a_k} - 1)$ is a subset of Ω_k which will be specified below and $1 > x > 1 - \alpha$. Note that $|\eta_s|$ (for $s \in \Omega_k$) grows linearly and is bounded by $2|\rho_k|k^{-x/1-\alpha}$. It follows that (for $n \in \Omega_k$)

$$|\phi_n| \leq \sum_{\Omega_k} \varphi_s |\eta_s| \leq c(k^{-\alpha/1-\alpha}|\rho_k|)(k^{-x/1-\alpha}|\rho_k|),\ c > 0.$$

Thus

$$\sum_k \sum_{n\in\Omega_k}{}' \varphi_n \left| \sum_{s\in\Omega_k}^{n}{}' \varphi_s \phi_s \right|$$

$$(6.4) \qquad \leq const. \sum_k (|\rho_k|k^{-\alpha/1-\alpha})^2 (|\rho_k|k^{-\alpha/1-\alpha}|\rho_k|K^{-x/1-\alpha})$$

$$= const \sum_k |\rho_k|^4 k^{-\frac{3\alpha+x}{1-\alpha}}.$$

On the other hand

$$\sum_s \varphi_s |\eta_s| \leq const. \sum_k (|\rho_k|k^{-\frac{\alpha}{1-\alpha}})$$

$$(6.5) \qquad (|\rho_k|k^{-x/1-\alpha}) = const. \sum_k |\rho_k|^2 k^{-\frac{x+\alpha}{1-\alpha}}.$$

Suppose that $|\rho_k| \asymp k^{\alpha\xi/1-\alpha}$, with a fixed $\xi \in (0, 1)$. Then the series $\sum_k |\rho_k|^4 k^{-\frac{3\alpha+x}{1-\alpha}}$ is convergent provided

$$(6.6) \qquad \xi < (4\alpha + x - 1)/4\alpha.$$

In turn the series $\sum_k |\rho_k|^2 k^{-\frac{x+\alpha}{1-\alpha}}$ is convergent if

$$(6.7) \qquad \xi < (2\alpha + x - 1)/2\alpha.$$

Consequently for $\xi \in (1 + \frac{x-1}{2\alpha}, 1 + \frac{x-1}{4\alpha})$ (6.6) holds but (6.7) doesn't, and this provides the constructive examples of Jacobi with nonegative weights.

Theorem 6.1 *There are examples of Jacobi matrices with weigths λ_n such that $\Delta_n \geq 0$ for all n (non-oscillating case), which satisfy all conditions of Theorem* 5.1 *but don't fulfil* (6.2).

Proof: Take Δ_n as in (6.3) with $1 > x > \frac{\alpha+3}{4}$ and ρ_k as above. Then $\Delta_n \in l^2$ and so (a) holds. For ξ in the interval $(1 + \frac{x-1}{2\alpha}, 1 + \frac{x-1}{4\alpha})$ we know that (b) is satisfied but (6.2) is not. Finally both ϕ_{a_k-1} and ω_k are estimated by $const.\, k^{\frac{\alpha-x}{1-\alpha}}$ which implies that (c) is fulfilled due to inequality $\frac{x-\alpha}{1-\alpha} > 3/4$. Finally condition (d) holds true because $\eta_n = O(\frac{1}{n^{x-\alpha\xi}})$. But $x - \alpha\xi > \frac{1-\alpha}{2}$ as $\xi < 1$ and $\alpha < 1$.

□

Examples of Jacobi matrices with oscillating perturbations in weights. Now we wish to present another examples illustrating new possibilities of application of Theorem 5.1. Oposite to the previous examples (see Theorem 6.1) this construction is based on the oscillating properties of perturbation Δ_n. This proves again usefulness of Cesaro averaging procedure which is one of crucial points of the paper. Without loss of generality we can assume that $a_k/4$ is an integer. Additionally (again without loss of generality for "quantization condition" (2.11)) we embed between any Ω_{k-1} and Ω_k new intervals of length four where $\Delta_n \equiv 0$. We use this only for the reason that our formulae look a bit simpler. This makes no problem in our considerations. Define Δ_n by a picewise formula (within Ω_k):

$$\Delta_n = \begin{cases} 0, & \text{for odd } n \\ (-1)^{n/2} c_k, & \text{for even } n \text{ and } n \neq a_{k-1} \\ c_{k/2}, & n = a_{k-1} \end{cases} . \tag{6.8}$$

Straightforward calculation shows that η_n is given (again within Ω_k) by the following description. First element η_n, $n \in \Omega_k$, equals to zero. The next two values of η_n are equal $(-c_k/2)$, the following two values of η_n coincide again and are equal $c_k/2$, and so on. Therefore we have oscillating behaviour of the sequence η_n. Concernig the real sequence $\{c_k\}$ it will be chosen below. Now let us estimate the sequence ϕ_n. Repeating the reasoning given above (3.9) in Section 3 we can replace in our estimates of ϕ_n the coefficients φ_s by their mean value $O(k^{-\alpha/1-\alpha})$ within the block Ω_k. Really

$$\begin{aligned} \varphi_n - \varphi_{a_{k-1}} &= O\left(\frac{|\Omega_k|}{n^{\alpha+1}}\right) + O(n^{-\alpha} max_{s\in\Omega_k}|\Delta_s|) \\ &= O(k^{-\frac{1}{1-\alpha}}) + O(c_k k^{-\frac{\alpha}{1-\alpha}}) \end{aligned}$$

if $\{c_k\} \in l^\infty$. Therefore ($n \in \Omega_k$)

$$\begin{aligned} |\phi_n| &= |\varphi_{a_{k-1}} \sum_{s\in\Omega_k}^{n} {}' \eta_s| + (O(k^{-\frac{1}{1-\alpha}}) + O(c_k k^{-\frac{\alpha}{1-\alpha}}))|\Omega_k| max_{s\in\Omega_k}|\eta_s| \\ &= O(k^{-\frac{\alpha}{1-\alpha}})O(c_k) + O\left(\frac{1}{k}\right)O(c_k) + O(c_k^2) \\ &= O\left(\frac{c_k}{k}\right) + O(c_k^2), \text{ as } \alpha > 1/2. \end{aligned}$$

Hence condition (b) of Theorem 5.1. reads as follows

$$\sum_k c_k^2 < +\infty. \tag{6.9}$$

Condition (a) is equivalent to

$$\sum_k c_k^2 k^{\frac{\alpha}{1-\alpha}} < +\infty. \tag{6.10}$$

In turn condition (c) means that

$$|c_k| = O(k^{-3/4-\omega}),\ \omega > 0. \tag{6.11}$$

Condition (d) is equivalent to the estimate $c_k = O(k^{-\frac{1}{2}-\omega})$, $\omega > 0$. This is surely weaker than (6.11). On the other hand condition (6.2) can be written as

$$\sum_k |c_k| < \infty \tag{6.12}$$

Comparing (6.12) with (6.9) we obtain the following

Theorem 6.2 *There are examples of Jacobi matrices with "oscillating" weights given by* (6.8) *such that conditions* (b), (c) *and* (d) *of Theorem* 5.1 *are satisfied but condition* (6.2) *is not.*

Actually for this example is not important the oscilation property of $\{\Delta_n\}$ itself but the oscilation of the sequence $\{\eta_n\}$ **around zero**. It should be mentioned that introducing the "zero" intervals between Ω_{k-1} and Ω_k creates only l^1 errors (in k and n simultaneously) because its length is constant, and therefore it follows from the convergence of the series $\sum_k \frac{1}{k^{\alpha/1-\alpha}}$, $\alpha > 1/2$.

Remark 6.3 It is clear that in our construction condition (6.12) (and so (6.2)) follows from (6.10) (i.e. from condition (a)) by Schwarz inequality as $\alpha > 1/2$. Hence the last example does not fulfil all assumptions of Theorem 5.1. The reason for this is too delicate oscilation behaviour of weights Δ_n. It forced that $|\eta_n|$ did not increase within one block Ω_k which was almost occasional. A slight change of weights in our example will lead to the linear growth of $|\eta_n|$ within Ω_k and therefore the condition (6.2) looks as follows

$$\sum_k k^{\frac{\alpha}{1-\alpha}} |c_k| < \infty. \tag{6.13}$$

Hence it is already possible to fulfil condition (a) of Theorem 5.1 (i.e. (6.10)) but do not condition (6.2) (now (6.13)). Concerning condition (c) it will have now the form

$$c_k = k^{-\frac{\alpha}{1-\alpha}} O(k^{-\frac{3}{4}-\omega}),\ \omega > 0.$$

Condition (d) will be satisfied if

$$c_k = k^{-\frac{\alpha}{1-\alpha}} O(k^{-\frac{1}{2}-\omega}),\ \omega > 0.$$

That it follows from the previous estimate. The most delicate condition (b) by repeating the reasoning above (6.9) holds provided

$$\sum_k k^{-\frac{2\alpha}{1-\alpha}} \max_{n\in\Omega_k} \left| \sum_{s\in\Omega_k}^{n}{}' \sum_{p\in\Omega_k}^{s}{}' \eta_p \right| < \infty. \tag{6.14}$$

This is surely possible if we choose our oscilating weights such that $|\eta_p|$ grows in average linearly within Ω_k but due to different signs of numbers η_p the sums $\sum_{s\in\Omega}^{n}{}' \sum_{p\in\Omega_k}^{s}{}' \eta_p$ would grow with n much more slowly. That we could also add in Theorem 6.2 the condition (a) of Theorem 5.1. However in our opinion it is too cumbersome and not essential to give all details of this reconstruction of the last example.

Acknowledgement

The research of the first author was supported by grant PB 2 PO3A 002 13 of the *Komitet Badań Naukowych, Warsaw.* The authors thank Pani Bozena Skoczylas for her valuable help in preparation of the manuscript.

References

[1] Yu.M. Berezanskii, *Expansions in Eigenfunctions of Selfadjoint Operators*, Naukova Dumka, Kiev (1965) (in Russian).

[2] H. Behncke, *Absolute Continuity of Hamiltonians with von Neumann-Wigner Potentials* II, Manuscripta Math. **71** (1991), 163–181.

[3] M. Christ and A. Kiselev, *Absolutely Continuous Spectrum for One-Dimensional Schrödinger Operators with Slowly Decaying Potentials*: Some Optimal Results (preprint 1997).

[4] J. Dombrowski, Cyclic Operators, Commutators, and Absolutely Continuous Measures, *Proc. Amer. Math. Soc.* vol. **100**, no. 3 (1987), 457–462.

[5] J. Dombrowski, *Spectral measures, orthogonal polynomials, and absolute continuity*, SIAM J. Math. Anal. vol. **19**, no. 4 (1988), 939–943.

[6] J. Dombrowski, *Absolutely continuous measures for systems of orthogonal polynomials with unbounded recurrence coefficients*, Constr. Approx. **8** (1992), 161–167.

[7] W.A. Harris and D.A. Lutz, *Asymptotic integration of adiabatic oscillator*, J. Math. Anal. Appl. **51** (1975), 76–93.

[8] J. Janas and S. Naboko, *Jacobi Matrices with Absolutely Continuous Spectrum*, Proc. Amer. Math. Soc. (to appear).

[9] J. Janas and S. Naboko, *Jacobi matrices with power like weights-grouping in blocks approach* (submitted).

[10] S. Khan and D.B. Parson, *Subordinacy and Spectral Theory for Infinite Matrices*, Helv. Phys. Acta, **65** (1992), 505–527.

[11] A. Kiselev, *Absolute Continuous Spectrum of One-dimensional Schrödinger Operators and Jacobi Matrices with Slowly Decreasing Potentials*, Comm. Math. Phys. **179** (1996), 377–400.

[12] C. Remling, *The Absolutely Continuous Spectrum of One-Dimensional Schrödinger Operators with Decaying Potentials* (to appear in Comm. Math. Phys).

[13] G. Stolz, *Spectral Theory for Slowly Oscillating Potentials* I. Jacobi Matrices, Manuscripta Math. **84** (1994), 245–260.

[14] G. Stolz, *Spectral Theory for Slowly Oscillating Potentials* II. Schrödinger Operators, Math. Nachr. **183** (1997), 275–294.

[15] R.E.A.C. Paley, *Some theorems on orthonormal functions*, Studia Math. **3** (1931), 226–245.
[16] A. Peyerimboff, Lectures on summability, Berlin, Springer, 1969.
[17] D.J. Gilbert and D.B. Pearson, *On subordinacy and analysis of the spectrum of one-dimensional Schrödinger operators*, J. Math. Anal. Appl. **128** (1987), 30–56.

Institute of Mathematics
Polish Academy of Sciences
Cracow Branch
Sw. Tomasza 30
31-027 Kraków
Poland
najanas@cyf-kr.edu.pl

Department of Mathematical
Physics, Institute of Physics
St. Petersburg University
Ulianovskaia 1, 198904
St. Petergoff, S.-Petersburg
Russia
naboko@snoopy.phys.spbu.ru

Operator Theory:
Advances and Applications, Vol. 117
© 2000 Birkhäuser Verlag Basel/Switzerland

Functional Means, Convolution Operators and Semigroups

S.V. Koshkin

The generalizations of *Levy's functional means* are considered, which are the limits of integral means over the infinitely-divisible product-measures. Convolution operators with the family of such means form the C_0-semigroup generated by the non-Gaussian generalization of the Levy-Laplacian. The concentration of means near the sphere of a certain radius is verified. Behaviour of this radius in time is studied and its relation to the analytical properties of the operator semigroup. By way of application the action of convolutions on the finite-supported functions is described.

Introduction

This paper is devoted to the C_0-semigroups of positive convolution operators acting in the spaces of integral functionals. Convolution is taken with respect to some functional mean irreducible to the σ-additive measure on the functional space or its extension. Considered C_0-semigroups are the limits of the finite-dimensional Markov semigroups. Unlike to the finite-dimensional analogs these ones display some properties of the shifts along the vector fields: multiplicativity and the preservation of finiteness of initial condition. At last the semigroups also enjoy some 'essentially infinite-dimensional' features as the trivial action on the cylindrical functions and vanishing at the finite times.

The first example of such a semigroup was considered by E.M. Polishchuk (see [8]) as generated by the Laplace-Levy functional differential operator introduced earlier by P. Levy [6]. Levy's semigroup is the limit of Markovian semigroups for the simpliest diffusion processes and was studied by many authors (see [1, 2, 8] and references therein). Applications of the theory to the white-noise analysis of the generalized Brownian functionals were developed by T. Hida and K. Saito [3, 9].

At [4] there was constructed the class of the semigroups acting on the functionals over $L_p([0,1],\mathbf{R})$ and analogous to Levy's one. Constructed semigroups are the limit Markovian semigroups for the processes more general then diffusion ones. Here we are aiming to generalize the results on the vector-valued spaces, thoroughly investigate the localization properties of the semigroups and its action on the finite-supported functions. The consideration of C_0-semigroups is perfected by the use of spaces with order unit introduced by M.G. Krein. The class of functionals is close to one appearing in the white-noise analysis and the results may be applied to the non-diffusion Ito's formula (cf. [9]).

1 Preliminaries

In this section we recall the general setting of the functional means theory and state some results from [4], which may be directly transferred on the vector-valued case considered here.

V is the real Banach space with the norm $|\cdot|$, $L_p([0,1],V)$ – the Banach space of V-valued integrable at the power p functions $x:[0,1]\to V$ with the norm $\|x\| = (\int_{[0,1]}|x(\tau)|^p d\tau)^{\frac{1}{p}}$. $St^N([0,1],V)\subset L_p([0,1],V)$ is the subspace of step-functions on the uniform partition of $[0,1]$, i.e. $St^N \ni x(\tau) = \text{const}$ on $[\frac{i-1}{N},\frac{i}{N}]$, $i=\overline{1,N}$. Evidently $St^N \simeq V^N = \underbrace{V\times\cdots\times V}_{N\ times}$. Let us take some probability measure μ on V and the continuous functional $\Phi: L_p([0,1],V)\to \mathbf{R}$ to define $I_\mu^N(\Phi)=\int_{St^N}\Phi(x)d\mu^N(x)$, where $\mu^N = \underbrace{\mu\otimes\cdots\otimes\mu}_{N\ times}$.

Definition 1 Suppose that the limit $I_\mu(\Phi) := \lim_{N\to\infty} I_\mu^N(\Phi)$ exists. Then it is called the continual product-mean (integral) of Φ generated by the measure μ.

Surely to make the definition meaningful one has to present the class of functionals for which the mean does exist. We start with the following functionals (see also [8]):

$$\Phi(x) = \int_{[0,1]^n}\varphi(\tau_1,\ldots,\tau_n;x(\tau_1),\ldots,x(\tau_n))d\tau_1\ldots d\tau_n, \tag{1}$$

where $\varphi(\tau,u)$ is Borel, continuous on u and symmetrical kernel such that

$$|\varphi(\tau,u)| \le C\prod_{i=1}^n (h(\tau_i)+|x_i|^p)$$

for some $0\le h(\tau)\in L_1([0,1],\mathbf{R})$. Symmetricity means that

$$\varphi(\tau_{i_1},\ldots,\tau_{i_n};x_{i_1},\ldots,x_{i_n}) = \varphi(\tau_1,\ldots,\tau_n;x_1,\ldots,x_n)$$

for an arbitrary rearrangement $i_1,\ldots,i_n$ of $1,\ldots,n$. Under these conditions Φ is continuous functional on L_p submitted to the estimate $|\Phi(x)|\le C(1+\|x\|^{np})$ [5, 8]. The functionals (1) form the linear space.

Definition 2 The space of these functionals is denoted $\Gamma^n(L_p)$. The following embeddings are valid: $\Gamma^1\subset\Gamma^2\ldots\subset\Gamma^n\subset\ldots$ and let $\Gamma^\infty=\cup_n\Gamma^n$. Γ^n is closed with respect to shifts on vectors of L_p and Γ^∞ is the linear algebra by the pointwise operations.

We call the measure μ **fast decreasing** if it enjoys the finite absolute moments of any order $\int_V |u|^p d\mu(u)<\infty$, $\forall p\ge 0$. The next result was established in [4] for the scalar-valued function spaces and may be generalized on the vector-valued case without the difficulties.

Theorem 1 *μ is the fast decreasing probability measure on V. Then the product-mean I_μ exists for $\Phi \in \Gamma^n$ and is expressed by the formula:*

$$I_\mu(\Phi) = \int_{[0,1]^n \times V^n} \varphi(\tau_1, \dots, \tau_n; u_1, \dots, u_n) \, d\tau_1, \dots, d\tau_n \, d\mu(u_1) \dots d\mu(u_n) \tag{2}$$

Example 1 Denote $m_1 = \int_V u d\mu(u)$. We may interpret it also as the constant function from $L_p([0,1], V)$. Take $l(\tau) \in L_q([0,1], V^*) \subset L_p^*([0,1], V)$, $p^{-1} + q^{-1} = 1$ so that $\langle l, x\rangle = \int_{[0,1]} \langle l(\tau), x(\tau)\rangle d\tau \in \Gamma^1$. Then according to the formula (2) $I_\mu(\langle l, \cdot\rangle) = \langle l, m_1\rangle$. If $m_1 = 0$ and $L_q([0,1], V^*) \simeq L_p^*([0,1], V)$ it implies that the mean vanishes on all the linear functionals and hence in the non-trivial case it *can not* be generated by some σ-additive measure on L_p. It also follows from the formula (2) that I_μ is *multiplicative* on Γ^∞. One more feature of the considered means is that though μ may be dispersed over the whole V the generated mean is in some sense finite-supported.

Definition 3 We say that I_μ is localized in some $D \subset L_p$ if

$$\lim_{N \to \infty} \int_{St^N \cap D} \Phi(x) d\mu^N(x) = \lim_{N \to \infty} \int_{St^N} \Phi(x) d\mu^N(x)$$

for an arbitrary $\Phi \in \Gamma^\infty$, i.e. the mean depends only on the values of Φ on D.

It follows from the definition that the intersection of the localizers is again the localizer and any set containing the localizer is the localizer itself. The proof of the next theorem is also analogous to the scalar-valued case [4].

Theorem 2 *μ is the fast decreasing probability measure on V.*

(i) *D is the localizer if and only if $\mu^N(St^N \backslash D) \to 0$.*

(ii) *Let $\Psi \in \Gamma^\infty$ and $\varepsilon > 0$. Then the set $M_\Psi(\varepsilon) = \{x \in L_p | \, |\Psi(x) - I_\mu(\Psi)| \le \varepsilon\}$ is the localizer of I_μ.*

Note that I_μ is localized in *any* neighbourhood of some level surface of the functional (1) though *not* on the surface itself.

Example 2 Let $l(\tau) \in L_q([0,1], V^*)$. Then $M_{\langle l, \cdot\rangle}(\varepsilon)$ is the neighbourhood of the plain of codimension 1 $\{y | \langle l, y - m_1\rangle| \le \varepsilon\}$. Next $\Psi(x) = \|x\|^p = \int_{[0,1]} |x(\tau)|^p d\tau \in \Gamma^1$ and $M_\Psi(\varepsilon)$ is the neighbourhood of the sphere with the radius

$$R_\mu = I_\mu(\| \cdot \|^p)^{\frac{1}{p}} = \left(\int_V |u|^p d\mu(u) \right)^{\frac{1}{p}} \tag{3}$$

Taking into account that intersection of the localizers is also the localizer we may conclude that the mean generated by the **centered** measure is concentrated in the arbitrary small neighbourhood of the sphere of any finite codimension and the radius (3). The radius R_μ expressed by the formulas (3) is called the **localization radius.**

Denote by $B_R = \{x \in L_p | \, \|x\| \leq R\}$ the ball of the radius R and $\|\Phi\|_{B_R} = \sup_{x \in B_R} |\Phi(x)|$. From the definition of localization we obtain $|I_\mu(\Phi)| \leq \|\Phi\|_{B_{R_\mu + \varepsilon}}$ for an arbitrary $\varepsilon > 0$. Therefore the mean is continuous in the topology of the uniform convergence on the bounded sets in L_p. We denote by Γ the completion of Γ^∞ in this topology and I_μ is extended on this completion as the densely defined continuous functional. The item (ii) of Theorem 2 also holds true for $\Psi \in \Gamma$. The next theorem collects some well-known properties of Γ [4, 8].

Theorem 3 (i) *Γ is the topological algebra with* 1 *closed with respect to shifts in L_p and I_μ is multiplicative on it.*

(ii) *If $\Phi_1, \ldots, \Phi_n \in \Gamma$ and $f : \mathbf{R}^N \to \mathbf{R}$ is continuous then $f(\Phi_1, \ldots, \Phi_n) \in \Gamma$ and $I_\mu(f(\Phi_1, \ldots, \Phi_n)) = f(I_\mu(\Phi_1), \ldots, I_\mu(\Phi_n))$* (**functional property**).

Corollary 1 *Since I_μ commutes with the continuous functional symbol we may rewrite the formula* (3) *in the form $R_\mu = I_\mu(\| \cdot \|)$.*

Corollary 2 *Let $\Phi(x) = f(\langle l_1, x \rangle, \ldots, \langle l_n, x \rangle)$ be the continuous cylindrical function and l_i be as in the Example* 1. *Then combining the result from this example with the Theorem* 3 (ii) *we obtain $I_\mu(\Phi) = \Phi(m_1)$, i.e. the mean coincides on these functions with the Dirac's delta.*

2 Banach Spaces $EU\Gamma^\alpha$ and the Convolution Operator

The algebra Γ is not Banach and is 'too large' for the most of analytical purposes. Here we introduce the scale of its Banach subspaces $EU\Gamma^\alpha$, which are the semiordered spaces with order unit initially introduced by M.G. Krein (see [7]). The point is that one has the developed theory of the positive operator C_0-semigroups in these spaces. $EU\Gamma^\alpha$ are useful for the description of the convolution semigroups later. To the same end we also introduce the convolution operator with the mean and find the connection formula between its norm and the localization radius. At last we investigate the action of the convolution operator on some special subspaces of Γ-cylindrical and finite-supported functionals. These results generalize those in [2] obtained for Levy's case.

Definition 4 The functional $\Phi \in \Gamma$ belongs to $EU\Gamma^\alpha$, $\alpha \geq 0$ if the functional $e^{-\alpha\|x\|}\Phi(x)$ is uniformly continuous on the whole L_p and $\|\Phi\|_{E_\alpha} = \sup_{x \in L_p} e^{-\alpha\|x\|} \, |\Phi(x)| < \infty$. In particular, $\Phi \in EU\Gamma^\alpha$ is uniformly continuous on any bounded set.

We recall that the semiordered Banach space X is called the space with order unit if there exists an element E such that for any $x \in X$ and some $\lambda > 0$ depending on x : $-\lambda E \leq x \leq \lambda E$. The infimum of such $\lambda > 0$ states the norm on X called the norm of order unit [7]. It follows from our definition that $E_\alpha(x) = e^{\alpha\|x\|}$ is the order unit in $EU\Gamma^\alpha$ and the introduced norm coincides with the norm of order unit.

The inclusions $EU\Gamma^\alpha \subset EU\Gamma^\beta, \alpha \leq \beta$ are obvious. The inclusion $\Gamma^\infty(L_p) \subset EU\Gamma^\alpha(L_{p+\delta})$ with $\alpha, \delta > 0$ follows from the fact that functional (1) continuous on the balls in L_p becomes uniformly continuous on the balls in the space with any greater indice [5]. Though this spaces are not algebras but their union is as $\Phi \in EU\Gamma^\alpha,\ \Psi \in EU\Gamma^\beta \Rightarrow \Phi\Psi \in EU\Gamma^{\alpha+\beta}$.

Since the shifted functional $\Phi_x(y) = \Phi(x+y)$ belongs to Γ together with Φ the linear operator $(T_\mu\Phi)(x) = I_\mu(\Phi_x)$ is properly defined on the whole Γ. Moreover one can show that operator T_μ leaves the spaces Γ^n invariant and the induced kernel operator is $(T_\mu^n\varphi)(\tau, u) = \int_{V^n} \varphi(\tau, u + v)d\mu^n(v)$ [4]. Via continuity Γ is also invariant under its action and T_μ inherits the multiplicativity of I_μ.

Definition 5 We call $T_\mu : \Gamma \to \Gamma$ the convolution operator and $T_\mu\Phi$ the convolution of functional Φ with the mean I_μ.

Proposition 1 *$T_\mu(EU\Gamma^\alpha) \subset EU\Gamma^\alpha$ and the narrowed operator on $EU\Gamma^\alpha$ is bounded and $\|T_\mu\|_{E_\alpha} = e^{\alpha R_\mu}$.*

Proof: First of all let us check that $T_\mu\Phi$ is uniformly continuous together with Φ:

$$\left|T_\mu\Phi(x) - T_\mu\Phi(y)\right| = \left|I_\mu(\Phi_x - \Phi_y)\right| \leq \sup_{z\in B_{R_\mu+\varepsilon}} |\Phi(x+z) - \Phi(y+z)| \longrightarrow 0,$$

since $\Phi \in EU\Gamma^\alpha$ is uniformly continuous on the balls.

It is known from the theory of spaces with order unit that $\|T_\mu\|_{E_\alpha} = \|T_\mu E_\alpha\|_{E_\alpha}$ [7]. Let us also recall that by the Corollary 1 $R_\mu = I_\mu(\|\cdot\|)$ and hence $e^{\alpha R_\mu} = I_\mu(e^{\alpha\|\cdot\|})$ by the functional property of I_μ. Therefore:

$$\begin{aligned} E_\alpha^{-1}(x)|T_\mu E_\alpha(x)| &= e^{-\alpha\|x\|} I_\mu(e^{\alpha\|x+\cdot\|}) \leq e^{-\alpha\|x\|} I_\mu(e^{\alpha\|x\|} e^{\alpha\|\cdot\|}) \\ &= I_\mu(e^{\alpha\|\cdot\|}) = e^{\alpha R_\mu}. \end{aligned}$$

And on the other hand the inequality turns into the equality for $x = 0$. So finally we obtain $\|T_\mu E_\alpha\|_{E_\alpha} = \sup_{x\in L_p} E_\alpha^{-1}(x)|T_\mu E_\alpha(x)| = e^{\alpha R_\mu}$. □

It follows from the Corollary 2 that $(T_\mu\Phi)(x) = \Phi(x + m_1)$ for Φ cylindrical, i.e T_μ degenerates on the cylindrical functionals into the convolution with the Dirac measure. Another interesting subset of Γ involves the finite-supported functionals. Actually it was noticed in [2] that T_μ does not enlarge the convex support of Φ and moreover maps the functional into 0, when the support is sufficiently small.

We are going to generalize the results on the considered case for the centered generating measure.

Definition 5 Φ is a functional on L_p. Its convex support c-suppΦ is the minimal closed convex set out of which $\Phi(x) = 0$. It also coincides with the convex closure of the usual support of Φ.

Proposition 2 *$\Phi \in \Gamma$ is finite-supported, μ is the centered measure and L_q $([0, 1], V^*) \simeq L_p^*([0, 1], V)$. Then*

(i) *c-supp$(T_\mu \Phi) \subset$ c-suppΦ*

(ii) *c-supp$\Phi \subset B_R$ and $R < R_\mu/2 \Rightarrow T_\mu \Phi \equiv 0$*

Proof: (i) First of all we note that c-suppΦ_x = (c-supp$\Phi - x$). Assume that $x \notin$ c-suppΦ and hence $0 \notin$ c-suppΦ_x. According to the well-known corollary of the Kchan-Banach theorem about the separation of convex sets there exists $l \in L_p^*$: $\langle l, 0\rangle < \varepsilon < \langle l, y\rangle$ $\forall y \in$ c-suppΦ and $\langle l, \cdot\rangle \in \Gamma^1$ by our assumption (see Example 1). But then I_μ is localized on the set $M_{\langle l,\cdot\rangle}(\varepsilon) = \{y \in L_p \mid |\langle l, y\rangle| \leq \varepsilon\}$, while (c-supp$\Phi_x$) $\cap M_{\langle l,\cdot\rangle}(\varepsilon) = 0$. Hence for $x \notin$ c-suppΦ : $T_\mu \Phi(x) = I_\mu(\Phi_x) = 0$ and c-supp$(T_\mu \Phi) \subset$ c-suppΦ since the latter is convex.

(ii) $T_\mu \Phi(x) = 0$ for $x \notin B_R$ follows from (i). Next c-suppΦ_x = (c-suppΦ $-$ $x) \subset (B_R - x)$ and for $x \in B_R$, $y \in$ c-suppΦ_x : $\|y\| \leq \|x\| + R \leq 2R$. But I_μ is localized near the sphere of the radius $R_\mu > 2R$ and one may choose so small its neighbourhood that it does not intersects with c-suppΦ_x. Hence $T_\mu \Phi(x) = I_\mu(\Phi_x) = 0$ on B_R and then everywhere. □

Our assumption $L_q([0, 1], V^*) \simeq L_p^*([0, 1], V)$, $p^{-1} + q^{-1} = 1$ is unavoidable to apply the Kchan-Banach theorem. It surely fulfills for V finite-dimensional or V Hilbert and $p = 2$.

Example 3 When $p = 2$ and V is Hilbert $\mathcal{H} = L_2([0, 1], V)$ is also Hilbert space and the estimates from the Proposition 2 may be improved. Take $x \in \mathcal{H}$ and the subspace of $\mathcal{H}$ orthogonal to x (arbitrary subspace of codimension 1, when $x = 0$). I_μ is localized near the sphere of the radius R_μ in this subspace. On account of orthogonality the distance from x to any element of the sphere is not less then $\|x\|^2 + R_\mu^2$. Let Φ be localized in B_R. The intersection of c-suppΦ_x with the neighbourhood of the localization sphere may be non-empty only if $\|x\|^2 + R_\mu^2 \leq R^2$ or $\|x\|^2 \leq R^2 - R_\mu^2$. So we may conclude that $T_\mu \Phi$ is localized in $B_{\sqrt{R^2 - R_\mu^2}}$ whenever Φ is localized in B_R. In particular $T_\mu \Phi \equiv 0$ already for $R < R_\mu$ (cf. [2]).

3 The Semigroup of Convolutions and its Localization Radius

In this section we consider the semigroups of measures on V and use the convolution operator defined above to construct the semigroups of linear operators in

$EU\Gamma^\alpha$. The analytical properties of these semigroups essentially depend on the asymptotical behaviour of the localization radius at 0 and ∞. Motivated by this fact we investigate the latter thoroughly for V finite-dimensional and establish that the localization radius behaves similarly for the different μ and p. In the end of the section we apply the results to describe the action of the semigroups on the finite-supported functions.

Let us take two different fast decreasing probability measures on V and consider the usual convolution $(\mu * \nu)(A) = \int_V \mu(A - v)d\nu(v)$ for the Borel set A. The new measure $\mu * \nu$ is again probability and fast decreasing and $I_{\mu*\nu}$, $T_{\mu*\nu}$ are properly defined. Using the properties of the finite-dimensional convolutions we may notice that $T^n_{\mu*\nu} = T^n_\mu T^n_\nu$ for all n and by passing to limit that generally on $\Gamma : T_{\mu*\nu} = T_\mu T_\nu$.

Suppose now that μ is the infinitely-divisible probability measure on V and $\chi_\mu(\xi) = \int_V e^{i\langle\xi,u\rangle}d\mu(u)$ is its Fourier transform. Then $(\chi_\mu)^t, t \geq 0$ is also Fourier transform of some probability measure μ_t and $\mu_t * \mu_s = \mu_{t+s}$, $\mu_0 = \delta^0$ (Dirac measure at 0). Thus $\{\mu_t\}$ form the semigroup under convolution and correspondingly $T_\mu(t) = T_{\mu_t}$ is the semigroup of the bounded linear operators in $EU\Gamma^\alpha$ since $T_\mu(t+s) = T_{\mu_{t+s}} = T_{\mu_t}T_{\mu_s} = T_\mu(t)T_\mu(s)$ and $T_{\delta^0} = I$.

In the theory of operator semigroups the operator $A_\mu\Phi = \lim_{t\to 0}\frac{1}{t}(T_\mu(t)\Phi - \Phi)$ defined on the submanifold of functions, where the limit exists is called the generator of the semigroup $T_\mu(t)$ [7]. It is closed and densely defined if the semigroup is the C_0-semigroup, i.e. $\lim_{t\to 0}\|T_\mu(t)\Phi - \Phi\| = 0$ for an arbitrary element Φ. We want to connect some inner properties of $T_\mu(t)$ with the behaviour of mean's localization radius.

Proposition 3 (i) $R_\mu(t) \equiv 0 \Leftrightarrow T_\mu(t) \equiv I$

(ii) $\lim_{t\to 0} R_\mu(t) = 0 \Leftrightarrow T_\mu(t)$ *is the* C_0*-semigroup in* $EU\Gamma^\alpha$.

(iii) *If the semigroup is* C_0 *one and* $\alpha > 0$*:*

$$\lim_{t\to\infty} t^{-1}R_\mu(t) = 0 \Longleftrightarrow Re\sigma(A_\mu) \subset \mathbf{R}^-.$$

Proof: (i) $R_\mu(t) \equiv 0$ means that $I_{\mu_t}(\Phi)$ depends only on values of Φ at the 0, i.e. $I_{\mu_t}(\Phi) \equiv \Phi(0)$ since I_{μ_t} is the mean. Hence $(T_\mu(t)\Phi)(x) = I_{\mu_t}(\Phi_x) = \Phi(x)$. The converse assertion follows directly from the formula: $R_\mu(t) = R_{\mu_t} = \alpha^{-1}\ln\|T_\mu(t)\|_{E_\alpha}$ (see Proposition 1).

(ii) For the positive C_0-semigroups in the spaces with order unit $\|T(t)\|$ is continuous with respect to t. Therefore $R_{\mu_t} = \alpha^{-1}\ln\|T_\mu(t)\|_{E_\alpha}$ is also continuous and $\ln\|T_\mu(0)\|_{E_\alpha} = \ln\|I\|_{E_\alpha} = 0$. On the other hand note that

$$\inf_{y\in(x+B_{R_\mu(t)+t})}\Phi(y) \leq (T_\mu(t)\Phi)(x) \leq \sup_{y\in(x+B_{R_\mu(t)+t})}\Phi(y)$$

by the definition of T_μ. Expressions in the sides of the inequality converge to $\Phi(x)$ since $R_\mu(t) \to 0$ due to the uniform continuity on the balls and hence the middle expression also converges to $\Phi(x)$.

(iii) For the generators of positive C_0-semigroups in the spaces with order unit it is known that $\sup_{\lambda\in\sigma(A_\mu)} \mathrm{Re}\lambda = \lim_{t\to\infty} t^{-1} \ln \|T_\mu(t)\|_{E_\alpha} = \alpha \lim_{t\to\infty} t^{-1} R_\mu(t)$, what provides the necessary fact. □

Denote $m_1 = \int_V u d\mu(u)$ for the first moment of the generating measure and δ^{m_1} the corresponding Dirac measure on V. μ may be decomposed into convolution of δ^{m_1} and the centered measure $d\widehat{\mu}(u) =: d\mu(u + m_1)$: $\mu = \widehat{\mu} * \delta^{m_1}$ and consequently $T_\mu(t) = T_{\widehat{\mu}}(t)T_{\delta^{m_1}}(t)$. Since the second multiplier is trivial by action $(T_{\delta^{m_1}}(t)\Phi)(x) = \Phi(x + m_1 t)$ we restrict ourselves in the sequel only with the **centered** measures (cf. Proposition 2). By the Corollary 2 the centered multiplier $T_{\widehat{\mu}}(t)$ trivializes on the cylindrical functions.

The Proposition 3 demonstrates that the structure of the semigroup is connected with the asymptotical behaviour of the localization radius at 0 and ∞. That is why we are going to study it in detail. For V finite-dimensional the problem may be solved with the help of the cumulants. The infinite-dimensional case is more complicated though our method seems to generalize on some separable spaces.

To consider $R_\mu(t)$ one has to work with the integrals $\int_V |u|^p d\mu_t(u)$. The natural approach is to use the connections between moments and cumulants as the latter depend on time linearly. To this end we adopt the notations of [10]: $\lambda = (\lambda^1, \ldots, \lambda^l)$ is the integer-valued vector of the length l, $|\lambda| = \lambda^1 + \cdots + \lambda^l$, $\lambda! = \lambda^1! \ldots \lambda^l!$. Next if $j_1, \ldots, j_l$ is the set of integers (the integers may repeat) then $m_\lambda^{j_1,\ldots,j_l}(\mu) = \int_V u_{j_1}^{\lambda^1} \ldots u_{j_l}^{\lambda^l} d\mu(u)$ and $\kappa_\lambda^{j_1,\ldots,j_l}(\mu) = \frac{1}{i^{|\lambda|}} \frac{\partial^{|\lambda|}}{\partial\xi_{j_1}^{\lambda^1} \ldots \partial\xi_{j_l}^{\lambda^l}} \ln \chi_\mu(\xi)|_{\xi=0}$ are respectively the moments and cumulants of μ. The connection formula holds true [10]:

$$m_\lambda^{j_1,\ldots,j_l} = \sum_{n=1}^{|\lambda|} \frac{1}{n!} \left[\sum_{\lambda_{(1)}+\cdots+\lambda_{(n)}=\lambda} \frac{\lambda!}{\lambda_{(1)}! \ldots \lambda_{(n)}!} \prod_{k=1}^{n} \kappa_{\lambda_{(k)}}^{j_1,\ldots,j_l} \right],$$

where summation in the inner sum is taken over all sets of the integer-valued vectors with $|\lambda_{(k)}| \geq 1$. Since $\ln \chi_{\mu_t} = \ln(\chi_\mu)^t = t \ln \chi_\mu$: $\kappa_\lambda^{j_1,\ldots,j_l}(\mu_t) = t\kappa_\lambda^{j_1,\ldots,j_l}(\mu)$. One finds for the even powers of $|u| = (\sum_{j=1}^d u_j^2)^{\frac{1}{2}}$:

$$\int_V |u|^{2l} d\mu(u) = \sum_{j_1,\ldots,j_n=1}^{d} \int_V u_{j_1}^2, \ldots, u_{j_l}^2 d\mu(u) = \sum_{j_1,\ldots,j_l=1}^{d} m_\lambda^{j_1,\ldots,j_l}(\mu),$$

where $\lambda = (2, \ldots, 2)$. And finally combining the latter representation with the connection formula and changing the order of summation we obtain:

$$\begin{aligned} &\int_V |u|^{2l} d\mu_t(u) \\ &= \sum_{n=1}^{|\lambda|} \frac{t^n}{n!} \left[\sum_{\lambda_{(1)}+\cdots+\lambda_{(n)}=\lambda} \frac{\lambda!}{\lambda_{(1)}! \ldots \lambda_{(n)}} \left[\sum_{j_1,\ldots,j_n=1}^{d} \prod_{k=1}^{n} \kappa_{\lambda_{(k)}}^{j_1,\ldots,j_l}(\mu) \right]\right] \end{aligned} \tag{4}$$

Theorem 4 *$V = \mathbf{R}^d$, μ is the centered measure on V and $p \in 2\mathbf{N}$. Then $R_\mu(t) = \sqrt[p]{P_r(t)}$, where $P_r(t)$ is the polynomial on t of degree r and either $r = p/2$ or $P_r(t) \equiv 0$.*

Proof: When $p \in 2\mathbf{N}$ and $l = p/2$ $R_\mu^p(t) = P_r(t)$ by the formula (4), where at list $r \leq |\lambda| = 2l = p$. Our first task is to show that actually $r \leq p/2$. Suppose that in some decomposition of $\lambda = (2, \ldots, 2)$ there exists $\lambda_{(s)} : \ |\lambda_{(s)}| = 1$. By the definition of $|\lambda_{(s)}|$ there implies that this vector has only one non-zero (equal 1) component. Let $\lambda_{(s)}^1 = 1$ without the loss of generality. Then

$$\kappa_{\lambda_{(s)}}^{j_1,\ldots,j_l}(\mu) = \frac{1}{i}\frac{\partial}{\partial \xi_{j_1}} \ln \chi_\mu(\xi)\Big|_{\xi=0} = \frac{1}{i}\frac{\partial}{\partial \xi_{j_1}} \chi_\mu(\xi)\Big|_{\xi=0} = \int_V u_{j_1} d\mu(u) = 0$$

as μ is centered by assumption. Hence $\prod_{k=1}^n \kappa_{\lambda_{(k)}}^{j_1,\ldots,j_l} = 0$ for any set $(j_1, \ldots, j_l)$ and the sum in the inner brackets of (4) vanishes for decompositions $\lambda_{(1)}, \ldots, \lambda_{(n)}$ with some $|\lambda_{(k)}| = 1$. Consequently in the non-zero addends $|\lambda_{(k)}| \geq 2 \ \forall k$ and $p = |\lambda| = |\lambda_{(1)}| + \cdots + |\lambda_{(r)}| \geq 2r \Rightarrow r \leq p/2$.

It remains to refute the strict inequality $r < p/2$. It would mean that *all* addends in the outer brackets with $|\lambda_{(k)}| \equiv 2$ vanish. Since the measure is centered $|\lambda_{(s)}| = 2$ implies that there is only one non-zero (equal 2) component in $\lambda_{(s)}$ for the definiteness $\lambda_{(s)}^1 = 2$. Hence via the centricity of μ:

$$\kappa_{\lambda_{(s)}}^{j_1,\ldots,j_l}(\mu) = \frac{1}{i^2}\frac{\partial^2}{\partial \xi_{j_1}^2} \ln \chi_\mu(\xi)|q_{\xi=0} = \int_V u_{j_1}^2 d\mu(u) \geq 0.$$

Thus the whole sum in the inner brackets contains only non-negative addends and it equals 0 if and only if each of them equals 0. In particular, put $j_1 = \cdots = j_l = j$:

$$\prod_{k=1}^n \kappa_{\lambda_{(k)}}^{j_1,\ldots,j_l}(\mu) = \left(-\frac{\partial^2}{\partial \xi_j^2} \ln \chi_\mu(\xi)|_{\xi=0}\right)^n = \left(\int_V u_j^2 d\mu(u)\right)^n = 0,$$

and $\sum_{j=1}^d \int_V u_j^2 d\mu(u) = \int_V |u|^2 d\mu(u) = 0$. It implies that μ is concentrated at $u = 0$. Being the probability measure $\mu = \delta^0$ and $T_\mu(t) \equiv I, \ R_\mu(t) \equiv 0$. □

Formulating the Corollaries of the Theorem 4 we adopt the customary notation: $f(t) \sim g(t) \Leftrightarrow \exists C_1, C_2 > 0 : \ C_1 g(t) \leq f(t) \leq C_2 g(t)$.

Corollary 3 *Under the conditions of Theorem* 4 ($\mu \neq \delta^0$)

$$t \longrightarrow 0 \ : \ R_\mu(t) \leq C t^{\frac{1}{p}}$$
$$t \longrightarrow \infty \ : \ R_\mu(t) \sim t^{\frac{1}{2}}.$$

Proof: It suffices to note that $R_\mu(t) = \sqrt[p]{P_{p/2}(t)}$ and at the 0 the lower power of $P_{p/2}(t)$ dominates, while at ∞ the upper one does. □

Now we are going to generalize these asymptotical estimates on the rest of indices p. To this end we recall two classical inequalities [10]. Denote $M_p(\mu) = \int_V |u|^p d\mu(u)$ then $M_s^{\frac{1}{s}} \leq M_p^{\frac{1}{p}}$, $0 \leq s \leq p$ and $M_p \leq M_s^\tau M_r^{1-\tau}$, $0 \leq s \leq p \leq r$, $p = \tau s + (1-\tau)r$.

Corollary 4 *$V = \mathbf{R}^d$, $\mu \neq \delta^0$ is the centered measure on V. Then*

$$\begin{aligned}
&\text{(i) } t \to 0: && R_\mu(t) \leq C t^{\frac{1}{p}},\ p \geq 2 \\
& && R_\mu(t) \leq C t^{\frac{1}{2}},\ 1 \leq p < 2 \\
&\text{(ii) } t \to \infty: && R_\mu(t) \sim t^{\frac{1}{2}},\ p \geq 2 \\
& && R_\mu(t) \leq C t^{\frac{p}{4}},\ 1 \leq p < 2
\end{aligned}$$

Proof: (i) For any $p \geq 2$ there exist such $r, s \in 2\mathbf{N}$, $r - s = 2$ that $2 \leq s \leq p \leq r$, $p = \tau s + (1-\tau)r$. Note that $M_p(\mu_t) = R_\mu^p(t)$. Therefore for $t \to 0$:

$$R_\mu^p(t) = M_p(\mu_t) \leq (M_s(\mu_t))^\tau M_r(\mu_t))^{1-\tau} \leq (C_1 t^{\frac{1}{s}})^{s\tau} (C_2 t^{\frac{1}{r}})^{r(1-\tau)} \leq C^p t$$

and $R_\mu(t) \leq C t^{\frac{1}{p}}$. Analogously $R_\mu(t) \leq C t^{\frac{1}{2}}$, $t \to \infty$. The lower bound for $t \to \infty$ follows from the inequality $C t^{\frac{1}{2}} \leq M_2^{\frac{1}{2}} \leq M_p^{\frac{1}{p}}$.

(ii) The proof is completely analogous to (1) with the only difference that now $0 \leq p \leq 2$, while $M_0(\mu_t) = \int_V d\mu_t(u) \equiv 1$. □

Combining this Corollary with the Proposition 3 one obtains:

Corollary 5 *If μ is centered measure on V finite-dimensional $T_\mu(t)$ is always the C_0-semigroup and $(\lambda I - A_\mu)^{-1}$ is bounded for $Re\lambda > 0$.*

Note that the asymptotical behaviour of $R_\mu(t)$ on ∞ is *universal* for $p \geq 2$, i.e. it does not depend neither on the nature of the measure μ no on the indice of the space. It is interesting to obtain the abstract derivation of such a growth.

For $p = 2$ we may state our result a bit preciser. It follows from the Corollary 4 that $R_\mu(t) \leq C t^{\frac{1}{2}}$, $t \to 0$. But actually $R_\mu(t) \sim t^{\frac{1}{2}}$ or elseworth $t^{-1} P_r(t) = t^{-1} P_1(t) = 0 \Rightarrow P_1(t) \equiv 0$ and $\mu = \delta^0$. Thus for the Hilbert space $\mathcal{H} = L_2([0,1], V)$ and V finite-dimensional $R_\mu(t) \sim t^{\frac{1}{2}}$ on the whole half-axis for an arbitrary measure.

Example 4 For the classical Levy's semigroup $V = \mathbf{R}$, $p = 2$ with $d\mu_t(u) = \frac{1}{2\sqrt{\pi t}} e^{-\frac{u^2}{4t}} du$. A_μ in that case is known as the functional Levy-Laplacian [8]. As acting on kernels in Γ^n it coincides with the n-dimensional Laplacian. In the general case some factorizable pseudodifferential operator is obtained (see [4] for details). It is known also that $R_\mu(t) = \sqrt{2t}$. We may conclude now that such type of behaviour preserves under the very wide conjectures about μ and V.

Now we are going to combine the results about the behaviour of the localization radius with the conclusions of Proposition 2.

Corollary 6 *V is finite-dimensional, $\mu \neq \delta^0$ and the functional Φ is finite-supported. Then $T_\mu(t)\Phi \equiv 0$ for all $t \geq t_0$.*

Proof: Since $\lim_{t\to\infty} R_\mu(t) = \infty$, $R_\mu(t) > 2R$ beginning from some t_0. It remains to apply Proposition 2 (ii). □

Moreover we may denote $\Gamma_R = \{\Phi \in \Gamma | \operatorname{supp}\Phi \subset B_R\}$ and $T_\mu(t)\Gamma_R \subset \Gamma_R$ by the Proposition 2 (i). Corollary 6 then gives:

Corollary 7 *The narrowing $T_\mu(t)|_{\Gamma_R}$ is the finite-time C_0-semigroup.*

As it is obvious from the integral representation the Levy-like semigroups are the certain limits of finite-dimensional parabolic semigroups. Meanwhile their prototypes disperse the initial condition over the whole space and never vanish in consequence of analyticity. On the first glance the preservation of finiteness and finite-time vanishing reminds the hyperbolic finite-dimensional semigroups. Those however usually shift the support of functions instead of narrowing it and vanish by 'sweeping' of initial condition into infinity not by 'collapsing' it at the bounded domain. So we deal with some new 'essentially infinite-dimensional' type of behaviour.

References

1. L. Accardi and O.G. Smolyanov, *The Gaussian process generated by the Levy Laplacian and associated Feynmann-Kac formula*, Preprint of the Vito Volterra centre (1994), no. 199, pp. 1–7.
2. Yu.V. Bogdansky and Yu.L. Dalecky, *Cauchy problem for the simpliest parabolic equation with essentially infinite-dimensional elliptic operator*, Suppl. to Yu.L. Dalecky, S.V. Fomin. Measures and differential equations in infinite-dimensional space, Kluwer Acad. Publ. (1991), pp. 309–322.
3. T. Hida, *White noise and Levy's functional analysis*, Lect. Notes in Math. **695** (1978), pp. 155–163.
4. S. Koshkin, *Levy-like continual means on the spaces L_p*, Methods of Functional Analysis and Topology **4** (1998), no. 2, pp. 53–65.
5. M. Krasnoselsky, et al. *Integral operators in spaces of summable functions*, Int. Publ. Leiden, 1976.
6. P. Levy, *Problemes concrets d'analyse fonctionelle*, Gauthier-Villars, Paris, 1951.

7. *One-parameter semigroups*, Ph. Clement et al., Springer-Verlag, New York-Berlin, 1987.
8. E.M. Polishchuk, *Continual means and boundary value problems in functional spaces*, Akademie-Verlag, Berlin, 1988.
9. K. Saito, *Ito's formula and Levy's Laplacian*, Nagoya J. Math. **108** (1987), pp. 67–76.
10. A.N. Shiryayev, *Probability*, Springer-Verlag, Berlin etc., 1984.

S.V. Koshkin
Postgraduate Student
National Technical University of Ukraine
'Kiev Polytechnic Institute'
Chair of Mathematical Methods of System Analysis
pr. Pobedy, 37, Kiev
Ukraine

1991 Mathematics Subject Classification. 46G05, 46F25, 60G60

Operator Theory:
Advances and Applications, Vol. 117
© 2000 Birkhäuser Verlag Basel/Switzerland

The Inverse Spectral Problem for First Order Systems on the Half Line

Matthias Lesch and Mark M. Malamud

Dedicated to the memory of M. G. Krein on the occasion of the 90th anniversary of his birth

On the half line $[0, \infty)$ we study first order differential operators of the form

$$B\frac{1}{i}\frac{d}{dx} + Q(x),$$

where $B := \begin{pmatrix} B_1 & 0 \\ 0 & -B_2 \end{pmatrix}$, $B_1, B_2 \in \mathrm{M}(n, \mathbb{C})$ are self-adjoint positive definite matrices and $Q : \mathbb{R}_+ \to \mathrm{M}(2n, \mathbb{C})$, $\mathbb{R}_+ := [0, \infty)$, is a continuous self-adjoint off-diagonal matrix function.

We determine the self-adjoint boundary conditions for these operators. We prove that for each such boundary value problem there exists a unique matrix spectral function σ and a generalized Fourier transform which diagonalizes the corresponding operator in $L^2_\sigma(\mathbb{R}, \mathbb{C})$.

We give necessary and sufficient conditions for a matrix function σ to be the spectral measure of a matrix potential Q. Moreover we present a procedure based on a Gelfand-Levitan type equation for the determination of Q from σ. Our results generalize earlier results of M. Gasymov and B. Levitan.

We apply our results to show the existence of $2n \times 2n$ Dirac systems with purely absolute continuous, purely singular continuous and purely discrete spectrum of multiplicity p, where $1 \leq p \leq n$ is arbitrary.

1 Introduction

We consider the differential operator

$$L := B\frac{1}{i}\frac{d}{dx} + Q(x), \tag{1.1}$$

where

$$B := \begin{pmatrix} B_1 & 0 \\ 0 & -B_2 \end{pmatrix},$$

$B_1, B_2 \in \mathrm{M}(n, \mathbb{C})$ are self-adjoint positive definite matrices and $Q : \mathbb{R}_+ \to \mathrm{M}(2n, \mathbb{C})$, $\mathbb{R}_+ := [0, \infty)$, is a continuous self-adjoint matrix function. If $B_1 = B_2 = I_n$ then (1.1) is a Dirac operator.

It turns out that the operator (1.1) subject to the boundary condition

$$f_2(0) = Hf_1(0) \quad \text{with} \quad B_1 = H^* B_2 H \tag{1.2}$$

generates a self-adjoint extension L_H of the minimal operator corresponding to L. Here, $f_1(0)$, $f_2(0)$ denote the first resp. last n components of the vector $f(0)$.

Let $Y(x,\lambda)$ be the $2n \times n$ matrix solution of the initial value problem

$$LY = \lambda Y, \quad Y(0,\lambda) = \begin{pmatrix} I \\ H \end{pmatrix}. \tag{1.3}$$

We will prove that there exists a unique increasing right-continuous $n \times n$ matrix function $\sigma(\lambda)$, $\lambda \in \mathbb{R}$, (spectral function or spectral measure) such that we have the symbolic identity

$$\int_{\mathbb{R}} Y(x,\lambda) d\sigma(\lambda) Y(t,\lambda)^* = \delta(x-t) I_{2n}. \tag{1.4}$$

The main purpose of this paper is to investigate the inverse spectral problem for the operator L_H. This means to find necessary and sufficient conditions for a $n \times n$ matrix function σ to be the spectral function of the boundary value problem (1.1), (1.2).

For a Sturm-Liouville operator this problem has been posed and completely solved by I. Gelfand and B. Levitan in the well-known paper [10] (see also [16], [23], [26]). Later on M. Gasymov and B. Levitan proved similar results for 2×2 Dirac systems [9], [23, Chap. 12] (see also [8] and [17]).

We note that in [23, Chap. 12] the determination of a potential Q with prescribed spectral function σ is incomplete. The self-adjointness of Q is not proved.

The paper is organized as follows. In Section 2 we present some auxiliary results. In particular we prove the self-adjointness of the operator L_H.

In Section 3 we introduce the generalized Fourier transform

$$(\mathcal{F}_{H,Q} f)(\lambda) := \int_0^\infty Y(x,\lambda)^* f(x) dx$$

(see (3.5)) for $f \in L^2_{\text{comp}}(\mathbb{R}_+, \mathbb{C}^{2n})$ and establish the existence of an $n \times n$ matrix (spectral) measure σ such that the Parseval equality

$$(f,g)_{L^2(\mathbb{R}_+,\mathbb{C}^{2n})} = (\mathcal{F}_{H,Q} f, \mathcal{F}_{H,Q} g)_{L^2_\sigma(\mathbb{R})}, \tag{1.4'}$$

which is equivalent to (1.4), holds. In the proof we follow Krein's method of directing functionals [14], [15]. Moreover we show that $\mathcal{F}_{H,Q}$ is a unitary transformation from $L^2(\mathbb{R}_+, \mathbb{C}^{2n})$ onto $L^2_\sigma(\mathbb{R})$ which diagonalizes the operator L_H. Namely, $\mathcal{F}_{H,Q} L_H \mathcal{F}_{H,Q}^{-1} = \Lambda$ where $\Lambda : L^2_\sigma(\mathbb{R}) \to L^2_\sigma(\mathbb{R})$ denotes the multiplication operator by the function $\lambda \mapsto \lambda$. Similar results (with similar proofs) hold for Sturm-Liouville operators as well as for higher order differential operators.

In Section 4 we introduce (under the additional assumptions on B) a triangular transformation operator $I + K$ and present a sketch of proof of the representation $Y(\cdot,\lambda) = ((I+K)e_0)(\cdot,\lambda)$ where $e_0(x,\lambda)$ is the solution of (1.3) with $Q = 0$.

Then we derive the linear Gelfand-Levitan equation

$$F(x,t)+K(x,t)+\int_0^x K(x,s)F(s,t)ds=0, \quad x>t, \tag{1.5}$$

with $F(x,t)$ defined by (4.34). F is the analog of the so-called transition function (cf. [16]). We present two proofs of (1.5). The proof after Theorem 4.8 is close to the proofs in [10] and [23, Chap. 12]. The second one is relatively short. It is based on simple identities for kernels of Volterra operators (see (4.17)–(4.23)). In Proposition 4.6 we derive two representations (4.31) and (4.34) for $F(x,t)$ which easily imply (1.5). In other words, this proof derives the linear equation (1.5) directly from the nonlinear Gelfand-Levitan equation (4.31). This proof seems to be new and is essential in the sequel.

Furthermore, in Section 5 we solve the inverse problem (Theorem 5.2). Namely, starting with the transition matrix function $F(x,t)$ of the form (5.1′) we prove the existence of the unique solution $K(x,t)$ of (1.5). Conversely, starting with $K(x,t)$ we determine the matrix potential $Q(x)=iBK(x,x)-iK(x,x)B$ and we prove that $Y(\cdot,\lambda):=((I+K)e_0)(\cdot,\lambda)$ satisfies the initial value problem (1.3).

We present several criteria for the prerequisites of Theorem 5.2 to hold.

Finally, in Section 6 we present some generalizations and improvements of the main result. The degenerate Gelfand-Levitan equation is also considered here. We point out that we have obtained a sufficient condition for an increasing matrix function σ to be the spectral function of the operator L. In the special case that $B=(\lambda_1 I_n,-\lambda_2 I_n)$ (or more generally for the class (T_B), cf. Section 4) our conditions are also necessary. Finally, we prove the existence of $2n\times 2n$ systems with purely absolute continuous, purely singular continuous, and purely discrete spectrum of any given multiplicity p, $1\le p\le n$.

In conclusion we mention some recent publications close to our work. D. Alpay and I. Gohberg [2], [3] have constructed some explicit formulas for the matrix potential of a Dirac system (1.1) from the rational spectral function. Their approach is based on the results of minimal factorizations and realizations of matrix functions [4].

A new approach to inverse spectral problems for one-dimensional Schrödinger operators with partial information on the potential as well as to different kinds of uniqueness problems on the half-line has been recently proposed by F. Gesztesy and B. Simon (see [12], [13] and references therein). Furthermore, we mention the recent paper F. Gesztesy and H. Holden [11] on trace formulas for Schrödinger-type operators.

The results of this paper have been announced in [20], a preliminary version of this paper has been published in [19].

Acknowledgements

The first named author gratefully acknowledges the hospitality and financial support of the Erwin-Schrödinger Institute, Vienna, where part of this work was completed. Furthermore, the first named author was supported through the Gerhard Hess Program and the Sonderforschungsbereich 288 of Deutsche Forschungsgemeinschaft.

The second named author gratefully acknowledges the hospitality and financial support of the Humboldt University, Berlin, where the part of this work was done.

2 Preliminaries

We consider again the operator (1.1) from the introduction. In the sequel for a vector $v \in \mathbb{C}^{2n}$ the vectors $v_1, v_2 \in \mathbb{C}^n$ will denote the first resp. last n components of v. In this paper scalar products will be antilinear in the first and linear in the second argument. This is necessary since we will be dealing with vector measures (see (3.1) below).

L is a formally self-adjoint operator acting on $H^1_{\text{comp}}((0,\infty), \mathbb{C}^{2n}) \subset L^2(\mathbb{R}_+, \mathbb{C}^{2n})$. We denote by L^* the adjoint of L in $L^2(\mathbb{R}_+, \mathbb{C}^{2n})$. To obtain self-adjoint extensions we impose boundary conditions of the form

$$H_2 f_2(0) = H_1 f_1(0). \tag{2.1}$$

Here, $H_1, H_2 \in \mathrm{M}(n, \mathbb{C})$ and $f_1(0), f_2(0) \in \mathbb{C}^n$ denote the first n resp. last n components of $f(0)$, where $f \in H^1_{\text{comp}}(\mathbb{R}_+, \mathbb{C}^{2n})$.

Proposition 2.1 *Let L_{H_1,H_2} be the operator L^* restricted to the domain*

$$\mathcal{D}(L_{H_1,H_2}) := \{f \in \mathcal{D}(L^*) \mid H_2 f_2(0) = H_1 f_1(0)\}.$$

Then the operator L_{H_1,H_2} is self-adjoint iff the matrices H_1, H_2 are invertible and $B_1 = H^* B_2 H$, where $H := H_2^{-1} H_1$.

Consequently we have $L_{H_1,H_2} = L_{H_2^{-1}H_1,I} = L_{H,I}$. From now on we will denote $L_{H,I}$ by L_H and we will write the boundary condition always in the form

$$f_2(0) = H f_1(0). \tag{2.2}$$

Proof: Since Q is continuous we have $\mathcal{D}(L^*) \subset H^1_{\text{loc}}(\mathbb{R}_+, \mathbb{C}^{2n})$. Now choose a sequence of functions $\chi_m \in C_0^\infty(\mathbb{R})$ with the following properties:

(i) $\chi_m | (-\infty, m] = 1$,

(ii) $0 \le \chi_m \le 1$,

(iii) $|\chi_m'| \le \frac{1}{m}$.

If $f \in \mathcal{D}(L^*)$ then $\chi_m f \to f$ in $L^2(\mathbb{R}_+, \mathbb{C}^{2n})$ and

$$L\chi_m f = B\frac{1}{i}\chi_m' f + \chi_m L f \longrightarrow Lf \tag{2.3}$$

in $L^2(\mathbb{R}_+, \mathbb{C}^{2n})$. Thus $\chi_m f \to f$ in $\mathcal{D}(L^*)$.

For $f, g \in \mathcal{D}(L^*)$ we then find

$$(2.4)\quad \begin{aligned}(L^* f, g) - (f, L^* g) &= \lim_{k\to\infty}\lim_{l\to\infty} (L^* \chi_k f, \chi_l g) - (\chi_k f, L^* \chi_l g) \\ &= -i \lim_{k\to\infty}\lim_{l\to\infty} \langle B\chi_k f(0), \chi_l g(0)\rangle_{\mathbb{C}^{2n}} \\ &= -i\langle Bf(0), g(0)\rangle_{\mathbb{C}^{2n}}.\end{aligned}$$

Hence, $g \in \mathcal{D}(L^*_{H_1,H_2})$ iff for all $f \in \mathcal{D}(L_{H_1,H_2})$

$$(2.5)\quad 0 = \langle Bf(0), g(0)\rangle_{\mathbb{C}^{2n}}.$$

This shows that any self-adjoint extension of L is given by a Lagrangian subspace V of the symplectic vector space $\mathbb{C}^{2n}$ with symplectic form

$$\omega(v, w) := \langle Bv, w\rangle = \langle B_1 v_1, w_1\rangle - \langle B_2 v_2, w_2\rangle.$$

Lagrangian means that $\dim V = n$ and $\omega|V = 0$. The domain of such an extension then is

$$\{f \in \mathcal{D}(L^*) \mid f(0) \in V\}.$$

Now let V be a Lagrangian subspace of $\mathbb{C}^{2n} = \mathbb{C}^n_+ \oplus \mathbb{C}^n_-$. We denote by π_1, π_2 the orthogonal projections onto the first resp. second factor. Since the symplectic form ω is positive resp. negative definite on $\ker \pi_1$ resp. $\ker \pi_2$ and since $\dim V = n$ the maps π_1, π_2 restricted to V are isomorphisms

$$(2.6)\quad \tilde{\pi}_1 : V \longrightarrow \mathbb{C}^n_+, \quad \tilde{\pi}_2 : V \longrightarrow \mathbb{C}^n_-.$$

Hence $V = \{(x, \tilde{\pi}_2 \circ \tilde{\pi}_1^{-1} x) \mid x \in \mathbb{C}^n_+\}$. Put $H := \tilde{\pi}_2 \circ \tilde{\pi}_1^{-1}$. Then $\omega|V = 0$ immediately implies $B_1 = H^* B_2 H$. This proves the proposition. □

Remark 2.2

1. The previous proposition shows that the deficiency indices $n_\pm(L)$ are equal to n, i.e. $n_\pm(L) = n$. This means that at infinity we do not have to impose a boundary condition. Thus infinity is always in the 'limit point case', which essentially distinguishes first order systems from Sturm-Liouville operators and higher order differential operators ([27, 7]).
2. For scalar Dirac systems ($n = B_1 = B_2 = 1$) another proof of Proposition 2.1 has been obtained earlier by B.M. Levitan [23, Theorem 8.6.1].

   The present proof is adapted from the standard proof of the essential self-adjointness of Dirac operators on complete manifolds (see e.g. [18, Theorem II.5.7]).

3. At the same time as our preprint [19] the paper Sakhnovich [30] appeared. Following Levitan's method he obtained some sufficient conditions for a canonical system to be selfadjoint. This is a system

$$J\frac{dy(x,\lambda)}{dx} = iH(x)y(x,\lambda), \quad J = \begin{pmatrix} 0 & I_n \\ -I_n & 0 \end{pmatrix}, \tag{2.7}$$

where $H(x)$ is a continuous nonnegative $2n \times 2n$ matrix function. The method of proof of Proposition 2.1 can be extended to arbitrary first order systems, in particular to generalize the recent result from [30] for canonical systems. Details will be given in a subsequent publication.

4. Another proof of the previous proposition could be given using the uniqueness of the solution of the Goursat problem for the hyperbolic system $\frac{du}{dt} = \pm iL_H^* u$ in $\mathbb{R}_+^2$. This method (see [5]) was also used to prove the essential self-adjointness of all powers of the Dirac operator on a complete manifold (cf. [6]). Sakhnovich's result [30] mentioned before also follows from the hyperbolic system method.

For the problem considered here we prefered to present an elementary direct proof.

From now on we will assume

$$B_1 = H^* B_2 H. \tag{2.8}$$

Note that this implies that H is invertible.

We first discuss in some detail the case $Q = 0$. Let $A \in M(n, \mathbb{C})$ be a positive definite matrix. Then we put for $f \in L^2(\mathbb{R}, \mathbb{C}^n)$

$$\mathcal{F}_A f(\lambda) := \int_{\mathbb{R}} e^{-iA^{-1}x\lambda} f(x)dx. \tag{2.9}$$

Then we have for $f, g \in L^2(\mathbb{R}, \mathbb{C}^n)$ the Parseval equality

$$(f, g) = \frac{1}{2\pi} \int_{\mathbb{R}} (\mathcal{F}_A f)(\lambda)^* A^{-1} (\mathcal{F}_A g)(\lambda) d\lambda. \tag{2.10}$$

To prove (2.10) we may assume A to be diagonal, i.e. $A = \operatorname{diag}(a_1, \ldots, a_n)$, because if $A = U\widetilde{A}U^*$ with a unitary matrix U then $(\mathcal{F}_A f)(\lambda) = U(\mathcal{F}_{\widetilde{A}} U^* f)(\lambda)$. Now $\mathcal{F}_A f(\lambda) = (\mathcal{F} f_j(\lambda/a_j))_{j=1,\ldots,n}$ and (2.10) follows easily from the Parseval equality for the Fourier transform.

Now let

$$e_0(x,\lambda) := e^{i\lambda B^{-1}x} \begin{pmatrix} I \\ H \end{pmatrix} = \begin{pmatrix} e^{i\lambda B_1^{-1}x} \\ e^{-i\lambda B_2^{-1}x} H \end{pmatrix} \tag{2.11}$$

and put

$$\mathcal{F}_{H,0} f(\lambda) := \int_0^\infty e_0(x,\lambda)^* f(x)dx = \mathcal{F}_{B_1} \widetilde{f}_1(\lambda) + H^* \mathcal{F}_{B_2} \widetilde{f}_2(-\lambda), \tag{2.12}$$

where $\widetilde{f}_j$ denotes the extension by 0 of f_j to $\mathbb{R}$.

If $f, g \in L^2(\mathbb{R}_+, \mathbb{C}^{2n})$ then the integrals

$$\begin{aligned} &\int_{\mathbb{R}} \mathcal{F}_{B_1}\widetilde{f}_1(\lambda)^* B_1^{-1} H^* \mathcal{F}_{B_2}\widetilde{g}_2(-\lambda)d\lambda, \\ &\int_{\mathbb{R}} \mathcal{F}_{B_2}\widetilde{f}_2(-\lambda)^* H B_1^{-1} \mathcal{F}_{B_1}\widetilde{g}_1(\lambda)d\lambda \end{aligned} \tag{2.13}$$

are sums of scalar products of the form

$$\int_{\mathbb{R}} \overline{\mathcal{F}\widetilde{\varphi}(-\lambda)}\mathcal{F}\widetilde{\psi}(\lambda)d\lambda, \tag{2.14}$$

where $\varphi, \psi \in L^2(\mathbb{R}_+)$. These scalar products vanish and hence we end up with the Parseval equality in the case of $Q = 0$

$$\begin{aligned} &\frac{1}{2\pi}\int_{\mathbb{R}} \mathcal{F}_{H,0}f(\lambda)^* B_1^{-1} \mathcal{F}_{H,0}g(\lambda)d\lambda \\ &= \frac{1}{2\pi}\int_{\mathbb{R}} \mathcal{F}_{B_1}\widetilde{f}_1(\lambda)^* B_1^{-1} \mathcal{F}_{B_1}\widetilde{g}_1(\lambda)d\lambda \\ &+ \frac{1}{2\pi}\int_{\mathbb{R}} \mathcal{F}_{B_2}\widetilde{f}_2(-\lambda)^* H B_1^{-1} H^* \mathcal{F}_{B_2}\widetilde{g}_2(-\lambda)d\lambda = (f, g), \end{aligned} \tag{2.15}$$

in view of (2.8) and (2.10).

3 The Spectral Measure

In this section we prove the existence of a spectral measure function for the self-adjoint operator L_H based on Krein's method of directing functionals [14], [15]. For the convenience of the reader we recall Krein's result.

Definition 3.1 ([14], [15]). Let A be a symmetric operator in a separable Hilbert space H and let E be a dense linear subspace of H containing $\mathcal{D}(A)$.

The system $\{\Phi_j\}_1^p$ of linear functionals defined on E and depending on $\lambda \in \mathbb{R}$ is called a directing system of functionals for A in E if the following three conditions are fulfilled:

1. $\Phi_j(f; \lambda)$, $j = 1, \dots, p$, is an analytic function of $\lambda \in \mathbb{R}$, for each $f \in E$;
2. the functionals $\Phi_j(\cdot; \lambda_0)$ are linearly independent for some $\lambda_0 \in \mathbb{R}$;
3. for each $f_0 \in E$ and $\lambda_0 \in \mathbb{R}$ the equation $Ag - \lambda_0 g = f_0$ has a solution in E if and only if

$$\Phi_j(f_0; \lambda_0) = 0 \quad \text{for all } 1 \le j \le p.$$

Theorem 3.2 ([14], [15]). *Let A be a symmetric operator in H with $\mathcal{D}(A) \subset E \subset H$ which has a directing system of functionals $\{\Phi_j(\cdot; \lambda)\}_1^p$ in E. Then*

1. *there exists an increasing* $p \times p$ *matrix function* $\sigma(\lambda) = (\sigma_{jk}(\lambda))_{j,k=1}^{p}$ *such that the equality*

$$(g, f) = \sum_{j,k=1}^{p} \int_{\mathbb{R}} \overline{\Phi_j(g;\lambda)} \Phi_k(f;\lambda) d\sigma_{jk}(\lambda)$$

holds for each $f, g \in E$.

2. *If* σ *is normalized by requiring it to be right-continuous with* $\sigma(0) = 0$ *then it is unique if and only if* $n_+(A) = n_-(A)$, *where* $n_\pm(A) := \dim\ker(A^* \mp i)$ *denote the deficiency indices of* A.

Definition 3.3 Let $\sigma(\lambda) = (\sigma_{ij}(\lambda))_{i,j=1}^{n}$ be an increasing $n \times n$ matrix function. On the space $C_0(\mathbb{R}, \mathbb{C}^n)$ of continuous $\mathbb{C}^n$-valued functions with compact support we introduce the scalar product

$$(3.1) \quad (f, g)_{L^2_\sigma} := \int_{\mathbb{R}} f(\lambda)^* d\sigma(\lambda) g(\lambda) := \sum_{i,j=1}^{n} \int_{\mathbb{R}} \overline{f_i(\lambda)} g_j(\lambda) d\sigma_{ij}(\lambda).$$

We denote by $L^2_\sigma(\mathbb{R})$ (cf. [27]) the Hilbert space completion of this space.

Remark 3.4 From now on we will consider - without saying this explicitly - only right - continuous $n \times n$ matrix functions which map 0 to the 0-matrix. Such a function σ is determined by its corresponding matrix measure $d\sigma$.

We turn to general Q. For future reference we state the boundary value problem for L:

$$(3.2) \qquad Lf = \lambda f, \quad f_2(0) = H f_1(0), \quad \text{where} \quad B_1 = H^* B_2 H.$$

Proposition 3.5 *Let* $Y : \mathbb{R}_+ \times \mathbb{C} \to \mathrm{M}(2n \times n, \mathbb{C})$ *be the unique solution of the initial value problem*

$$(3.3) \qquad LY(x,\lambda) = \lambda Y(x,\lambda), \quad Y(0,\lambda) = \begin{pmatrix} I \\ H \end{pmatrix}.$$

Then:

1. *There exists an increasing* $n \times n$ *matrix function* $\sigma(\lambda)$, $\lambda \in \mathbb{R}$, *(spectral function) such that the map*

$$(3.4) \qquad \begin{aligned} \mathcal{F}_{H,Q} \ &: \ L^2_{\text{comp}}(\mathbb{R}_+, \mathbb{C}^{2n}) \ni f \longmapsto (\mathcal{F}_{H,Q} f)(\lambda) := F(\lambda) \\ &:= \int_0^\infty Y(x,\lambda)^* f(x) dx \end{aligned}$$

extends by continuity to an isometric transformation from $L^2(\mathbb{R}_+, \mathbb{C}^{2n})$ into the space $L^2_\sigma(\mathbb{R})$, i.e. for $f, g \in L^2(\mathbb{R}_+, \mathbb{C}^{2n})$ we have the Parseval equation

$$\int_0^\infty f^*(t)g(t)dt = \int_{\mathbb{R}} F^*(\lambda)d\sigma(\lambda)G(\lambda) \tag{3.5}$$

with F, G being the $\mathcal{F}_{H,Q}$-transforms of f, g.

2. *If σ is normalized by requiring it to be right-continuous with $\sigma(0) = 0$ then it is unique.*

Proof: 1. Let $b \in \mathbb{R}_+$ be a fixed point and let L_b be the operator L^* restricted to the domain

$$\mathcal{D}(L_b) = \{f \in H^1([0, b], \mathbb{C}^{2n}) \mid f_2(0) = Hf_1(0),\ f_1(b) = f_2(b) = 0\}. \tag{3.6}$$

It is clear that L_b is a symmetric operator and $L_b^* = (L_b)^*$ is a restriction of L^* to the domain

$$\mathcal{D}(L_b^*) = \{f \in H^1([0, b], \mathbb{C}^{2n}) \mid f_2(0) = Hf_1(0)\}. \tag{3.7}$$

We consider $\mathcal{D}(L_b)$ as a subset of $H^1(\mathbb{R}_+, \mathbb{C}^{2n})$ identifying each function $f \in \mathcal{D}(L_b)$ with its continuation by zero to $\mathbb{R}_+$.

Since L_b is a regular differential operator on a finite interval, each $\lambda \in \mathbb{C}$ is a regular type point for L_b, i.e. $\|(L_b - \lambda)f\| \geq \epsilon\|f\|$ for all $f \in \mathcal{D}(L_b)$ with some $\epsilon > 0$. In particular, $L_b - \lambda$ has closed range. Hence, for a fixed $\lambda \in \mathbb{R}$ and $f \in L^2([0, b]; \mathbb{C}^{2n})$ the equation

$$Lg - \lambda g = f, \quad \lambda \in \mathbb{R}, \tag{3.8}$$

has a solution $g \in L^2([0, b], \mathbb{C}^{2n})$ if and only if f is orthogonal to the kernel $\ker(L_b^* - \lambda)$, that is if $\int_0^b Y^*(x, \lambda)f(x)dx = 0$.

Denoting by Y_i the i-th column of Y, on rewrites the last equation as

$$\begin{aligned} \Phi_i(f; \lambda) &:= \int_0^b \langle Y_i(x, \lambda), f(x)\rangle dx \\ &= \sum_{j=1}^{2n} \int_0^b \overline{Y_{ji}(x, \lambda)} f_j(x)dx = 0, \quad 1 \leq i \leq n. \end{aligned} \tag{3.9}$$

It is clear that the functionals Φ_i on $L^2_{\text{comp}}(\mathbb{R}_+, \mathbb{C}^{2n})$, defined by the left-hand side of (3.9), are linearly independent and holomorphic in $\lambda \in \mathbb{R}$. Thus the conditions 1. and 2. of Definition 3.1 are satisfied. Since $E := L^2_{\text{comp}}(\mathbb{R}_+, \mathbb{C}^{2n})$ is dense in $L^2(\mathbb{R}_+, \mathbb{C}^{2n})$ the functionals $\Phi_i(f, \lambda)$ thus form a directing system of functionals for the operator $A := L_H \restriction \mathcal{D}(L_H) \cap L^2_{\text{comp}}(\mathbb{R}_+, \mathbb{C}^{2n})$. By Krein's Theorem 3.2 there exists $\sigma(\lambda)$ such that (3.5) holds for arbitrary $f, g \in L^2_{\text{comp}}(\mathbb{R}_+, \mathbb{C}^{2n})$.

2. In view of Proposition 2.1 the operator $A = L_H \restriction \mathcal{D}(L_H) \cap L^2_{\text{comp}}(\mathbb{R}_+, \mathbb{C}^{2n})$ is essentially selfadjoint and consequently $n_+(A) = n_-(A) = 0$. Thus the uniqueness of $\sigma(\lambda)$ follows from the assertion 2. of Krein's Theorem 3.2. □

Remark 3.6

1. The Parseval identity may be symbolically rewritten as

$$\int_{\mathbb{R}} Y(x,\lambda)d\sigma(\lambda)Y(t,\lambda)^* = \delta(x-t)I_{2n}. \tag{3.5'}$$

   To obtain (3.5′) from (3.5) it suffices to set in (3.5) $f(\xi) = \delta_x(\xi)\otimes e_i$, $g(\xi) = \delta_t(\xi)\otimes e_j$, $1\le i,j\le 2n$ and to note that $(\mathcal{F}_{H,Q}f)(\lambda) = Y(x,\lambda)^*e_i$, $(\mathcal{F}_{H,Q}g)(\lambda) = Y(t,\lambda)^*e_j$.
2. Another proof of Proposition 3.5 based on the approximation method proposed independently by B.M. Levitan [23, Chap. 8] and N. Levinson [7, Chap. 9] was given in the preliminary version of this paper [19].

For convenience we denote the extension of $\mathcal{F}_{H,Q}$ to $L^2(\mathbb{R}_+,\mathbb{C}^{2n})$ by the same letter. Next we prove the surjectivity of $\mathcal{F}_{H,Q}$.

Theorem 3.7 *Under the assumptions of Proposition* 3.5 *the mapping* $\mathcal{F}_{H,Q}$ *is surjective, that is* $\mathcal{F}_{H,Q}$ *maps* $L^2(\mathbb{R}_+,\mathbb{C}^{2n})$ *onto* $L^2_\sigma(\mathbb{R})$.

Proof: So far we have proved that $\mathcal{F}_{H,Q}: L^2(\mathbb{R}_+,\mathbb{C}^{2n})\to L^2_\sigma(\mathbb{R})$ is an isometry. To prove surjectivity we mimick the proof of [7, Sec. 9.3] for second order operators.

Note first that for $f\in\mathcal{D}(L_H)$ we have

$$(\mathcal{F}_{H,Q}L_Hf)(\lambda) = -Y(0,\lambda)^*\frac{1}{i}Bf(0)+\lambda(\mathcal{F}_{H,Q}f)(\lambda) = \lambda(\mathcal{F}_{H,Q}f)(\lambda) \tag{3.10}$$

since in view of (3.2) and (3.3) $Y(0,\lambda)^*Bf(0)=0$.

For $f\in L^2_{\text{comp}}(\mathbb{R}_+,\mathbb{C}^{2n})\cap\mathcal{D}(L_H)$ formula (3.10) follows from integration by parts. For arbitrary $f\in\mathcal{D}(L_H)$ it follows from the fact that $L^2_{\text{comp}}(\mathbb{R}_+,\mathbb{C}^{2n})\cap\mathcal{D}(L_H)$ is a core for L_H. The latter follows from the proof of Proposition 2.1.

Next we construct the adjoint of $\mathcal{F}_{H,Q}$: we put for $g\in L^2_{\sigma,\text{comp}}(\mathbb{R})$

$$(\mathcal{G}_Hg)(x) := \int_{\mathbb{R}} Y(x,\lambda)d\sigma(\lambda)g(\lambda). \tag{3.11}$$

Then for $f\in L^2_{\text{comp}}(\mathbb{R}_+,\mathbb{C}^{2n})$

$$\begin{aligned}(\mathcal{G}_Hg,f)_{L^2(\mathbb{R}_+,\mathbb{C}^{2n})} &= \int_0^\infty (\mathcal{G}_Hg)(x)^*f(x)dx\\ &= \int_0^\infty\int_{\mathbb{R}} g(\lambda)^*d\sigma(\lambda)Y(x,\lambda)^*f(x)dx\\ &= (g,\mathcal{F}_{H,Q}f)_{L^2_\sigma(\mathbb{R})}.\end{aligned} \tag{3.12}$$

From the estimate

(3.13) $\left|(\mathcal{G}_H g, f)_{L^2(\mathbb{R}_+,\mathbb{C}^{2n})}\right| = \left|(g, \mathcal{F}_{H,Q} f)_{L^2_\sigma(\mathbb{R})}\right| \le \|g\|_{L^2_\sigma(\mathbb{R})} \|f\|_{L^2(\mathbb{R}_+,\mathbb{C}^{2n})}$

we infer that $\mathcal{G}_H$ extends by continuity for $L^2_\sigma(\mathbb{R})$. Moreover, it equals the adjoint of $\mathcal{F}_{H,Q}$, i.e.

(3.14) $$\mathcal{G}_H = \mathcal{F}^*_{H,Q}.$$

Since $\mathcal{F}_{H,Q}$ is an isometry it remains to prove injectivity of $\mathcal{F}^*_{H,Q}$.

It follows from (3.10) and Proposition 2.1

(3.15) $$\mathcal{F}_{H,Q}(L_H - \zeta)^{-1} = (\Lambda - \zeta)^{-1}\mathcal{F}_{H,Q}$$

for $\zeta = \nu + i\varepsilon \in \mathbb{C}\backslash\mathbb{R}$. Here $\Lambda : L^2_\sigma(\mathbb{R}) \to L^2_\sigma(\mathbb{R})$, $(\Lambda g)(\lambda) := \lambda g(\lambda)$ denotes the operator of multiplication by λ. Therefore $(L_H - \bar{\zeta})^{-1}\mathcal{F}^*_{H,Q} = \mathcal{F}^*_{H,Q}(\Lambda - \bar{\zeta})^{-1}$ and hence we have the implication:

(3.16) $\Phi \in \ker \mathcal{F}^*_{H,Q} \Longrightarrow (\Lambda - \zeta)^{-1}\Phi \in \ker \mathcal{F}^*_{H,Q}$ for all $\zeta \in \mathbb{C} \setminus \mathbb{R}$.

We put

(3.17) $$\widetilde{Y}(x,\lambda) := \int_0^x Y(t,\lambda)dt.$$

Note that the i-th row $\widetilde{Y}_i(x,\lambda)$ of $\widetilde{Y}$ satisfies

(3.18) $$\widetilde{Y}_i(x,\lambda)^* = \int_0^x Y(t,\lambda)^* e_i dt = (\mathcal{F}_{H,Q}(1_{[0,x]} \otimes e_i))(\lambda),$$

where e_i denotes the i-th unit vector in $\mathbb{C}^{2n}$.

Now let $\Phi \in \ker \mathcal{F}^*_{H,Q}$.

In particular $\widetilde{Y}_i(x,\cdot)^* \in L^2_\sigma(\mathbb{R})$ and thus in view of (3.12) and (3.16) we have for $x \ge 0$, $\varepsilon > 0$, $\nu \in \mathbb{R}$

(3.19) $$\begin{aligned} 0 &= \left(1_{[0,x]} \otimes e_i, \mathcal{F}^*_{H,Q}\frac{\varepsilon}{(\Lambda - \nu)^2 + \varepsilon^2}\Phi\right) \\ &= \left(\widetilde{Y}^*_i, \frac{\varepsilon}{(\Lambda - \nu)^2 + \varepsilon^2}\Phi\right), \end{aligned}$$

thus

(3.20) $$0 = \int_{\mathbb{R}} \widetilde{Y}(x,\lambda)\frac{\varepsilon}{(\lambda - \nu)^2 + \varepsilon^2} d\sigma(\lambda)\Phi(\lambda).$$

Since $\widetilde{Y}(x,\cdot)d\sigma(\cdot)\Phi(\cdot)$ is L^1 the dominated convergence theorem implies for $\alpha, \beta \in \mathbb{R}$

(3.21) $$\begin{aligned} 0 &= \lim_{\varepsilon\to 0}\int_\alpha^\beta \int_{\mathbb{R}} \widetilde{Y}(x,\lambda)\frac{\varepsilon}{(\lambda - \nu)^2 + \varepsilon^2} d\sigma(\lambda)\Phi(\lambda)d\nu \\ &= \int_{\mathbb{R}} \widetilde{Y}(x,\lambda) \lim_{\varepsilon\to 0}\int_\alpha^\beta \frac{\varepsilon}{(\lambda - \nu)^2 + \varepsilon^2} d\nu d\sigma(\lambda)\Phi(\lambda) \\ &= \pi \int_\alpha^\beta \widetilde{Y}(x,\lambda)d\sigma(\lambda)\Phi(\lambda). \end{aligned}$$

Differentiating by x and putting $x = 0$ yields for $\alpha, \beta \in \mathbb{R}$

$$0 = \int_\alpha^\beta Y(0,\lambda)d\sigma(\lambda)\Phi(\lambda).$$

Since $Y(0,\lambda) = \binom{I}{H}$ and H is invertible we have

$$\int_\alpha^\beta d\sigma(\lambda)\Phi(\lambda) = 0$$

for all $\alpha, \beta \in \mathbb{R}$. This implies $\Phi = 0$ in $L^2_\sigma(\mathbb{R})$. □

Remark 3.8

1. We note that another proof of the uniqueness of the spectral function in Proposition 3.5 can be given using Theorem 3.7. To prove the uniqueness statement we assume we had another increasing right continuous $n \times n$ matrix function ϱ, $\varrho(0) = 0$, such that $\mathcal{F}_{H,\varrho}$ is a unitary transformation from $L^2(\mathbb{R}_+, \mathbb{C}^{2n})$ onto $L^2_\varrho(\mathbb{R})$. Then in view of (3.5) we have for all $F, G \in L^2_{\mathrm{comp}}(\mathbb{R}, \mathbb{C}^n)$

   $$\int_{\mathbb{R}} F(\lambda)^* d\sigma(\lambda)G(\lambda) = \int_{\mathbb{R}} F(\lambda)^* d\varrho(\lambda)G(\lambda),$$

   and hence the two Radon vector measures $d\sigma$ and $d\varrho$ coincide. By the right-continuity and the normalization $\varrho(0) = \sigma(0) = 0$ this implies $\sigma = \varrho$.
2. For $n = 1$ Proposition 3.5 follows from [15, Theorem 4]. We also note that a generalization of Krein's theorem to the case $n > 1$ may be obtained by a slight modification of the proof of Proposition 3.5.
3. In [29, Chap. 3] the existence of the spectral function for a canonical system (2.7) is stated. For nonsingular Hamiltonians this fact follows from Krein's Theorem 3.2 in just the same way as Proposition 3.5.

   We note also that for a singular Hamiltonian similar results may be obtained by the corresponding generalization of Krein's Theorem 3.2 for linear relations.

Example 3.9 (2.15) shows that in the case $Q = 0$ we can choose for σ the function $\sigma_0(\lambda) := \frac{1}{2\pi} B_1^{-1}\lambda$.

4 Transformation Operator and Gelfand-Levitan Equation

1. We present a special case of [24, Theorem 7.1], (see also [25, Theorem 1.2]). In the sequel we assume B to be a diagonal matrix, which can be achieved by conjugating L with an appropriate unitary matrix.

Let

$$(4.1)\quad \begin{aligned} B &= \operatorname{diag}(\lambda_1 I_{n_1}, \ldots, \lambda_r I_{n_r}), \\ n_1 &= \min\{n_i \mid 1 \le i \le r\}, \quad n_1 + n_2 + \cdots + n_r = 2n. \end{aligned}$$

Furthermore, we put

$$(4.2)\quad \Omega := \{(x,t) \in \mathbb{R}^2 \mid 0 \le t \le x\}.$$

Theorem 4.1 *Let B be as in* (4.1) *and let $Q = (Q_{ij})_{i,j=1}^r : \mathbb{R}_+ \to \mathrm{M}(2n, \mathbb{C})$ continuous, where Q_{ij} denotes the block-matrix decomposition with respect to the orthogonal decomposition $\mathbb{C}^{2n} = \oplus_{i=1}^r \mathbb{C}^{n_i}$. Moreover, we assume that Q is off-diagonal, i.e.*

$$(4.3)\quad Q_{ii} = 0, \quad i = 1, \ldots, r.$$

Let Y be the solution of the equation (3.3) *satisfying*

$$(4.4)\quad \begin{aligned} &Y(0,\lambda) = A := \operatorname{col}(A_1, \ldots, A_r), \\ &A_j \in \mathrm{M}(n_j \times n_1, \mathbb{C}), \quad \operatorname{rank} A_j = n_1. \end{aligned}$$

Then there exists a continuous function $K : \Omega \to \mathrm{M}(2n, \mathbb{C})$ such that we have

$$(4.5)\quad Y(x,\lambda) = Y_0(x,\lambda) + \int_0^x K(x,t) Y_0(t,\lambda)\,dt,$$

where

$$Y_0(x,\lambda) = e^{i\lambda B^{-1} x} A$$

is the solution of the equation (3.2) *with $Q = 0$ and satisfying the same initial conditions* (4.4).

If $Q \in C^1(\mathbb{R}_+, \mathrm{M}(2n, \mathbb{C}))$ then $K \in C^1(\Omega, \mathrm{M}(2n, \mathbb{C}))$ and it satisfies

$$(4.6a)\quad B\partial_x K(x,t) + \partial_t K(x,t) B + iQ(x) K(x,t) = 0,$$

$$(4.6b)\quad BK(x,x) - K(x,x)B + iQ(x) = 0,$$

$$(4.6c)\quad K(x,0)BA = 0.$$

If $Q \in C(\mathbb{R}_+, \mathrm{M}(2n, \mathbb{C}))$ then K is the generalized continuous solution of (4.6).

Conversely, if K is a (generalized) solution of (4.6) then $Y(x,\lambda)$ defined by (4.5) *is the (generalized) solution of the initial value problem* (3.3).

Sketch of proof: i) Suppose that $K \in C^1(\mathbb{R}_+, \mathrm{M}(2n, \mathbb{C}))$ and that formula (4.5) holds. Substituting (4.5) into (3.3) and integrating by parts one obtains

$$(4.7)\quad \begin{aligned} &[BK(x,x) - K(x,x)B + iQ(x)] Y_0(x,\lambda) + K(x,0) B Y_0(0,\lambda) \\ &+ \int_0^x [B\partial_x K(x,t) + \partial_t K(x,t) B + iQ(x) K(x,t)] Y_0(t,\lambda)\,dt = 0. \end{aligned}$$

Since $Y_0(0,\lambda) = A$ does not depend on λ one concludes from (4.7) and the Riemann-Lebesgue Lemma that (4.7) is equivalent to (4.6). Thus in this case the representation (4.5) is equivalent to the solvability of the problem (4.6).

ii) Next we prove the existence of a (not unique) solution of the problem (4.6a)–(4.6b). Let $R(x,t)$ be one of them. Using the block-matrix representation $R(x,t) = (R_{ij}(x,t))_{i,j=1}^{r}$ we rewrite the problem (4.6a)–(4.6b) as

$$\begin{aligned}\lambda_i\partial_x R_{ij}(x,t) + \lambda_j\partial_t R_{ij}(x,t)\\ = -\sqrt{-1}\sum_{p=1}^{r} Q_{ip}(x)R_{pj}(x,t),\ 1\le i,j\le r,\end{aligned} \tag{4.8}$$

$$R_{ij}(x,x) = -\sqrt{-1}(\lambda_i-\lambda_j)^{-1}\,Q_{ij}(x),\quad 1\le i\ne j\le r. \tag{4.9}$$

It is clear that the system (4.8) is hyperbolic with real characteristics $l_{ij} : x = k_{ij}t + c(k_{ij} = \lambda_j\lambda_i^{-1})$. Thus, in $\Omega = \{0 \le t \le x < \infty\}$ we have the incomplete characteristic Cauchy problem (4.8), (4.9) with $(2n)^2 - n_1^2 - \cdots - n_r^2$ scalar conditions (4.9). Fixing $x_0 \in \mathbb{R}_+$ and setting

$$k_{\min} = \max\{k_{ij} \mid k_{ij} \in (0,1),\ 1 \le i,j \le r\},\quad k_{\max} = k_{\min}^{-1},$$

we consider the triangle Δ_{ABC} confined by the lines $AB : x = t,\ AC : x - x_0 = k_{\min}t,\ BC : x - x_0 = k_{\max}t$. We preserve the notation $Q(x)$ for a continuous extension to $\mathbb{R}$ of the function $Q(x)$ with the same norm. Furthermore, we denote by a and b the abscissas of the points A and B respectively. Now we impose the following $n_1^2 + \cdots + n_r^2$ conditions on the characteristic line AC:

$$R_{jj}(x,(x-1)k_{\min}) = 0,\quad \text{for}\quad x \in [a,x_0],\quad j \in 1,\ldots,r. \tag{4.10}$$

Thus, we arrive at the Goursat problem (4.8)–(4.10) for the hyperbolic system (4.8) in the triangle Δ_{ABC}. Integrating the system (4.8) along the characteristics and using (4.9), (4.10) one deduces the system of integral equations

$$\begin{aligned}\lambda_i R_{ij}(x,t) &= \frac{\lambda_i Q_{ij}(\xi_{ij}(x,t))}{\lambda_i-\lambda_j}\\ &\quad -\sqrt{-1}\int_{\xi_{ij}(x,t)}^{x}\sum_{p=1}^{r} Q_{ip}(\xi)R_{pj}(\xi,(\xi-x)k_{ij}+t)d\xi,\end{aligned} \tag{4.11}$$

where for brevity it is set $\lambda_i(\lambda_i-\lambda_j)^{-1}Q_{ij}(\xi_{ij}(x,t)) = 0$ for $i = j$ and

$$\xi_{ij}(x,t) = \begin{cases}(\lambda_j x - \lambda_i t)(\lambda_j-\lambda_i)^{-1}, & i \ne j,\\ a + (1-a)(x-t), & i = j.\end{cases}$$

For $Q \in C^1(\mathbb{R}, \mathrm{M}(n,\mathbb{C})$ the system (4.11) is equivalent to the Goursat problem (4.8)–(4.10). The solvability (and uniqueness) of the solution of (4.11) is proved by the method of successive approximations.

For $Q \in C(\mathbb{R}, \mathrm{M}(n, \mathbb{C})) \setminus C^1(\mathbb{R}, \mathrm{M}(n, \mathbb{C}))$ we understand the solution of (4.8)–(4.10) as a solution of (4.11).

iii) To finish the proof, starting with the solution $R(x, t)$ of the Goursat problem (4.8)–(4.10) we introduce a convolution operator

$$\Phi : f \longrightarrow \int_0^x \Phi(x-t) f(t) dt$$

with $\Phi(x) = \mathrm{diag}(\Phi_1(x), \ldots, \Phi_r(x))$ being a block-diagonal $2n \times 2n$ matrix function, consisting of $n_j \times n_j$ blocks Φ_j and define the operator K by the equality $I + K = (I + R)(I + \Phi)$. It is clear that K is a Volterra operator with the kernel

$$K(x,t) = R(x,t) + \Phi(x-t) + \int_0^x R(x,s)\Phi(s-t) ds. \tag{4.12}$$

Since the operator $I + R$ intertwines the restrictions L_0 and $-iB \otimes D_0$ of the operators L and $-iB \otimes D$ onto $\{f \in H^1([0,1], \mathbb{C}^{2n}) \mid f(0) = 0\}$, that is $L_0(I + R) = (I + R)(-iB \otimes D_0)$, so is $I + K$. This fact amounts to saying that $K(x, t)$ satisfies the problem (4.6a)–(4.6b). To satisfy the condition (4.6c) it suffices (in view of (4.12)) to choose $\Phi(x)$ as the solution of the equation

$$\Phi(x)BA + \int_0^x R(x,s)\Phi(s)BA ds = -R(x,0)BA. \tag{4.13}$$

Since $\mathrm{rank} A_j = n_1, 2 \le j \le r$, the Volterra equation (4.13) is of the second kind and therefore has the unique solution $\Phi \in C([0, \infty), \mathrm{M}(2n, \mathbb{C}))$. Thus $K(x, t)$ is the required solution of (4.6a)–(4.6c). □

Corollary 4.2 *Under the assumptions of Theorem* 4.1 *let* $B = (\lambda_1 I_n, \lambda_2 I_n)$ *(that is* $r = 2$.)

Then there exists a continuous function $K : \Omega \to \mathrm{M}(2n, \mathbb{C})$ *such that we have*

$$Y(x,\lambda) = e_0(x,\lambda) + \int_0^x K(x,t) e_0(t,\lambda) dt, \tag{4.14}$$

where $e_0(x, \lambda)$ *was defined in* (2.11).

If $Q \in C^1(\mathbb{R}_+, \mathrm{M}(2n, \mathbb{C}))$ *then* $K \in C^1(\Omega, \mathrm{M}(2n, \mathbb{C}))$ *and it satisfies*

$$B\partial_x K(x,t) + \partial_t K(x,t) B + iQ(x)K(x,t) = 0, \tag{4.15a}$$

$$BK(x,x) - K(x,x)B + iQ(x) = 0, \tag{4.15b}$$

$$K(x,0)B\begin{pmatrix} I \\ H \end{pmatrix} = 0. \tag{4.15c}$$

If $Q \in C(\mathbb{R}_+, \mathrm{M}(2n, \mathbb{C}))$ *then* K *is the generalized continuous solution of (4.6).*

Conversely, if K *is a (generalized) solution of (4.15) then* $Y(x, \lambda)$ *defined by* (4.14) *is the (generalized) solution of the initial value problem* (3.3). *This last statement holds even for general* B *of the form* (1.1).

2. We continue with some general remarks about Volterra operators:
For any continuous matrix function $K : \Omega \to \mathrm{M}(2n, \mathbb{C})$ we obtain a Volterra operator

$$Kf(x) := \int_0^x K(x,t)f(t)dt \tag{4.16}$$

acting on $C(\mathbb{R}_+, \mathbb{C}^{2n})$ or $L^2([0,a], \mathbb{C}^{2n})$ for any $a > 0$. By slight abuse of notation we will use the same symbol for the operator and its kernel. The set of operators $I + K$ with K being a Volterra operator forms a group. The operator

$$R := (I + K)^{-1} - I \tag{4.17}$$

is again a Volterra operator with continuous kernel $R(x,t), t \le x$. From the equation

$$I = (I + R)(I + K) = (I + K)(I + R) \tag{4.18}$$

we deduce

$$RK = KR = -R - K. \tag{4.19}$$

Put

$$F := R + R^* + RR^*. \tag{4.20}$$

The kernel of F obviously is

$$F(x,t) = \begin{cases} R(x,t) + \displaystyle\int_0^t R(x,s)R(t,s)^* ds, & x > t, \\ R(t,x)^* + \displaystyle\int_0^x R(x,s)R(t,s)^* ds, & x < t. \end{cases} \tag{4.21}$$

Furthermore, using (4.19) we conclude

$$F + K + KF = R + K + R^* + RR^* + KR + KR^* + KRR^* = R^* \tag{4.22}$$

thus we have the "Gelfand-Levitan equation"

$$F + K - R^* + KF = 0. \tag{4.23}$$

Proposition 4.3 *Let $K : \Omega \to \mathrm{M}(2n, \mathbb{C})$ be continuous and let $R : \Omega \to \mathrm{M}(2n, \mathbb{C})$ be the continuous kernel of the Volterra operator $(I + K)^{-1} - I$. Then the function $F : \mathbb{R}_+^2 \to \mathrm{M}(2n, \mathbb{C})$ defined by* (4.21) *satisfies the "Gelfand-Levitan equation"*

$$F(x,t) + K(x,t) + \int_0^x K(x,s)F(s,t)ds = 0, \quad x > t, \tag{4.24}$$

$$F(x,t) - R(t,x)^* + \int_0^x K(x,s)F(s,t)ds = 0, \quad x < t. \tag{4.25}$$

Conversely, if $F_1 : \Omega \to \mathrm{M}(2n, \mathbb{C})$ is continuous and satisfies (4.24) *then $F_1 = F|\Omega$.*

Proof: It only remains to prove the assertion about F_1. The difference $F(x,t) - F_1(x,t)$ satisfies the equation

$$F(x,t) - F_1(x,t) + \int_0^x K(x,s)[F(s,t) - F_1(s,t)]ds = 0, \quad 0 \le t \le x.$$

For each fixed $t \in [0,x]$ this is a homogeneous Volterra equation of the second kind and consequently has only the trivial solution $F(x,t) - F_1(x,t) = 0$. □

We turn back to the system (3.2).

Definition 4.4 We say that the system (3.2) (resp. the operator L) belongs to the class (T_B) if for this system there exists a transformation operator.

This means that the solution $Y(x,\lambda)$ of the initial value problem (3.3) admits a representation (4.14) with a continuous function $K : \Omega \to M(2n;\mathbb{C})$. Corollary 4.2 says that the system is of class (T_B) if $B = (\lambda_1 I_n, \lambda_2 I_n)$.

It follows easily from Proposition 4.6 below that for an operator L_H of class (T_B) the transformation operator $I + K$ is unique, i.e. the representation (3.3) for $Y(x,\lambda)$ is unique.

If the system is of class (T_B) then we denote by K the unique Volterra operator with continuous kernel satisfying (4.15). As before R denotes the Volterra operator defined by $R := (I+K)^{-1} - I$.

In particular we have in view of (4.15)

$$(4.26) \qquad e_0(x,\lambda) = ((I+R)Y(\cdot,\lambda))(x) = Y(x,\lambda) + \int_0^x R(x,t)Y(t,\lambda)dt.$$

Lemma 4.5 *Let L be of class* (T_B).

1. *Let σ be the spectral function of the boundary value problem* (3.2) *and let $g \in L^2_{\mathrm{comp}}(\mathbb{R}_+, \mathbb{C}^{2n})$. Put*

$$(4.27) \qquad G_0(\lambda) := (\mathcal{F}_{H,0}g)(\lambda) = \int_0^\infty e_0(x,\lambda)^* g(x)dx.$$

Then $G_0 \in L^2_\sigma(\mathbb{R})$ and if

$$(4.28) \qquad \int_{\mathbb{R}} G_0(\lambda)^* d\sigma(\lambda) G_0(\lambda) = 0$$

then $g = 0$.

2. *We have $\mathcal{F}_{H,0}(L^2_{\mathrm{comp}}(\mathbb{R}_+, \mathbb{C}^{2n})) = \mathcal{F}_{H,Q}(L^2_{\mathrm{comp}}(\mathbb{R}_+, \mathbb{C}^{2n}))$.*

Proof: In view of (4.26) we have

$$(4.29)\qquad \begin{aligned} G_0(\lambda) &= \int_0^\infty \left[Y(x,\lambda)^* + \int_0^x Y(t,\lambda)^* R(x,t)^* dt \right] g(x)dx \\ &= \int_0^\infty Y(x,\lambda)^* \left[g(x) + \int_x^\infty R(t,x)^* g(t)dt \right] dx, \end{aligned}$$

hence $G_0(\lambda)$ is also the $\mathcal{F}_{H,Q}$-transform of the function

$$(4.30)\qquad \widetilde{g}(x) := ((I+R^*)g)(x) = g(x) + \int_x^\infty R(t,x)^* g(t)dt.$$

Since $g \in L^2_{\rm comp}(\mathbb{R}_+, \mathbb{C}^{2n})$ we also have $\widetilde{g} \in L^2_{\rm comp}(\mathbb{R}_+, \mathbb{C}^{2n})$. This shows the inclusion $\mathcal{F}_{H,0}(L^2_{\rm comp}(\mathbb{R}_+, \mathbb{C}^{2n})) \subset \mathcal{F}_{H,Q}(L^2_{\rm comp}(\mathbb{R}_+, \mathbb{C}^{2n}))$. The converse inclusion is proved analogously using (4.5) instead of (4.26).

In view of the Parseval equality (Proposition 3.5) we find

$$\int_{\mathbb{R}} G_0(\lambda)^* d\sigma(\lambda) G_0(\lambda) = \int_0^\infty \widetilde{g}(x)^* \widetilde{g}(x) dx,$$

which by assumption (4.28) implies $\widetilde{g} = 0$. Since g has compact support (4.30) is a Volterra equation and thus $g = 0$. □

Proposition 4.6 *Let σ be the spectral function of the boundary value problem* (3.2) *and let $\sigma_0 = \frac{1}{2\pi} B_1^{-1}\lambda$ be the corresponding spectral function for $Q = 0$. We abbreviate $\Sigma := \sigma - \sigma_0$.*

1. *Let L be of class (T_B) and let $I + R$ be the transformation operator of the form* (4.26)*. Furthermore, let F be the $2n \times 2n$ matrix function defined by* (4.21)*, i.e.*

$$(4.31)\qquad F(x,t) := \begin{cases} R(x,t) + \int_0^t R(x,s)R(t,s)^* ds, & x > t > 0, \\ R(t,x)^* + \int_0^x R(x,s)R(t,s)^* ds, & 0 < x < t. \end{cases}$$

   Then we have for all $f, g \in L^2_{\rm comp}(\mathbb{R}_+, \mathbb{C}^{2n})$

$$(4.32)\qquad \int_{\mathbb{R}} F_0(\lambda)^* d\Sigma(\lambda) G_0(\lambda) = \int_0^\infty \int_0^\infty f(x)^* F(x,t) g(t) dx dt,$$

   where F_0, G_0 denote the $\mathcal{F}_{H,0}$-transforms of f, g.

2. *Again assuming L to be of class (T_B) we put*

$$(4.33)\qquad \widetilde{e}_0(x,\lambda) := \int_0^x e_0(t,\lambda)dt.$$

Then the function

$$\widetilde{F}(x,t) := \int_{\mathbb{R}} \widetilde{e}_0(x,\lambda)d\Sigma(\lambda)\widetilde{e}_0(t,\lambda)^* \tag{4.34}$$

exists and has a continuous mixed second derivative which coincides with $F(x,t)$, *i.e.* $\frac{\partial^2}{\partial x \partial t}\widetilde{F}(x,t) = F(x,t)$.

3. *Conversely, given any increasing* $n \times n$ *matrix function* σ *put* $\Sigma := \sigma - \sigma_0$. *If the integral* (4.34) *exists and has a continuous mixed second derivative* $F_1(x,t) := \frac{\partial^2}{\partial x \partial t}\widetilde{F}(x,t)$ *then* (4.32) *holds for all* $f, g \in L^2_{\text{comp}}(\mathbb{R}_+, \mathbb{C}^{2n})$ *with* F_1 *instead of* F.

Remark 4.7 We emphasize that 3. holds for arbitrary L of the form (3.2) not necessarily being of class (T_B).

We note that the identity (4.32) characterizes the spectral function of the problem (3.2). More precisely, if ϱ is an increasing (normalized) $n \times n$ matrix function such that (4.32) holds with $\Sigma_\varrho := \varrho - \sigma_0$ then $\varrho = \sigma$.

Indeed from (4.32) we infer

$$\int_{\mathbb{R}} F_0(\lambda)^* d\sigma(\lambda) G_0(\lambda) = \int_{\mathbb{R}} F_0(\lambda)^* d\varrho(\lambda) G_0(\lambda), \tag{4.35}$$
$$F_0 := \mathcal{F}_{H,0}f, \quad G_0 := \mathcal{F}_{H,0}g,$$

for all $f, g \in L^2_{\text{comp}}(\mathbb{R}_+, \mathbb{C}^{2n})$. By Theorem 3.7 and Lemma 4.5, 2. this implies that (4.35) holds for all $F_0, G_0 \in L^2_\varrho(\mathbb{R})$, in particular it holds for all $F_0, G_0 \in C(\mathbb{R}, \mathbb{C}^n)$ with compact support. Thus the vector measures $d\sigma, d\varrho$ and hence the right-continuous functions ϱ, σ coincide.

Proof:

1. In view of (4.30) F_0 is the $\mathcal{F}_{H,\varrho}$-transform of

$$\widetilde{f}(x) = f(x) + \int_x^\infty R(t,x)^* f(t)dt, \tag{4.30'}$$

thus the Parseval equality (3.5) gives

$$\begin{aligned}
&\int_{\mathbb{R}} F_0(\lambda)^* d\Sigma(\lambda) G_0(\lambda) \\
&\quad = \int_{\mathbb{R}} F_0(\lambda)^* d\sigma(\lambda) G_0(\lambda) - (f,g) = (\widetilde{f}, \widetilde{g}) - (f,g) \\
&\quad = \int_0^\infty \left(f(x) + \int_x^\infty R(t,x)^* f(t)dt \right)^* \left(g(x) + \int_x^\infty R(t,x)^* g(t)dt \right) \\
&\qquad dx - (f,g) = \int_0^\infty \int_0^\infty f(x)^* F(x,t) g(t) dx dt
\end{aligned}$$

by a straightforward calculation.

2. For $x, t \geq 0$ and $f_0, g_0 \in \mathbb{C}^{2n}$ we apply 1. with $f(u) := 1_{[0,x]}(u) f_0$, $g(v) := 1_{[0,t]}(v) g_0$ and find

$$f_0^* \int_R \widetilde{e}_0(x,\lambda) d\Sigma(\lambda) \widetilde{e}_0(t,\lambda)^* g_0 = f_0^* \int_0^x \int_0^y F(u,v) du dv g_0, \tag{4.36}$$

which implies the first assertion.

3. To prove the converse statement we note that now we have (4.36) with $F_1(x,t) = \frac{\partial^2}{\partial x \partial t} \widetilde{F}(x,t)$. This identity implies (4.32) with F_1 instead of F for step functions

$$f = \sum_{j=1}^{n} f_j 1_{[a_j,b_j]}, \quad g = \sum_{j=1}^{n} g_j 1_{[c_j,d_j]}, \quad f_j, g_j \in \mathbb{C}^{2n}. \tag{4.37}$$

There is a slight subtlety since Σ is not necessarily increasing. However, we conclude from (4.32) and the Parseval equality that for all step functions f, g

$$(F_0, G_0)_{L^2_\sigma(\mathbb{R})} = (f,g)_{L^2(\mathbb{R}_+, \mathbb{C}^{2n})} + \int_0^\infty \int_0^\infty f(x)^* F(x,t) g(t) dx dt. \tag{4.38}$$

Since σ is increasing the assertion now follows from the denseness of the step functions in $L^2_{\text{comp}}(\mathbb{R}_+, \mathbb{C}^{2n})$. To complete the proof it remains to note that the equality $F(x,t) = F_1(x,t)$ is a consequence of (4.32) and (4.38).

□

Combining Propositions 4.3 and 4.6 one immediately obtains the following theorem.

Theorem 4.8 *Assume that the system* (3.2) *is of class* (T_B). *Let* σ *be its spectral measure function and* $\sigma_0(\lambda) = \frac{1}{2\pi} B_1^{-1} \lambda$. *Then with* F *defined by* (4.34) *we have the Gelfand-Levitan equation*

$$F(x,t) + K(x,t) + \int_0^x K(x,s) F(s,t) ds = 0, \quad t < x. \tag{4.39}$$

Remark 4.9 Note that by Proposition 4.6, 2. the function F is continuous also on the diagonal. In view of (4.21) the continuity of F at the diagonal implies $R(x,x) = R(x,x)^*$.

Proof: We present a second proof of the Gelfand-Levitan equation based on the formula (4.34) for F, which is similar to [10] and [23, Chap. 12].

For $f, g \in L^2_{\text{comp}}(\mathbb{R}_+, \mathbb{C}^{2n})$ we consider

$$\mathrm{I}(f,g) := \int_{\mathbb{R}} \int_0^\infty dx \int_0^\infty dt f(x)^* Y(x,\lambda) d\sigma(\lambda) e_0(t,\lambda)^* g(t). \tag{4.40}$$

Substituting (4.5) for Y we find using the Parseval equality and Lemma 4.6

$$\begin{aligned} \mathrm{I}(f,g) &= (f,g) + \int_0^\infty \int_0^\infty f(x)^* F(x,t) g(t) dx dt \\ &+ \underbrace{\int_{\mathbb{R}} \int_0^\infty dx \int_0^\infty dt f(x)^* \int_0^\infty K(x,s) e_0(s,\lambda) ds d\sigma(\lambda) e_0(t,\lambda)^* g(t)}_{\mathrm{II}(f,g)}, \end{aligned} \tag{4.41}$$

$$\mathrm{II}(f,g) = \int_{\mathbb{R}} \int_0^\infty dx \int_0^\infty ds \int_0^\infty f(x)^* K(x,s) dx e_0(s,\lambda) d\sigma(\lambda) e_0(t,\lambda)^* g(t).$$

Writing $d\sigma = d\Sigma + d\sigma_0$ and using Lemma 4.6 we find

$$\begin{aligned} \mathrm{II}(f,g) &= \int_0^\infty \int_0^\infty f(x)^* K(x,t) g(t) dx dt \\ &+ \int_0^\infty \int_0^\infty \int_0^\infty f(x)^* K(x,s) F(s,t) g(t) dx dt ds, \end{aligned} \tag{4.42}$$

hence

$$\begin{aligned} &\mathrm{I}(f,g) = (f,g) + \int_0^\infty \int_0^\infty \\ &f(x)^* \left[F(x,y) + K(x,y) + \int_0^x K(x,t) F(t,y) dt \right] g(y) dx dy. \end{aligned} \tag{4.43}$$

Now if $\operatorname{supp} f \subset [b,\infty)$, $\operatorname{supp} g \subset [0,a]$, $a < b$, then $(f,g) = 0$ and

$$\int_0^\infty e_0(x,\lambda)^* g(x) dx$$

is the $\mathcal{F}_{H,Q}$-transform of

$$g(x) + \int_x^\infty R(t,x)^* g(t) dt$$

which also has support in $[0,a]$, hence by the Parseval equality $\mathrm{I}(f,g) = 0$. This implies the assertion. □

5 The Inverse Problem

5.1 The Main Result

Proposition 5.1 *Let $B = \operatorname{diag}(B_1, -B_2)$ be an arbitrary nonsingular self-adjoint matrix of signature 0. Let $\sigma(\lambda)$ be a $n \times n$ matrix function satisfying:*

1. *If $g \in L^2_{\mathrm{comp}}(\mathbb{R}_+, \mathbb{C}^{2n})$ and if*

$$\int_{\mathbb{R}} G_0(\lambda)^* d\sigma(\lambda) G_0(\lambda) = 0,$$

where G_0 is the $\mathcal{F}_{H,0}$-transform of g, then $g = 0$.

2. *The function*

$$\widetilde{F}(x,t) := \int_{\mathbb{R}} \widetilde{e}_0(x,\lambda)d\Sigma(\lambda)\widetilde{e}_0(t,\lambda)^* \tag{5.1}$$

with $\Sigma = \sigma - \sigma_0$ *exists, and has a continuous mixed second derivative*

$$F(x,t) := \frac{\partial^2}{\partial x \partial t}\widetilde{F}(x,t). \tag{5.1'}$$

Then the Gelfand-Levitan equation (4.39) *has a unique continuous solution* $K : \Omega \to \mathrm{M}(2n, \mathbb{C})$.

Moreover, if $F(x,t)$ *is continuously differentiable, then so is* $K(x,t)$.

Proof: Since for fixed x equation (4.39) is a Fredholm equation it suffices to show that the dual equation

$$k(t) + \int_0^x k(s)F(t,s)^* ds = 0, \tag{5.2}$$

where $k : [0,x] \to M(2n, \mathbb{C})$ is square integrable, has only the zero solution. Looking at the individual columns in (5.2) it suffices to show that

$$g(t)^* + \int_0^x g(s)^* F(t,s)^* ds = 0, \quad g \in L^2([0,x], \mathbb{C}^{2n}) \tag{5.3}$$

implies $g = 0$. Extending g by 0 to $\mathbb{R}_+$ we may consider g as an element of $L^2_{\mathrm{comp}}(\mathbb{R}_+, \mathbb{C}^{2n})$ and (5.3) implies in view of 2. and Proposition 4.6, 3.

$$\begin{aligned} 0 &= \|g\|^2 + \int_0^\infty \int_0^\infty g(s)^* F(s,t) g(t) ds dt \\ &= \|g\|^2 + \int_{\mathbb{R}} G_0(\lambda)^* d\Sigma(\lambda) G_0(\lambda) = \int_{\mathbb{R}} G_0(\lambda)^* d\sigma(\lambda) G_0(\lambda) \end{aligned}$$

and thus $g = 0$ by 1.

The proof of C^1-smoothness of $K(x,t)$ is similar to that used in [23] and [10] and is omitted. □

Next we prove the main result of this paper:

Theorem 5.2 *Let* $B = \mathrm{diag}(B_1, -B_2)$ *be an arbitrary nonsingular self-adjoint matrix of signature* 0 *as in* (1.1). *Let* $\sigma(\lambda)$ *be an increasing (right-continuous,* $\sigma(0) = 0$*)* $n \times n$ *matrix function satisfying the conditions* 1. *and* 2. *of Proposition* 5.1.

Then there exists a unique continuous $2n \times 2n$ *matrix potential* Q *satisfying* (4.3) *such that the corresponding system* (3.2) *is of class* (T_B) *and such that* σ *is its spectral measure function.* $Q(x)$ *has* p *continuous derivative iff* $D_x^p D_t^p F(x,t)$ *is continuous.*

Conversely, if σ is the spectral measure function of the boundary value problem (3.2) *of class* (T_B) *then the conditions* 1. *and* 2. *of Proposition* 5.1 *hold.*

Proof: The necessity was proved in Lemma 4.5 and Proposition 4.6.

To prove the sufficiency we assume that the conditions 1. and 2. of Proposition 5.1 hold:

i) Starting with $\sigma(\lambda)$ we define $\widetilde{F}$, F by (5.1) and (5.1′). Then we consider the Gelfand-Levitan equation (4.39)

$$\text{(5.4)} \quad \Phi(x,t) := F(x,t) + K(x,t) + \int_0^x K(x,s)F(s,t)ds = 0, \quad x > t.$$

By Proposition 5.1 this equation has a unique continuous solution $K : \Omega \to \mathrm{M}(2n, \mathbb{C})$.

Then F also equals the right hand side of (4.21): namely, starting with K we consider the operator R of the form (4.17) and introduce F_1 by (4.21). According to Proposition 4.3 F_1 and K are connected by equation (4.24). Thus F defined by (5.1) and F_1 defined by (4.21) satisfy the equation (5.4) and therefore we infer from Proposition 4.3 that $F = F_1$.

We collect further properties of F: in view of (4.21) we have

$$\text{(5.5a)} \qquad F(x,t) = F(t,x)^*.$$

By continuity, the equation (4.21) also holds for $x = t$ and consequently $R(x,x)$ is self-adjoint. Therefore, so is $K(x,x) = -R(x,x)$. Furthermore,

$$\text{(5.5b)} \qquad \partial_t F(x,t)B = -B\partial_x F(x,t),$$

where this equality holds in the distributional sense if F is only continuous. To see this let $f, g \in C_0^\infty((0,\infty), \mathbb{C}^{2n})$. In view of (4.32) and (3.10) applied with $Q = 0$ we calculate

$$\begin{aligned}
&\int_0^\infty \int_0^\infty f(x)^* \partial_t F(x,t) B g(t) dx dt \\
&= -i \int_0^\infty \int_0^\infty f(x)^* F(x,t) \frac{1}{i} B \partial_t g(t) dx dt \\
&= -i \int_{\mathbb{R}} (\mathcal{F}_{H,0} f)(\lambda)^* d\Sigma(\lambda) \lambda (\mathcal{F}_{H,0} g)(\lambda) \\
&= -i \int_{\mathbb{R}} \left(\mathcal{F}_{H,0} \frac{1}{i} B f' \right)(\lambda)^* d\Sigma(\lambda) (\mathcal{F}_{H,0} g)(\lambda) \\
&= - \int_0^\infty \int_0^\infty f(x)^* B \partial_x F(x,t) g(t) dx dt.
\end{aligned}$$

Moreover, it follows from (5.1) and (2.11) that with some matrix function $T(t)$ we have

$$\text{(5.5c)} \qquad F(0,t) = \begin{pmatrix} I \\ H \end{pmatrix} T(t).$$

We now define (cf. (4.5))

$$Y(x,\lambda)=e_0(x,\lambda)+\int_0^x K(x,t)e_0(t,\lambda)dt \tag{5.6}$$

and we will show that the properties (5.5a–c) imply that $Y(x,\lambda)$ satisfies the initial value problem

$$B\frac{1}{i}\frac{dY(x,\lambda)}{dx}+Q(x)Y(x,\lambda)=\lambda Y(x,\lambda),\quad Y(0,\lambda)=e_0(0,\lambda)=\begin{pmatrix} I\\ H\end{pmatrix}, \tag{5.7}$$

where

$$Q(x):=iBK(x,x)-iK(x,x)B. \tag{5.8}$$

Note that since $K(x,x)$ is self-adjoint $Q(x)$ is self-adjoint, too. Moreover, from (5.8) we also conclude that $Q(x)$ is off-diagonal, i.e. $Q_{ii}=0$.

It follows from (5.5c) that

$$F(x,0)BF(0,t)=T(x)^*[B_1-H^*B_2H]T(t)=0. \tag{5.9}$$

Plugging (5.9) into the Gelfand-Levitan equation (5.4) gives

$$K(x,0)B\begin{pmatrix} I\\ H\end{pmatrix}=0\quad\text{for}\quad x\in[0,\infty). \tag{5.10}$$

ii) For the moment we assume in addition that F is continuously differentiable. Then by Proposition 5.1 K also is continuously differentiable. Differentiating (5.4) we obtain

$$\begin{aligned} B\partial_x\Phi(x,t) &= B\partial_xF(x,t)+B\partial_xK(x,t)+BK(x,x)F(x,t)\\ &\quad+\int_0^x B\partial_xK(x,s)F(s,t)ds=0, \end{aligned} \tag{5.11}$$

$$\begin{aligned} \partial_t\Phi(x,t)B &= \partial_tF(x,t)B+\partial_tK(x,t)B\\ &\quad+\int_0^x K(x,s)\partial_tF(s,t)Bds=0. \end{aligned} \tag{5.12}$$

Integrating by parts and using (5.5b) and (5.10) we obtain

$$\begin{aligned} \int_0^x K(x,s)\partial_tF(s,t)Bds &= -\int_0^x K(x,s)B\partial_sF(s,t)ds\\ &=\int_0^x \partial_sK(x,s)BF(s,t)ds-K(x,x)BF(x,t). \end{aligned} \tag{5.13}$$

Adding up (5.11) and (5.12) and using (5.13) and the Gelfand-Levitan equation (5.4) we obtain

$$\begin{aligned} &B\partial_xK(x,t)+\partial_tK(x,t)B+iQ(x)K(x,t)\\ &\quad+\int_0^x [B\partial_xK(x,s)+\partial_sK(x,s)B+iQ(x)K(x,s)]F(s,t)ds=0. \end{aligned}$$

Since the homogeneous integral equation corresponding to the Gelfand-Levitan equation (5.4) has only the trivial solution (see the proof of Proposition 5.1) we infer from (5.5) that

$$B\partial_x K(x,t) + \partial_t K(x,t)B + iQ(x)K(x,t) = 0. \tag{5.14}$$

Since K satisfies the relations (5.10), (5.8) and (5.14) it follows from Theorem 4.1 that $Y(x,\lambda)$ (cf. (5.6)) satisfies the initial value problem (5.7).

iii) We now assume that F is just continuous. Assume for the moment that for $\delta > 0$ we have a continuously differentiable matrix function $F^\delta : \mathbb{R}_+^2 \to \mathrm{M}(2n,\mathbb{C})$ with the properties:

(5.15a) F^δ converges to F as $\delta \to 0$ uniformly on compact subsets of $\mathbb{R}_+^2$.

(5.15b) F^δ satisfies (5.5a–c).

We fix $x_0 > 0$. For $0 < x \le x_0$ let T_F be the integral operator in $C([0,x],\mathbb{C}^{2n})$ defined by $(T_F f)(t) := \int_0^x f(s)F(s,t)ds$. The proof of Proposition 5.1 shows that $-1 \notin \operatorname{spec} T_F$. Thus for $\delta \le \delta_0(x_0)$ we have $-1 \notin \operatorname{spec} T_{F^\delta}$ and the Gelfand-Levitan equations

$$((I + T_{F^\delta})K_\delta(x,.))(t) = K_\delta(x,t) + \int_0^x K_\delta(x,s)F^\delta(s,t)ds = -F^\delta(x,t) \tag{5.16}$$

have (for each fixed $x \in (0,x_0]$) unique solutions $K_\delta(x,t)$, $(x,t) \in [0,x_0]^2$, which converge to K as $\delta \to 0$ uniformly on $[0,x_0]^2$. Since F^δ is C^1 it can be shown (cf. the proof of Proposition 5.1) that K_δ is C^1, too.

Moreover, K_δ satisfies (5.10) for $0 < x \le x_0$ which follows from (5.15b) and (5.16). Now part ii) of this proof shows that K_δ also satisfies (5.14) with $Q_\delta(x) := iBK_\delta(x,x) - iK_\delta(x,x)B$, $x \in [0,x_0]$. Hence, K_δ satisfies (4.6a–c) (on $[0,x_0]^2$) and therefore,

$$Y_\delta(x,\lambda) := e_0(x,\lambda) + \int_0^x K_\delta(x,t)e_0(t,\lambda)dt, \quad 0 \le x \le x_0,$$

satisfies the initial value problem (5.7) with Q_δ instead of Q.

Since $F^\delta(x,t)^* = F^\delta(t,x)$ one concludes as in part i) of this proof that $Q_\delta(x)^* = Q_\delta(x)$.

Since K_δ converges to K as $\delta \to 0$ uniformly on $[0,x_0]^2$, Q_δ converges to Q uniformly on $[0,x_0]$. Thus $Y(x,\lambda)$ satisfies the initial value problem (5.7) on $[0,x_0]$. Since x_0 was arbitrary $Y(x,\lambda)$ satisfies (5.7) on $\mathbb{R}_+$.

It remains to prove the existence of the sequence F^δ:

Let $F(x,t) := (F_{ij}(x,t))_{i,j=1}^r$ be the block-matrix representation with respect to the orthogonal decomposition $\mathbb{C}^{2n} = \oplus_{i=1}^r \mathbb{C}^{n_i}$.

It follows from (5.1) and (5.1′) that

$$F_{ij}(x,t) = f_{ij}(\mu_i x - \mu_j t), \qquad f_{ij}(\xi) = H_i g(\xi) H_j^*, \tag{5.17}$$

with $\mu_i = \lambda_i^{-1}$, $1 \le i \le r$, and $H_1 := I_{n_1} = I_n$. Here the map $g : \mathbb{R} \to M(n \times n, \mathbb{C})$ is continuous and satisfies $g(\xi)^* = g(-\xi)$. Therefore the maps $f_{ij} : \mathbb{R} \to M(n_i \times n_j, \mathbb{C})$ are continuous and satisfy $f_{ij}(\xi)^* = f_{ji}(-\xi)$. We note that if the measure $\Sigma(\lambda)$ is finite, that is $\int_{\mathbb{R}} |d\Sigma(\lambda)| \in M(n, \mathbb{C})$, then $g(\xi) = \int_{\mathbb{R}} e^{i\lambda\xi} d\Sigma(\lambda)$.

We put

$$
\begin{aligned}
g^\delta(\xi) &:= \frac{1}{2\delta} \int_{\xi-\delta}^{\xi+\delta} g(s)ds, \\
f_{ij}^\delta(\xi) &:= H_i g^\delta(\xi) H_j^*, \quad F_{ij}^\delta(x,t) := f_{ij}^\delta(\mu_i x - \mu_j t),
\end{aligned} \tag{5.18}
$$

and $F^\delta(x,t) := (F_{ij}^\delta(x,t))_{i,j=1}^r$.

Obviously, F^δ is continuously differentiable and satisfies (5.15a). It is clear from (5.17) that

$$
\begin{aligned}
g^\delta(\xi)^* &= \frac{1}{2\delta} \int_{\xi-\delta}^{\xi+\delta} g^\delta(s)^* ds = \frac{1}{2\delta} \int_{\xi-\delta}^{\xi+\delta} g^\delta(-s) ds \\
&= \frac{1}{2\delta} \int_{-\xi-\delta}^{-\xi+\delta} g^\delta(s) ds = g^\delta(-\xi),
\end{aligned} \tag{5.19}
$$

and thus $f_{ij}^\delta(\xi)^* = f_{ji}^\delta(-\xi)$.

In view of (5.18) and (5.19) F^δ satisfies (5.5a, b). To prove the property (5.5c) for F^δ we note that in view of (5.17) and (5.18) $F_{ij}^\delta(0,t) = f_{ij}^\delta(-\mu_j t) = H_i g^\delta(-\mu_j t) H_j^*$ and consequently

$$
F^\delta(0,t) = (F_{ij}^\delta(0,t))_{i,j=1}^r = (H_i g^\delta(-\mu_j t) H_j^*)_{i,j=1}^r =: \begin{pmatrix} I \\ H \end{pmatrix} T^\delta(t), \tag{5.20}
$$

where $T^\delta(t) = (g^\delta(-\mu_1 t) H_1^*, g^\delta(-\mu_2 t) H_2^*, \ldots, g^\delta(-\mu_r t) H_r^*)$.

This proves that F^δ satisfies (5.5c). Summing up, we have proved that F^δ satisfies (5.15a, b).

iv) Starting with an increasing $n \times n$ matrix function $\sigma(\lambda)$ satisfying the conditions 1. and 2. of Proposition 5.1 we have constructed the boundary value problem (3.2) resp. (5.7). To complete the proof it remains to show that $\sigma(\lambda)$ is, in fact, the spectral function for the problem (5.7).

Let $\varrho(\lambda)$ be the spectral function of the problem (5.7). Starting with $\Sigma_\varrho := \varrho - \sigma_0$ we define F_ϱ by (5.1′). Then by Theorem 4.8 K satisfies the Gelfand-Levitan equation (4.39) with F_ϱ. On the other hand, in view of (5.4) K satisfies the Gelfand-Levitan equation with F instead of F_ϱ. From Proposition 4.3 we infer $F = F_\varrho$. By Remark 4.7 this implies $\varrho = \sigma$. □

Remark 5.3

1. The case $n = 1$ and $B_1 = B_2 = 1$, i.e. the case of a 2×2 Dirac system, is due to M. Gasymov and B. Levitan [9], [23, Chap. 12]. We note, however, that the proof in [23, Chap. 12] is incomplete, since the self-adjointness of Q is not proved.
2. We also note that following Krein's method [17] L. Sakhnovich [29, Chap. 3, §3] has obtained some (implicit) sufficient conditions for a matrix measure to be the spectral function of a canonical system.

5.2 Some Complements to the Main Result

Next we will discuss several other criteria which imply conditions 1. or 2. of Proposition 5.1. For brevity, in the sequel we will address them just as "condition 1./2.". As in the proof of Theorem 3.7 we denote by $\Lambda : L^2_\sigma(\mathbb{R}) \to L^2_\sigma(\mathbb{R})$, $(\Lambda g)(\lambda) := \lambda g(\lambda)$ the operator of multiplication by λ. Furthermore, we denote by $\mu_T(\lambda_0)$ the multiplicity of the spectrum of a self-adjoint operator T at the point λ_0. We first note some simple facts:

Remark 5.4

1. If $\Sigma(\lambda)$ is increasing then condition 1. is trivially fulfilled.

   For let $g \in L^2_{\text{comp}}(\mathbb{R}_+, \mathbb{C}^{2n})$ with

$$\begin{aligned}0 &= \int_{\mathbb{R}} G_0(\lambda)^* d\sigma(\lambda) G_0(\lambda) \\ &= \int_{\mathbb{R}} G_0(\lambda) d\sigma_0(\lambda) G_0(\lambda) + \int_{\mathbb{R}} G_0(\lambda) d\Sigma(\lambda) G_0(\lambda),\end{aligned} \tag{5.21}$$

   where G_0 is the $\mathcal{F}_{H,0}$-transform of g. Since $\Sigma(\lambda)$ is assumed to be increasing both summands on the right hand side of (5.21) are nonnegative and hence 0. Then Proposition 3.5 implies $g = 0$.
2. Assume that the matrix measure $\Sigma(\lambda)$ is finite, i.e. $\int_{\mathbb{R}} |d\Sigma(\lambda)| \in \mathrm{M}(n, \mathbb{C})$. Then condition 2. is obviously fulfilled.

Recall that a subset $X \subset \mathbb{R}$ is said to have finite density (cf. [21]) if

$$\limsup_{R \to \infty} \frac{1}{R} |\{x \in X \mid |x| \leq R\}| < \infty. \tag{5.22}$$

Otherwise, X is said to have infinite density.

Proposition 5.5 *Let $B = (B_1, -B_2)$ be as in* (1.1). *For an increasing $n \times n$ matrix function σ the condition*

1′. *The set* $\operatorname{supp}_n(d\sigma) := \{\lambda \in \mathbb{R} \mid \mu_\Lambda(\lambda) = n\}$ *has infinite density implies condition* 1.

Proof: Let $\sigma(\lambda) = (\sigma_{ij}(\lambda))_{i,j=1}^n$ and $\varrho(\lambda) := \operatorname{tr}\sigma(\lambda) = \sigma_{11}(\lambda) + \cdots + \sigma_{nn}(\lambda)$. From the inequality

$$|\sigma_{ij}(\lambda) - \sigma_{ij}(\mu)| \le \sqrt{\sigma_{ii}(\lambda) - \sigma_{ii}(\mu)}\sqrt{\sigma_{jj}(\lambda) - \sigma_{jj}(\mu)}, \quad \mu < \lambda,$$

we infer that $d\sigma_{ij}(\lambda)$ is absolutely continuous with respect to $d\varrho(\lambda)$. Hence, by the Radon-Nikodym Theorem there exists a density matrix

$$\Phi(\lambda) = (\phi_{ij}(\lambda))_{i,j=1}^n, \tag{5.23}$$

such that

$$\sigma(\lambda) = \begin{cases} \int_{(0,\lambda]} \Phi(t) d\varrho(t), & \lambda \ge 0, \\ -\int_{(\lambda,0]} \Phi(t) d\varrho(t), & \lambda < 0. \end{cases}$$

Obviously, $\Phi(\lambda) \ge 0$ and thus we have

$$\begin{aligned} \operatorname{supp}_n(d\sigma) &= \operatorname{supp}(\det \Phi d\varrho) \\ &= \left\{\lambda \in \operatorname{supp}(d\varrho) \,\middle|\, \int_{\lambda-\varepsilon}^{\lambda+\varepsilon} \det \Phi(\lambda) d\varrho(\lambda) > 0 \text{ for all } \varepsilon > 0\right\}. \end{aligned} \tag{5.24}$$

To check condition 1. let $g \in L^2([0,b], \mathbb{C}^{2n})$ with

$$0 = \int_{-\infty}^{\infty} G_0(\lambda)^* d\sigma(\lambda) G_0(\lambda) = \int_{-\infty}^{\infty} G_0(\lambda)^* \Phi(\lambda) G_0(\lambda) d\varrho(\lambda).$$

Then we have $G_0(\lambda) = 0$ for all $\lambda \in \operatorname{supp}_n(d\sigma)$.

On the other hand $G_0(\lambda)$ is an entire (vector) function of strict order one and hence (cf. [21]) either $G_0 = 0$ or the set of its zeros is of finite density. But since $\operatorname{supp}_n(d\sigma)$ is assumed to have infinite density, we must have $G = 0$. □

Remark 5.6

1. Condition 1.′ of the previous proposition is satisfied if $\operatorname{supp}_n(d\sigma)$ has at least one finite limit point.
2. Note that if $n = 1$ then $\operatorname{supp}_n(d\sigma) = \operatorname{supp}(d\sigma)$ equals the support of the Radon measure $d\sigma$.

Corollary 5.7 *Let* B *be as before and* $\sigma_0(\lambda) = \frac{1}{2\pi} B_1^{-1}\lambda$. *If*

(i) *the measure* $d\sigma$ *is discrete,*

(ii) $\operatorname{supp}_n(d\sigma) = \operatorname{supp}(d\varrho)$,

(iii) *the measure* $d\Sigma = d\sigma - d\sigma_0$, *is finite, i.e.* $\int_{\mathbb{R}} |d\Sigma(\lambda)| \in M(n, \mathbb{C})$.

Then σ *satisfies condition* 1.′ *of Proposition* 5.5 *and hence condition* 1.

Proof: Since $d\Sigma(\lambda)$ is finite we have

$$\sigma(\lambda) = \frac{1}{2\pi} B_1^{-1}\lambda + C_\pm + o(1), \quad \lambda \longrightarrow \pm\infty, \tag{5.25}$$

where $C_\pm := \Sigma(\pm\infty) = C_\pm^*$. Hence

$$\varrho(\lambda) = c_0\lambda + c_1 + o(1), \quad \lambda \longrightarrow +\infty. \tag{5.26}$$

Since $d\sigma$ and hence $d\varrho$ is discrete, we infer from (5.26) that the set of discontinuities of ϱ has infinite density. Hence, by (ii) the set $\mathrm{supp}_n(d\sigma)$ has infinite density. □

Remark 5.8 Note that this proof only uses the asymptotic relation (5.25) which is slightly weaker than the finiteness of the measure $d\Sigma$ (since the $o(1)$ need not be of bounded variation).

Finally, we give a criterion for the condition 2.

Proposition 5.9 *The condition* 2. *is fulfilled if the limit*

$$\lim_{\Lambda\to\infty} \int_{|\lambda|\le\Lambda} e_0(x,\lambda) d\Sigma(\lambda) e_0(t,\lambda)^* \tag{5.27}$$

exists locally uniformly in x, t. Then indeed $F(x,t)$ is given by (5.27).

This is the case if the matrix function $\Sigma(\lambda)$ is integrable with respect to Lebesgue measure and satisfies $\lim_{\lambda\to\pm\infty} \Sigma(\lambda) = 0$.

Proof: If the limit (5.27) exists locally uniformly in x, t then we have

$$\begin{aligned}\widetilde{F}(x,t) &:= \int_{\mathbb{R}} \widetilde{e}_0(x,\lambda) d\Sigma(\lambda) \widetilde{e}_0(t,\lambda)^* \\ &= \int_0^x \int_0^t \lim_{\Lambda\to\infty} \int_{|\lambda|\le\Lambda} e_0(x',\lambda) d\Sigma(\lambda) e_0(t',\lambda)^* dx' dt'\end{aligned}$$

and we reach the first assertion.

If $\Sigma(\lambda)$ is integrable with respect to Lebesgue measure and satisfies $\lim_{\lambda\to\pm\infty} \Sigma(\lambda) = 0$, then we apply integration by parts for Lebesgue-Stieltjes integrals to obtain

$$\begin{aligned}\int_{|\lambda|\le\Lambda} e_0(x,\lambda) d\Sigma(\lambda) e_0(t,\lambda)^* &= e_0(x,\lambda)\Sigma(\lambda)e_0(t,\lambda) \Big|_{\lambda=-\Lambda}^{\lambda=\Lambda} \\ &- \int_{|\lambda|\le\Lambda} (\partial_\lambda e_0(x,\lambda))\Sigma(\lambda)e_0(t,\lambda) + e_0(x,\lambda)\Sigma(\lambda)(\partial_\lambda e_0(t,\lambda)) d\lambda.\end{aligned}$$

Since $e_0(x,\lambda)$ is uniformly bounded and $\partial_\lambda e_0(x,\lambda)$ is uniformly bounded for $|x| \le R$ for each R, we reach the conclusion. □

Corollary 5.10 *The spectral function can be prescribed on an arbitrary finite interval. More precisely, there exists a boundary value problem* (3.2) *with continuous Q satisfying* (4.3) *and such that its spectral function $\sigma(\lambda)$ coincides on an arbitrary finite interval with a prescribed increasing $n \times n$ spectral measure.*

Proof: This follows from the fact that if $\Sigma(\lambda)$ is constant outside a compact interval then it satisfies condition 2. by Remark 5.4 (or the previous proposition) and it satisfies condition 1. by Proposition 5.5. □

Conjecture 5.11 *We conjecture that condition* 1. *in Theorem* 5.2 *is obsolete in general.*

6 Some Generalizations, Comments, Examples

6.1 Generalization of the Main Result Theorem 5.2

Before we have investigated an operator L of the form (3.2) starting with the operator L_0 (with $Q = 0$). This has an obvious generalization. Namely, we may investigate two operators $L_1 := L_{1,H}, L_2 := L_{2,H}$ and consider L_2 as a perturbation of L_1. More precisely, let

$$L_j = \frac{1}{i} B \frac{d}{dx} + Q_j, \tag{6.1}$$

and

$$\mathcal{D}(L_j) = \{f \in \mathcal{D}(L_j^*) \mid f_2(0) = H f_1(0)\}, \quad B_1 = H^* B_2 H. \tag{6.2}$$

Furthermore, let Y_j be the $2n \times n$ matrix solution of the initial value problem (3.3) (with L_j instead of L). If both operators L_j are of class (T_B) then Y_j admits the representation $Y_j(., \lambda) = (I + K_j)e_0(., \lambda)$, where K_j is a Volterra operator with kernel $K_j(x, t)$. Therefore

$$Y_2(x, \lambda) = ((I + K)Y_1(., \lambda))(x) = Y_1(x, \lambda) + \int_0^x K(x, t)Y_1(t, \lambda)dt, \tag{6.3}$$

where

$$I + K = (1 + K_2)(I + K_1)^{-1}. \tag{6.4}$$

Repeating the arguments used in the proof of Theorem 4.1 one concludes that if $Q_1, Q_2 \in C^1(\mathbb{R}_+, \mathrm{M}(2n, \mathbb{C}))$ then $K \in C^1(\Omega, \mathrm{M}(2n, \mathbb{C}))$ and, moreover, K satisfies the following Goursat problem

$$B\partial_x K(x, t) + \partial_t K(x, t)B + iQ_2(x)K(x, t) - iK(x, t)Q_1(t) = 0, \tag{6.5a}$$

$$BK(x, x) - K(x, x)B = i(Q_1(x) - Q_2(x)), \tag{6.5b}$$

$$K(x, 0)B\begin{pmatrix} I \\ H \end{pmatrix} = 0. \tag{6.5c}$$

We also note that (6.5a)–(6.5c) may be deduced directly from (6.4) and (4.6) for K_1, K_2. For example (6.5b) follows from (4.6b) and the identity $K(x,x) = K_2(x,x) - K_1(x,x)$.

Putting $R := (I+K)^{-1} - I$ we obtain from (6.3)

$$Y_1(x,\lambda) = ((I+R)Y_2(.,\lambda))(x) = Y_2(x,\lambda) + \int_0^x R(x,t)Y_2(t,\lambda)dt. \tag{6.6}$$

Since Proposition 4.3 remains valid in the case under consideration, the following result, being a complete analog of Proposition 4.6, may be obtained in the same way as Proposition 4.6.

Proposition 6.1 *Let $\sigma_j(\lambda)$ be the $n\times n$ spectral function (cf. Proposition 3.5) of the operator L_j, $j=1,2$, and $\Sigma := \sigma_2 - \sigma_1$.*

1. *Let L_j be of class (T_B) and let $F(x,t)$ be defined by (4.31) with $R(x,t)$ being the kernel of the transformation operator (6.6). Then we have for all $f, g \in L^2_{\mathrm{comp}}(\mathbb{R}_+, \mathbb{C}^{2n})$*

$$\int_{\mathbb{R}} F_1(\lambda)^* d\Sigma(\lambda) G_1(\lambda) = \int_0^\infty \int_0^\infty f(x)^* F(x,t) g(t)dxdt, \tag{6.7}$$

where F_1 and G_1 are the $\mathcal{F}_{H,Q_1}$-transforms of f and g respectively.

2. *Again assuming L_j to be of class (T_B) we put*

$$\widetilde{Y}_1(x,\lambda) := \int_0^x Y_1(t,\lambda)dt.$$

Then the function

$$\widetilde{F}(x,t) := \int_{\mathbb{R}} \widetilde{Y}_1(x,\lambda) d\Sigma(\lambda) \widetilde{Y}_1(t,\lambda)^* \tag{6.8}$$

exists and has a continuous mixed second derivative which coincides with $F(x,t)$, i.e. $\frac{\partial^2}{\partial x \partial t}\widetilde{F}(x,t) = F(x,t)$.

3. *Conversely, given any increasing $n\times n$ matrix function σ_2 put $\Sigma := \sigma_2 - \sigma_1$. If the integral (6.8) exists and has a continuous mixed second derivative $F_1(x,t) := \frac{\partial^2}{\partial x \partial t}\widetilde{F}(x,t)$ then (6.7) holds for all $f, g \in L^2_{\mathrm{comp}}(\mathbb{R}_+, \mathbb{C}^{2n})$ with F_1 instead of F.*

Again, we emphasize that 3. holds for arbitrary L of the form (3.2) not necessarily being of class (T_B).

Combining Propositions 6.1 and 4.3 we arrive at the Gelfand-Levitan equation:

Proposition 6.2 *Let L_j be of class (T_B) and let σ_j be the spectral function of the problem (3.2) with $Q_j = Q_j^*$, $j=1,2$, instead of Q. Then with F defined by (6.8) we have the Gelfand-Levitan equation (4.39).*

Now we are ready to present a generalization of the main result (Theorem 5.2).

Theorem 6.3 *Let $\sigma_1(\lambda)$ be the spectral function of the operator L_1 of the form* (6.1). *For an increasing $n \times n$ matrix function $\sigma(\lambda)$ to be the spectral function of the boundary value problem* (3.2) *with (unique) continuous $2n \times 2n$ matrix potential Q satisfying* (4.3) *it is sufficient that the following conditions hold:*

1. *If $g \in L^2_{\text{comp}}(\mathbb{R}_+, \mathbb{C}^{2n})$ and if*

$$\int_{\mathbb{R}} G_1(\lambda)^* d\sigma(\lambda) G_1(\lambda) = 0,$$

   where G_1 is the $\mathcal{F}_{H,Q_1}$-transform of g, then $g = 0$.

2. *The function*

$$\widetilde{F}(x,t) := \int_{\mathbb{R}} \widetilde{Y}_1(x,\lambda) d\Sigma(\lambda) \widetilde{Y}_1(t,\lambda)^*, \quad \widetilde{Y}_1(x,\lambda) := \int_0^x Y_1(t,\lambda) dt, \tag{6.9}$$

   with $\Sigma = \sigma - \sigma_1$ exists and has a continuous mixed second derivative

$$F(x,t) := \frac{\partial^2}{\partial x \partial t} \widetilde{F}(x,t). \tag{6.9'}$$

   Moreover Q has m continuous derivatives if and only if $D_x^m D_t^m F(x,t)$ exists and is continuous.

   Again, if we content ourselves to operators of class (T_B) then the conditions 1. *and* 2. *are also necessary.*

Sketch of Proof: The necessity is proved in just the same way as Lemma 4.5 and Proposition 4.6. □

Sufficiency: Starting with $\sigma(\lambda)$ we define $\widetilde{F}$, F by (6.8) with $\Sigma(\lambda) := \sigma(\lambda) - \sigma_1(\lambda)$. Then we consider the Gelfand-Levitan equation

$$F(x,t) + K(x,t) + \int_0^x K(x,s)F(s,t)ds = 0, \quad t < x. \tag{6.10}$$

with F defined by (6.8). Following the proof of Proposition 5.1 one concludes that (6.10) has a continuous solution $K : \Omega \to \mathrm{M}(2n, \mathbb{C})$. Next we define $Y(x,\lambda)$ setting $Y(.,\lambda) = (I + K)Y_1(.,\lambda)$ and show that $Y(x,\lambda)$ satisfies the initial value problem (5.7) with

$$Q(x) = Q_1(x) + iBK(x,x) - iK(x,x)B. \tag{6.11}$$

Since Q_1 satisfies (4.3) we infer from (6.11) that Q also satisfies (4.3). Moreover the self-adjointness of Q may be proved as in the proof of Theorem 5.1.

Furthermore, we note that if F is continuously differentiable it satisfies the equality

$$BD_xF(x,t) + D_tF(x,t)B = -iQ_1(x)F(x,t) + iF(x,t)Q_1(t) \tag{6.12}$$

and according to Proposition 5.1 K is continuously differentiable, too. If F is just continuous then (6.12) still holds in the distributional sense. This is shown similar to (5.5b).

Since $Y_1(0,\lambda) = \binom{I}{H}$ we may argue exactly as in (5.5c), (5.9), (5.10) to obtain

$$K(x,0)B\begin{pmatrix} I \\ H \end{pmatrix} = 0, \quad \text{for} \quad x \in [0,\infty). \tag{6.13}$$

In view of (6.10)–(6.13) the relation (6.5a) for K is proved along the same lines as part ii) of the proof of Theorem 5.2.

Thus K satisfies the initial value problem (6.5a)–(6.5c). Therefore $Y(x,\lambda)$ satisfies the initial value problem (5.7) with Q defined by (6.11).

If now F is just continuous then one proceeds as in part iii) of the proof of Theorem 5.2.

That σ is indeed the spectral function of the problem (5.7) with Q from (6.11) is shown as part iv) of Theorem 5.2. Instead of (5.1), Theorem 4.8 (5.4), and Proposition 4.6 one uses (6.8), Proposition 6.2 (6.10), and Proposition 6.1. □

6.2 The Degenerate Gelfand-Levitan Equation

We discuss solutions of the Gelfand-Levitan equation in the special case where $\Sigma(\lambda)$ is a step function:

We consider the situation of Theorem 6.3 and fix an operator L_1 of the form (6.1) with spectral function $\sigma_1(\lambda)$.

Let $A \in \mathrm{M}(n,\mathbb{C})$ be a hermitian nonnegative matrix and

$$\Sigma(\lambda) := A\ 1_{[a,\infty)}(\lambda) \tag{6.14}$$

an increasing step function with one jump of "height" A.

We show that

$$\sigma := \sigma_1 + \Sigma \tag{6.15}$$

is the spectral function of the boundary value problem (3.2) for some (unique) continuous self-adjoint $2n \times 2n$-matrix potential Q satisfying (4.3).

Since jumps of the spectral function correspond to eigenvalues this shows in particular that for a given potential Q_1 and given real number a there is a potential Q such that

$$\mathrm{spec}(L_1 + Q - Q_1) = \mathrm{spec}(L_1) \cup \{a\}. \tag{6.16}$$

For the proof we have to verify the conditions 1. and 2. of Theorem 5.3. By Remark 5.4 condition 1. is fulfilled since A is nonnegative. To verify 2. we calculate

$$\begin{aligned}\widetilde{F}(x,t) &= \int_{\mathbb{R}} \widetilde{Y}_1(x,\lambda)d\Sigma(\lambda)\widetilde{Y}_1(t,\lambda)^* \\ &= \widetilde{Y}_1(x,a)A\widetilde{Y}_1(t,a)^*.\end{aligned}$$

Obviously, this has a continuous mixed second derivative, namely

$$F(x,t) := \frac{\partial^2}{\partial x \partial t}\widetilde{F}(x,t) =: Y_1(x,a)AY_1(t,a)^*. \tag{6.17}$$

In this case we can solve the Gelfand-Levitan equation explicitly. First we introduce for $x > 0$

$$T(x) := \int_0^x Y_1(s,a)^* Y_1(s,a)ds. \tag{6.18}$$

From

$$Y_1(0,a)^* Y_1(0,a) = I + H^*H \geq I$$

we infer that $T(x) > 0$ is positive definite for $x \geq 0$.

We put for $t \leq x$

$$\begin{aligned}K(x,t) &:= -Y_1(x,a)AY_1(t,a)^* \\ &\quad +Y_1(x,a)AT(x)A^{1/2}(I + A^{1/2}T(x)A^{1/2})^{-1}A^{1/2}Y_1(t,a)^* \\ &= -Y_1(x,a)A^{1/2}(I + A^{1/2}T(x)A^{1/2})^{-1}A^{1/2}Y_1(t,a)^*.\end{aligned} \tag{6.19}$$

Note that $(I + A^{1/2}T(x)A^{1/2}) \geq I$ is positive definite, thus invertible. We abbreviate $S(x) := A^{1/2}(I + A^{1/2}T(x)A^{1/2})^{-1}A^{1/2}$. If A is positive definite then we simply have $S(x) = (A^{-1} + T(x))^{-1}$.

One immediately checks that $K(x,t)$ solves the Gelfand-Levitan equation (6.10) corresponding to F and consequently determines Q by means of (6.11).

Summarizing the previous considerations we arrive at the following proposition.

Proposition 6.4 *Let L_1 be an operator of the form* (6.1), (6.2) *with the spectral function $\sigma_1(\lambda)$ and let $\Sigma(\lambda)$ be of the form* (6.14). *Then $\sigma = \sigma_1 + \Sigma$ is the spectral function of the boundary value problem* (3.2) *with $2n \times 2n$ matrix potential*

$$Q(x) = Q_1(x) + i\{Y_1(x,a)S(x)Y_1^*(x,a)B - BY_1(x,a)S(x)Y_1^*(x,a)\}.$$

Corollary 6.5 *Under the assumptions of the previous Proposition* 6.4 *let $Q_1 = 0$ (i.e. $L_1 = -iB\frac{d}{dx}$). Then the $2n \times 2n$ matrix potential corresponding to the spectral function $\sigma(\lambda) = \frac{1}{2\pi\lambda_1}\lambda I_n + \Sigma(\lambda)$ with one jump of "height" A is given by*

$$Q(x) = ie^{iaB^{-1}x}\{\widetilde{S}(x)B - B\widetilde{S}(x)\}e^{-iaB^{-1}x},$$

where $\widetilde{S}(x) := \binom{I}{H}A^{1/2}(I + xA^{1/2}(I + H^*H)A^{1/2})^{-1}A^{1/2}(I, H^*)$.

If Σ is a general increasing step function then the Gelfand-Levitan equation is still solvable. However, we do not have such an explicit formula:

Proposition 6.6 *Let L_1 be an operator of the form* (6.1), (6.2) *with spectral function $\sigma_1(\lambda)$. Furthermore, let $-\infty < a_1 < \cdots < a_r < \infty$ be real numbers and $A_j \in \mathrm{M}(n, \mathbb{C})$ nonnegative matrices.*

Then for the increasing step function

$$\Sigma(\lambda) := \sum_{j=1}^{r} A_j 1_{[a_j,\infty)}(\lambda)$$

there exists a unique continuous matrix potential Q satisfying (4.3) *such that $\sigma_1 + \Sigma$ is the spectral function of $L_1 + Q - Q_1$.*

Namely, Q is uniquely determined by (6.11) *with $K(x,t)$ being the solution of the Gelfand-Levitan equation* (6.10).

Proof: This follows by induction from Theorem 6.3 and the preceding discussion.

The conditions 1. and 2. of Theorem 6.3 can also immediately be checked directly: namely, condition 1. is fulfilled in view of Remark 5.4 since Σ is increasing. Condition 2. follows immediately from

$$\widetilde{F}(x,t) = \sum_{j=1}^{r} \widetilde{Y}_1(x,a_j) A_j \widetilde{Y}_1(t,a_j)^* \text{ and } F(x,t) = \sum_{j=1}^{r} Y_1(x,a_j) A_j Y_1(t,a_j)^*.$$

□

6.3 On Unitary Invariants of $2n \times 2n$ Systems

It is well-known that a selfadjoint operator A in a Hilbert space is uniquely determined (up to unitary equivalence) by the spectral type $[E_A]$ and the multiplicity function N_{E_A}. In this section we will show that there exist potentials Q such that the corresponding operator L_H has constant multiplicity one and $[E]$ is of pure type (absolute continuous, singular continuous, pure point).

Definition 6.7 An increasing function $\mu : \mathbb{R} \to \mathbb{R}$ on the real line will be called p-admissible if there exists a strictly increasing sequence of real numbers, $(x_\nu)_{\nu \in \mathbb{Z}}$, such that

1. $x_0 = 0$,
2. the sequence $(x_{\nu+1} - x_\nu)$ is square summable,
3. $\lim_{\nu \to \pm\infty} x_\nu = \pm\infty$,
4. $\mu(a_{\nu j}) < \mu(a_{\nu,j+1})$, where $a_{\nu j} := x_\nu + j 2^{-n-p}(x_{\nu+1} - x_\nu)$, $j = 0, \ldots, 2^{n+p} - 1$.

In particular, a strictly increasing function μ is p-admissible for any p.

We will show that for a p-admissible increasing function μ there exists an operator L_H of the form (3.2) such that its spectral measure $E := E_{L_H}$ satisfies

$$[E] = [d\mu], \qquad N_E(x) = p \text{ for } \mu\text{- a.e. } x \in \mathbb{R}.$$

In particular there exist $2n \times 2n$ systems such that each point in the spectrum has multiplicity one. To prove this result we will use the criteria from the end of Section 5.

Proposition 6.8 *Let* $B = (B_1, -B_2) \in \mathrm{M}(2n, \mathbb{C})$ *be a matrix as in* (1.1) *and* $H \in \mathrm{M}(n, \mathbb{C})$ *as in* (2.8). *Let* μ *be a p-admissible increasing function on the real line,* $1 \leq p \leq n$. *Then there exists a continuous potential* Q *satisfying* (4.3) *and such that the corresponding operator* L_H *is unitary equivalent to the operator* $\Lambda_p = \oplus_1^p \Lambda_1$, *where*

$$\Lambda_1 : L^2_\mu(\mathbb{R}) \longrightarrow L^2_\mu(\mathbb{R}), \quad \Lambda_1 f(\lambda) = \lambda f(\lambda).$$

Proof: Let (ψ_j) be the Rademacher functions [31, Sec. I.3], i.e. $\psi_1 : \mathbb{R} \to \mathbb{R}$ is a function of period one, such that

$$\psi_1(x) = \begin{cases} 1, & 0 < x \leq \frac{1}{2}, \\ -1, & \frac{1}{2} < x \leq 1, \end{cases} \tag{6.20}$$

and

$$\psi_j(x) = \psi_1(2^j x). \tag{6.21}$$

(In [31, Sec. I.3] one puts $\psi_1(1/2) = 0$). ψ_j takes values ± 1. The set (ψ_j) is orthonormal in $L^2[0,1]$ and

$$\int_0^1 \psi_j(x)dx = 0. \tag{6.22}$$

We put

$$\psi_0(\lambda) := \begin{pmatrix} \psi_1(\lambda) & \psi_2(\lambda) & \dots & \psi_n(\lambda) \\ \psi_2(\lambda) & \psi_3(\lambda) & \dots & \psi_{n+1}(\lambda) \\ \dots & \dots & \dots & \dots \\ \psi_p(\lambda) & \psi_{p+1}(\lambda) & \dots & \psi_{n+p-1}(\lambda) \end{pmatrix}, \tag{6.23}$$

$$\psi(\lambda) := \psi_0\left(\frac{\lambda - x_\nu}{x_{\nu+1} - x_\nu}\right), \quad \text{for } x_\nu \leq \lambda < x_{\nu+1}, \tag{6.24}$$

$$\Phi(\lambda) := p^{-1}\frac{1}{2\pi}B_1^{-1/2}\psi(\lambda)^*\psi(\lambda)B_1^{-1/2}. \tag{6.25}$$

$\Phi(\lambda)$ is a symmetric nonnegative matrix of rank p for each $\lambda \in \mathbb{R}$.

Furthermore, let

$$f(x) := \frac{a_{\nu,j+1} - a_{\nu j}}{\mu(a_{\nu,j+1}) - \mu(a_{\nu j})}, \quad a_{\nu j} < x \le a_{\nu,j+1}, \tag{6.26}$$

and put

$$\varrho(\lambda) := \begin{cases} \int_{(0,\lambda]} f(t)d\mu(t), & \lambda > 0, \\ -\int_{(\lambda,0]} f(t)d\mu(t), & \lambda \le 0. \end{cases} \tag{6.27}$$

Obviously the measures $d\varrho$ and $d\mu$ are mutually equivalent and

$$\varrho(a_{\nu j}) = a_{\nu j}. \tag{6.28}$$

Finally we put

$$\sigma(\lambda) := \begin{cases} \int_{(0,\lambda]} \Phi(t)d\varrho(t), & \lambda > 0, \\ -\int_{(\lambda,0]} \Phi(t)d\varrho(t), & \lambda \le 0. \end{cases} \tag{6.29}$$

Note that in view of (6.22) and the orthonormality of the Rademacher functions we have $\sigma(x_\nu) = \frac{1}{2\pi}B_1^{-1}x_\nu$. Again, by the orthonormality of the Rademacher functions and the fact that the entries of ψ are constant on the intervals $(a_{\nu j}, a_{\nu,j+1}]$, we have for $x_\nu < \lambda \le x_{\nu+1}$

$$\left\| \sigma(\lambda) - \frac{1}{2\pi}B_1^{-1}x_\nu \right\| \le \int_{(x_\nu,\lambda]} \|\Phi(t)\| d\varrho(t) \le C(x_{\nu+1} - x_\nu), \tag{6.30}$$

thus

$$\begin{aligned} \int_{x_\nu}^{x_{\nu+1}} \left\| \sigma(\lambda) - \frac{1}{2\pi}B_1^{-1}\lambda \right\| d\lambda &\le C' \int_{x_\nu}^{x_{\nu+1}} (x_{\nu+1} - x_\nu) d\lambda \\ &= C'(x_{\nu+1} - x_\nu)^2. \end{aligned} \tag{6.31}$$

Since $(x_{\nu+1} - x_\nu)_\nu$ is square summable, we infer that the function

$$\Sigma(\lambda) := \sigma(\lambda) - \sigma_0(\lambda)$$

is integrable with respect to Lebesgue measure and $\lim_{\lambda\to\pm\infty} \Sigma(\lambda) = 0$. In view of Proposition 5.9 it satisfies condition 2.

To show that it satisfies condition 1. let $g \in L^2([0,b], \mathbb{C}^{2n})$ with

$$0 = \int_{-\infty}^{\infty} G_0(\lambda)^* d\sigma(\lambda) G_0(\lambda) = \int_{-\infty}^{\infty} G_0(\lambda)^* \Phi(\lambda) G_0(\lambda) d\varrho(\lambda).$$

In view of 4. of the definition of admissibility we infer that in each interval $(x_\nu, x_{\nu+1}]$ there exist points $\lambda_{\nu j}$, $j = 0, \dots, 2^{n+p} - 1$, such that the vector $B_1^{-1/2}$ $G_0(\lambda_{\nu j})$ lies in the null space of the matrix $\psi_0(j2^{-n-p})$. Since $(x_{\nu+1} - x_\nu)$ is square summable, each of the sequences $(\lambda_{\nu j})_\nu$ is a sequence of infinite density. Noting that $G_0(\lambda)$ is an entire (vector) function of strict order one and of finite type we infer as in the proof of Proposition 5.5 that $B_1^{-1/2}G_0(\lambda)$ lies in the null space of the matrix $\psi_0(j2^{-n-p})$, $j = 0, \dots, 2^{n+p} - 1$, for each λ. It is easy to check that the intersection of these null spaces is 0, hence $G_0(\lambda) = 0$.

By Theorem 5.2 there exists Q satisfying (4.3) such that σ is the spectral measure function of the corresponding operator L_H. This proves the theorem. □

Corollary 6.9 *For* $1 \leq p \leq n$ *there exist continuous potentials* $Q : \mathbb{R}_+ \to \mathrm{M}(n, \mathbb{C})$ *satisfying* (4.3) *such that the corresponding operator* L_H *has*

i) *absolute continuous spectrum of multiplicity* p,

ii) *singular continuous spectrum of multiplicity* p,

iii) *pure point spectrum of multiplicity* p.

Proof: We only have to note that there exist admissible increasing functions $\mu_{\mathrm{ac}}, \mu_{\mathrm{sc}}, \mu_{\mathrm{pp}}$ such that the measures $d\mu_{\mathrm{ac}}, d\mu_{\mathrm{sc}}, d\mu_{\mathrm{pp}}$ are absolute continuous, singular continuous, discrete, respectively. Such measures obviously exist. □

References

[1] Z.S. Agranovich and V.A. Marchenko, *The inverse problem of scattering theory.* Gordon and Breach, New York, 1963.

[2] D. Alpay and I. Gohberg, *Inverse spectral problem for differential operators with rational scattering matrix functions.* J. Differ. Equations **118** (1995), 1–19.

[3] D. Alpay and I. Gohberg, *Potentials associated to rational weights.* In: Operator Theory, Advances and Applications (I. Gohberg, ed.), vol. **98**, 1997, Birkhäuser, Basel, pp. 23–40.

[4] H. Bart, I. Gohberg and M.A. Kaashoek, *Minimal factorization of matrix and operator functions.* In: Operator Theory, Advances and Applications, vol. **1**, 1997, Birkhäuser, Basel.

[5] Y.M. Berezanskii, *Expansions in Eigenfunctions of Selfadjoint Operators.* AMS, Providence, 1968.

[6] P.R. Chernoff, *Essential self-adjointness of powers of generators of hyperbolic equations.* J. Funct. Anal. **12** (1973), 401–414.

[7] E. Coddington and N. Levinson, *Theory of ordinary differential equations.* McGraw Hill, New York, 1955.

[8] H. Dym and A. Jacob, *Positive definite extensions, canonical equations and inverse problems.* In: Operator theory, advances and applications, vol. **12**, 1984, Birkhäuser, Basel, pp. 141–240.

[9] M.G. Gasymov and B.M. Levitan, *The inverse problem for a Dirac system.* Sov. Math. Dokl. **7** (1966), 495–499.

[10] I.M. Gel'fand and B.M. Levitan, *On the determination of a differential equation from its spectral function.* Izv. Akad. Nauk SSSR **15** (1951), 309–360 (Russian); Amer. Math. Transl. **1** (1955), 253–304 (Engl. transl.).

[11] F. Gesztesy and H. Holden, *On trace formulas for Schroedinger-type operators.* In: D.G. Truhlar et al. (ed.), *Multiparticle quantum scattering with applications to nuclear, atomic and molecular physics.* IMA Vol. Math. Appl. 89, 1997, Springer, New York, pp. 121–145.

[12] F. Gesztesy and B. Simon, *Uniqueness theorems in inverse spectral theory for one-dimensional Schrödinger operators.* Trans. Amer. Math. Soc. **348** (1996), 349–373.

[13] F. Gesztesy and B. Simon, *Inverse spectral analysis with partial information on the potential. I: The case of an a.c. component in the spectrum.* Helv. Phys. Acta **70** (1997), 66–71.

[14] M.G. Krein, *On a general method of decomposing Hermite-positive nuclei into elementary products.* Sov. Math. Dokl. **53** (1946), 3–6.

[15] M.G. Krein, *On Hermitian operators with direct functionals.* (Russian) Sbornik trudov Instituta Matemetiki AN USSR **10** (1948), 83–105.

[16] M.G. Krein, *On transition function of one-dimensional second order boundary value problem.* Sov. Math. Dokl. **88** (1953), 405–408.

[17] M.G. Krein, *Continuous analogs of theorems on polynomials orthogonal on the unit circle.* Sov. Math. Dokl. **105** (1955), 637–640.

[18] H.B. Lawson and M.L. Michelsohn, *Spin Geometry.* Princeton University Press, Princeton, N.J., 1989.

[19] M. Lesch and M.M. Malamud, *The inverse spectral problem for first order systems on the half line.* SFB 288, Preprint no. 322, Berlin, 1998, 30p.; math. SP/9805033, http://xxx.lanl.gov/abs/math/9805033.

[20] M. Lesch and M.M. Malamud, *The inverse spectral problem for systems on the half line.* Uspekhi Matem. Nauk **53** (1998), 157.

[21] B.Ya. Levin, *Distribution of Zeros of Entire functions.* Transl. Math. Monographs vol. **5**, AMS, Providence, 1964.

[22] B.M. Levitan, *Inverse Sturm-Liouville problems.* VNU Science Press, Utrecht, 1987.

[23] B.M. Levitan and I.S. Sargsjan, *Sturm-Liouville and Dirac operators.* Kluwer, Dordrecht, 1991.

[24] M.M. Malamud, *Spectral analysis of Volterra operators and inverse problems for systems of ordinary operators.* SFB 288, Preprint no. 269, Berlin, 1997.

[25] M.M. Malamud, *Uniqueness questions in inverse problems for first order systems on a finite interval.* Trans. Moscow Math. Soc. **60** (1999), 204–262.

[26] V.A. Marchenko, *Sturm-Liouville operators and applications.* Birkhäuser, Basel, 1986.

[27] M.A. Naimark, *Linear differential operators.* Frederick Ungar Publ., New York, 1967.

[28] R.G. Newton and R. Jost, *The construction of potentials from the S-matrix for systems of Differential Equations.* Nuovo Cimento **1** (1955), 590–622.

[29] L.A. Sakhnovich, *Factorization problems and operator identities.* Uspekhi Matem. Nauk **41** (1986), 3–55 (Russian); Russ. Math. Surv. **41** (1986), 1–64 (English translation).

[30] L.A. Sakhnovich, *Defekt indices of a first order system of differential equations.* Sibirskii matematicheskii Jurnal **38** (1997), 1360–1361.

[31] A. Zygmund, *Trigonometric series.* Second Edition, Cambridge University Press, Cambridge, 1959.

Humboldt Universität zu Berlin
Institut für Mathematik
Unter den Linden 6
D–10099 Berlin
Germany

Department of Mathematics
University of Donetsk
Donetsk
Ukraine
mmm@univ.donetsk.ua

Current address:
Universität Bonn
Mathematisches Institut
Beringstr. 1
D–53115 Bonn
lesch@mathematik.hu-berlin.de
lesch@math.uni-bonn.de
URL: http://spectrum.mathematik.hu-berlin.de/~lesch

1991 Mathematics Subject Classification. 34A25 (Primary), 34L (Secondary)

Operator Theory:
Advances and Applications, Vol. 117
© 2000 Birkhäuser Verlag Basel/Switzerland

Exact Solution of the Marchenko Equation Relevant to Inverse Scattering on the Line

*Cornelis van der Mee**

This paper presents explicit solutions of the Marchenko equation relevant to the solution of the inverse scattering problem of determining the real potential $Q(x)$ in the 1-D Schrödinger equation on the line from the reflection coefficient $R(k)$, which is assumed to be rational. The reflection coefficient is written in the form $i\,\mathcal{C}(k - i\mathcal{A})^{-1}\mathcal{B}$. State space methods are applied to solve the Marchenko equation, both without and with bound states.

1 Introduction

Consider the Schrödinger equation

$$\psi''(k,x) + k^2\psi(k,x) = Q(x)\,\psi(k,x), \qquad x \in \mathbf{R}, \tag{1.1}$$

where $\mathbf{R}$ is the real line, the prime denotes the derivative with respect to the spatial coordinate x, k is the wavenumber, k^2 is energy, and $Q(x)$ is a real potential that is at least integrable on the line. Then the Jost solution from the left $f_l(k,x)$ and the Jost solution from the right $f_r(k,x)$ are the solutions of (1.1) satisfying the boundary conditions

$$f_l(k,x) = \begin{cases} e^{ikx} + o(1), & x \longrightarrow +\infty, \\ \dfrac{1}{T(k)}e^{ikx} + \dfrac{L(k)}{T(k)}e^{-ikx} + o(1), & x \longrightarrow -\infty, \end{cases} \tag{1.2a}$$

$$f_r(k,x) = \begin{cases} \dfrac{1}{T(k)}e^{-ikx} + \dfrac{R(k)}{T(k)}e^{ikx} + o(1), & x \longrightarrow +\infty, \\ e^{-ikx} + o(1), & x \longrightarrow -\infty, \end{cases} \tag{1.2b}$$

where $0 \neq k \in \mathbf{R}$, $T(k)$ is the transmission coefficient and $R(k)$ and $L(k)$ are the reflection coefficients from the right and from the left, respectively. The scattering matrix $\mathbf{S}(k)$ is given by

$$\mathbf{S}(k) = \begin{bmatrix} T(k) & R(k) \\ L(k) & T(k) \end{bmatrix}, \qquad k \in \mathbf{R}.$$

*This material is based on work supported by C.N.R., MURST, and a University of Cagliari Coordinated Research Project.

Apart from the scattering solutions, one should also consider the nontrivial solutions of (1.1) that are square integrable on the real line, its so-called bound-state solutions. Such solutions occur at the values of k in the open upper half-plane $\mathbf{C}^+$ where the Jost solutions from the left and from the right are linearly dependent. When $(1+|x|)Q(x)$ is integrable, there are at most finitely many such k, all of which are purely imaginary. Moreover, for such potentials the reflection and transmission coefficients are continuous in $k \in \mathbf{R}$. For later use, let $L^1_q(I)$ denote the set of measurable functions f on the interval I such that $\|f\|_{1,q} := \int_I dx\,(1+|x|)^q\,|f(x)|$ is finite.

The direct and inverse scattering theory for the Schrödinger equation (1.1) has been studied intensively [13, 15–18]. The inverse problem that one usually considers, consists of the determination of a real potential $Q \in L^1_1(\mathbf{R})$ from the reflection coefficient $R(k)$, the bound state poles $i\kappa_j$ of the transmission coefficient $T(k)$, and the bound state constants $c_j = f_r(i\kappa_j, x)/f_l(i\kappa_j, x)$ ($j = 1, \ldots, \mathcal{N}$; $\mathcal{N}$ finite, where there might not be any bound states). Necessary and sufficient conditions to construct $Q(x)$ from these data have been given in [15] for $Q \in L^1_2(\mathbf{R})$ and in [29] for $Q \in L^1_1(\mathbf{R})$. For quite general $R(k)$, an inversion algorithm based on the Marchenko integral equation was given by Faddeev [18] (see also [15, 16], and [13], Chapter XVII). The unique solvability of the Marchenko integral equation was proved in [16, 18], also when bound state information is incorporated.

When using the so-called Faddeev functions

$$m_l(k,x) = e^{-ikx} f_l(k,x), \qquad m_r(k,x) = e^{ikx} f_r(k,x),$$

one obtains the Riemann-Hilbert problem

$$\begin{bmatrix} m_l(-k,x) \\ m_r(-k,x) \end{bmatrix} = \begin{bmatrix} T(k) & -R(k)e^{2ikx} \\ -L(k)e^{-2ikx} & T(k) \end{bmatrix} \begin{bmatrix} m_r(k,x) \\ m_l(k,x) \end{bmatrix}, \qquad k \in \mathbf{R}. \tag{1.3}$$

Since there exist integrable functions $B_l(x,\cdot)$, $B_r(x,\cdot)$, $\hat{R}$ and $\hat{L}$ such that

$$\begin{aligned} m_l(k,x) &= 1 + \int_0^\infty dy\, e^{iky} B_l(x,y), \quad m_r(k,x) = 1 + \int_0^\infty dy\, e^{iky} B_r(x,y); \\ R(k) &= \int_{-\infty}^\infty dz\, e^{-ikz} \hat{R}(z), \quad L(k) = \int_{-\infty}^\infty dz\, e^{-ikz} \hat{L}(z), \end{aligned}$$

when $Q \in L^1_1(\mathbf{R})$ [15], one can, by Fourier transformation and some calculus of residues, convert the Riemann-Hilbert problem (1.3) into the Marchenko integral equations

$$\begin{aligned} &B_l(x,y) + \int_0^\infty dz\, \hat{S}_l(2x+y+z) B_l(x,z) \\ &\quad = -\hat{S}_l(2x+y), \quad y > 0; \end{aligned} \tag{1.4a}$$

$$(1.4b)\qquad \begin{aligned} &B_r(x,y) + \int_0^\infty dz\, \hat{S}_r(-2x+y+z)B_r(x,z) \\ &= -\hat{S}_r(-2x+y), \quad y>0. \end{aligned}$$

Here $\hat{S}_l$ and $\hat{S}_r$ coincide with $\hat{R}$ and $\hat{L}$, respectively, when there are no bound states. Otherwise

$$\hat{S}_l(z) = \hat{R}(z) - i\sum_{j=1}^{\mathcal{N}} t_j c_j e^{-\kappa_j z}, \qquad \hat{S}_r(z) = \hat{L}(z) - i\sum_{j=1}^{\mathcal{N}} \frac{t_j}{c_j} e^{-\kappa_j z},$$

where $i\kappa_1, \dots, i\kappa_{\mathcal{N}}$ are the poles of the transmission coefficient $T(k)$, $t_1, \dots, t_{\mathcal{N}}$ are the residues of $T(k)$ at these (simple) poles, and $c_j = f_r(i\kappa_j, x)/f_l(i\kappa_j, x)$ ($j = 1, \dots, \mathcal{N}$) are the bound state constants. It can be shown that $-it_jc_j > 0$ ($j = 1, \dots, \mathcal{N}$) [see Section 3; also [16]). The potential $Q(x)$ is easily found from the solution of either (1.4a) or (1.4b), since we have

$$(1.5)\qquad B_l(x,0^+) = \frac{1}{2}\int_x^\infty dt\, Q(t), \qquad B_r(x,0^+) = \frac{1}{2}\int_{-\infty}^x dt\, Q(t).$$

Jump discontinuities in either of $B_l(x,0^+)$ and $B_r(x,0^+)$ lead to Dirac delta function terms in $Q(x)$ and hence to an extension of the class of potentials considered. Derivations of (1.4a) and (1.4b) can be found in [13, 16].

The inverse scattering problem for rational reflection coefficients is easily solved in an ad hoc way by computing $B_l(x,y)$ for $x \geq 0$ [$B_r(x,y)$ for $x \leq 0$, respectively] from (1.3) using calculus of residues, where the values of $m_l(k,x)$ at the poles of $R(k)$ in $\mathbf{C}^+$ when $x \geq 0$ [the values of $m_r(k,x)$ at the poles of $L(k)$ in $\mathbf{C}^+$ when $x \leq 0$, respectively] follow by solving a linear system of equations (cf. [2] and references therein). However, in this article we apply state space methods (see, e.g., [9]) which allow us to derive explicit formulas for thet solution of the above inverse scattering problem when $R(k)$ is a rational function without real poles that is real-valued for purely imaginary k, vanishes as $k \to \infty$, and satisfies $|R(k)| < 1$ for $0 \neq k \in \mathbf{R}$ and $R(0) \in [-1, 1)$. We then have the state space realization

$$(1.6)\qquad R(k) = i\,\mathcal{C}(k - i\mathcal{A})^{-1}\mathcal{B} = i((k - i\mathcal{A})^{-1}\beta, \gamma),$$

where $\mathcal{A}$ is a real $n \times n$ matrix without imaginary eigenvalues, $\mathcal{B}$ is a real $n \times 1$ matrix coinciding with the column vector β, $\mathcal{C}$ is a real $1 \times n$ matrix coinciding with the row vector γ^T, and $(\cdot,\cdot)$ denotes the usual scalar product on $\mathbf{C}^n$. Writing

$$(1.7)\qquad (k - i\mathcal{A})^{-1} = i\int_{-\infty}^\infty dt\, e^{-ikt} E(t; -\mathcal{A}),$$

the solution $Q(x)$ of the inverse problem follows from the solution $B_l(x,y)$ of the Marchenko integral equation

$$(1.8)\qquad \begin{aligned} &B_l(x,y) - \int_0^\infty dz\, \mathcal{C}E(2x+y+z; -\mathcal{A})\mathcal{B}B_l(x,z) \\ &= \mathcal{C}E(2x+y; -\mathcal{A})\mathcal{B}, \quad y>0, \end{aligned}$$

by means of the formula [cf. (1.5)]

$$Q(x) = -2\frac{d}{dx}B_l(x, 0^+), \tag{1.9}$$

when there are no bound states. In fact, we shall prove the following theorem.

Theorem 1.1 *Suppose there are no bound states. Then for $x > 0$ the unique solution of* (1.8) *is given by*

$$B_l(x, y) = \mathcal{C}E(2x + y; -\mathcal{A})\Lambda(x)^{-1}\mathcal{B}, \tag{1.10}$$

and the potential $Q(x)$ is given by

$$\begin{aligned} Q(x) &= 4\mathcal{C}E(2x; -\mathcal{A})\Lambda(x)^{-1}\mathcal{A}\Lambda(x)^{-1}\mathcal{B} \\ &= 4(E(2x; -\mathcal{A})\Lambda(x)^{-1}\mathcal{A}\Lambda(x)^{-1}\beta, \gamma), \end{aligned} \tag{1.11}$$

where

$$\Lambda(x) = I - \mathcal{D}E(2x; -\mathcal{A}) \tag{1.12}$$

and

$$\mathcal{D} = \int_0^\infty du\, E(u; -\mathcal{A})\mathcal{B}\mathcal{C}E(u; -\mathcal{A}). \tag{1.13}$$

When there are bound states, i.e., when $T(k)$ has finitely many (simple) poles at $i\kappa_j$ with residue t_j and bound state constant c_j with $d_j = -it_jc_j > 0$ $(j = 1, \ldots, \mathcal{N})$, the Marchenko integral equation is given by (1.4a), where

$$\hat{S}_l(z) = -\mathcal{C}E(z; -\mathcal{A})\mathcal{B} + \sum_{j=1}^{\mathcal{N}} d_j e^{-\kappa_j z} = -\breve{C}E(z; -\breve{A})\breve{B}, \tag{1.14}$$

and $\breve{A}$, $\breve{B}$ and $\breve{C}$ are the real $(n + \mathcal{N}) \times (n + \mathcal{N})$, $(n + \mathcal{N}) \times 1$ and $1 \times (n + \mathcal{N})$ matrices given by

$$\begin{aligned} \breve{\mathcal{A}} &= \mathcal{A} \oplus \operatorname{diag}(\kappa_1, \ldots, \kappa_{\mathcal{N}}), \quad \breve{\mathrm{B}} = [\mathcal{B}\ -\mathrm{d}_1\ \ldots\ -\mathrm{d}_{\mathcal{N}}]^{\mathrm{T}}, \\ \breve{\mathcal{C}} &= [\mathcal{C}\ 1\ \ldots\ 1]. \end{aligned} \tag{1.15}$$

We also write $\breve{\beta} = [\beta\ -d_1\ \ldots\ -d_{\mathcal{N}}]^T$ and $\breve{\gamma} = [\gamma\ 1\ \ldots\ 1]^T$. We then have the following theorem.

Theorem 1.2 *For $x > 0$ the unique solution of the Marchenko equation* (1.4a) *is given by*

$$B_l(x, y) = \breve{\mathcal{C}}E(2x + y; -\breve{\mathcal{A}})\breve{\Lambda}(x)^{-1}\breve{\mathcal{B}},$$

and the potential $Q(x)$ *is given by*

$$\begin{aligned} Q(x) &= 4\breve{C}E(2x;-\breve{A})\breve{\Lambda}(x)^{-1}\breve{A}\breve{\Lambda}(x)^{-1}\breve{B} \\ &= 4(E(2x;-\breve{\mathcal{A}})\breve{\Lambda}(x)^{-1}\breve{\mathcal{A}}\breve{\Lambda}(x)^{-1}\breve{\beta},\breve{\gamma}), \end{aligned}$$

where

$$\breve{\Lambda}(x) = I - \breve{\mathcal{D}}E(2x;-\breve{\mathcal{A}}), \qquad \breve{\mathcal{D}} = \int_0^\infty du\, E(u;-\breve{\mathcal{A}})\breve{\mathcal{B}}\breve{C}E(u;-\breve{\mathcal{A}}).$$

Since $E(t;-\mathcal{A})$ is a so-called bisemigroup [10, 11] – we shall give its definition below –, (1.8) is an integral equation with a separable kernel when $x > 0$, which makes it trivial to solve. When $x < 0$, there are two approaches. The first is to solve (1.4b) for $x < 0$, where $\hat{S}_r(z)$ is written as in (1.14) and which is an integral equation with separable kernel. The other approach is to continue studying (1.8) which leads to

$$B_l(x,y) = \mathcal{C}E(2x+y;-\mathcal{A})\omega(x), \qquad y > -2x,$$

for a suitable vector $\omega(x)$. Then for $0 < y < -2x$, (1.8) is first written as a system of integral equations for the vector with components $B_l(x,y)$ and $\tilde{B}_l(x,y) = B_l(x,-2x-y)$. Since this system has a so-called semi-separable kernel, it could in principle be solved using the methods of [21], Chapter IX. Instead, we will reduce this coupled set of integral equations to a linear system governed by a first order linear differential equation and solve it by elementary means, yielding $Q(x)$ for $x < 0$ as the final result. This program will first be carried out if there are no bound states (Section 2) and then when bound states are taken into account (Section 3). The potential $Q(x)$ will contain the additional term $(\lim_{k\to\infty} 2ikR(k))\delta(x) = -2\mathcal{C}\mathcal{B}\delta(x)$ where $\delta(x)$ is the Dirac delta function [see (1.6), (1.8), and (1.9)]; this term vanishes if $R(k) = o(1/k)$ as $k \to \infty$. Finally, in the Appendix we give a concise proof of the unique solvability of the Marchenko integral equations.

The inverse problem for the Schrödinger equation (1.1) on the half-line $x \in (0,+\infty)$ has basically been solved in the 1950's [1, 17, 19, 20, 27, 28]. This problem consists of the determination of a real potential $Q(x)$ from the spectral function of the differential operator $-(d^2/dx^2) + Q(x)$ with Dirichlet boundary condition at $x = 0$. In recent years state space methods have been used to derive the exact solution of this problem with rational spectral function and of the problem of computing the potential from a rational scattering function. This was done for the usual Schrödinger equation [24] and for the so-called canonical differential operators [3–7, 22, 23].

When the research leading to the present article was completed, the author learned of the existence of [8], where the state space method is applied to the inverse scattering problem for the $n \times n$ matrix Schrödinger equation on the full line with selfadjoint potential $Q(x)$ and rational reflection coefficient $R(k) = o(1/k)$ as $k \to \infty$. In the present paper a different method is used to arrive at explicit solutions of the inverse problem. Our paper is not based on [34] (which is in turn based on [35]), where the inverse scattering theory of the $n \times n$ matrix Schrödinger equation has not been fully developed. The articles [8, 34, 35] will be discussed in more detail in a later publication.

2 Solution of the Inverse Problem without Bound States

In this paper we distinguish between the generic and the exceptional case [15, 18]. Generically $f_l(0,x)$ and $f_r(0,x)$ are linearly independent and $T(k) \sim ick, k \to 0$ in $\overline{\mathbf{C}^+}$, where $0 \neq c \in \mathbf{R}$. We then have

$$T(0) = 0, \qquad R(0) = L(0) = -1.$$

In the so-called exceptional case these two functions are linearly dependent and $T(k) = T(0) + o(1)$, $k \to 0$ in $\overline{\mathbf{C}^+}$, for some $0 \neq T(0) \in \mathbf{R}$ ([15, 18] if $Q \in L^1_2(\mathbf{R})$; [26] if $Q \in L^1_1(\mathbf{R})$). Thus the modulus of $T(k)$ is known; the final form of $T(k)$ now depends on the presence of bound states or not.

A. Basic Concepts

Let $\mathcal{A}$ be an arbitrary real $n \times n$ matrix without zero or imaginary eigenvalues and let $\sigma(\mathcal{A})$ denote its set of eigenvalues. Let $\Gamma^{(+)}$ and $\Gamma^{(-)}$ be positively oriented simple Jordan contours in the right and left half-plane enclosing all of the eigenvalues of $\mathcal{A}$ in the open right and left half-plane, respectively. Let $P^{(+)}_{\mathcal{A}}$ and $P^{(-)}_{\mathcal{A}}$ be the spectral projections of $\mathcal{A}$ corresponding to its eigenvalues in the right and left half-plane, respectively. For $0 \neq t \in \mathbf{R}$ we then define the *bisemigroup* generated by $\mathcal{A}$ by

$$E(t; -\mathcal{A}) = \begin{cases} e^{-t\mathcal{A}} P^{(+)}_{\mathcal{A}} = \dfrac{1}{2\pi i} \displaystyle\int_{\Gamma^{(+)}} e^{-tz}(z - \mathcal{A})^{-1}\, dz, & t > 0 \\ -e^{-t\mathcal{A}} P^{(-)}_{\mathcal{A}} = -\dfrac{1}{2\pi i} \displaystyle\int_{\Gamma^{(-)}} e^{-tz}(z - \mathcal{A})^{-1}\, dz, & t < 0. \end{cases}$$

Then (1.7) holds true. For later use, let $\mathcal{J}^{(\pm)}_{\mathcal{A}}$ be the natural imbedding of the range of $P^{(\pm)}_{\mathcal{A}}$ into $\mathbf{C}^n$ and $\pi^{(\pm)}_{\mathcal{A}}$ the unique operator from $\mathbf{C}^n$ onto the range of $P^{(\pm)}_{\mathcal{A}}$ such that $\mathcal{J}^{(\pm)}_{\mathcal{A}} \pi^{(\pm)}_{\mathcal{A}} = P^{(\pm)}_{\mathcal{A}}$.

B. Reflection and Transmission Coefficients

Let $R(k)$ be a rational function of k without real poles vanishing at infinity and satisfying $R(-k) = \overline{R(k)}$ for $k \in \mathbf{R}$. Then there exist real matrices $\mathcal{A}$ (of size $n \times n$), $\mathcal{B}$ (of size $n \times 1$ and thus representable as the column vector β) and $\mathcal{C}$ (of size $1 \times n$ and thus representable as the row vector γ^T) such that (1.6) holds. In that case

$$\hat{R}(t) = \frac{1}{2\pi} \int_{-\infty}^{\infty} dk\, e^{ikt} R(k) = -\mathcal{C}E(t; -\mathcal{A})\mathcal{B} = -\left(E(t; -\mathcal{A})\beta, \gamma\right).$$

We require n to be minimal. We put $\mathcal{C}_\pm = \mathcal{C}\mathcal{J}_{\mathcal{A}}^{(\pm)}$, $\mathcal{A}_\pm = \pi_{\mathcal{A}}^{(\pm)}\mathcal{A}\mathcal{J}_{\mathcal{A}}^{(\pm)}$ and $\mathcal{B}_\pm = \pi_{\mathcal{A}}^{(\pm)}\mathcal{B}$.

Since the scattering matrix $\mathbf{S}(k)$ is unitary for $k \in \mathbf{R}$, we now easily compute

$$|T(k)|^2 = T(k)T(-k) = 1 - i\,[\mathcal{C} \quad 0] \begin{bmatrix} k - i\mathcal{A} & i\,\mathcal{B}\mathcal{C} \\ 0 & k + i\mathcal{A} \end{bmatrix}^{-1} \begin{bmatrix} 0 \\ \mathcal{B} \end{bmatrix}.$$

Moreover,

$$\frac{1}{|T(k)|^2} = \frac{1}{T(k)T(-k)} = 1 + i\,[\mathcal{C} \quad 0](k - i\mathcal{E})^{-1} \begin{bmatrix} 0 \\ \mathcal{B} \end{bmatrix},$$

where

$$\mathcal{E} = \begin{bmatrix} \mathcal{A} & -\mathcal{B}\mathcal{C} \\ \mathcal{B}\mathcal{C} & -\mathcal{A} \end{bmatrix}. \tag{2.1}$$

Then $M\mathcal{E}M = -\mathcal{E}$ for $M = \begin{bmatrix} 0 & I \\ I & 0 \end{bmatrix}$, so that the spectrum of $\mathcal{E}$ is symmetric with respect to the origin.

Since $R(-k) = \overline{R(k)}$ for $k \in \mathbf{R}$ and the order of the matrix $\mathcal{A}$ in the realization (1.6) is minimal, there exists a unique nonsingular J such that $J\mathcal{A} = \mathcal{A}^* J$, $J\mathcal{B} = \mathcal{C}^*$ and $\mathcal{C} = \mathcal{B}^* J$, where the asterisk denotes the conjugate transpose. Then J is selfadjoint, while

$$(J \oplus (-J))\mathcal{E} = \mathcal{E}^*(J \oplus (-J))$$

implies that the spectrum $\sigma(\mathcal{E})$ of $\mathcal{E}$ is symmetric with respect to the real line. In the generic case $\mathcal{E}$ is nonsingular, whereas in the exceptional case $\mathcal{E}$ has a double zero eigenvalue.

The following well-known result [12, 32, 33] is needed to compute integrals of the form (1.13) above. We will omit the proof.

Proposition 2.1 *Let $\Gamma^{(+)}$ and $\Gamma^{(-)}$ be simple positively oriented Jordan contours enclosing all of the eigenvalues of $\mathcal{A}$ and $-\mathcal{A}$ in the right and left half-plane, respectively, and let*

$$\begin{aligned} \mathcal{D} &= \frac{1}{2\pi i} \int_{\Gamma^{(+)}} dz\,(z - \mathcal{A})^{-1}\mathcal{B}\mathcal{C}(z + \mathcal{A})^{-1} \\ &= \frac{-1}{2\pi i} \int_{\Gamma^{(-)}} dz\,(z - \mathcal{A})^{-1}\mathcal{B}\mathcal{C}(z + \mathcal{A})^{-1}. \end{aligned}$$

Then $\mathcal{D}$ maps the range of $P_{\mathcal{A}}^{(+)}$ into the range of $P_{\mathcal{A}}^{(+)}$, $\mathcal{D}^\bullet = \pi_{\mathcal{A}}^{(+)}\mathcal{D}\mathcal{J}_{\mathcal{A}}^{(+)}$ is the unique solution of the matrix equation

$$\mathcal{A}_+\mathcal{D}^\bullet + \mathcal{D}^\bullet\mathcal{A}_+ = \pi_{\mathcal{A}}^{(+)}\mathcal{B}\mathcal{C}\mathcal{J}_{\mathcal{A}}^{(+)}$$

where $\mathcal{A}_+ = \pi_{\mathcal{A}}^{(+)}\mathcal{A}\mathcal{J}_{\mathcal{A}}^{(+)}$, and $\mathcal{D}$ is also given by (1.13).

C. Solution of the Marchenko Equation for $x > 0$

Reducing (1.8) with $x > 0$ to an integral equation for $C_l(x) = \int_0^\infty dz\, E(z; -\mathcal{A})\mathcal{B}B_l(x, z)$, we find that the solution of (1.8) for $x > 0$ is given by

$$B_l(x, y) = \mathcal{C}E(2x + y; -\mathcal{A})\Lambda(x)^{-1}\mathcal{B},$$

where $\Lambda(x)$ given by (1.12) is nonsingular. Indeed, assuming $\Lambda(x)$ singular would lead to a solution of the homogeneous counterpart to (1.8), which contradicts Theorem A.1. Using (1.5) and $\frac{d}{dx}\Lambda(x)^{-1} = -\Lambda(x)^{-1}(\frac{d}{dx}\Lambda(x))\Lambda(x)^{-1}$, we get (1.11) for $x > 0$.

D. Solution of the Marchenko Equation for $x < 0$

To solve (1.8), put

$$C_l(x, y) = E(2x + y; -\mathcal{A})\mathcal{B} + \int_0^\infty dz\, E(2x + y + z; -\mathcal{A})\mathcal{B}B_l(x, z). \tag{2.2}$$

Then

$$B_l(x, y) = \mathcal{C}C_l(x, y) \tag{2.3}$$

and

$$\begin{aligned} C_l(x, y) - \int_0^\infty dz\, E(2x + y + z; -\mathcal{A})\mathcal{B}\mathcal{C}C_l(x, z) \\ = E(2x + y; -\mathcal{A})\mathcal{B}, \quad y > 0. \end{aligned} \tag{2.4}$$

For $y > -2x$ we easily obtain $(\partial/\partial y)C_l(x, y) = -\mathcal{A}C_l(x, y)$, where the entries of $C_l(x, y)$ are exponentially decreasing as $y \to +\infty$. Hence,

$$C_l(x, y) = E(2x + y; -\mathcal{A})\omega(x)$$

for some vector $\omega(x)$ in the range of $P_{\mathcal{A}}^{(+)}$ to be determined shortly.

Now consider (2.4) for $0 < y < -2x$. Putting $\tilde{C}_l(x, y) = C_l(x, -2x - y)$ and using the identity [cf. (1.13)]

$$\int_{-2x}^\infty dz\, E(2x + z; -\mathcal{A})\mathcal{B}\mathcal{C}E(2x + z; -\mathcal{A})\omega(x) = \mathcal{D}\omega(x),$$

we obtain the coupled system of integral equations

$$C_l(x,y) - \int_0^{-2x} dz\, E(y-z;-\mathcal{A})\mathcal{B}\mathcal{C}\tilde{C}_l(x,z) = E(2x+y;-\mathcal{A})\mathcal{B} + E(y;-\mathcal{A})\mathcal{D}\omega(x); \tag{2.5a}$$

$$\tilde{C}_l(x,y) - \int_0^{-2x} dz\, E(z-y;-\mathcal{A})\mathcal{B}\mathcal{C}C_l(x,z) = E(-y;-\mathcal{A})\mathcal{B} + E(-2x-y;-\mathcal{A})\mathcal{D}\omega(x), \tag{2.5b}$$

where $0 < y < -2x$. As a result, we get

$$\frac{\partial}{\partial y}\begin{bmatrix} C_l(x,y) \\ \tilde{C}_l(x,y) \end{bmatrix} = -\mathcal{E}\begin{bmatrix} C_l(x,y) \\ \tilde{C}_l(x,y) \end{bmatrix}, \qquad 0 < y < -2x, \tag{2.6}$$

where $\mathcal{E}$ is given by (2.1). Putting $\mathcal{Q}_\mathcal{A} = P_\mathcal{A}^{(+)} \oplus P_\mathcal{A}^{(-)}$, one may write the boundary conditions in the concise form

$$\mathcal{Q}_\mathcal{A}\begin{bmatrix} C_l(x,0^+) \\ \tilde{C}_l(x,0^+) \end{bmatrix} = \begin{bmatrix} \mathcal{D}\omega(x) \\ -P_\mathcal{A}^{(-)}\mathcal{B} \end{bmatrix},$$

$$(I-\mathcal{Q}_\mathcal{A})\begin{bmatrix} C_l(x,(-2x)^-) \\ \tilde{C}_l(x,(-2x)^-) \end{bmatrix} = \begin{bmatrix} -P_\mathcal{A}^{(-)}\mathcal{B} \\ \mathcal{D}\omega(x) \end{bmatrix},$$

implying

$$(\mathcal{Q}_\mathcal{A} + (I-\mathcal{Q}_\mathcal{A})e^{2x\mathcal{E}})\begin{bmatrix} C_l(x,0^+) \\ \tilde{C}_l(x,0^+) \end{bmatrix} = \begin{bmatrix} I \\ I \end{bmatrix}(\mathcal{D}\omega(x) - P_\mathcal{A}^{(-)}\mathcal{B}).$$

Now note that the matrix $(\mathcal{Q}_\mathcal{A} + (I-\mathcal{Q}_\mathcal{A})e^{2x\mathcal{E}})$ is nonsingular; otherwise the homogeneous counterpart to the system (2.5) would have a nontrivial solution, which would contradict Corollary A.2. As a result,

$$\begin{bmatrix} C_l(x,0^+) \\ \tilde{C}_l(x,0^+) \end{bmatrix} = (\mathcal{Q}_\mathcal{A} + (I-\mathcal{Q}_\mathcal{A})e^{2x\mathcal{E}})^{-1}\begin{bmatrix} I \\ I \end{bmatrix}(\mathcal{D}\omega(x) - P_\mathcal{A}^{(-)}\mathcal{B}),$$

which implies

$$C_l(x,y) = [I \quad 0]e^{-y\mathcal{E}}(\mathcal{Q}_\mathcal{A} + (I-\mathcal{Q}_\mathcal{A})e^{2x\mathcal{E}})^{-1}\begin{bmatrix} I \\ I \end{bmatrix}(\mathcal{D}\omega(x) - P_\mathcal{A}^{(-)}\mathcal{B}).$$

It remains to determine $\omega(x)$. From (2.2) we derive the identity

$$C_l(x,(-2x)^+) - C_l(x,(-2x)^-) = \mathcal{B},$$

which implies [with the help of $P_{\mathcal{A}}^{(+)}\omega(x) = \omega(x) = C_l(x, (-2x)^+)$]

$$\begin{aligned}\omega(x) &= \mathcal{B} + C_l(x, (-2x)^-) \\ &= \mathcal{B} + [I \quad 0](\mathcal{Q}_{\mathcal{A}} e^{-2x\mathcal{E}} + (I - \mathcal{Q}_{\mathcal{A}}))^{-1} \begin{bmatrix} I \\ I \end{bmatrix} (\mathcal{D}\omega(x) - P_{\mathcal{A}}^{(-)}\mathcal{B}),\end{aligned}$$

yielding

$$\begin{aligned}\omega(x) &= \left(I - [I \quad 0](\mathcal{Q}_{\mathcal{A}} e^{-2x\mathcal{E}} + (I - \mathcal{Q}_{\mathcal{A}}))^{-1} \begin{bmatrix} I \\ I \end{bmatrix} \mathcal{D}\right)^{-1} \\ &\quad \times \left(I - [I \quad 0](\mathcal{Q}_{\mathcal{A}} e^{-2x\mathcal{E}} + (I - \mathcal{Q}_{\mathcal{A}}))^{-1} \begin{bmatrix} I \\ I \end{bmatrix} P_{\mathcal{A}}^{(-)}\right) \mathcal{B},\end{aligned}$$

where the inverse operator in the right-hand side exists. In fact, reasoning as above, we see that the unique solvability of (1.8) [see Theorem A.1] implies the nonsingularity of the two matrices $\mathcal{Q}_{\mathcal{A}} + (I - \mathcal{Q}_{\mathcal{A}})e^{2x\mathcal{E}}$ and $I - [I \quad 0](\mathcal{Q}_{\mathcal{A}} e^{-2x\mathcal{E}} + (I - \mathcal{Q}_{\mathcal{A}}))^{-1} \begin{bmatrix} I \\ I \end{bmatrix} \mathcal{D}$. Finally,

$$B_l(x, 0^+) = \mathcal{C}[I \quad 0](\mathcal{Q}_{\mathcal{A}} + (I - \mathcal{Q}_{\mathcal{A}})e^{2x\mathcal{E}})^{-1} \begin{bmatrix} I \\ I \end{bmatrix} (\mathcal{D}\omega(x) - P_{\mathcal{A}}^{(-)}\mathcal{B}), \tag{2.7}$$

yielding $Q(x) = -2\frac{d}{dx} B_l(x, 0^+)$, which decreases exponentially as $x \to -\infty$. We omit the rather cumbersome expression for $Q(x)$ with $x < 0$.

When $R(k)$ is analytic in $\mathbf{C}^+$ and hence the matrix $\mathcal{A}$ in its realization (1.6) has only eigenvalues in the open left half-plane, we have $E(t; -\mathcal{A}) = -e^{-t\mathcal{A}}$ for $t < 0$ and $E(t; -\mathcal{A}) = 0$ for $t > 0$. For $x < 0$ the Marchenko integral equation (1.8) reduces to $B_l(x, y) = 0$ for $y > -2x$ and this leads to significant simplifications in the above inversion algorithm. Defining $C_l(x, y)$ as in (2.2), we get (2.3) and

$$\begin{aligned} C_l(x, y) + \int_0^{-2x-y} dz\, e^{-(2x+y+z)\mathcal{A}} \mathcal{B}\mathcal{C} C_l(x, z) \\ = -e^{-(2x+y)\mathcal{A}}\mathcal{B}, \qquad 0 < y < -2x. \end{aligned} \tag{2.8}$$

Next we get the linear system of equations

$$\begin{aligned} C_l(x, y) + \int_y^{-2x} dz\, e^{(z-y)\mathcal{A}} \mathcal{B}\mathcal{C} \tilde{C}_l(x, z) &= -e^{-(2x+y)\mathcal{A}}\mathcal{B}; \\ \tilde{C}_l(x, y) + \int_0^{y} dz\, e^{(y-z)\mathcal{A}} \mathcal{B}\mathcal{C} C_l(x, z) &= -e^{y\mathcal{A}}\mathcal{B}, \end{aligned}$$

where $0 < y < -2x$ and $\tilde{C}_l(x, y) = C_l(x, -2x - y)$. We then get the system of differential equations (2.6) where $\tilde{C}_l(x, 0^+) = C_l(x, (-2x)^-) = -\mathcal{B}$. Hence

$$\begin{bmatrix} C_l(x, y) \\ \tilde{C}_l(x, y) \end{bmatrix} = e^{-y\mathcal{E}} \begin{bmatrix} h_1 \\ h_2 \end{bmatrix}, \tag{2.9}$$

where

$$\begin{bmatrix} [e^{2x\mathcal{E}}]_{11} & [e^{2x\mathcal{E}}]_{12} \\ 0 & I \end{bmatrix} \begin{bmatrix} h_1 \\ h_2 \end{bmatrix} = \left(\begin{bmatrix} 0 & 0 \\ 0 & I \end{bmatrix} + \begin{bmatrix} I & 0 \\ 0 & 0 \end{bmatrix} e^{2x\mathcal{E}} \right)$$

(2.10)

$$\begin{bmatrix} h_1 \\ h_2 \end{bmatrix} = \begin{bmatrix} -\mathcal{B} \\ -\mathcal{B} \end{bmatrix}.$$

Now note that the matrix $[e^{2x\mathcal{E}}]_{11}$ is nonsingular; otherwise one could construct a nontrivial solution of the homogeneous counterpart of (2.8), which would contradict Corollary A.2. As a result,

$$\begin{bmatrix} h_1 \\ h_2 \end{bmatrix} = \begin{bmatrix} [e^{2x\mathcal{E}}]_{11}^{-1} & -[e^{2x\mathcal{E}}]_{11}^{-1}[e^{2x\mathcal{E}}]_{12} \\ 0 & I \end{bmatrix} \begin{bmatrix} -\mathcal{B} \\ -\mathcal{B} \end{bmatrix}.$$

Consequently,

$$\begin{aligned} B_l(x, 0^+) &= \mathcal{C}C_l(x, 0^+) \\ &= -\mathcal{C}[I \quad 0] \begin{bmatrix} [e^{2x\mathcal{E}}]_{11}^{-1} & -[e^{2x\mathcal{E}}]_{11}^{-1}[e^{2x\mathcal{E}}]_{12} \\ 0 & I \end{bmatrix} \begin{bmatrix} I \\ I \end{bmatrix} \mathcal{B} \\ &= -([e^{2x\mathcal{E}}]_{11}^{-1}(I - [e^{2x\mathcal{E}}]_{12})\beta, \gamma). \end{aligned}$$

Using (1.5) we finally obtain for $x < 0$

$$Q(x) = -4([e^{2x\mathcal{E}}]_{11}^{-1}\{[\mathcal{E}e^{2x\mathcal{E}}]_{11}[e^{2x\mathcal{E}}]_{11}^{-1}(I - [e^{2x\mathcal{E}}]_{12}) + [\mathcal{E}e^{2x\mathcal{E}}]_{12}\}\beta, \gamma). \tag{2.11}$$

When $R(k)$ is analytic in $\mathbf{C}^-$ and hence the matrix $\mathcal{A}$ in (1.6) has its eigenvalues in the open right half-plane, the situation is somewhat more complicated. For $x > 0$, $Q(x)$ is given by (1.11). For $x < 0$ we define $C_l(x, y)$ as in (2.2), and derive (2.3) and the integral equation

$$C_l(x, y) - \int_{\max(0,-2x-y)}^{\infty} dz\, e^{-(2x+y+z)\mathcal{A}}\mathcal{B}\mathcal{C}C_l(x, z)$$

(2.12)

$$= \begin{cases} e^{-(2x+y)\mathcal{A}}\mathcal{B}, & y > -2x \\ 0, & 0 < y < -2x. \end{cases}$$

Since $C_l(x,y) = e^{-(2x+y)\mathcal{A}}\omega(x)$ for $y > -2x$, we obtain

$$C_l(x,y) - \int_{-2x-y}^{-2x} dz\, e^{-(2x+y+z)\mathcal{A}}\mathcal{B}\mathcal{C}C_l(x,z) = e^{-y\mathcal{A}}\mathcal{D}\omega(x), \qquad 0 < y < -2x,$$

where $\mathcal{D}$ is given by (1.13). Putting $\tilde{C}_l(x,y) = C_l(x,-2x-y)$ we find

$$C_l(x,y) - \int_0^y dz\, e^{(z-y)\mathcal{A}}\mathcal{B}\mathcal{C}\tilde{C}_l(x,z) = e^{-y\mathcal{A}}\mathcal{D}\omega(x),$$

$$\tilde{C}_l(x,y) - \int_y^{-2x} dz\, e^{(y-z)\mathcal{A}}\mathcal{B}\mathcal{C}C_l(x,z) = e^{(2x+y)\mathcal{A}}\mathcal{D}\omega(x),$$

implying (2.6), where $C_l(x,0^+) = \tilde{C}_l(x,(-2x)^-) = \mathcal{D}\omega(x)$. We thus get (2.9), where

$$\begin{bmatrix} I & 0 \\ [e^{2x\mathcal{E}}]_{21} & [e^{2x\mathcal{E}}]_{22} \end{bmatrix}\begin{bmatrix} h_1 \\ h_2 \end{bmatrix} = \left(\begin{bmatrix} I & 0 \\ 0 & 0 \end{bmatrix} + \begin{bmatrix} 0 & 0 \\ 0 & I \end{bmatrix} e^{2x\mathcal{E}}\right)$$

$$\begin{bmatrix} h_1 \\ h_2 \end{bmatrix} = \begin{bmatrix} \mathcal{D}\omega(x) \\ \mathcal{D}\omega(x) \end{bmatrix}.$$

Now note that the matrix $[e^{2x\mathcal{E}}]_{22}$ is nonsingular; otherwise the homogeneous counterpart of (2.12) would have a nontrivial solution, which would contradict Theorem A.1. In that case

$$\begin{bmatrix} h_1 \\ h_2 \end{bmatrix} = \begin{bmatrix} I & 0 \\ -[e^{2x\mathcal{E}}]_{22}^{-1}[e^{2x\mathcal{E}}]_{21} & [e^{2x\mathcal{E}}]_{22}^{-1} \end{bmatrix}\begin{bmatrix} I \\ I \end{bmatrix}\mathcal{D}\omega(x).$$

Consequently, $B_l(x,0^+) = \mathcal{C}C_l(x,0^+) = \mathcal{D}\omega(x)$, where

$$\omega(x) = C_l(x,(-2x)^-) = k + \mathcal{B} = [e^{2x\mathcal{E}}]_{22}^{-1}(I - [e^{2x\mathcal{E}}]_{21})\mathcal{D}\omega(x) + \mathcal{B}.$$

As a result, we obtain for $x < 0$

$$\begin{aligned} B_l(x,0^+) &= \mathcal{D}\omega(x) = U(x)\mathcal{B} \\ &= (I - \mathcal{D}[e^{2x\mathcal{E}}]_{22}^{-1}(I - [e^{2x\mathcal{E}}]_{21}))^{-1}\mathcal{B}; \\ Q(x) &= 4U(x)\mathcal{D}[e^{2x\mathcal{E}}]_{22}^{-1} \\ &\quad \{[\mathcal{E}e^{2x\mathcal{E}}]_{22}[e^{2x\mathcal{E}}]_{22}^{-1}(I - [e^{2x\mathcal{E}}]_{21}) + [\mathcal{E}e^{2x\mathcal{E}}]_{21}\}U(x)\mathcal{B}, \end{aligned} \tag{2.13}$$

where $U(x) = (I - \mathcal{D}[e^{2x\mathcal{E}}]_{22}^{-1}(I - [e^{2x\mathcal{E}}]_{21}0))^{-1}$ is nonsingular; otherwise a nonunique $\omega(x)$ may result.

We now present two illustrative examples.

Example 2.1 Let $R(k) = i\xi/(k+i\eta)$ where $\eta > 0$ and $0 \neq \xi \in [-\eta, \eta)$. Then $Q(x) \equiv 0$ for $x > 0$. Moreover, $\mathcal{A} = [-\eta]$, $\mathcal{B} = [\xi]$ and $\mathcal{C} = [1]$. Further,

$$\mathcal{E} = \begin{bmatrix} -\eta & -\xi \\ \xi & \eta \end{bmatrix},$$

so that $\mathcal{E}^2 = \gamma^2 I$ with $\gamma = \sqrt{\eta^2 - \xi^2}$. Using the identity

$$e^{y\mathcal{E}} = \begin{cases} \cosh(y\gamma)I + \dfrac{\sinh(y\gamma)}{\gamma}\mathcal{E}, & 0 < |\xi| < \eta \\ I + y\mathcal{E}, & \xi = -\eta, \end{cases}$$

we get for $0 < |\xi| < \eta$

$$e^{y\mathcal{E}} = \begin{bmatrix} \cosh(\gamma y) - \eta\dfrac{\sinh(\gamma y)}{\gamma} & -\xi\dfrac{\sinh(\gamma y)}{\gamma} \\ \xi\dfrac{\sinh(\gamma y)}{\gamma} & \cosh(\gamma y) + \eta\dfrac{\sinh(\gamma y)}{\gamma} \end{bmatrix}$$

and for $\xi = -\eta$, $e^{y\mathcal{E}} = \begin{bmatrix} 1-\eta y & \eta y \\ -\eta y & 1+\eta y \end{bmatrix}$. From (2.10) and using that $C_l(x, y) = B_l(x, y)$ when $x < 0$, we get (2.9), where

$$\begin{bmatrix} \cosh(2\gamma x) - \eta\dfrac{\sinh(2\gamma x)}{\gamma} & -\xi\dfrac{\sinh(2\gamma x)}{\gamma} \\ 0 & 1 \end{bmatrix} \begin{bmatrix} h_1 \\ h_2 \end{bmatrix} = \begin{bmatrix} -\xi \\ -\xi \end{bmatrix},$$

and therefore, as $B_l(x, 0^+) = h_1$, we get for $0 < |\xi| < \eta$ and $x < 0$

$$B_l(x, 0^+) = \frac{-\xi[\gamma - \xi\sinh(-2\gamma x)]}{\gamma\cosh(-2\gamma x) + \eta\sinh(-2\gamma x)},$$

$$Q(x) = 4\xi\gamma^2 \frac{\xi + \eta\cosh(-2\gamma x) + \gamma\sinh(-2\gamma x)}{[\gamma\cosh(-2\gamma x) + \eta\sinh(-2\gamma x)]^2} - 2\xi\delta(x),$$

where $\delta(x)$ is Dirac's delta function. For $\xi = -\eta$ we get $h_1 = h_2 = \eta$ and $B_l(x, 0^+) = \eta$, so that $Q(x) = 2\eta\delta(x)$.

One may also compute $B_r(x, y)$ from (1.4b), using the fact that (1.4b) has a separable kernel when $x < 0$. As a result, we obtain

$$B_r(x, y) = \frac{2\xi\gamma e^{-\gamma(y-2x)}}{\eta + \gamma - \xi e^{2\gamma x}}.$$

We find the same potential $Q(x)$, because $B_l(x, y) + B_r(x, y) = \eta - \gamma$ for $x < 0$.

Example 2.2 Let $R(k) = \xi\eta^2/(k^2+\eta^2)$ where $\eta > 0$ and $0 \neq \xi \in [-1, 1)$ and bound states are absent. Then

$$\mathcal{A} = \begin{bmatrix} \eta & 0 \\ 0 & -\eta \end{bmatrix}, \quad \mathcal{B} = -\frac{\eta\xi}{2}\begin{bmatrix} 1 \\ -1 \end{bmatrix}, \quad \mathcal{C} = [1 \quad 1], \quad \mathcal{D} = -\frac{\xi}{4}\begin{bmatrix} 1 & 0 \\ 0 & 0 \end{bmatrix},$$

while $\hat{R}(z) = (\xi\eta/2)e^{-\eta|z|}$. For $x > 0$ we easily solve (1.4a) and obtain

$$B_l(x, y) = \frac{-2\xi\eta\, e^{-(2x+y)\eta}}{4 + \eta\, e^{-2x\eta}}, \qquad Q(x) = \frac{-32\xi\eta^2 e^{-2x\eta}}{[4 + \xi\, e^{-2x\eta}]^2}.$$

Letting E_{24} be the matrix obtained from the 4×4 unit matrix by interchanging the second and fourth columns, we obtain

$$E_{24}\mathcal{E}E_{24} = \eta T_\xi = \eta \begin{bmatrix} 1 & \xi/2 & \xi/2 & 0 \\ \xi/2 & 1 & 0 & \xi/2 \\ -\xi/2 & 0 & -1 & -\xi/2 \\ 0 & -\xi/2 & -\xi/2 & -1 \end{bmatrix},$$

so that $E_{24}\mathcal{E}^2E_{24} = \eta^2[S_\xi \oplus S_\xi]$, where $S_\xi = \begin{bmatrix} 1 & \xi \\ \xi & 1 \end{bmatrix}$. Then

$$\begin{aligned} E_{24}e^{y\mathcal{E}}E_{24} &= [\cosh(y\eta S_\xi^{1/2}) \oplus \cosh(y\eta S_\xi^{1/2})] \\ &\quad + [\sinh(y\eta S_\xi^{1/2}) \oplus \sinh(y\eta S_\xi^{1/2})][S_\xi^{-1/2} \oplus S_\xi^{-1/2}]T_\xi, \end{aligned} \tag{2.14}$$

where for $a_\xi = \sqrt{1+\xi}$, $b_\xi = \sqrt{1-\xi}$ and $j = 0, 1, 2, \ldots$

$$S_\xi^{j/2} = \frac{1}{2}\begin{bmatrix} a_\xi^j + b_\xi^j & a_\xi^j - b_\xi^j \\ a_\xi^j - b_\xi^j & a_\xi^j + b_\xi^j \end{bmatrix}, \quad S_\xi^{-j/2} = \frac{1}{2a_\xi^j b_\xi^j}\begin{bmatrix} a_\xi^j + b_\xi^j & b_\xi^j - a_\xi^j \\ b_\xi^j - a_\xi^j & a_\xi^j + b_\xi^j \end{bmatrix}.$$

Applying the similarity $\frac{1}{\sqrt{2}}\left(\begin{bmatrix} 1 & 1 \\ -1 & 1 \end{bmatrix} \oplus \begin{bmatrix} 1 & 1 \\ -1 & 1 \end{bmatrix}\right)$ to (2.14), we find

$$e^{y\mathcal{E}} = \frac{1}{2}\begin{bmatrix} T_{11}+T_{44} & T_{11}-T_{44} & T_{13}-T_{24} & T_{13}+T_{24} \\ T_{11}-T_{44} & T_{11}+T_{44} & T_{13}+T_{24} & T_{13}-T_{24} \\ -T_{13}+T_{24} & -T_{13}-T_{24} & T_{33}+T_{22} & T_{33}-T_{22} \\ -T_{13}-T_{24} & -T_{13}+T_{24} & T_{33}-T_{22} & T_{33}+T_{22} \end{bmatrix},$$

where

$$\begin{cases} T_{11} = \cosh(y\eta a_\xi) + \left(1+\frac{\xi}{2}\right)\frac{\sinh(y\eta a_\xi)}{a_\xi}, & T_{44} = \cosh(y\eta b_\xi) - \left(1-\frac{\xi}{2}\right)\frac{\sinh(y\eta b_\xi)}{b_\xi}; \\ T_{22} = \cosh(y\eta b_\xi) + \left(1-\frac{\xi}{2}\right)\frac{\sinh(y\eta b_\xi)}{b_\xi}, & T_{33} = \cosh(y\eta a_\xi) - \left(1+\frac{\xi}{2}\right)\frac{\sinh(y\eta a_\xi)}{a_\xi}; \\ T_{13} = \frac{\xi}{2}\frac{\sinh(y\eta a_\xi)}{a_\xi}, & T_{24} = \frac{\xi}{2}\frac{\sinh(y\eta b_\xi)}{b_\xi}, \end{cases}$$

and the y-dependence has not been written. Using $\mathcal{Q}_{\mathcal{A}} = \operatorname{diag}(1,0,0,1)$ and $I - \mathcal{Q}_{\mathcal{A}} = \operatorname{diag}(0,1,1,0)$, we obtain the invertibility of the matrix

$$\mathcal{Q}_{\mathcal{A}} + (I - \mathcal{Q}_{\mathcal{A}})e^{2x\mathcal{E}}$$

(2.15)

$$= \begin{bmatrix} 1 & 0 & 0 & 0 \\ \frac{1}{2}(T_{11}-T_{44}) & \frac{1}{2}(T_{11}+T_{44}) & \frac{1}{2}(T_{13}+T_{24}) & \frac{1}{2}(T_{13}-T_{24}) \\ \frac{1}{2}(-T_{13}+T_{24}) & -\frac{1}{2}(T_{13}+T_{24}) & \frac{1}{2}(T_{33}+T_{22}) & \frac{1}{2}(T_{33}-T_{22}) \\ 0 & 0 & 0 & 1 \end{bmatrix},$$

where $T_{ij} = T_{ij}(2x)$, because the determinant of the 2×2 middle block

$$\Delta(x,\eta,\xi) = \frac{1}{2}\left[\cosh(2x\eta a_\xi)\cosh(2x\eta b_\xi) + \frac{\sinh(2x\eta a_\xi)\sinh(2x\eta b_\xi)}{a_\xi b_\xi}\right] > 0.$$

Writing $[S(x,\eta,\xi)]_{i,j=1}^{4}$ for the inverse of the matrix in (2.15), we obtain

$$B_l(x,0^+) = \begin{bmatrix} 1 & 1 \end{bmatrix} \begin{bmatrix} S_{11}(x,\eta,\xi)+S_{13}(x,\eta,\xi) & S_{12}(x,\eta,\xi)+S_{14}(x,\eta,\xi) \\ S_{21}(x,\eta,\xi)+S_{23}(x,\eta,\xi) & S_{22}(x,\eta,\xi)+S_{24}(x,\eta,\xi) \end{bmatrix} \begin{bmatrix} [h(x)]_1 \\ [h(x)]_2 \end{bmatrix},$$

where $h(x) = \mathcal{D}\omega(x) - P_{\mathcal{A}}^{(-)}\mathcal{B}$. Since $S_{11} = 1$ and $S_{12} = S_{13} = S_{14} = 0$, we get

$$\begin{aligned} B_l(x,0^+) &= -\frac{\xi}{4}(1 + S_{21}(x,\eta,\xi) + S_{23}(x,\eta,\xi))\,[\omega(x)]_1 \\ &\quad -\frac{\eta\xi}{2}(S_{22}(x,\eta,\xi) + S_{24}(x,\eta,\xi)), \end{aligned}$$

where it suffices to compute the first component $[\omega(x)]_1$ of $\omega(x)$. Then $Q(x)$ will follow using (1.9) for $x < 0$. Since $R(k) = o(1/k)$ as $k \to \infty$, there will not be any delta function terms in the potential [15]. We omit the rather cumbersome computation of $[\omega(x)]_1$.

3 Solution of the Inverse Problem with Bound States

A. Preliminaries

Let the (simple) poles of $T(k)$ occur at $i\kappa_j$, let the corresponding residues and bound state constants be t_j and c_j ($j = 1, \ldots, \mathcal{N}$), and let $\kappa_1 > \cdots > \kappa_{\mathcal{N}}$. Considering $T(k)$ on the positive imaginary axis where $T(k) \to 1$ as $k \to +i\infty$, $T(k) = \overline{T(-\bar{k})} \in \mathbf{R}$ and $-it_j = \lim_{\kappa \to \kappa_j} (\kappa - \kappa_j) T(i\kappa)$, we see that $(-1)^{j-1}(-it_j) > 0$, since for every pole $i\kappa_j$, κ_j is a sign change of $\kappa \mapsto 1/T(i\kappa)$. Further, $f_l(i\kappa_j, x) = e^{-\kappa_j x} + o(1)$ as $x \to +\infty$, $f_r(i\kappa_j x) = e^{\kappa_j x} + o(1)$ as $x \to -\infty$, $c_j = f_r(i\kappa_j, x)/f_l(i\kappa_j, x)$ and the fact that $f_l(i\kappa_j, x)$ has $j - 1$ simple zeros (which follows from the usual oscillation theorems [14]), imply that $(-1)^{j-1} c_j > 0$. Consequently,

$$d_j = -it_j c_j > 0, \qquad j = 1, \ldots, \mathcal{N}. \tag{3.1}$$

B. Marchenko Equation for Reflectionless Potentials

When $\hat{R}(z) \equiv 0$, (1.4a) reduces to

$$B_l(x, y) = -\sum_{j=1}^{\mathcal{N}} d_j e^{-(2x+y)\kappa_j} [1 + u_j(x)],$$

where

$$\begin{aligned} u_j(x) &= \int_0^\infty dz\, e^{-\kappa_j z} B_l(x, z); \\ \sum_{j=1}^{\mathcal{N}} \left\{ \delta_{ij} + \frac{d_j e^{-2x\kappa_j}}{\kappa_i + \kappa_j} \right\} u_j(x) &= -\sum_{j=1}^{\mathcal{N}} d_j \frac{e^{-2x\kappa_j}}{\kappa_i + \kappa_j}, \qquad i = 1, \ldots, \mathcal{N}. \end{aligned} \tag{3.2}$$

We now easily see that $(d_j e^{-2\kappa_j}/(\kappa_i + \kappa_j))_{i,j=1}^{\mathcal{N}}$ is similar to the real symmetric matrix $M(x)$ with entries

$$[M(x)]_{ij} = \frac{\sqrt{d_i d_j}\, e^{-(\kappa_i + \kappa_j)x}}{\kappa_i + \kappa_j}.$$

This matrix is positive semidefinite, because for every $\xi = (\xi_1, \ldots, \xi_{\mathcal{N}}) \in \mathbf{C}^{\mathcal{N}}$

$$(M(x)\xi, \xi) = \int_0^\infty dz \left| \sum_{i=1}^{\mathcal{N}} \xi_i \sqrt{d_i}\, e^{-\kappa_i (z+x)} \right|^2 \geq 0.$$

Hence (3.2) is uniquely solvable and

$$B_l(x, y) = \sum_{i,j=1}^{\mathcal{N}} \sqrt{d_i d_j}\, e^{-\kappa_i y} e^{-(\kappa_i + \kappa_j)x} [(I + M(x))^{-1}]_{ij}.$$

Consequently,

$$Q(x) = 2\sum_{i,j=1}^{\mathcal{N}} \sqrt{d_i d_j}\, e^{-(\kappa_i+\kappa_j)x} + \Bigg\{(\kappa_i+\kappa_j)[(I+M(x))^{-1}]_{ij} + \left[(I+M(x))^{-1}\left(\frac{d}{dx}M(x)\right)(I+M(x))^{-1}\right]_{ij}\Bigg\}.$$

C. Formalism and Solution of the Marchenko Equation

The integral kernel function $\hat{S}_l(z)$ can be written in the form (1.14), where $\breve{\mathcal{A}}$, $\breve{\mathcal{B}}$ and $\breve{\mathcal{C}}$ are given by (1.15), $\breve{\beta} = [\beta - d_1 \dots - d_{\mathcal{N}}]^T$ and $\breve{\gamma} = [\gamma 1 \dots 1]^T$, and $d_1, \dots, d_{\mathcal{N}}$ are positive constants [cf. (3.1)]. Then

$$P^{(+)}_{\breve{\mathcal{A}}} = P^{(+)}_{\mathcal{A}} \oplus I_{\mathbf{C}^{\mathcal{N}}}, \qquad P^{(-)}_{\breve{\mathcal{A}}} = P^{(-)}_{\mathcal{A}} \oplus 0_{\mathbf{C}^{\mathcal{N}}}.$$

We define $\breve{\mathcal{D}}$ as in the statement of Theorem 1.2. With these definitions, one can repeat Subsections 2.c and 2.d when there are $\mathcal{N}$ bound states, where $\mathcal{A}$, $\mathcal{B}$, $\mathcal{C}$, $\mathcal{D}$, β and γ are to be replaced by $\breve{\mathcal{A}}$, $\breve{\mathcal{B}}$, $\breve{\mathcal{C}}$, $\breve{\mathcal{D}}$, $\breve{\beta}$ and $\breve{\gamma}$, respectively. As there is no conceptual problem, it will not be discussed at great length.

We limit ourselves to a few remarks. Instead of (1.8), we now have the Marchenko equation

$$\begin{aligned} B_l(x,y) - \int_0^\infty dz\, \breve{\mathcal{C}}E(2x+y+z; -\breve{\mathcal{A}})\breve{\mathcal{B}}B_l(x,z) \\ = \breve{\mathcal{C}}E(2x+y; -\breve{\mathcal{A}})\breve{\mathcal{B}}, \quad y > 0, \end{aligned} \tag{3.3}$$

where the potential $Q(x)$ follows from (1.9). When $x > 0$, (3.3) has a separable kernel and its solution $B_l(x,y)$ and the corresponding potential $Q(x)$ are given in the statement of Theorem 1.2. On the other hand, when $x < 0$, the solution of (3.3) for $y = 0^+$ is given by (2.7), where $\mathcal{A}$, $\mathcal{B}$, $\mathcal{C}$, and $\mathcal{D}$ are to be replaced by $\breve{\mathcal{A}}$, $\breve{\mathcal{B}}$, $\breve{\mathcal{C}}$, and $\breve{\mathcal{D}}$, respectively; these replacements are also to be made in the expressions for $\mathcal{E}$ [thus converting (2.1) into (3.4) below], $\mathcal{Q}_{\mathcal{A}}$ and $\omega(x)$. The potential $Q(x)$ follows by applying (1.9). When $x < 0$ and $R(k)$ is analytic in $\mathbf{C}^-$, the solution of (3.3) for $y = 0^+$ and the potential $Q(x)$ are given by

$$\begin{aligned} B_l(x,0^+) &= (I - \breve{\mathcal{D}}[e^{2x\breve{\mathcal{E}}}]_{22}^{-1}(I - [e^{2x\breve{\mathcal{E}}}]_{21}))^{-1}\breve{\mathcal{B}}; \\ Q(x) &= 4\breve{U}(x)\breve{\mathcal{D}}[e^{2x\breve{\mathcal{E}}}]_{22}^{-1} \\ &\quad \{[\breve{\mathcal{E}}e^{2x\breve{\mathcal{E}}}]_{22}[e^{2x\breve{\mathcal{E}}}]_{22}^{-1}(I - [e^{2x\breve{\mathcal{E}}}]_{21}) + [\breve{\mathcal{E}}e^{2x\breve{\mathcal{E}}}]_{21}\}\breve{U}(x)\breve{\mathcal{B}}, \end{aligned}$$

where $\breve{U}(x) = (I - \breve{\mathcal{D}}[e^{2x\breve{\mathcal{E}}}]_{22}^{-1}(I - [e^{2x\breve{\mathcal{E}}}]_{21}))^{-1}$ is nonsingular. Here

$$\breve{\mathcal{E}} = \begin{bmatrix} \breve{\mathcal{A}} & -\breve{\mathcal{B}}\breve{\mathcal{C}} \\ \breve{\mathcal{B}}\breve{\mathcal{C}} & -\breve{\mathcal{A}} \end{bmatrix}. \tag{3.4}$$

When $x < 0$ and $R(k)$ is analytic in $\mathbf{C}^+$, one does not find an expression analogous to (2.11), because $\breve{A}$ has eigenvalues in both the left and the right half-plane.

Appendix A. Well-posedness of the Marchenko Equations

The main result of the Appendix is well-known [16, 18]. We give a short proof.

For real $\hat{\rho} \in L^1(\mathbf{R}^+)$, define

$$(\mathcal{L}_{\hat{\rho}} f)(y) = \int_0^\infty dz\, \hat{\rho}(y+z) f(z).$$

Then $\mathcal{L}_{\hat{\rho}}$ is a compact operator on each of the Banach spaces $L^p(\mathbf{R}^+)$ $(1 \leq p \leq +\infty)$ and on $BC([0, +\infty))$, the space of bounded continuous complex-valued functions on $[0, +\infty)$ with supremum norm. Moreover, $\mathcal{L}_{\hat{\rho}}$ maps $L^\infty(\mathbf{R}^+)$ into $BC([0, +\infty))$. Let $H^p_\pm(\mathbf{R})$ be the closed subspaces of $L^p(\mathbf{R})$ consisting of those L^p-functions that have an analytic continuation to $\mathbf{C}^\pm$ (cf. [25]). Then $H^p_+(\mathbf{R}) \oplus H^p_-(\mathbf{R}) = L^p(\mathbf{R})$ $(1 < p < +\infty)$, where the direct sum is orthogonal when $p = 2$. Letting $\tau^{(H)}_\pm : H^2_\pm(\mathbf{R}) \to L^2(\mathbf{R})$ denote the natural embeddings and $\pi^{(H)}_\pm : L^2(\mathbf{R}) \to H^2_\pm(\mathbf{R})$ the orthogonal projections, $(\mathcal{F}h)(k) = \int_{-\infty}^\infty dt\, e^{ikt} h(t)$ the Fourier transform, and $(Jh)(k) = h(-k)$ the sign inversion, we have on $L^2(\mathbf{R}^+)$

$$\mathcal{L}_{\hat{\rho}} = \mathcal{F}^{-1} \pi^{(H)}_+ J M_\rho \tau^{(H)}_+ \mathcal{F},$$

where $(M_\rho h)(k) = \rho(k) h(k)$ is the operator of multiplication by $\rho(k) = \int_{-\infty}^\infty dt\, e^{-ikt} \hat{\rho}(t)$. By Nehari's theorem [30, 31],

$$\left\| \mathcal{L}_{\hat{\rho}} \right\|_{L^2(\mathbf{R}^+)} = \inf_{u \in H^\infty_+(\mathbf{R})} \|\rho - u\|_\infty. \tag{A.1}$$

The following result implies the unique solvability of the Marchenko equations (1.4a) and (1.4b) on a variety of function spaces.

Theorem A.1 *On the Banach spaces $BC([0, +\infty))$ and $L^p(\mathbf{R}^+)$ $(1 \leq p \leq +\infty)$, the integral operators with real symmetric kernels $\hat{S}_l(2x + y + z)$ and $\hat{S}_r(-2x + y + z)$ have their eigenvalues in the interval* $(-1, +\infty)$.

Proof: Without bound states, $|R(k)| < 1$ for $0 \leq k \in \mathbf{R}$ and $R(0) \in [-1, 1)$ imply that $|R(k) + \varepsilon| < 1$ for $k \in \mathbf{R}$ for some suitable $\varepsilon \geq 0$ not depending on k. Then (A.1) implies that $\mathcal{L}_{\hat{\rho}}$ is a strict contraction on $L^2(\mathbf{R}^+)$. Since $d_1, \ldots, d_{\mathcal{N}} > 0$, the integral kernel with kernel $\sum_{j=1}^{\mathcal{N}} d_j e^{-\kappa_j(2x+y+z)}$ is positive selfadjoint on $L^2(\mathbf{R}^+)$. Hence, when bound states are taken into account, $\mathcal{L}_{\hat{S}_l}$ has its eigenvalues in $(-1, +\infty)$. The proof for $\mathcal{L}_{\hat{S}_r}$ is similar.

A simple Fredholm argument yields the same results in any of the other function spaces mentioned above. □

Corollary A.2 *Let $c > 0$. On the Banach spaces $L^p(0,c)$ $(1 \leq p \leq +\infty)$ and $BC([0,c])$, the integral operators with real symmetric kernels $\hat{S}_l(2x+y+z)$ and $\hat{S}_r(-2x+y+z)$ have their eigenvalues in the interval $(-1,+\infty)$.*

Proof: Let $\tau_c : L^2(0,c) \to L^2(\mathbf{R}^+)$ denote the natural embedding and $\pi_c : L^2(\mathbf{R}^+) \to L^2(0,c)$ the orthogonal projection. Then these integral operators can be written as $\pi_c \mathcal{L}_{\hat{S}_l} \tau_c$ and $\pi_c \mathcal{L}_{\hat{S}_r} \tau_c$, respectively, and hence when adding the identity one gets positive selfadjoint operators.

A simple Fredholm argument again yields the same results in any of the other Banach spaces of functions on $(0,c)$. □

References

1. Z.S. Agranovich and V.A. Marchenko, *The Inverse Problem of Scattering Theory*. Kharkov Univ. Press, Kharkov, 1960 [English Translation: Gordon & Breach, New York, 1963]. MR **28**, #5696.
2. T. Aktosun, M. Klaus and C. van der Mee, *Explicit Wiener-Hopf factorization for certain non-rational matrix functions*. Integral Equations and Operator Theory **15** (1992), 879–900. MR 93j:47030.
3. D. Alpay and I. Gohberg, *Inverse spectral problems for differential operators with rational scattering matrix functions*. J. Diff. Eqs. **118** (1995), 1–19. MR 96f:34121.
4. D. Alpay and I. Gohberg, *Inverse scattering problem for differential operators with rational scattering matrix functions*. In: Böttcher, A. and Gohberg, I. (Eds.), Singular Integral Operators and Related Topics. Joint German-Israeli Workshop, Tel Aviv, March 1–10, 1995. Birkhäuser OT **90**, Basel, 1996; pp. 1–18. MR 97j:34112.
5. D. Alpay and I. Gohberg, *Potentials associated to rational weights*. In: Gohberg, I. and Lyubich, Yu. (Eds.), New Results in Operator Theory and its Applications. The Israel M. Glazman Memorial Volume, Birkhäuser OT **98**, Basel, 1997; pp. 23–40. MR 99a:34218.
6. D. Alpay and I. Gohberg, *State space method for inverse spectral problems*. Progress in Systems and Control Theory **22** (1997), 1–16 MR 98d:34113.
7. D. Alpay, I. Gohberg and L. Sakhnovich, *Inverse scattering problem for continuous transmission lines with rational reflection coefficient function*. In: Gohberg, I., Lancaster, P. and Shivakumar, P. N. (Eds.), Recent Developments in Operator Theory and its Applications. International Conference in Winnipeg, October 2–6, 1994. Birkhäuser OT **87**, Basel, 1996. MR 97f:340721.
8. D. Alpay and I. Gohberg, *Inverse problem for Sturm-Liouville operators with rational reflection coefficient*. Integral Equations and Operator Theory **30** (1998), 317–325. MR 1608685.
9. H. Bart, I. Gohberg and M.A. Kaashoek, *Minimal Factorization of Matrix and Operator Functions*. Birkhäuser OT **1**, Basel, 1979. MR 81a:47001.
10. H. Bart, I. Gohberg and M.A. Kaashoek, *Exponentially dichotomous operators and inverse Fourier transforms*. Report 8511/M, Econometric Institute, Erasmus University of Rotterdam, The Netherlands, 1985.
11. H. Bart, I. Gohberg and M.A. Kaashoek, *Wiener-Hopf factorization, inverse Fourier transforms and exponentially dichotomous operators*. J. Funct. Anal. **68** (1986), 1–42. MR 88d:47052.

12. R. Bhatia, *Matrix Analysis*. Graduate Texts in Mathematics **169**, Springer, New York, 1997. MR 98i:15003.
13. K. Chadan and P. Sabatier, *Inverse Problems in Quantum Scattering Theory*. Texts and Monographs in Physics, 2nd ed., Springer, New York, 1989. MR 90b:81002.
14. E.A. Coddington, and N. Levinson, *Theory of Ordinary Differential Equations*. McGraw-Hill, New York, 1955. MR **16**, #1022b.
15. P. Deift and E. Trubowitz, *Inverse scattering on the line*. Comm. Pure Appl. Math. **32** (1979), 121–251. MR 80e:34011.
16. W. Eckhaus and A. van Harten, *The Inverse Scattering Transformation and the Theory of Solitons*. North-Holland, Math. Studies **50** (1981). MR 83c:35101.
17. L.D. Faddeev, *The inverse problem in the quantum theory of scattering*. Uspekhi Matem. Nauk **14** (1959), 57–119 [English translation (in collaboration with B. Seckler): J. Math. Phys. **4** (1963), 72–104]. MR **22**, #1344 (Uspekhi); MR **26**, #7328 (JMP).
18. L.D. Faddeev, *Properties of the S-matrix of the one-dimensional Schrödinger equation*. Amer. Math. Soc. Transl. **2** (1964), 139–166 [Trudy Mat. Inst. Steklova **73** (1964), 314–336 (Russian)]. MR **31**, #2446.
19. I.M. Gel'fand and B.M. Levitan, *On the determination of a differential equation from its spectral function*. Dokl. Akad. Nauk SSSR **77**(4) (1951), 557–560 [Russian]. MR **13**, 240f.
20. I.M. Gel'fand and B.M. Levitan, *On the determination of a differential equation from its spectral function*. Izv. Akad. Nauk SSSR **15** (1951), 309–360 [Russian]. MR **13**, 558f.
21. I. Gohberg, S. Goldberg and M.A. Kaashoek, *Classes of Linear Operators*. vol. I. Birkhäuser OT **49**, Basel, 1990. MR 93d:47002.
22. I. Gohberg, M.A. Kaashoek and A.L. Sakhnovich, *Canonical systems with rational spectral densities: Explicit formulas and applications*. Math. Nachr. **194** (1998), 93–125. MR 1653082.
23. I. Gohberg, M.A. Kaashoek and A.L. Sakhnovich, *Pseudocanonical systems with rational Weyl functions: Explicit formulas and applications*. J. Diff. Eqs. **146** (1998), 375–398. MR 1631291.
24. I. Gohberg, M.A. Kaashoek and A.L. Sakhnovich, *Sturm-Liouville systems with rational Weyl functions: Explicit formulas and applications*. Integral Equations and Operator Theory **30** (1998), 338–377. MR 1608648.
25. K. Hoffman, *Banach Spaces of Analytic Functions*. Prentice-Hall, Englewood Cliffs, 1962; also: Dover Publ., New York, 1988. MR **24**, #A2844.
26. M. Klaus, *Low-energy behaviour of the scattering matrix for the Schrödinger equation on the line*. Inverse Problems **4** (1988), 505–512. MR 89k:81185.
27. M.G. Krein, *On the determination of a potential from its S-function*. Dokl. Akad. Nauk SSSR **105**(3) (1955), 433–436 [Russian]. MR **17**, 1210b.
28. V.A. Marchenko, *Recovery of the potential energy from the phases of the scattered waves*. Dokl. Akad. Nauk SSSR **104**(5) (1955), 695–698 [Russian]. MR **17**, 740d.
29. A. Melin, *Operator methods for inverse scattering on the real line*. Comm. in Partial Differential Equations **10**(7) (1985), 677–766. MR 86f:35177.
30. J.R. Partington, *An Introduction to Hankel Operators*. London Math. Soc. Student Texts, vol. **13**, Cambridge Univ. Press, Cambridge, 1988. MR 90c:47047.
31. S.C. Power, *Hankel Operators on Hilbert Space*. Research Notes in Mathematics **64**, Pitman, Boston, 1982. MR 84e:47037.

32. M. Rosenblum, *On the operator equation $BX - XA = Q$*, Duke Math. J. **23** (1956), 263–269. MR **18**, 54d.
33. W.E. Roth, *The equations $AX - YB = C$ and $AX - XB = C$ in matrices*. Proc. Amer. Math. Soc. **3** (1952), 392–396. MR **13**, 900c.
34. M. Wadati, *Generalized matrix form of the inverse scattering method*. In Bullough, R.K. and Caudry, P. J. (Eds.), Solitons, Topics in current physics, vol. **17**, pp. 287–299, Springer, Berlin, 1980.
35. M. Wadati and T. Kamijo, *On the extension of inverse scattering method*. Progress of Theor. Phys. **52** (1974), 397–414. MR **53**, #9964.

Cornelis van der Mee
Dipartimento di Matematica
Università di Cagliari
Via Ospedale 72
09124 Cagliari
Italy
corneliskrein.unica.it
http://krein.unica.it/~cornelis

MSC Primary 34A55, 81U40

Operator Theory:
Advances and Applications, Vol. 117
© 2000 Birkhäuser Verlag Basel/Switzerland

An Arbitrary Oriented Crack in the Box Shell

V.I. Migdalski and V.V. Reut

Stresses are considered in the box shell formed by two semi-infinite plates joined at right angles. The plates are similar but their thicknesses are different. The crack is an arbitrarily oriented and goes to the shell's tip. The edges of the crack are loaded in a plane of box shell's plate. It has been known that numerical methods of solving of stress state problems in the box shell weakened by a crack coming out on the tip possess bad convergence through the necessity of taking into account real singularities near the crack. The problem is solved on the assumption that the plates have small thickness with respect to the length of crack and this makes possible the consideration of problem in an asymptotical formulation, (see Popov, Reut [4]).

The initial problem is reduced to the combined state of plane and bending stress of an imaginary plate with intersecting defects whose roles are performed by the crack and the tip of box.

Using the generalized integral transforms method this problem is reduced to the one dimensional discontinuous boundary value problem in the Mellin transforms. Its solution is constructed using Green's functions. The boundary conditions on the edge of the box are satisfied exactly. After the inversion of the Mellin transforms the original problem is reduced to the system of two singular integral equation over a finite interval with the kernel of the Mellin convolution type with respect to the displacement jumps of crack's edges.

An exact solution has been obtained is quadratures by reducing the original problem to the Riemannian problem for a vector. The solution asymptotic near the point of intersecting of crack and the box's shell is investigated. Numerical results for the stress intensity factors in the apex of crack depending on the angle formed by the crack and the tip of box are given.

Under the defect we mean the line while crossing which the stresses and displacement of box have discontinuities. The advantages of this approach are on the one hand is the diminution of number of the solving equations and the quantity of boundary conditions to be fulfilled,on the other hand the solving methods for such problem are well worked out, (see Popov, Reut [3]).

1

In the case of load which is symmetric relatively to crack the mathematical formulation of this problem is given by equations of the plane elasticity theory together with the conditions on the box's tip:

$$\begin{aligned} \langle v \rangle = 0, \langle \tau_{r\theta} \rangle &= K_\theta \tau_{r\theta}|_{\theta+0}, \sigma_\theta(r, \theta+0) = \sigma_\theta(r, \theta-0) = 0 \\ \theta &= 0, \theta = \pi, 0 < r < \infty \end{aligned} \tag{1.1}$$

and conditions on the crack

$$\begin{aligned} \sigma_\theta &= f_1(r), \tau_{r\theta} = f_2(r) \\ \theta &= \alpha, 0 < r < 1 \end{aligned} \tag{1.2}$$

Here $\langle A \rangle = A(\theta-0) - A(\theta+0)$, $\sigma_\theta, \sigma_r, \tau_{r\theta}$ are the normal and tangential stresses, u, v are the radial and tangential displacements respectively, $\theta > 0$ corresponds to the plate with a crack, α is the angle between the crack and the tip of box.

$$K_0 = 1 - h_1/h_2;\ K_\pi = 1 - h_2/h_1$$

where h_1, h_2 are thicknesses of shell's plates.

Let us denote

$$\begin{aligned} \langle v(r,\theta) \rangle|_{\theta=\alpha} &= \chi(r), \\ \langle u(r,\theta) \rangle|_{\theta=\alpha} &= \mu(r), \\ 0 < r < \infty \end{aligned} \tag{1.3}$$

where $\chi(r) \equiv 0$, $\mu(r) \equiv 0$ when $r > 1$.

Using the generalized integral transforms method (see Popov [1]) and introducing into consideration the Mellin transformations

$$[\sigma_{\theta p}, \sigma_{rp}, u_p, v_p, \tau_{r\theta p}] = M[\sigma_\theta, \sigma_r, u, vr^{-1}, \tau_{r\theta} r^{-1}] \tag{1.4}$$

where

$$M[A(r)] = \int_0^\infty A(r) r^p dr\ (p \epsilon \Lambda)$$

and Λ is a contour lying in the strip $\{-1 < Re\ p < 0, -\infty < Im\ p < \infty\}$. Our problem can be reduced to the boundary-value problem

$$\begin{aligned} L(\sigma_{\theta p}) &= 0 \\ \sigma_{\theta p}|_{\theta=0,\theta=\alpha} &= 0 \end{aligned} \tag{1.5}$$

Here

$$L \equiv \partial^4/\partial\theta^4 + [(p+1)^2 + (p-1)^2]\partial^2/\partial\theta^2 + (p+1)^2(p-1)^2 \tag{1.6}$$

The solution of one-dimensional discontinuity boundary-value problems is constructed by means of Green's functions method [1]

$$\sigma_{\theta p}(\theta) = -\omega_3 G^*_{(j)}(\theta, \alpha) + \omega_2 \partial/\partial\xi G^*_{(j)}(\theta, \alpha) \quad (1.7)$$

where $G^*_{(1)}(\theta, \xi)$- the Green's function of boundary-value problem (1.5); the case: j = 1 is the case of equal thickness, j = 2 is the case of different thickness. Here

$$G^*_{(1)}(\theta, \xi) = G(\theta, \xi) + A(\xi)G(\theta, 0) + B(\xi)G(\theta, \pi) \quad (1.8)$$

where

$$\begin{aligned} A(\xi) &= -(\sin \pi p)^{-1} [p \sin\xi \cos(\pi - \xi)p + \cos\xi \sin(\pi - \xi)p], \\ B(\xi) &= -(\sin \pi p)^{-1}[p \sin\xi \cos\xi p - \cos\xi \sin\xi p], \\ G(\theta, \xi) &= -(8p(p^2 - 1))^{-1} \sin \pi p\{(p+1)\cos(p-1)[\pi - |\theta - \xi|] \\ &\quad -(p-1)\cos(p+1)[\pi - |\theta - \xi|]\}; \\ G^*_{(2)}(\theta, \xi) &= G^*_{(1)}(\theta, \xi) + \gamma(\xi)S[G(\theta, 0)] + \beta(\xi)S[G(\theta, \pi)] \end{aligned} \quad (1.9)$$

$$\begin{aligned} \gamma(\xi) &= 2K_0(2 + K_0)^{-1}(4p \sin \pi p)^{-1} \sin\xi \cos(\pi - \xi)p, \\ \beta(\xi) &= -2K_\pi(2 - K_\pi)^{-1}(4p \sin \pi p)^{-1} \sin\xi \cos\xi p, \end{aligned} \quad (1.10)$$

$$S \equiv \partial^2/\partial\theta^2 + [(p+1)^2 + (p-1)^2] \quad (1.11)$$

Satisfying the boundary conditions (1.1)–(1.3) and using the relations between stresses and displacements by solving the system of linear equations with respect to ω_2 and ω_3, we can write the following:

$$[\sigma_{\theta p}, \tau_{r\theta p}]^T = Ep^2 W(p, \alpha)[X_p, M_p]^T, \quad (1.12)$$

where

$$[X_p, M_p]^T = M[\chi(r)r^{-1}, \mu(r)]^T,$$

E is the Young's modulus of elasticity, $W(p, a) = \|w_{ij}\|$, (i, j = 1, 2) is a matrix-function

$$\begin{aligned} w_{11} &= (p^2 - 1)G^*_{(j)}(\alpha, \alpha); \; w_{12} = -(p-1)\partial/\partial\xi G^*_{(j)}(\alpha, \alpha); \\ w_{21} &= (p+1)\partial/\partial\theta G^*_{(j)}(\alpha, \alpha); \; w_{22} = \partial^2/\partial\theta\partial\xi G^*_{(j)}(\alpha, \alpha) \end{aligned} \quad (1.13)$$

Denoting

$$F(p) = [F_1(p), F_2(p)]^T = M[f_1(r)r^{-1}, f_2(r)]^T$$

and employing the conditions on the crack (1.2) we get a boundary value problem for a vector

$$Ep^2 W(p, \alpha)[X_p, M_p]^T = F(p), \qquad (p \in \Lambda) \quad (1.14)$$

After the inversion of Mellin transforms (1.14) can be represented as a system of two singular integral equation in the finite interval with a kernel of the Mellin convolution type with respect to $\chi(r)$ and $\mu(r)$-displacement of crack's edges.

2

Let us introduce the vector

$$\phi(r) = [(\phi)_1(r), (\phi)_2(r)]^T = \frac{1}{4}E\partial/\partial r[\langle v\rangle, \langle u\rangle]^T, 0 < r < 1 \tag{2.1}$$

$$\Phi_-(p) = M[(\phi)(r)] \tag{2.2}$$

Then, by the completing the definition of (1.14) we can write the Riemann boundary-value problem for a vector

$$G(p)\Phi_-(p) = F(p) + \Psi(p), \qquad (p \in \Lambda) \tag{2.3}$$

where $G(p) = pW(p, \alpha)$ is the coefficient of Riemann problem.

By means of (1.7), (1.10) and (1.13) the matrix function G(p) can be written in the following form:

$$G(p) = s(p)I(p) + b(p)\left\| \begin{matrix} I(p) & m(p) \\ n(p) & -I(p) \end{matrix} \right\| \tag{2.4}$$

where s(p) is the half-trace of matrix G(p) i.e. the half-sum of diagonal elements, b(p), I(p), m(p), n(p) are the polynomials.

The further solving of (2.4) is realized by constructing of the factorization for the coefficient G(p) which structure assumes the matrix factorization by the Khrapkov's scheme (see Khrapkov [2]).

As the final result it allow us to write the representation for the Mellin transformation $\Phi_-(p)$.

$$\Phi_-(p) = -Q_-(p)X^-(p)/(D^-(p)D^0(p)) \tag{2.5}$$

$$Q_-(p) = (2\pi i)^{-1}\int_\Lambda [X^+(s)]^{-1}F(s)(D^+(s)(s-p)))^{-1}ds \tag{2.6}$$

$$G(p) = ctg\,\pi p V(p) \tag{2.7}$$

$$ctg\pi p = D^+(p)D^-(p)D^0(p),\ D^0(p) = \left(p + \frac{1}{2}\right)/p \tag{2.8}$$

$$\begin{aligned} D^+(p) &= \Gamma(1-p)/\Gamma\left(\frac{1}{2} - p\right), D^-(p) \\ &= \Gamma(p+1)/\Gamma\left(p + \frac{1}{2} + 1\right) \end{aligned} \tag{2.9}$$

$$\begin{aligned} V(p) &= X^+(p)[X^-(p)]^{-1} \\ &= [X^-(p)]^{-1}[X^-(p)]^{-1}X^+(p), (p \in \Lambda) \end{aligned} \tag{2.10}$$

Conversing the Mellin transformations we obtain in quadratures the vector $\phi(r)$, which was defined in (2.1).

3

On the base of the Tauberian theorems the asymptotic behaviour of $\phi(r)$ when $r \to 1-0$ and $r \to 1+0$ is determine by the behaviour of $\Phi_-(p)$ when $|p| \to \infty$.

Taking into account (2.5) and (2.8) when $r \to 1-0$ we have

$$\begin{aligned}\phi(r) &= (\sqrt{(-\pi \ln r)})^{-1}(2\pi i)^{-1} \int_\Lambda [X^+(s)]^{-1} F(s)(D^+(s))^{-1} ds \\ &\quad + O(\sqrt{(-\pi \ln r)})^{-1}\end{aligned}$$

In the case when singularities of F(p) are different from σ_i-roots of the characteristics equation $\det G(p) = 0$ for $r \to +0$ it follows

$$\begin{aligned}\phi(r) &= \sum_{i=1}^{\infty} Res\,(p = \sigma_i)[r^{-p-1}(G(p))^{-1}(F(p) \\ &\quad - X^+(p)Q_+(p)D^+(p))]\ f_i(r) = \eta_i \exp(\gamma_i \ln r), \quad (i = 1, 2) \qquad (3.1) \\ F_1(p) &= \eta_1/(p+\gamma_1), \qquad F_2(p) = \eta_2/(p+\gamma_2+1)\end{aligned}$$

Using the formulas for the stress intensity factors [5]

$$K_1 = [k_1, k_2]^T = \frac{1}{4} E \lim_{r \to 1+0} \phi(r)\sqrt{(2\pi(r-1))}$$

we get

$$\begin{aligned}[k_1, k_2]^T &= \sqrt{2}\{\eta_1[X^+_{22}(-\gamma_1), -X^+_{12}(-\gamma_1)]^T/(D^+(-\gamma_1)\det X^+(-\gamma_1) \\ &\quad + \eta_2[-X^+_{21}(-\gamma_2-1), X^+_{11}(\gamma_2-1)]^T/(D^+(-\gamma_2-1) \\ &\quad \det X^+(-\gamma_2-1)\end{aligned}$$

where X^+_{ij} are the elements of the matrix $X^+(p)$.

There were studied the numerical values of the stress intensity factors as functions of the angle α formed by the crack and the tip of box, and loads ($f_i, i = 1, 2$).

References

[1] G.Ya. Popov, The elastic stress concentration near stamps, cuts, thin inclusions and reinforcements. Moscow, Nauka, 1982, p. 342.

[2] A.A. Khrapkov, Some cases of elastic equilibrium of infinite wedge with an unsymmetrical cut in the apex by the action of concentrated force, PMM, vol. **35** (1971), N4 pp. 677–689.

[3] G.Ya. Popov and V.V. Reut, The calculation of the box shells.Thesises of XIV All-Union conference on the theory of plates and shells, vol. **2** Kutaisy, 1987, pp. 327–333.

[4] G.Ya. Popov and V.V. Reut, The asymptotic approach to the problems of the state of stress in elastic boxed bodies. Annotation of reports of VII All-Union congress on theoretical and applied mechanics. Moskow, 1991, pp. 291–292.

[5] V.I. Migdalski and V.V. Reut, On the stress state of the box shell which is weaked by the crack//Vestnik Odesskogo universiteta, 1998, N3, pp. 43–49.

Institute Mathematics, Economics and Mechanics
Odessa State University
270057 Dvoryanskaya str. 2, Odessa
Ukraine

Operator Theory:
Advances and Applications, Vol. 117
© 2000 Birkhäuser Verlag Basel/Switzerland

Homogeneity of a String having Three Unperturbed Spectra

Vyacheslav N. Pivovarchik

We prove that there does not exist a smooth inhomogeneous string such the spectra of three particular boundary value problems describing its small vibrations coincide with those for a homogeneous string. These three boundary value problems are the Dirichlet problem on $[0, \frac{l}{2}]$, the Dirichlet problem on $[\frac{l}{2}, l]$, and the Dirichlet-Neumann problem on $[0, l]$.

Consider an inhomogeneous string with the left end fixed and the right one free. Denote by $B(s)$ the density of the string, by l its length and by $u(s, t)$ the transverse displacement. Then we deal with the problem

$$\frac{\partial^2 u}{\partial s^2} - B(s)\frac{\partial^2 u}{\partial t^2} = 0, \tag{1}$$

$$u(0) = 0, \tag{2}$$

$$\left.\frac{\partial u}{\partial s}\right|_{s=l} = 0. \tag{3}$$

Consider the same string with fixed ends clamped at an interior point $l_1 \in (0, l)$. Then we obtain equation (1) for $s \in (0, l_1)$ with the boundary conditions

$$u(0) = u(l_1) = 0, \tag{4}$$

and equation (1) for $s \in (l_1, l)$ with the boundary conditions

$$u(l_1) = u(l) = 0. \tag{5}$$

Let the string be smooth enough, namely $B(s) \in W_2^2(0, l)$, $B(s) \geq \varepsilon > 0$, and let $\left.\frac{dB(s)}{ds}\right|_{s=l} = 0$, and $\int_0^{l_1} B^{\frac{1}{2}}(s)ds = \int_{l_1}^{l} B^{\frac{1}{2}}(s)ds$. Then substituting $u(s, t) = v(\lambda, s)e^{i\lambda t}$ and applying the Liouville transform [1]

$$\begin{aligned} x(s) &= \int_0^s B^{\frac{1}{2}}(s')ds', \\ y(x) &= B^{\frac{1}{4}}(s(x))\, v(\lambda, s(x)), \end{aligned}$$

we reduce the above problems to the following Sturm-Liouville problems

$$y'' + (\lambda^2 - q(x))y = 0, \tag{6}$$

$$y(0) = y'(a) = 0, \tag{7}$$

$$\begin{aligned} y'' + (\lambda^2 - q(x))y = 0,\\ y(0) = y\left(\frac{a}{2}\right) = 0, \end{aligned} \tag{8}$$

$$\begin{aligned} y'' + (\lambda^2 - q(x))y = 0,\\ y\left(\frac{a}{2}\right) = y(a) = 0, \end{aligned} \tag{9}$$

where

$$a = \int_0^l B^{\frac{1}{2}}(s)ds, \tag{10}$$

$$\begin{aligned} \int_0^{l_1} B^{\frac{1}{2}}(s)ds &= \int_{l_1}^{l} B^{\frac{1}{2}}(s)ds = \frac{a}{2},\\ q(x) &= B^{-\frac{1}{4}}(s(x))\frac{d^2}{dx^2}B^{\frac{1}{4}}(s(x)). \end{aligned} \tag{11}$$

Denote by $\{\xi_k\}_{k=-\infty, k\neq 0}^{\infty}$ the spectrum of problem (6), (7) and by $\{\nu_k\}_{k=-\infty, k\neq 0}^{\infty}$ $\left(\{\nu_k^{(1)}\}_{k=-\infty, k\neq 0}^{\infty}\right)$ the spectrum of problem (6), (8) (problem (6), (9)). It is well known that the three spectra consist of real simple eigenvalues only (due to the condition $B(s) \geq \varepsilon > 0$). We enumerate the eigenvalues in the following way: $\xi_{-k} = -\xi_k$, $\xi_{k+1} > \xi_k$, $\nu_{-k} = -\nu_k$, $\nu_{k+1} > \nu_k$, $\nu_{-k}^{(1)} = -\nu_k^{(1)}$, $\nu_{k+1}^{(1)} > \nu_k^{(1)}$, $k \in \mathbb{N}$. We call unperturbed the spectrum of a problem corresponding to $q(x) \overset{a.e.}{=} 0$. We shall make use of the following theorem of [4] in the form adapted for our case. Denote by $\{\vartheta_k\}_{k=-\infty, k\neq 0}^{\infty}$ the spectrum of the following problem

$$\begin{aligned} y'' + (\lambda^2 - q(x))y &= 0,\\ y(0) = y(a) &= 0. \end{aligned}$$

Theorem (Borg): Let $q(x) \in L_1(0, a)$ be real-valued. Then the two spectra $\{\vartheta_k\}_{k=-\infty, k\neq 0}^{\infty}$ and $\{\xi_k\}_{k=-\infty, k\neq 0}^{\infty}$ determine uniquely $q(x)$.

New Theorem: Let $\xi_k = \frac{\pi}{a}(k - \frac{1}{2})$, $\nu_k = \nu_k^{(1)} = \frac{2\pi k}{a}$, $(k \in \mathbb{N})$, and let $q(x) \in L_2(0, a)$ be real-valued. Then $q(x) \overset{a.e.}{=} 0$.

Proof: Denote by $s(\lambda, x)$ $(c(\lambda, x))$ the solution of (6) satisfying the conditions $s(\lambda, 0) = s'(\lambda, 0) - 1 = 0$ $(c(\lambda, 0) - 1 = c'(\lambda, 0) = 0)$. Denote by $s_1(\lambda, x)$

($c_1(\lambda, x)$) the solution of (6) satisfying the conditions $s_1(\lambda, \frac{a}{2}) = s_1'(\lambda, \frac{a}{2}) - 1 = 0$ ($c_1(\lambda, \frac{a}{2}) - 1 = c_1'(\lambda, \frac{a}{2}) = 0$). The set of zeroes of $s'(\lambda, a)$ coincides with $\{\xi_k\}_{\substack{k\neq 0\\ k=-\infty}}^{\infty}$ and the set of zeroes of $s(\lambda, \frac{a}{2})$ ($s_1(\lambda, a)$) coincides with $\{\nu_k\}_{\substack{k\neq 0\\ k=-\infty}}^{\infty}$ $\left(\{\nu_k^{(1)}\}_{\substack{k\neq 0\\ k=-\infty}}^{\infty}\right)$. Using formula (1.3.11) of [2], we obtain

$$s'(\lambda, a) = \cos\lambda a + \frac{F\sin\lambda a}{\lambda} + \frac{g(\lambda)}{\lambda}, \tag{12}$$

where $F \in \mathbb{R}$, $g(\lambda) = \int_0^a \tilde{g}(t)\sin\lambda t dt$, $\tilde{g}(t) \in L_2(0, a)$. Then Lemma 3.4.2 of [2] implies

$$s\prime(\lambda, a) = \prod_{k=1}^{\infty}(k - \frac{1}{2})^{-2}\left(\frac{a^2\left(\xi_k^2 - \lambda^2\right)}{\pi^2}\right),$$

where $\{\xi_k\}_{\substack{k\neq 0\\ k=-\infty}}^{\infty}$ is the set of zeroes of $s'(\lambda, a)$ and, what is the same, the spectrum of problem (6), (7). In our case $\xi_k = \frac{\pi}{a}(k - \frac{1}{2})$ and consequently

$$s'(\lambda, a) = \prod_{k=1}^{\infty}\left(1 - \left(\frac{\lambda a}{\pi(k - \frac{1}{2})}\right)^2\right) = \cos\lambda a. \tag{13}$$

In the same way we apply Lemma 3.4.2 of [2] to the problems (6), (8) and (6), (9) and obtain

$$s\left(\lambda, \frac{a}{2}\right) = s_1(\lambda, a) = \frac{a}{2}\prod_{k=1}^{\infty}\left(1 - \left(\frac{\lambda a}{2\pi k}\right)^2\right) = \frac{\sin\frac{\lambda a}{2}}{\lambda}. \tag{14}$$

Tedious calculations show that

$$s(\lambda, a) = s'\left(\lambda, \frac{a}{2}\right)s_1(\lambda, a) + s\left(\lambda, \frac{a}{2}\right)c_1(\lambda, a), \tag{15}$$

$$s'(\lambda, a) = s'\left(\lambda, \frac{a}{2}\right)s_1'(\lambda, a) + s\left(\lambda, \frac{a}{2}\right)c_1'(\lambda, a). \tag{16}$$

Multiplying (15) by $s_1'(\lambda, a)$ and using the identity

$$c_1(\lambda, a)s_1'(\lambda, a) = 1 + s_1(\lambda, a)c_1'(\lambda, a)$$

we obtain

$$\begin{aligned} s(\lambda, a)s_1'(\lambda, a) &= s_1(\lambda, a)\left(s'\left(\lambda, \frac{a}{2}\right)s_1'(\lambda, a) + s\left(\lambda, \frac{a}{2}\right)c_1'(\lambda, a)\right) \\ &\quad + s\left(\lambda, \frac{a}{2}\right) = s_1(\lambda, a)s'(\lambda, a) + s\left(\lambda, \frac{a}{2}\right). \end{aligned} \tag{17}$$

Substituting (16) into (17) and using (13) and (14) we obtain

$$s(\lambda, a)s_1'(\lambda, a) = \frac{\sin\lambda a}{\lambda}\cos\frac{\lambda a}{2}.$$

Let us prove that $s_1'(\lambda, a) = \cos\frac{\lambda a}{2}$. If for some $k \in \mathbb{N}$, $\lambda = \frac{\pi}{a}(2k-1)$ is not a zero of $s_1'(\lambda, a)$, then being a double zero of $s(\lambda, a)s_1'(\lambda, a)$, it has to be a double zero of $s(\lambda, a)$. But all the zeroes of $s(\lambda, a)$ are simple (see for example [3]). A contradiction. Hence, due to the symmetry of the problem all zeroes of $\cos\frac{\lambda a}{2}$ are those of $s_1'(\lambda, a)$. Let $s_1'(\lambda, a)$ have zeroes not equal to $\lambda = \frac{\pi}{a}(2k-1)$. Then

$$s_1'(\lambda, a) = C\cos\frac{\lambda a}{2}\prod_{k=1}^{n}\left(1-\left(\frac{a\lambda}{2\pi k_p}\right)^2\right), \tag{18}$$

where $C \in \mathbb{R}\backslash\{0\}$, $\{k_p\}_{p\neq 0, -n}^{n} \subset \{\pm 1, \pm 2, \ldots\}$, and $n \in \mathbb{N}\cup\{\infty\}$. The comparison of (18) and the analog of formula (12) written for $s_1'(\lambda, a)$ gives

$$C\prod_{k=1}^{n}\left(1-\left(\frac{a\lambda}{2\pi k_p}\right)^2\right) = 1 + \frac{F_1\tan\frac{\lambda a}{2}}{\lambda} + \frac{g_1(\lambda)}{\lambda\cos\frac{\lambda a}{2}}, \tag{19}$$

where $F_1 \in \mathbb{R}\backslash\{0\}$, $g_1(\lambda) = \int_{\frac{a}{2}}^{a}\tilde{g}_1(t)\sin\lambda(\frac{a}{2}-t)dt$, $\tilde{g}_1(t) \in L_2(\frac{a}{2}, a)$. The left-hand side of (19) is an entire function, and hence so is the right-hand side. If $F_1 \neq 0$, then the right-hand side has poles at $\lambda = \frac{\pi}{a}(2k-1)$ for k large enough, a contradiction. If $F_1 = 0$ and $g_1(\lambda) = 0$ for all $\lambda = \frac{\pi}{a}(2k-1)$, then, due to the completeness of the set $\{\sin(\frac{\pi}{a}(2k-1)(\frac{a}{2}-t))\}_{-\infty}^{\infty}$ in $L_2(\frac{a}{2}, a)$ we obtain $\tilde{g}_1(t) \overset{a.e.}{=} 0$, i.e. the right-hand side of (19) is equal to 1. A contradiction. Hence, $s_1'(\lambda, a) = \cos\frac{\lambda a}{2}$, and, consequently,

$$s(\lambda, a) = \frac{\sin\lambda a}{\lambda}. \tag{20}$$

Now in accordance with Borg's theorem [4] explained above, the formulae (13), (20) imply $q(x) \overset{a.e.}{=} 0$. The theorem is proved.

If $q(x) \overset{a.e.}{=} 0$ then the formula (12) with the condition $\left.\frac{dB(s)}{ds}\right|_{s=l} = 0$ taken into account implies $B(s) = B(s(x)) = C = const$. Substituting $B(s) = C$ into (10) we obtain

$$B = \frac{a}{l}, \tag{21}$$

where the parameter a can be found from the asymptotics

$$a = \left(\lim_{n\to\infty}\frac{\xi_n}{\pi n}\right)^{-1}.$$

Hence, we see that the string is homogeneous if the three spectra are unperturbed. To find the density we need to know the length l of the string in addition to the spectra. It is a consequence of the invariance of the problems (1)–(3); (1), (4) and (1), (5) with respect to the transformation $s' := rs$, $B'(s') = B(s)r^{-2}$, $l' = rl$.

□

It should be mentioned that the first results on the three-spectra inverse problem were obtained in [5, 6].

Acknowledgements

It is pleasure to acknowledge useful conversations with Professor Fritz Gesztesy.

This work is partly supported by Grant UM1-298 of the Ukrainian Government and Civil Research and Development Foundation (USA).

References

[1] R. Courant and D. Hilbert, Methods of Mathematical Physics, vol. **1**, Interscience, New York, 1953.

[2] V.A. Marchenko, Sturm-Liouville operators and applications (in Russian), Kiev, Naukova Dumka, 1977, p. 331.

[3] F.V. Atkinson, Discrete and Continuous Boundary Problems, Academic Press, New York-London, 1964.

[4] G. Borg, Uniqueness theorems in the spectral theory of $y'' + (\lambda - q(x))y = 0$, Proc. 11th Scandinavian Congress of Mathematicians, Johan Grundt Tanums Forlag, Oslo, (1952), 276–287.

[5] V.N. Pivovarchik, An Inverse Sturm-Liouville problem by Three Spectra, to appear in Integral Equations and Operator Theory.

[6] F. Gesztesy and B. Simon, On the Determination of a Potential from Three Spectra, to appear in Birman Birthday Volume in Advances in Mathematical Sciences.

V.N. Pivovarchik
Preobrazhenskaya str, 59/61, a.17
270045, Odessa
Ukraine
v.pivovarchik@paco.net

MSC Primary 34A55, 34B24, 34B10, 34L20, Secondary 73K03

Operator Theory:
Advances and Applications, Vol. 117
© 2000 Birkhäuser Verlag Basel/Switzerland

On the Integro-differential Equation of a Torsion of an Elastic Medium Including a Cylindrical Crack

G. Popov and B. Kebli

We have in mind such a problem: an unbounded $(0 < r < \infty, -\pi < \varphi < \pi, -\infty < z < \infty)$ elastic medium including a crack coinciding with the surface:

$$r = R, -\pi < \varphi < \pi, a \leq z \leq b \tag{1.1}$$

has been exposed by the deformation of the torsion from an arbitrary load, which causes stress $\tau^0_{r\varphi}(r, z)$ and displacement $u^0_\varphi(r, z)$ in the elastic medium without the crack to demand to determine the stress intensity factor.

Constructing the discontinuous solution of the elasticity equations for the defect (1.1) using the scheme [1, 2] we reduce the problem to integro-differential equation

$$\frac{d^2}{dz^2}\frac{1}{4}\int_a^b \langle V(R,\xi)\rangle k\left(\frac{z-\xi}{R}\right) d\xi = \tau^0_{r\varphi}(R, z), \; a \leq z \leq b \tag{1.2}$$

where (G is the shear modulus)

$$\langle V(R, z)\rangle = 2G[u_\varphi(R-0, z) - u_\varphi(R+0, z)] \tag{1.3}$$

is the unknown jump of the displacement on the crack,

$$\begin{aligned} k(y) &= \frac{2}{\pi}\int_0^\infty I_2(\alpha)K_2(\alpha)\cos\alpha y d\alpha \\ &= \frac{1}{2\pi}\int_{-\infty}^\infty 2I_2(|\alpha|)K_2(|\alpha|)e^{-i\alpha y} d\alpha \end{aligned} \tag{1.4}$$

$I_n(z), K_n(z)$ are the modified Bessel functions.

By means of the substitution

$$\begin{aligned} z &= R(c_+ + c_- x), \zeta = R(c_+ + c_-\xi); 2Rc_\pm = b \pm a \\ &< V(R, Rc_+ + Rc_-\xi) = \varphi(\zeta), \\ &-4Rc_-\tau^0_{r\varphi}(R, Rc_+ + Rc_- x) = f(x) \end{aligned} \tag{1.5}$$

the equation (1.2) is reduced to

$$-\frac{d^2}{dx^2}\int_{-1}^1 k(c_- x - c_-\zeta)\varphi(\zeta)d\zeta = f(x), |x| \leq 1 \tag{1.6}$$

We would explore the analytical property of the kernel of obtained equation and segregate from it the discontinuous part, using (1.4). For this purpose we are using the formula [3]:

$$I_2(\alpha)K_2(\alpha) = \frac{1}{2}\int_0^\infty \frac{J_4(\alpha\tau)d\tau}{\sqrt{1+(\frac{1}{2}\tau)^2}}$$

It enables to lead the first integral from (1.4) to the form

$$k(y) = \frac{1}{\pi}\int_0^\infty \frac{d\tau}{\sqrt{1+(\frac{1}{2}\tau)^2}}\int_0^\infty \cos\alpha y J_4(\alpha\tau)d\tau \tag{1.7}$$

The last integral is known [4] and after the obvious transforms we obtain:

$$\begin{aligned} k(y) &= 2\int_0^{\frac{\pi}{2}} \frac{8\cos^4 t - 8\cos^2 t + 1}{\pi\,|y|\sqrt{1+4y^{-2}\sin^2 t}}dt = \frac{1}{|y|}\sum_{j=1}^{3} c_j F\left(\frac{1}{2},\frac{1}{2};j;-\frac{4}{y^2}\right) \\ & c_1 = 1, c_2 = -4, c_3 = 3 \end{aligned} \tag{1.8}$$

The second equation in (1.8) follows from the formula 3.681(1) in [5]. To segregate from the function (1.8) the discontinuous part we need to continue the Gauss hypergeometric function containing there to a neighbourhood of the zero, but needed for this purpose formula 9.132(2) from [5] in this case does not work because the third parameter in Gauss function is the integer. For the purpose to overcome this difficulty we act so: that parameter we make equal j+ε and use the indicated formula. As a result we have then ε has been tended to the zero. As a result we have:

$$\begin{aligned} F\left(\frac{1}{2},\frac{1}{2};j;-\frac{4}{y^2}\right) &= \frac{|y|\,\Gamma(j)}{2\sqrt{\pi}\Gamma(j-\frac{1}{2})}\left\{\sum_{m=0}^{\infty}\frac{(\frac{1}{2})_m(\frac{3}{2}-j)_m}{m!^2}\left(-\frac{y^2}{4}\right)^m\right. \\ & \left[2\Psi(1+m)-\Psi\left(\frac{1}{2}+m\right)-\Psi\left(\frac{3}{2}-j+m\right)\right] \\ & \left. -2\ln\frac{|y|}{2}\sum_{m=0}^{\infty}\frac{(\frac{1}{2})_m(\frac{3}{2}-j)_m}{m!^2}\left(-\frac{y^2}{4}\right)^m\right\} \end{aligned} \tag{1.9}$$

Having this formula we easy find the discontinuous part of kernel (1.8) explored and write the equation (1.6) in the form

$$-\frac{d^2}{dx^2}\frac{1}{\pi}\int_{-1}^{1}\left[\ln\frac{1}{|x-\zeta|}+R(c_-x-c_-\zeta)\right]\varphi(\zeta)d\zeta = f(x), |x|\le 1, \tag{1.10}$$

where the regular part of the kernel R(y), having the continuous first derivative defined by the formula

$$(1.11)\quad \begin{aligned} R(y) &= \ln\frac{2}{|y|}\sum_{m=1}^{\infty} A_m y^{2m} + \sum_{m=0}^{\infty} \Psi_m y^{2m} \\ A_m &= a_m\left(1+\frac{8}{3}m+\frac{4}{3}m^2\right), \\ \Psi_m &= A_m\left[\Psi(1+m)-\Psi\left(m-\frac{3}{2}\right)\right] \\ &\quad -\frac{8(2m-1)(7m-4)}{3(2m-3)}m!^2 a_m = (-1)^m\left(\frac{1}{2}\right)_m\left(-\frac{3}{2}\right)_m \end{aligned}$$

where $\Psi(z)$ is the Euler Ψ function.

2

We offer an efficient approximate method for solving the equation (1.10) based on the spectral relation [1] (U_n-Chebyshev's polynomials of the second kind)

$$(2.1)\quad \begin{aligned} &-\frac{d^2}{dx^2}\frac{1}{\pi}\int_{-1}^{1}\sqrt{1-\xi^2}U_{n-1}(\xi)\ln\frac{1}{|x-\xi|}d\xi \\ &= nU_{n-1}(x),\ |x|\le 1,\ n=\overrightarrow{1,\infty} \end{aligned}$$

According to (2.1), we shall construct a solution of the integral equation (1.10) in the form:

$$(2.2)\quad \varphi(\xi) = \sqrt{1-\xi^2}\sum_{n=1}^{\infty}\varphi_n U_{n-1}(\xi)$$

After the substitution (2.2) into (1.10), using (2.1) and this result:

$$(2.3)\quad \frac{d}{dz}\left[\sqrt{1-z^2}U_{n-1}(z)\right] = -\frac{nT_n(z)}{\sqrt{1-z^2}},\ n=\overline{1,\infty}$$

$T_n(z)$ is Chebyshev's polynomial of the first kind and carrying standard scheme of the orthogonal polynomials method [1], we reduce the integral equation (1.10) to an infinite system:

$$(2.4)\quad \begin{aligned} &\psi_k + \frac{2}{\pi^2}\sum_{n=1}^{\infty}\sqrt{kn}d_{kn}\psi_n = \frac{f_k}{\sqrt{k}};\ \psi_k = \sqrt{k}\varphi_k \\ &d_{kn} = \int_{-1}^{1} R(c_-x - c_-\xi)\frac{T_k(x)T_k(\xi)}{\sqrt{1-x^2}\sqrt{1-\xi^2}}dx d\xi \\ &f_k = \int_{-1}^{1} f(x)\sqrt{1-x^2}U_{k-1}(x)dx \end{aligned}$$

For calculating the integrals contained here we offer to use the Gauss-quadratureal formula [6]. That is

$$
(2.5)\qquad
\begin{aligned}
d_{kn} &= \frac{\pi^2}{l^2}\sum_{i=1}^{l}\sum_{j=1}^{l} R(c_-x_i - c_-x_j)T_k(x_i)T_n(x_j), l > k, n;\\
f_k &= \frac{\pi}{l+1}\sum_{i=1}^{l}\sin^2\frac{i\pi}{l+1} f(x_i^*)U_{k-1}(x_i^*),\\
x_i &= \cos\frac{2i-1}{2l}\pi, x_i^* = \cos\frac{i\pi}{l+1}
\end{aligned}
$$

The infinite system (2.4) to be solved approximately by the reduction method for the basis of which to be required [7] to prove, the following series converge

$$
(2.6)\qquad S_1 = \sum_{k=1}^{\infty}\sum_{n=1}^{\infty} knd_{kn}^2, S_2 = \sum_{k=1}^{\infty}\frac{f_k^2}{k}
$$

This may be done using the scheme [1].

3

The intensity factors we find by passing to the limit

$$
N_{\mp} = \lim\frac{\sqrt{2\pi(a-z)}}{\sqrt{2\pi(z-b)}}\tau_{r\varphi}(R, z), z \longrightarrow a-0, z \longrightarrow b+0
$$

or taking (1.5) into account

$$
(3.1)\qquad
\begin{aligned}
N_{\mp} &= \lim\frac{\sqrt{\pi(b-a)(-x-1)}}{\sqrt{\pi(b-a)(x-1)}}\tau_{r\varphi}(R, c_+R + c_-Rx),\\
& x \longrightarrow -1-0, x \longrightarrow 1+0
\end{aligned}
$$

in this case by (1.2) and (1.10)

$$
(3.2)\qquad
\begin{aligned}
\tau_{r\varphi}(R, c_+R + c_-Rx) &= -\frac{d^2}{dx^2}\frac{1}{4\pi R^2}\int_{-1}^{1}\left[\ln\frac{1}{|x-\xi|} + R(c_-x - c_-\xi)\right]\\
&\qquad \varphi(\xi)d\xi + \tau_{r\varphi}^0(R, c_+R + c_-Rx), |x| > 1
\end{aligned}
$$

For the realization of the limit (3.1) it is necessary to continue the spectral relation (2.1) to the interval $|x| > 0$. For this purpose we have used the relation

$$\frac{d^{k+1}}{dx^{k+1}}\int_{-1}^{1}\ln\frac{1}{|x-s|}\frac{P_m^{\alpha,k-\alpha}(s)ds}{(1-s)^{-\alpha}(1+s)^{\alpha-k}}$$

$$= \frac{(-2)^{k+1}2^m\Gamma(1+\alpha+m)\Gamma(1+k+m-\alpha)}{m!(1+k+2m)!(m+k)!^{-1}(x-1)^{m+k+1}} \tag{3.3}$$

$$\times\; F(1+\alpha+m, m+k+1; 2+k+2m; (1-x)/2)$$

following from the results of the work [8]. For calculation N_- the Gauss function contained here has been continued to a neighbourhood $x = -1$, using formula 9.131(2) from [5] and then to set $k = 1, \alpha = \frac{1}{2}$. As result instead of (2.1) we have

$$-\frac{d^2}{dx^2}\frac{1}{\pi}\int_{-1}^{1}\ln\frac{1}{|x-s|}\sqrt{1-s^2}U_m(s)ds = \frac{(m+1)^2 2^{m+2}}{(x-1)^{m+2}}$$

$$\left[F\left(\frac{3}{2}+m, m+2; \frac{3}{2}; \frac{x+1}{x-1}\right) - \frac{m+1}{2}\sqrt{\frac{1-x}{-1-x}}\right. \tag{3.4}$$

$$\left. F\left(\frac{3}{2}+m, m+1; \frac{1}{2}; \frac{x+1}{x-1}\right)\right], x < -1$$

Having this relation N_- has been easy calculated. The using the formula (3.2), (2.2) and (3.4), (2.4) lead to the result

$$N_- = \sqrt{\frac{\pi c}{2}}\sum_{m=1}^{\infty}(-1)^{m+1}\sqrt{m}\psi_m \tag{3.5}$$

The similar formula has been obtained for the stress intensity factor N_+.

4

We now consider the case of the semi-infinite crack, when in (1.1) and (1.2) $b = \infty$. In this case to be made the substitutions

$$z = a + Rx, \zeta = a + R\xi; \langle V(R, a + \xi R)\rangle = \varphi(\zeta),$$

$$4R\tau_{r\varphi}^{0}(R, a + Rx) = -f(x) \tag{4.1}$$

As a result instead of (1.2) we have the equation

$$-\frac{d^2}{dx^2}\int_0^{\infty}k(x-\xi)\varphi(\xi)d\xi = f(x), 0 \leq x < \infty \tag{4.2}$$

which to be solved exactly by the method of factorization [9]. However in order to do this procedure more simple we will transform the equation (4.2) using the integration by parts:

$$-\int_0^\infty k(x-\xi)\varphi''(\xi)d\xi = f(x), 0 \leq x < \infty \tag{4.3}$$

According to the mechanical sense

$$\varphi''(\xi) = O(\xi^{-\frac{3}{2}}), \xi \longrightarrow 0$$

that is why we must solve the integral equation (4.3) in the class of function having non-integrable singularities and the integrals must be understood in the generalized (regularized) sense [10, 1]. In this sense must be understood the following relation [12, 1]

$$\begin{gathered}\int_a^b \left[\frac{h(\zeta)}{(\zeta-a)^\alpha(b-\zeta)^\beta}\right]' g(\zeta)d\zeta = -\int_a^b \left[\frac{h(\zeta)g'(\zeta)d\zeta}{(\zeta-a)^\alpha(b-\zeta)^\beta}\right] \\ Re(\alpha,\beta) < 1, h(\zeta), g'(\zeta) \in C([a,b])\end{gathered} \tag{4.4}$$

$$\int_0^\infty \varphi''(\xi)d\xi = 0 \tag{4.5}$$

For the reducing (4.2) to (4.3) it has been made the integration by parts, using (4.4) and has been supposed that the solution is built in the class of functions decreasing in the infinity together its first derivatives.

At first we have solved the equation (4.3) with the special right-hand part [9]

$$f(x) = e^{i\zeta x}, Im\zeta > 0 \tag{4.6}$$

It has been seen from (1.4), that the symbol of the kernel (1.4) is

$$\begin{aligned} K(\alpha) &= 2I_2(|\alpha|)K_2(|\alpha|) = \frac{th\pi\alpha}{\alpha}G(\alpha) \\ G(\alpha) &= 2\alpha I_2(|\alpha|)K_2(|\alpha|)cth\pi\alpha, \lim G(\alpha) = 1, |\alpha| \longrightarrow \infty \end{aligned} \tag{4.7}$$

Carrying standard scheme of the method of factorization we come to Riemann's problem [11]

$$\begin{aligned} -\Phi^+(\alpha)K(\alpha) &= i(\alpha+\zeta)^{-1} + Y^{-1}(\alpha), -\infty < \alpha < \infty \\ \Phi^+(\alpha) &= \int_0^\infty \varphi''(\xi)e^{i\alpha\xi}d\xi, Y^-(\alpha) \\ &= 4R\int_{-\infty}^0 \tau_{r\varphi}(R, a+xR)e^{i\alpha x}dx \end{aligned} \tag{4.8}$$

For using the known formula [11] for the solution Riemann's problem (4.8) it is necessary to prove its correctness for the case of the Cauchy-type integrals with the unintegrable density. It has been made in the work [12]. The factorization

$$K(\alpha) = X^+(\alpha)X^-(\alpha)$$

has been given by the formula [9, 11]

$$X^\pm(\alpha) = X_0^\pm(\alpha)X_1^\pm(\alpha),\ X_0^\pm(\alpha) = \Gamma^{-1}(1 \pm i\alpha)\Gamma\left(\frac{1}{2} \pm i\alpha\right) \tag{4.9}$$
$$X_1^\pm(\alpha) = \exp\left(\mp \int_{-\infty}^{\infty} \ln G(\beta)\frac{d\beta}{\beta - \alpha}\right)$$

and the solution of Riemann problem (4.8) is

$$\Phi^+(\alpha) = -\left[C_\zeta + \frac{1}{(\alpha+\zeta)X^-(-\zeta)}\right]\frac{1}{X^+(\alpha)}, \tag{4.10}$$
$$\frac{Y^-(\alpha)}{X^-(\alpha)} = C_\zeta - \frac{1}{(\alpha+\zeta)}\left(\frac{1}{X^-(\alpha)} - \frac{1}{X^-(-\zeta)}\right)$$

The arbitrary constant C_ς will be found from the condition (4.5). As a result we have

$$C_0 = -i[\zeta X^-(-\varsigma)]^{-1} \tag{4.11}$$

For the stress in the continuation of the crack we have formula

$$4R\tau_{r\varphi}^0(R, a + xR) = \frac{1}{2\pi}\int_{-\infty}^{\infty} Y^-(\alpha)e^{-i\alpha x}d\alpha,\ x < 0 \tag{4.12}$$

and accordingly for the stress intensity factor

$$N = \lim_{z\to a-0} \sqrt{2\pi(z-a)}\tau_{r\varphi}(R, z) = \lim_{x\to -0} \sqrt{2\pi Rx}\tau_{r\varphi}(R, a + Rx) \tag{4.13}$$

These formulae are correct for every loading of the elastic medium. However the function $Y^-(\alpha)$ determined by the formulae (4.10), (4.11) corresponds to loading which provides the equality (4.6). The intensity factor for this loading we denote N_ζ. Substituting the indicated expressions for $Y^-(\alpha)$ into (4.12) and passing to the limit (4.13), we find

$$N_\zeta = -\sqrt{\frac{2\pi}{R}}\frac{i}{4\zeta X^-(-\zeta)} \tag{4.14}$$

For calculating the stress intensity factor in the general case of the right-hand part of the equation (4.3) and giving by formula (4.1), we use the idea [9] that it can be represented in a look the continuous sum from the functions (4.6). Indeed suppose the integral

$$F(\alpha) = -4R \int_0^\infty \tau_{r\varphi}^0(R, a + xR)e^{i\alpha x}dx$$

is existing, then by the certain conditions we can write that

$$f(x) = \frac{1}{2\pi}\int_{-\infty}^{\infty} F(\alpha)e^{-i\alpha x}d\alpha = \frac{1}{2\pi}\int_{-\infty}^{\infty} F(-\zeta)e^{i\zeta x}d\zeta$$

and hence the stress intensity factor for the load $\tau_{r\varphi}^0(r, z)$ we shall find by the formula

$$N = \frac{1}{2\pi}\int_{-\infty}^{\infty} F(-\zeta)N_\zeta d\zeta \tag{4.15}$$

References

[1] G.Ya. Popov, *The Concentration of Elastic Stresses near Punches, cuts, Thin Inclusions and Reinforcments*. Nauka, Moscow, 1982.

[2] G.Ya. Popov, *Problems of stress concentration in the neighbourhood of a spherical defect*. Achiev. in Mech, vol. **15** N1–2 (1992), 71–110.

[3] G. Bateman and A. Erdelyi, *Higher Transtendentae Functions*, vol. **2**, Nauka, Moscow, 1966.

[4] G. Bateman, and A. Erdelyi, *Tables of Integral transforms*, vol. **1**, Nauka, Moscow, 1969.

[5] J.S. Gradshtein and I.M. Ryzhik, *Tables of Integrals, Sums, Series and Products*, Fizmatgiz, Moscow, 1962.

[6] V.J. Krylov, *Approximate Calculation of Integrals*. Fizmatgiz, Moscow, 1959.

[7] L.V. Kantorovich and G.P. Akilov, *Functional Analysis*, Nauka, Moscow, 1977.

[8] G.Ya. Popov, *About the one remarkable property of Yakobi's polynomials*. Ukr. mat. journal, vol. **20**, N4, 1968.

[9] M.G. Krein, *Integral equations on a half-line with a kernel that depends on the difference in the arguments*. Uspekhi Mat. Nauk, 1958, vol. **13**, N5, 1958.

[10] I.M. Gelfand and V.E. Shilov, *Generalized functions and Operations on them*. Fizmatgiz, Moscow, 1958.

[11] F.D. Gakhov, *Boundary-valued Problems*. Nauka, Moscow, 1977.

[12] O.V. Onishchuk, G.Ya. Popov and P.G. Forshite, Prikl. mat. and Mech. vol. **50**, N2, 1986.

Institute Mathematics, Economics and Mechanics
Odessa State University
270057 Dvoryanskaya str. 2
Odessa
Ukraine

Operator Theory:
Advances and Applications, Vol. 117
© 2000 Birkhäuser Verlag Basel/Switzerland

Green's Formula and Theorems on Isomorphisms for General Elliptic Problems for Douglis-Nirenberg Elliptic Systems*

Inna Roitberg and Yakov Roitberg

Green's formula is obtained for general elliptic boundary value problem for elliptic systems of Douglis-Nirenberg structure. The formally adjoint problem with respect to Green's formula is studied. The same results are obtained for parameter elliptic and parabolic problems for general systems of equations. The Green's formula permits us to prove various theorems on complete collections of isomorphisms for problems under consideration. Relations between these theorems are studied. Various applications of the isomorphisms theorems are considered. In particular, the question on smoothness of weak solutions of such problems in the whole domain is studied. These investigations were stimulated by works of M.S. Agranovich and A.N. Kozhevnikov in the spectral theory of such problems.

1 General Elliptic Boundary Value Problems for Systems of Equations

1.1

In the bounded domain $G \subset \mathbb{R}^n$ with the boundary $\partial G \in C^\infty$ we consider the elliptic boundary value problem

$$l(x, D)u(x) = f(x) \qquad (x \in G), \tag{1}$$

$$b(x, D)u(x) = \phi(x) \qquad (x \in \partial G). \tag{2}$$

Here

$$l(x, D) = \big(l_{rj}(x, D)\big)_{r,j=1,\dots,N}, \quad \operatorname{ord} l_{rj} \le s_r + t_j,$$

numbers $t_1, \dots, t_N, s_1, \dots, s_N$ are given integer, $|S| + |T| = s_1 + \cdots + s_N + t_1 + \cdots + t_N = 2m$;

$$l_{rj}(x, D) = \begin{cases} \sum_{|\alpha| \le s_r + t_j} a_\alpha(x) D^\alpha & \text{for } s_r + t_j \ge 0 \\ 0 & \text{for } s_r + t_j < 0 \end{cases}$$

$(\alpha = (\alpha_1, \dots, \alpha_n)$, $D^\alpha = D_1^{\alpha_1} \dots D_n^{\alpha_n}$, $D_j = i\partial/\partial x_j$; $|\alpha| = \alpha_1 + \cdots + \alpha_n)$. Since $s_k + t_j = (s_k - r) + (t_j + r)$, without restriction of generality, one can suppose that $\max\{s_j\} = 0$. Then the maximal order of differentiation of the

*This work has been partially supported by Grant INTAS-94-2187.

function u_j in the system (1) is not larger than t_j. In what follows we assume that $t_1 \geq \cdots \geq t_N \geq 0 = s_1 \geq \cdots \geq s_N$ (one can obtain this relation by changing both the numbers of the functions and the numbers of the equations). In addition,

$$b(x, D) = (b_{hj}(x, D))_{\substack{h=1,\dots,m \\ j=1,\dots,N}}, \quad \operatorname{ord} b_{hj} \leq \sigma_h + t_j,$$

$$b_{hj}(x, D) = \begin{cases} \sum_{|\alpha| \leq \sigma_h + t_j} b_\alpha(x) D^\alpha & \text{for } \sigma_h + t_j \geq 0, \\ 0 & \text{for } \sigma_h + t_j < 0, \end{cases}$$

the numbers $\sigma_1, \dots, \sigma_n$ are given integer.

We assume, for simplicity, that the coefficients of all the differential expressions are infinitely smooth.

Denote

$$\begin{aligned} \kappa &= \max\{0, \sigma_1 + 1, \dots, \sigma_m + 1\}, \tau = (\tau_1, \dots, \tau_N), \\ |\tau| &= \tau_1 + \cdots + \tau_N, \quad \tau_j = t_j + \kappa \ (j = 1, \dots, N). \end{aligned} \tag{3}$$

1.2 Functional Spaces

(see [1], Ch. 1–2). Let $p, p' \in (1, \infty)$, $1/p + 1/p' = 1$, and let $\Omega \subset \mathbb{R}^n$ be a domain with the boundary $\partial\Omega$. In the work we consider only the cases i) where $\Omega = G$ is a bounded domain with the boundary $\partial\Omega = \partial G$, and ii) where $\Omega = \mathbb{R}^n_- = \{x = (x', x_n) \in \mathbb{R}^n : x_n < 0\}$ (or $\Omega = \mathbb{R}^n_+ = \{x = (x', x_n) \in \mathbb{R}^n : x_n > 0\}$) with the boundary $\partial\Omega = \{x \in \mathbb{R}^n : x_n = 0\} = \mathbb{R}^{n-1}$.

Denote by $H^{s,p}(\Omega)$ $s \geq 0$ the space of Bessel potentials (Liouville classes); $H^{-s,p}(\Omega)$ denotes the space dual to $H^{s,p'}(\Omega)$ with respect to the extension $(\cdot, \cdot)$ of the scalar product in $L_2(\Omega)$. The norm in $H^{s,p}(\Omega)$ is denoted by $\|\cdot\|_{s,p} = \|\cdot, \Omega\|_{s,p}$, $s \in \mathbb{R}$.

Denote by $B^{s,p}(\partial\Omega)$, $s \in \mathbb{R}$, the Besov space with the norm $\langle\langle\cdot\rangle\rangle_{s,p} = \langle\langle\cdot, \partial\Omega\rangle\rangle_{s,p}$. The spaces $B^{s,p}(\partial\Omega)$ and $B^{-s,p'}(\partial\Omega)$ are dual to each other with respect to the extension $\langle\cdot, \cdot\rangle$ of the scalar product in $L_2(\partial\Omega)$.

In what follows we denote by $(\cdot, \cdot)$ and by $\langle\cdot, \cdot\rangle$ the scalar products (and their extensions) in $L_2(\Omega)$ and $L_2(\partial\Omega)$ respectively, and the scalar products (and their extensions) in the direct products $(L_2(\Omega))^N$ and $(L_2(\partial\Omega))^{|\tau|}$ as well.

Denote by $C_0^\infty(\overline{\Omega})$ the set of restrictions of functions from $C_0^\infty(\mathbb{R}^n)$ to $\overline{\Omega}$. It is clear that $C_0^\infty(\overline{G}) = C^\infty(\overline{G})$ for a bounded domain G.

Let r be a fixed natural number, $1 < p < \infty$, $s \in \mathbb{R}$, and $s \neq k + 1/p$ $(k = 0, \dots, r-1)$. By $\widetilde{H}^{s,p,(r)} = \widetilde{H}^{s,p,(r)}(\Omega)$ we denote the completion of $C_0^\infty(\overline{\Omega})$ in the norm

$$\begin{aligned} |||u|||_{s,p,(r)} &= |||u, \Omega|||_{s,p,(r)} \\ &= \left(\|u\|^p_{s,p} + \sum_{j=1}^{p} \langle\langle D_\nu^{j-1} u\rangle\rangle_{s-j+1-1/p,p} \right)^{1/p}, \end{aligned} \tag{4}$$

where $D_\nu = i\partial/\partial\nu$ and ν is the vector normal to $\partial\Omega$.

We identify every element $u \in C_0^\infty(\overline{\Omega})$ with the vector

(5) $u = (u_0, u_1, \dots, u_r), \quad u_0 = u\big|_{\overline{\Omega}}, \quad u_j = D_\nu^{j-1} u\big|_{\partial\Omega}, \quad j = 1, \dots, r.$

Then the space $\widetilde{H}^{s,p,(r)}(\Omega)$ is isometrically equivalent to the closure $\mathcal{H}_0^{s,p,(r)}(\Omega)$ of the space of these vectors $\{u = (u_0, u_1, \dots, u_r) \in C_0^\infty(\overline{\Omega})\}$ in the direct product

$$\mathcal{H}^{s,p,(r)}(\Omega) = H^{s,p}(\Omega) \times \prod_{j=1}^{r} B^{s-j+1-1/p,p}(\partial\Omega).$$

In particular, if $s < 1/p$ then $\widetilde{H}^{s,p,(r)}(\Omega) = \mathcal{H}^{s,p,(r)}(\Omega) = H^{s,p}(\Omega) \times \prod_{j=1}^{r} B^{s-j+1-1/p,p}(\partial\Omega)$. Hence, the space $\widetilde{H}^{s,p,(r)}(\Omega)$ consist of the vectors $u = (u_0, u_1, \dots, u_r)$ where $u_0 \in H^{s,p}(\Omega)$, $u_j \in B^{s-j+1-1/p,p}(\partial\Omega)$, $j = 1, \dots, r$, and $u_j = D_\nu^{j-1} u_0|_{\partial\Omega}$, $j : s - j + 1 - 1/p > 0$. For any sequence $u_k \in C_0^\infty(\overline{\Omega})$ convergent to $u = (u_0, u_1, \dots, u_r)$ in $\widetilde{H}^{s,p,(r)}(\Omega)$ the sequence $D_\nu^{j-1} u_k|_{\partial\Omega}$ convergent to u_j in $B^{s-j+1-1/p,p}(\partial\Omega)$. In this (strong) sense the equality $u_j = D_\nu^{j-1} u\big|_{\partial\Omega}$, $j = 1, \dots, r$, $u \in \widetilde{H}^{s,p,(r)}(\Omega)$, is true.

If $s > r - 1/p$ then norm (4) is equivalent to the norm $\|\cdot, \Omega\|_{s,p}$ and $\widetilde{H}^{s,p,(r)}(\Omega) = H^{s,p}(\Omega)$. In this case all components of the vector $(u_0, u_1, \dots, u_r) \in \widetilde{H}^{s,p,(r)}(\Omega)$ are defined by the first component $u_j = D_\nu^{j-1} u_0\big|_{\partial\Omega}$ $(j = 1, \dots, r)$.

For $s = k + 1/p$ $(k = 0, \dots, r-1)$ we define the space $\widetilde{H}^{s,p,(r)}$ and the norm (4) by the method of complex interpolation.

Finally, for $r = 0$ we set that $\widetilde{H}^{s,p,(0)}(\Omega) := H^{s,p}(\Omega)$, $\quad |||u|||_{s,p,(r)} := \|u\|_{s,p}$. Recall here ([1], Ch. 2) that if

(6) $$M(x, D) = \sum_{|\alpha| \le q} a_\alpha(x) D^\alpha, \qquad a_\alpha(x) \in C_0^\infty(\overline{\Omega}),$$

is a differential expression of order $q \le r$ then the closure M of the mapping $u \mapsto M(x, D)u$, $u \in C_0^\infty(\overline{\Omega})$, acts continuously in the pair of spaces $\widetilde{H}^{s,p,(r)}(\Omega) \to H^{s-q,p}(\Omega)$; and if $q \le r - 1$ then the closure $M_{\partial\Omega}$ of the mapping $u \mapsto M(x, D)u|_{\partial\Omega}$, $u \in C_0^\infty(\overline{\Omega})$, acts continuously in the pair of spaces $\widetilde{H}^{s,p,(r)}(\Omega) \to B^{s-q-1/p,p}(\partial\Omega)$. It implies that the closure M of the mapping $u \mapsto (Mu|_{\overline{\Omega}}, Mu|_{\partial\Omega}, \dots, D_\nu^{r-q-1} Mu|_{\partial\Omega})$, $u \in C_0^\infty(\overline{\Omega})$, acts continuously in the pair of spaces $\widetilde{H}^{s,p,(r)}(\Omega) \to \widetilde{H}^{s-q,p,(r-q)}(\Omega)$.

Let us rewrite expression (6) in the form of $M(x, D) = \sum_{j=0}^{q} M_j(x, D') D_\nu^j$, where $M_j(x, D')$ is a $(q-1)$th-order tangential operator. By integration by parts we obtain that

$$(M(x, D)u, v) = (u, M^+(x, D)v) - i \sum_{j=1}^{q} \sum_{k=1}^{j} \langle D_\nu^{k-1} u, D_\nu^{j-k} M_j^+ v \rangle,$$
$$u, v \in C_0^\infty(\overline{\Omega}).$$

Here the expressions M^+ and M_j^+ are formally adjoint to the expressions M and M_j, respectively. By passing to the limit it implies that if $u = (u_0, \dots, u_r) \in \widetilde{H}^{s,p,(r)}(\Omega)$ then $Mu = f \in H^{s-q,p}(\Omega)$ if and only if

$$(7) \qquad (f, v) = (u_0, M^+ v) - i \sum_{j=1}^{q} \sum_{k=1}^{j} \langle u_k, D_\nu^{j-k} M_j^+ v \rangle \quad (\forall v \in C_0^\infty(\overline{\Omega})).$$

Formula (7) gives us a rule of calculation of the element $Mu = f \in H^{s-q,p}(\Omega)$ according to the element $u = (u_0, \dots, u_r) \in \widetilde{H}^{s,p,(r)}(\Omega)$. Rewrite (7) in the form of

$$(M(x, D)u)_+ = M(x, D)u_{0+} - i \sum_{j=1}^{q} \sum_{k=1}^{j} M_j D_\nu^{j-k} u_k \times \delta(\partial\Omega).$$

Here $(M(x, D)u)_+$ and u_{0+} are the extensions by zero of the functions $M(x, D)u$ and u_0 onto $\mathbb{R}^n$, and $\delta(\partial\Omega)$ is the Dirac measure concentrated on $\partial\Omega$:

$$M_j D_\nu^{j-k}(u_k \times \delta(\partial\Omega)) := \langle u_k \times D_\nu^{j-k} M_j^+ v \rangle \quad (v \in C_0^\infty(\mathbb{R}^n)).$$

In the set of the expressions of the form $M(x, D) = \sum_{j=1}^{q} M_j(x, D') D_\nu^j$ we introduce the operator J such that

$$JM(x, D) = \begin{cases} 0 & \text{for } q = 0 \\ \sum_{j=1}^{q} M_j(x, D') D_\nu^{j-1} & \text{for } q \geq 1. \end{cases}$$

Now one can rewrite (7) in the form of

$$(7)' \qquad f_+ = (Mu)_+ = Mu_{0+} - i \sum_{k=1}^{q} J^k M(x, D)(u_k \times \delta(\Omega)).$$

Here f_+, $(Mu)_+$ and u_{0+} are the extensions by zero of the functions f, Mu and u_0 onto $\mathbb{R}^n$. The equality (7)$'$ considered in whole $\mathbb{R}^n$ gives us a formula for calculation of the element f_+ for the vector $(u_0, \dots, u_r)$.

In a similar way it is easy to verify that if $\sum_{j=0}^{q} M_j(x, D') u_{j+1} = \phi$ then $Mu|_{\partial\Omega} = \phi \in B^{s-q-1/p,p}(\partial\Omega)$, $q \leq r - 1$.

1.3

It follows from all mentioned in 1.2 that for any $s \in \mathbb{R}$ and any $p \in (1, \infty)$ the closure $A = A_{s,p}$ of the mapping $u \mapsto (lu|_{\overline{G}}, \{(l_j u|_{\partial G}, \dots, D_\nu^{\kappa - s_j - 1} l_u|_{\partial G}) : j = 1, \dots, N\}, bu|_{\partial G})$ $(u \in (C^\infty(\overline{G}))^N, l_j u = \sum_{k=1}^{N} l_{jk}(x, D) u_k)$ acts continuously in the pair of spaces

$$\widetilde{H}^{T+s,p,(\tau)} := \prod_{j=1}^{N} \widetilde{H}^{t_j+s,p,(\tau_j)} \longrightarrow K_{s,p} := \prod_{j=1}^{N} \widetilde{H}^{s-s_j,p,(\kappa-s_j)} \times \prod_{h=1}^{m} B^{s-\sigma_h-1/p,p}(\partial G). \tag{8}$$

Definition: The element $u = (u_1, \ldots, u_N) \in \widetilde{H}^{T+s,p,(\tau)}$, $u_j = (u_{j0}, \ldots, u_{j,\tau_j}) \in \widetilde{H}^{t_j+s,p,(\tau_j)}$, $j = 1, \ldots, N$, such that

$$\begin{aligned} Au &= F = (f, \phi_1, \ldots, \phi_m), \quad f = (f_1, \ldots, f_N), \\ f_j &= (f_{j0}, \ldots, f_{j,\kappa-s_j}) \end{aligned} \tag{9}$$

is called a generalized solution of the problem (1)–(2).

It turns out [1] that the theorem on complete collection of isomorphisms is true:

Theorem 1 *Let* (1) *be an elliptic problem. Then*
i) *for any* $s \in \mathbb{R}$ *and any* $p \in (1, +\infty)$ *the operator* $A = A_{s,p}$ *is Noetherian. It means that the kernel* $\mathfrak{N}$ *and the cokernel* $\mathfrak{N}^*$ *are finite dimensional and do not depend on s and p,*

$$\begin{aligned} \mathfrak{N} &= \{u \in (C^\infty(\overline{G}))^N : Au = 0\}, \\ \mathfrak{N}^* &\subset \{V = (V_1, \ldots, V_N, \psi_1, \ldots, \psi_m) : \\ V_j &= (V_{j0}, \ldots, V_{j,\kappa-s_j}) \in C^\infty(\overline{G}) \times (C^\infty(\partial G))^{\kappa-s_j}; \\ &\quad \psi_j \in C^\infty(\partial G)(j = 1, \ldots, m)\}. \end{aligned}$$

The problem (9) *is solvable in* $\widetilde{H}^{T+s,p,(\tau)}$ *if and only if the following relation holds:*

$$\begin{aligned} [F, V] := &\sum_{j=1}^{N} (f_{j0}, V_{j0}) + \sum_{r=1}^{N} \sum_{k=1}^{\kappa-s_r} \langle f_{rk}, V_{rk} \rangle \\ &+ \sum_{h=1}^{m} \langle \phi_h, \psi_h \rangle = 0 (\forall V \in \mathfrak{N}^*) \end{aligned}$$

ii) *The restriction* $\widehat{A}_{s,p}$ *of the operator* $A_{s,p}$ *realizes an isomorphism*

$$P\widetilde{H}^{T+s,p,(\tau)}(G) \longrightarrow Q^+ K_{s,p} \quad (s \in \mathbb{R},\ 1 < p < \infty).$$

Here $P\widetilde{H}^{T+s,p,(\tau)}(G) = \{u \in \widetilde{H}^{T+s,p,(\tau)} : (u_0, v) = 0\ (\forall v \in \mathfrak{N})\}$ *is a subspace of the space* $\widetilde{H}^{T+s,p,(\tau)}$, *and* $Q^+ K_{s,p} = \{F \in K^{s,p} : [F, V] = 0 \quad (\forall V \in \mathfrak{N}^*)\}$ *is a subspace of* $K_{s,p}$.

2 Green's Formula

2.1

Definition: Matrix $B(x, D)$ of the system of $|\tau| = \tau_1 + \cdots + \tau_N$ boundary conditions

$$B_h u(x) = \sum_{j=1}^{N} B_{hj}(x, D) u_j(x) \quad (h = 1, \ldots, |\tau|) \tag{10}$$

is called a τ-complete matrix ([2], [3]) if

a) $B_{hj}(x, D) = \sum_{s=1}^{\tau_j} \Lambda_{h,js}(x, D') D_\nu^{s-1}$ $(x \in \partial G,\ j = 1, \ldots, N,\ h = 1, \ldots, |\tau|)$, where $\Lambda_{h,js}(x, D')$ are tangential, generally speaking, pseudodifferential operators. Moreover there exist integer $\sigma_h < 0$ $(h = 1, \ldots, |\tau|)$ such that $\operatorname{ord} B_{hj} \leq \sigma_h + \tau_j$ for $\sigma_h + \tau_j > 0$, and $B_{hj} \equiv 0$ for $\sigma_h + \tau_j \leq 0$;

b) For every real vector $\gamma \neq 0$ tangential to ∂G at the point x the determinant of the square $|\tau| \times |\tau|$-matrix

$$(\Lambda^0_{h,js}(x, \gamma))_{\substack{h=1,\ldots,|\tau| \\ s=1,\ldots,|\tau_j|,\ j=1,\ldots,N}} \tag{11}$$

does not equal to zero at every point $x \in \partial G$. Here $\Lambda^0_{h,js}(x, D')$ is the principal part of the expression $\Lambda_{h,js}(x, D')$ which includes only $(\tau_j + \sigma_h - s + 1)$th-order differentiations. In addition $\Lambda^0_{h,js} = 0$ for $\tau_j + \sigma_h - s + 1 < 0$.

This definition implies that if the system (10) is a τ-complete matrix then the system of the equations

$$\sum_{j=1}^{N} \sum_{s=1}^{\tau_j} \Lambda_{h,js}(x, D') \eta_{js} = \phi_h \quad (h = 1, \ldots, |\tau|;\ x \in \partial G) \tag{12}$$

or, in short, $\Lambda \eta = \phi$, is an elliptic in Douglis-Nirenberg sense on the manifold ∂G.

In fact, if we put the number $\tau_{js} = \tau_j - s + 1$ into a correspondence to the function η_{js}, and if we put the number σ_h into a correspondence to the equation with the number h then we get that $\operatorname{ord} \Lambda_{h,js} \leq \tau_{js} + \sigma_h$ for $\tau_{js} + \sigma_h \geq 0$, and $\Lambda_{h,js} \equiv 0$ for $\tau_{js} + \sigma_h < 0$. Thus, the ellipticity of system (12) on ∂G follows from the fact that the determinant of matrix (11) does not equal to zero.

It is clear that the formally adjoint system

$$\sum_{h=1}^{|\tau|} \Lambda^+_{h,js}(x, D') \zeta_h = \psi_{js} \quad (s = 1, \ldots, \tau_j;\ j = 1, \ldots, N;\ x \in \partial G) \tag{13}$$

or, in short, $\Lambda^+ \zeta = \psi$ (where $\Lambda^+_{h,js}$ is the expression formally adjoint to the expression $\Lambda_{h,js}$), is a Douglis-Nirenberg elliptic system. In addition, the following Green's formula is true:

$$\langle \Lambda \eta, \zeta \rangle = \langle \eta, \Lambda^+ \zeta \rangle \qquad (\eta, \zeta \in (C^\infty(\partial G))^{|\tau|}). \tag{14}$$

For any $q \in \mathbb{R}$ the closure $\Lambda = \Lambda_q$ of the mapping $\eta \mapsto \Lambda\eta$ for $\eta \in (C^\infty(\partial G))^{|\tau|}$ acts continuously in the pair of spaces

$$(15) \quad U^{q,\tau,p} := \prod_{j=1}^{N}\prod_{s=1}^{\tau_j} B^{q+\tau_j-s,p}(\partial G) \longrightarrow V^{q,\sigma,p} := \prod_{h=1}^{|\tau|} B^{q-\sigma_h-1,p}(\partial G).$$

The kernel $\mathfrak{N}_\lambda$ of this operator is finite-dimensional and does not depend on q, i.e., $\mathfrak{N}_\Lambda \subset (C^\infty(\partial G))^{|\tau|}$. The range $\mathfrak{R}(\Lambda_q)$ of values of operator Λ_q is closed in the space $V^{q,\sigma,p}$ and has a finite codimension independent of q ([2]; [1], Ch. X).

The similar statement is true also for the operator Λ_q^+ defined by the problem (13) with the kernel $\mathfrak{N}_{\Lambda^+}$. In addition, the cokernel (the kernel) of the operator Λ is the kernel (the cokernel) of the operator Λ^+.

It easily follows from finite-dimensionality of $\mathfrak{N}_\Lambda$ that every element $\eta \in \cup_{q=-\infty}^{+\infty} U^{q,\tau,p}$ can be uniquely represented in the form of the sum

$$(16) \qquad \eta = \eta' + \eta'', \quad \eta'' \in \mathfrak{N}_\Lambda, \quad \langle \eta', \mathfrak{N}_\Lambda \rangle = 0.$$

It is clear ([3]; [1], Ch. V, Ch. X) that the restriction $\widehat{\Lambda}_q$ of the operator Λ_q onto $P_\Lambda U^{q,\tau,p}$ realizes an isomorphism $\widehat{\Lambda}_q : P_\Lambda U^{q,\tau,p} \to P_{\Lambda^+} V^{q,\sigma,p}$. The operator $\Pi_q = \widehat{\Lambda}_q^{-1} P_{\Lambda^+}$ acts continuously from the entire $V^{q,\sigma,p}$ into $U^{q,\tau,p}$. Since for $q_1 \le q_2$ the operators Λ_{q_1}, $\widehat{\Lambda}_{q_1}$ and Π_{q_2} are the extension in continuity of the operators Λ_{q_2}, $\widehat{\Lambda}_{q_2}$ and Π_{q_2} respectively, in what follows we will write Λ, $\widehat{\Lambda}$ and Π in place of Λ_q, $\widehat{\Lambda}_q$ and Π_q.

Note that the following statement is true.

Lemma 1 *For any $\phi \in V^{q,\sigma,p}$ there exists an element $u \in \widetilde{H}^{T+q,p,(\tau)}$ such that $Bu|_{\partial G} = \phi$ if and only if $\mathfrak{N}_{\Lambda^+} = 0$. In addition there exists a linear continuous operator $T_q : \phi \mapsto u$ acting in the pair of spaces $V^{q,\sigma,p} \to \widetilde{H}^{T+q,p,(\tau)}$. For $q_2 < q_1$ the operator T_{q_2} is an extension in continuity of the operator T_{q_1}. If $\phi \in (C^\infty(\partial G))^{|\tau|}$ then $u \in (C^\infty(\overline{G}))^N$.*

For any element $\phi \in (C^\infty(\overline{G}))^{|\tau|}$ there exists the element $u \in (C^\infty(\overline{G}))^N$ such that $Bu|_{\partial G} = \phi$ if and only if the same condition $\mathfrak{N}_{\Lambda^+} = 0$ holds.

2.2

Let us now obtain Green's formula. Assume that $l_{rj}(x, D) = \sum_{k=0}^{s_r+t_j} l_k^{rj}(x, D')D_\nu^k$, $r, j : s_r + t_j \ge 0$, in some neighborhood of the boundary ∂G in $\overline{G}$. Here $l_k^{rj}(x, D')$ are the tangential operators whose orders are not larger than $s_r + t_j - k$. By integration by parts and changing the the order of summation, we obtain that

$$(17) \qquad (lu, v) - (u, l^+v) = \sum_{j:t_j\ge 1}\sum_{s=1}^{t_j} \langle D_\nu^{s-1}u_j, M_j^s v\rangle, \; M_j^s v = -i \sum_{r:s_r+t_j\ge s}\sum_{k=s}^{s_r+t_j} D_\nu^{k-j}(l_k^{rj}(x, D'))^+ v_r.$$

We put the vector

$$\begin{aligned} \eta &= (\eta_1, \dots, \eta_N), \eta_j = (\eta_{j1}, \dots, \eta_{j,\tau_j}), \eta_{js} = D_\nu^{s-1} u_j\big|_{\partial G}, \\ s &= 1, \dots, \tau_j, j = 1, \dots, N, \end{aligned}$$

into a correspondence to the element $u = (u_1, \dots, u_N)$, and put the vector

$$\begin{aligned} Mv &= (\zeta_1, \dots, \zeta_N), \quad \zeta_j = (\zeta_{j1}, \dots, \zeta_{j,\tau_j}), \\ \zeta_{jk} &= \begin{cases} M_j^k v\big|_{\partial G} & \text{for } k = 1, \dots, t_j \\ 0 & \text{for } t_j < k \leq \tau_j \end{cases} \end{aligned}$$

into a correspondence to the element $v = (v_1, \dots, v_N)$. Then $Bu|_{\partial G} = \Lambda\eta$, and one can rewrite formula (18) in the form

$$(lu, v) - (u, l^+ v) = \langle \eta, Mv \rangle \quad (u, v \in (C^\infty(\overline{G}))^{|\tau|}). \tag{18}$$

By (16), using the fact that $\Pi\Lambda\eta = \eta'$, rewrite (18) in the form

$$\begin{aligned} (lu, v) - (u, l^+ v) &= \langle \Pi\Lambda\eta, Mv \rangle + \langle \eta'', Mv \rangle = \langle \Lambda\eta, \Pi^+ Mv \rangle + \langle \eta'', Mv \rangle \\ &= \langle \Lambda\eta, P_{\Lambda^+}\Pi^+ Mv \rangle + \langle \eta'', Mv \rangle. \end{aligned}$$

This proves the following theorem.

Theorem 2 *The Green's formula*

$$\begin{aligned} (lu, v) - (u, l^+ v) &= \langle Bu, B'v \rangle + \langle \eta'', Mv \rangle, \quad u, v \in (C^\infty(\overline{G})), \\ \eta'' &= \eta - P\eta \in N_\Lambda, \end{aligned} \tag{19}$$

is true. Here

$$B'v = P_{\Lambda^+}\Pi^+ Mv|_{\partial G} \tag{20}$$

is, generally speaking, a pseudodifferential matrix expression.

In particular, if $\mathfrak{N}_\Lambda = 0$ then the last term in the right-hand side of (19) is missing. It will be, for example, if the matrix B is a parameter elliptic matrix, or if B is a Dirichlet matrix ([4]). In the last case, if the matrix B is differential then the matrix B' is also differential.

2.3

Let us write the Green's formula in more convenient form. To do this, we complement the matrix $b(x, D)$ (see (2)) to the τ-complete system $B(x, D)$ by new rows.

Denote:

$$c(x, D) = (c_{hj}(x, D))_{\substack{h=1,\dots,m \\ j=1,\dots,N}}, \tag{21}$$

where $\operatorname{ord} c_{hj} \le \sigma_h^c + t_j$ for $\sigma_h^c + t_j > 0$, and $c_{hj} \equiv 0$ for $\sigma_h^c + t_j < 0$;

$$c_h(x, D) = (c_{h1}(x, D), \ldots, c_{hN}(x, D)), \quad h = 1, \ldots, m,$$

is a row with the number h of the matrix $c(x, D)$.

Let

$$e(x, D) = (D_\nu^{k-1} l_r(x, D))_{\substack{r=1,\ldots,M \\ k=1,\ldots,-s_r+\kappa}} . \tag{22}$$

Here $l_r(x, D) = (l_{r1}(x, D), \ldots, l_{rN}(x, D))$ is a row with the number r of the matrix $l(x, D)$.

The matrix $c(x, D)$ consists of m rows, and the matrix $e(x, D)$ consists of $|\tau| - 2m$ rows. It turns out that the considerations from [1], Ch. X, imply that the next statement true.

Lemma 2 *Let* (1)–(2) *be an elliptic problem. Then there exists the matrix* $c(x, D)$ *with* $\sigma_h^c < 0$, $h = 1, \ldots, m$ *such that the* $|\tau| \times N$*-matrix*

$$B(x, D) = \begin{pmatrix} b(x, D) \\ e(x, D) \\ c(x, D) \end{pmatrix} \tag{23}$$

is τ*-complete.*

Lemma 2 and Theorem 2 directly imply the validity of the following theorem.

Theorem 3 *Let the problem* (1)–(2) *be an elliptic, and let the matrix* $b(x, D)$ *be complemented to the* τ*-complete matrix* $B(x, D)$ (23) *by matrixes* (21) *and* (22). *Then the Green's formula*

$$\begin{aligned} &(lu, v) + \sum_{h=1}^{m} \langle b_h u, c_h' v \rangle + \sum_{r=1}^{N} \sum_{k=1}^{-s_r+\kappa} \langle D_\nu^{k-1} l_r u, e_{kr}' v \rangle \\ &= (u, l^+ v) + \sum_{h=1}^{m} \langle c_h u, b_h' v \rangle + \langle \eta'', M v \rangle, \\ &u, v \in (C^\infty(\overline{G}))^N, \quad \eta'' = \eta - P\eta \in \mathfrak{N}_\Lambda, \end{aligned} \tag{24}$$

is true.

To obtain formula (24), it is necessary to represent matrix (20) in the form of the matrix $\begin{pmatrix} -c'(x, D) \\ -e'(x, D) \\ b'(x, D) \end{pmatrix}'$, transposed to $\begin{pmatrix} -c'(x, D) \\ -e'(x, D) \\ b'(x, D) \end{pmatrix}$.

2.4

In this subsection let us additionally assume that $\mathfrak{N}_{\Lambda^+} = 0$. In other words, we set that for any $\phi \in (C^\infty(\partial G))^{|\tau|}$ there exists an element $u \in (C^\infty(\overline{G}))^N$ such that $Bu|_{\partial G} = \phi$.

The problem

$$l^+ v(x) = g(x), \qquad x \in G, \tag{25}$$

$$b'v\big|_{\partial G} = \phi, \tag{26}$$

$$\langle Mv, \mathfrak{N}_\Lambda \rangle = 0 \tag{27}$$

is called a formally adjoint to the problem (1)–(2) with respect to Green's formula (24).

Let

$$\mathfrak{N}^+ = \{v \in (C^\infty(\overline{G}))^N : l^+ v = 0,\ b'v\big|_{\partial G} = 0,\ \langle Mv, \mathfrak{N}_\Lambda \rangle = 0\}. \tag{28}$$

Theorem 4 *Under the conditions of Theorem* 3, *let* $\mathfrak{N}_{\Lambda^+} = 0$, *and let* $s \in \mathbb{R}$, $p \in (1, +\infty)$. *Then the problem* (9) *with* $F \in K_{s,p}$ *has a solution* $u \in \widetilde{H}^{T+s,p,(\tau)}$ *if and only if the relation*

$$(f_0, v) + \sum_{j=1}^{N} \sum_{r=1}^{\kappa - s_j} \langle f_{jr}, e'_{kr} v \rangle + \sum_{h=1}^{m} \langle \phi_h, c'_h v \rangle = 0, \qquad v \in \mathfrak{N}^+, \tag{29}$$

holds. In addition, $\mathfrak{N}^+$ *is finite-dimensional.*

The proof of the Theorem follows from the Green's formula (24) and Theorem 1.

2.5

Using the methods from the paper [2] and [3], we get the following theorem.

Theorem 5 *Let the problem* (1)–(2) *be elliptic, and let* $\mathfrak{N}_{\Lambda^+} = 0$. *Assume that for any* $\psi \in B^{T'-\sigma'-1/2,2}(\partial G)$ *there exists the element* $v_0 \in H^{T',2}(G)$ *such that* $b'v_0|_{\partial G} = \psi$ *and* $\langle Mv_0, \mathfrak{N}_\Lambda \rangle = 0$, *and the extension operator* $T : \psi \mapsto v_0$ *is continuous. Then the problem* (25)–(26), *formally adjoint to the problem* (1)–(2) *with respect to the Green's formula, is elliptic.*

2.6

In what follows we assume that

$$\mathfrak{N}_\Lambda = 0, \qquad \mathfrak{N}_{\Lambda^+} = 0. \tag{30}$$

It will be, for example, in the case where the matrix B is elliptic with a parameter, or where B is Dirichlet matrix ([4]). Under this assumption the term $\langle \eta'', Mv \rangle$

is missing in formulas (19), (24), and (28). Then the formally adjoint problem is defined by equalities (25)–(26), and all the considerations are rather simplified.

The case where $\mathfrak{N}_\Lambda \neq 0$ and $\mathfrak{N}_{\Lambda^+} \neq 0$ is considered in [2] for one equation and in [3] for Petrovskii elliptic system. It is based on the stadying of the set of various projection operators onto finite-dimensional subspaces.

3 On Various Isomorphisms Theorems for General Elliptic Boundary Value Problems for Systems of Equations

3.1

Let us describe two methods those permit us to obtain new theorems on isomorphisms from known isomorphism theorems.

Let B_1 and B_2 be Banach spaces and let T be a linear operator that isomorphically maps the space B_1 onto the space B_2. Let E_1 be a subspace of B_1, and let $E_2 = TE_1$. Then it is clear that the operator T naturally defines a linear operator T_1 that isomorphically maps the factor space B_1/E_1 onto the factor space B_2/E_2.

Further, let Q_2 be a Banach space, and let $Q_2 \subset B_2$ (the imbedding is algebraic and topological). Then $Q_1 = T^{-1}Q_2$ is a linear (generally speaking, nonclosed) subset of the space B_1. However, the space Q_1 becomes a Banach space (denoted by Q_{1T}) with respect to the graph norm $\|x\|_{Q_{1T}} = \|x\|_{B_1} + \|Tx\|_{Q_2}$, $x \in Q_1$. The restriction of the operator T onto Q_1 establishes the isomorphism $Q_{1T} \to Q_2$.

These simple procedures those are called **"Pasting" (or factorization) method** and **graph method** give a possibility to obtain in [5] and [6] all known theorems and a number of new theorems on isomorphisms for the case of one equation from the isomorphisms theorems from [7] and [2].

The obtained above results on Green's formula permit us to use these two methods to the case of general boundary value problems for Douglis-Nirenberg systems.

3.2

The norm $|||\cdot|||$ of the space $\widetilde{H}^{s,p,(r)}$ is defined by formula (4) for $s \neq k + 1/p$, $k = 0, \dots, r-1$. Therefore in what follows we shall consider such spaces for these values of $s \in \mathbb{R}$. For $s = k + 1/p$, $k = 0, \dots, r-1$ the theorems on isomorphisms is obtained by interpolation theorem.

It was already mentioned in 1.3 that the closure $A = A_{s,p}$ of the mapping $u \mapsto (lu, bu|_{\partial G})$, $u \in (C^\infty(\overline{G}))^N$, acts continuously in the pair of spaces (9). Moreover, according to Theorem 1, the operator $A = A_{s,p}$ is Noetherian, the restriction $\widehat{A} = \widehat{A}_{s,p}$ of the operator A realizes an isomorphism $P\widetilde{H}^{T+s,p,(\tau)} \to Q^+K_{s,p}$. For simplicity, in what follows we shall assume that the defect is missing, i.e., $\mathfrak{N} = 0$ and $\mathfrak{N}^* = 0$. Then the operator $A = A_{s,p}$ realizes an isomorphism

$$\widetilde{H}^{T+s,p,(\tau)} \longrightarrow K_{s,p}. \tag{31}$$

3.3

Relations (30) imply that the operator $\Lambda = \Lambda_q$ realizes an isomorphism between spaces (15). Therefore, since the matrix B is defined by formula (23), the norm

$$
\begin{aligned}
\|u\|_{\widetilde{H}^{T+s,p,(\tau)}} &= \left(\sum_{j=1}^{N} |||u_j|||^p_{t_j+s,p,(\tau_j)}\right)^{1/p} \\
&= \sum_{j=1}^{N}\left(\|u_j\|^p_{t_j+s,p} + \sum_{k=1}^{\tau_j}\langle\langle D_\nu^{k-1}u_j, \partial G\rangle\rangle^p_{t_j+s-k+1-1/p,p}\right)^{1/p}
\end{aligned} \tag{32}
$$

is equivalent to the norm

$$
\begin{aligned}
&\left(\sum_{j=1}^{N}\|u_j\|_{t_j+s,p} + \sum_{r=1}^{N}\sum_{k=1}^{-s_r+\kappa}\langle\langle D_\nu^{k-1}l_r u\rangle\rangle_{s-s_r-k+1-1/p,p}\right. \\
&\left. + \sum_{j=1}^{m}\langle\langle b_j u\rangle\rangle_{s-\sigma_j-1/p,p} + \sum_{h=1}^{m}\langle\langle c_h u\rangle\rangle_{s-\sigma_h^c-1/p,p}\right)^{1/p},
\end{aligned} \tag{33}
$$

and the space $\widetilde{H}^{T+s,p,(\tau)}$ coincides with the completion $\widetilde{H}_B^{T+s,p,(\tau)}$ of the space $(C^\infty(\overline{G}))^N$ in the norm (34). If $s \geq \kappa$ (4), then norms (33) and (34) are equivalent to the norm $\sum_{j=1}^{N}\|u_j\|_{t_j+s,p}$, and both the spaces $\widetilde{H}^{T+s,p,(\tau)}$ and $\widetilde{H}_B^{T+s,p,(\tau)}$ coincide with the space

$$\prod_{j=1}^{N} H^{t_j+s,p}(G) := H^{T+s,p}.$$

If $s < \kappa$ then norms (33) and (34) are equivalent to the norms

$$\|u\|_{H^{T+s,p}} + \sum_{k,j:t_j+s-k+1-1/p<0}\langle\langle D_\nu^{k-1}u_j\rangle\rangle_{t_j+s-k+1-1/p,p} \tag{32'}$$

and

$$
\begin{aligned}
&||u||_{H^{T+s,p}} + \sum_{r,k:s-s_r-k+1-1/p<0}\langle\langle D_\nu^{k-1}l_r u\rangle\rangle_{s-s_r-k+1-1/p,p} \\
&+ \sum_{j:s-s_j-1/p<0}\langle\langle b_j u\rangle\rangle_{s-s_j-1/p,p} + \sum_{h:s-\sigma_h^c-1/p<0}\langle\langle c_h u\rangle\rangle_{s-\sigma_h^c-1/p,p},
\end{aligned} \tag{33'}
$$

respectively.

The closure S of the mapping

$$u \longmapsto (u|_{\overline{G}}, \{D_\nu^{j-1}u_k | j,k : t_j + s - k + 1 - 1/p < 0\}), u \in (C^\infty(\overline{G}))^N,$$

is an isometry between the space $\widetilde{H}^{T+s,p,(\tau)}$ with the metric (32)′ and the space

$$(34)\quad \mathcal{H}^{T+s,p,(\tau)} := H^{T+s,p} \times \prod_{k,j:t_j+s-k+1-1/p<0} B^{t_j+s-k+1-1/p,p}(\partial G)$$

(see [8], [1]). Analogously, the closure S_B of the mapping

$$\begin{aligned} u \longmapsto\ & (u|_{\overline{G}}, \{D_\nu^{k-1} l_r u|_{\partial G} | r, k : s - s_r - k + 1 - 1/p < 0\},\\ & \{b_j u|_{\partial G} : s - \sigma_j - 1/p < 0\},\\ & \{c_h u|_{\partial G} : s - \sigma_h^c - 1/p < 0\}), \quad u \in (C^\infty(\overline{G}))^N \end{aligned}$$

is an isometry between the space $\widetilde{H}^{T+s,p,(\tau)}$ with metric (33)′ and the space

$$(35)\quad \begin{aligned} \mathcal{H}_B^{T+s,p,(\tau)} := H^{T+s,p} \times & \prod_{r,k:s-s_r-k+1-1/p<0} B^{s-s_r-k+1-1/p,p}(\partial G)\\ \times \prod_{j:s-\sigma_j-1/p<0} B^{s-\sigma_j-1/p,p}(\partial G) \times & \prod_{h:s-\sigma_h^c-1/p<0} B^{s-\sigma_h^c-1/p,p}(\partial G). \end{aligned}$$

This gives us a possibility to identify every element $u \in \widetilde{H}_B^{T+s,p,(\tau)}$ with the element

$$(36)\quad \begin{aligned} S_B u = \ & (u_0, \{u_{rk} : s - s_r - k + 1 - 1/p < 0\},\\ & \{u_j^b : s - \sigma_j - 1/p < 0\},\\ & \{u_h^c : s - \sigma_h^c - 1/p < 0\}) \in \mathcal{H}_B^{T+s,p,(\tau)}. \end{aligned}$$

We shall write $u = S_B u$ for any $u \in \widetilde{H}_B^{T+s,p(\tau)}$.

For any sequence $u_i \in (C^\infty(\overline{G}))^N$ that converges to u in $\widetilde{H}_B^{T+s,p(\tau)}$ as $i \to \infty$ the sequence $D_\nu^{k-1} l_r u_i|_{\partial G}$ converges to u_{rk} in $B^{s-s_r-k+1-1/p,p}(\partial G)$ $(r, k : s - s_r - k + 1 - 1/p < 0)$, the sequence $b_j u_i|_{\partial G}$ converges to u_j^b in $B^{s-\sigma_j-1/p,p}(\partial G)$, $s - \sigma_j^b - 1/p < 0$, the sequence $c_h u_i|_{\partial G}$ converges to u_h^c, in the space $B^{s-\sigma_h^c-1/p,p}(\partial G)$, $s - \sigma_h^c - 1/p < 0$. In this sense, $u_{rk} = D_\nu^{k-1} l_r u|_{\partial G}$, $u_j^b = b_j u|_{\partial G}$, $u_h^c = c_h u|_{\partial G}$.

Otherwise, if $s - s_r - k + 1 - 1/p > 0$, $u_i \in (C^\infty(\overline{G}))^N$ such that $u_i \to u$ in $\widetilde{H}^{T+s,p,(\tau)}$, then $D_\nu^{k-1} l_r u_i|_{\partial G} \to D_\nu^{k-1} l_r u_0|_{\partial G}$. If $s - \sigma_j - 1/p > 0$ then $b_j u_i|_{\partial G} \to b_j u_0|_{\partial G}$ in $B^{s-\sigma_j-1/p,p}(\partial G)$. If $s - \sigma_h^c - 1/p > 0$ then $c_h u_i|_{\partial G} \to c_h u_0|_{\partial G}$.

Thus, the expressions $D_\nu^{k-1} l_r u|_{\partial G}$, $b_j u|_{\partial G}$ and $c_h u|_{\partial G}$ are defined for every element $u \in \mathcal{H}_B^{T+s,p,(\tau)} \simeq \widetilde{H}^{T+s,p,(\tau)}$.

Green's formula (24) and equations (9) easily imply that element $u = Su \in \widetilde{H}^{T+s,p,(\tau)}$ is a generalized solution of the problem (1)–(2) if and only if

$$(37)\quad \begin{aligned} (u_0, l^+ v) + \sum_{h=1}^m \langle c_h u, b_h' v\rangle = (f_0, v) + \sum_{h=1}^m \langle \phi_h, c_h' v\rangle + \sum_{r=1}^N \sum_{k=1}^{-s_r+\kappa}\\ \langle f_{rk}, e_{kr}' v\rangle, \forall v \in (C^\infty(\overline{G}))^N. \end{aligned}$$

3.4

It is possible that the various elements of the space $\widetilde{H}_B^{T+s,p,(\tau)}$ have the same components $(u_0, c_1 u|_{\partial G}, \dots, c_m u|_{\partial G})$. "Pasting" these elements and doing the corresponding factorization in the space of images, we obtain new statement on isomorphisms from the isomorphism $\widetilde{H}_B^{T+s,p,(\tau)} \to K^{s,p}$ Formula (36) easily implies that after this pasting the space of pre-images

$$H^{T+s,p} \times \prod_{h:s-\sigma_h^c-1/p<0} B^{s-\sigma_h^c-1/p,p}(\partial G) \tag{38}$$

is the completion of $(C^\infty(\overline{G}))^N$ in the norm $\|u\|_{T+s,p} + \sum_{h=1}^m \langle\langle c_h u \rangle\rangle_{s-\sigma_h^c-1/p,p}$, and the space of images (see (9) and (38)) is the direct product

$$\left(\prod_{j=1}^N H^{s-s_j,p}(G) \times \prod_{h:s-\sigma_h-1/p>0} B^{s-\sigma_h-1/p,p}(\partial G) \right) \Big/ M_{s,p}^1, \tag{39}$$

where

$$\begin{aligned} M_{s,p}^1 = \Bigg\{ & (f, \{\phi_h\}) \in \prod_{j=1}^N H^{s-s_j,p}(G) \times \prod_{h:s-\sigma_h-1/p>0} B^{s-\sigma_h-1/p,p}(\partial G) : \\ & (f, v) + \sum_{h:s-\sigma_h-1/p>0} \langle \phi_h, c_h' v \rangle + \sum_{k,r:s-s_r-k+1-1/p>0} \\ & \langle D_\nu^{k-1} f_r, e_{kr}' v \rangle = 0, \forall v \in (C^\infty(\overline{G}))^N : \\ & c_h' v|_{\partial G} = 0, e_{kr}' v|_{\partial G} = 0, h : s - \sigma_h^c - 1/p < 0, k, r : \\ & s - s_r - k + 1 - 1/p < 0 \Bigg\} \end{aligned}$$

is a subspace of the direct product $\prod_{j=1}^N H^{s-s_j,p}(G) \times \prod_{h:s-\sigma_h-1/p>0} B^{s-\sigma_h-1/p,p}$ (∂G).

As a result, we obtain the following statement.

Theorem 6 *Let the problem* (1)–(2) *be elliptic, and let relations* (30) *hold. For simplicity, assume that $\mathfrak{N} = 0$ and $\mathfrak{N}^* = 0$. Then the closure $A_1 = A_{1sp}$ of the mapping*

$$u \longmapsto (lu, \{b_h u|_{\partial G} : s - \sigma_h - 1/p > 0\}), \quad u \in (C^\infty(\overline{G}))^N,$$

realizes an isomorphism from space (38) *into space* (39).

Corollary 1 *For $u \in (C^\infty(\overline{G}))^N$ the estimate*

$$\|u\|_{T+s,p} + \sum_{j=1}^{m} \langle\langle c_j u\rangle\rangle_{s-\sigma_j^c-1/p,p} \le c_1 \left(\|lu\|_{s-S,p} + \sum_{h:s-\sigma_h-1/p>0} \langle\langle b_h u\rangle\rangle_{s-\sigma_h-1/p,p} + \|u\|_{T+s-k,p} \right)$$

is valid. Here the number $k > 0$ may be chosen arbitrary large.

Note that if $\mathfrak{N} = 0$ then one can take, the term $\|u\|_{T+s-k,p}$ is absent.

3.5

The various elements of the space $\widetilde{H}^{T+s,p,(\tau)}$ may have the same components $(u_0, c_1u|_{\partial G}, \ldots, c_m u|_{\partial G}, b_1 u|_{\partial G}, \ldots, b_m u|_{\partial G})$. Pasting them and pasting the corresponding elements in the space of images we get below new theorem on isomorphisms.

Denote by $\widetilde{H}_{c,b}^{T+s,p,(\tau)}$ the completion of the space $C^\infty(\overline{G}))^N$ in the norm

$$|||u|||_{T+s,p,(c,b)} := \left(\|u\|^p_{T+s,p} + \sum_{j=1}^{m} \langle\langle c_j u\rangle\rangle^p_{s-\sigma_h^c-1/p,p} + \sum_{j=1}^{m} \langle\langle b_j u\rangle\rangle^p_{s-\sigma_j-1/p,p} \right)^{1/p}.$$

It follows from the consideration of subsection 3.3 that the space $\widetilde{H}_{c,b}^{T+s,p,(\tau)}$, that is the space of pre-images obtained by this pasting, is isomorphic to the direct product

$$H^{T+s,p} \times \prod_{h:s-\sigma_h^c-1/p<0} B^{s-\sigma_h^c-1/p,p}(\partial G) \times \prod_{h:s-\sigma_h-1/p<0} B^{s-\sigma_h-1/p,p}(\partial G)$$

The space of images is the factor-space

$$\left(H^{s-S,p} \times \prod_{h=1}^{m} B^{s-\sigma_h-1/p,p}(\partial G) \right) \Big/ M^2_{s,p} := K^2_{s,p},$$

where $H^{s-S,p} = \prod_{j=1}^{N} H^{s-s_j,p}(G)$, and

$$M^2_{s,p} = \Bigg\{ (f, \phi_1, \ldots, \phi_m) \in H^{s-S,p} \times \prod_{h=1}^{m} B^{s-\sigma_h-1/p,p}(\partial G) : (f, v) + \sum_{h=1}^{m} \langle \phi_h, c'_h v\rangle + \sum_{k,r:s-s_r-k+1-1/p>0} \langle D_\nu^{k-1} f_r, e'_{kr} v\rangle = 0, \forall v \in C^\infty(\overline{G}))^N : e'_{kr} v\big|_{\partial G} = 0, k, r : s - s_r - k + 1 - 1/p < 0) \Bigg\}$$

(see (8) and (36)).

Thus, we establish the validity of the following statement.

Theorem 7 *Under the conditions of Theorem 6, the closure $A_2 = A_{2sp}$ of the mapping $u \mapsto (lu, bu)$, $u \in (C^\infty(\overline{G}))^N$, realizes an isomorphism $\widetilde{H}_{c,b}^{T+s,p,(\tau)} \to K_{s,p}^2$.*

Corollary 2 *The estimate*

$$|||u|||_{T+s,p,(c,b)} \leq c_1 \left(\|lu\|_{s-S,p} + \sum_{j=1}^{m} \langle\langle b_j u\rangle\rangle_{s-\sigma_j-1/p,p} + \|u\|_{T+s-k,p} \right),$$
$$u \in (C^\infty(\overline{G}))^N,$$

is valid. Here the number $k > 0$ may be chosen arbitrary large.

Note that if $\mathfrak{N} = 0$ then one can take, the term $\|u\|_{T+s-k,p}$ is absent.

3.6

The various elements of the space $\widetilde{H}_B^{T+s,p,(\tau)}$ may have the same components $(u_0, b_1 u|_{\partial G}, \ldots, b_m u|_{\partial G})$. Paste them and make a corresponding factorization in the space of images.

Denote by $\widetilde{H}_b^{T+s,p,(\tau)}$ the completion of $(C^\infty(\overline{G}))^N$ in the norm

$$|||u|||_{T+s,p,(b)} = \left(\|u\|_{T+s,p}^p + \sum_{j=1}^{m} \langle\langle b_j u\rangle\rangle_{s-\sigma_j-1/p,p}^p \right)^{1/p}.$$

It is clear (see subsection 3.3) that the obtained space $\widetilde{H}_b^{T+s,p,(\tau)}$ of pre-images is isomorphic to the direct product $\widetilde{H}_b^{T+s,p,(\tau)} \simeq H^{T+s,p} \times \prod_{h:s-\sigma_h-1/p<0} B^{s-\sigma_h-1/p,p}(\partial G)$. The space of images is described by the formula $(H^{s-S,p} \times \prod_{h=1}^m B^{s-\sigma_h-1/p,p}(\partial G))/M_{s,p}^3 = K_{s,p}^3$, where

$$M_{s,p}^3 = \Big\{ (f, \phi_1, \ldots, \phi_m) \in H^{s-S,p} \times \prod_{h=1}^{m} B^{s-\sigma_h-1/p,p}(\partial G) :$$

$$(f, v) + \sum_{h=1}^{m} \langle \phi_h, c_h' v\rangle + \sum_{k,r:s-s_r-k+1-1/p>0} \langle D_\nu^{k-1} f_r, e_{kr}' v\rangle = 0, \tag{40}$$
$$\forall v \in (C^\infty(\overline{G}))^N : e_{kr}' v\big|_{\partial G} = 0,\ b_h' v\big|_{\partial G} = 0,\ k, r : s - s_r - k$$
$$+1 - 1/p < 0,\ h = 1, \ldots, m \Big\}$$

is the subspace of the direct product $H^{s-S,p} \times \prod_{h=1}^m B^{s-\sigma_h-1/p,p}(\partial G)$.

Theorem 8 *Under the assumption of Theorem 6, the closure $A_3 = A_{3sp}$ of the mapping $u \mapsto (lu, bu)$, $u \in (C^\infty(\overline{G}))^N$, realizes an isomorphism*

$$\widetilde{H}_b^{T+s,p,(\tau)} \longrightarrow K^3_{s,p}. \tag{41}$$

Corollary 3 *The estimate $|||u|||_{T+s,p,(b)} \leq c(|||lu||_{s-S,p} + \sum_{j=1}^m \langle\langle b_j u\rangle\rangle_{s-\sigma_j-1/p,p} + \|u\|_{T+s-k,p})$ is valid. Here the number $k > 0$ may be chosen arbitrary large.*

Note that if $\mathfrak{N} = 0$ then one can take, the term $\|u\|_{T+s-k,p}$ is absent.

Remark: If the matrix $\begin{pmatrix} c'(x,D) \\ b'(x,D) \\ e'(x,D) \end{pmatrix}$ is τ'-complete then it follows from definition (40) of the space $M^3_{s,p}$ that

$$\begin{aligned} M^3_{s,p} &= \Bigg\{(f, 0, \ldots, 0) \in H^{s-S,p} \times \prod_{j=1}^m B^{s-\sigma_j-1/p,p}(\partial G) : \\ &\qquad D_\nu^{k-1} f_r|_{\partial G} = 0, k, r : s - s_r - k + 1 - 1/p > 0, \\ (f, v) &= 0, \forall v \in (C^\infty(\overline{G}))^N : e'_{kr} v = 0,\ k, r : s - s_r - k + 1 - 1/p < 0, \\ &\qquad b'_h v|_{\partial G} = 0,\ h = 1, \ldots, m\Bigg\}. \end{aligned}$$

If we denote by $\widetilde{M}^3_{s,p}$ the subspace of the space $H^{s-S,p}$ consisting of the elements f such that $(f, 0, \ldots, 0) \in M^3_{s,p}$ then

$$K^3_{s,p} = H^{s-S,p} \Big/ \widetilde{M}^3_{s,p} \times \prod_{j=1}^m B^{s-\sigma_j-1/p,p}(\partial G) \tag{42}$$

3.7

Let $C^\infty_{(b)} = \{u \in (C^\infty(\overline{G}))^N : bu|_{\partial G} = 0\}$. Denote by $H^{T+s}_{(b)}$ the subspace of $\widetilde{H}_b^{T+s,p,(\tau)}$:

$$\begin{aligned} H^{T+s}_{(b)} &= \{u \in \widetilde{H}_b^{T+s,p,(\tau)} : bu|_{\partial G} = 0\} = \{u \in H^{T+s,p}(G) : \\ &\qquad b_h u|_{\partial G} = 0, h : s - \sigma_h - 1/p > 0\}. \end{aligned}$$

The space $H^{T+s,p}_{(b)}$ is the closure of $C^\infty_{(b)}$ in the space $H^{t+s,p}$. Theorem 8 directly implies the following assertion.

Theorem 9 *Under the condition of Theorem 6, the closure $A_{3(b)}$ of the mapping $u \mapsto lu, u \in C^{\infty}_{(b)}$, realizes an isomorphism $H^{T+s,p}_{(b)} \to H^{s-S,p}/M^4_{s,p}$, where*

$$M^4_{s,p} = \Bigg\{ f \in H^{s-S,p} : (f, v) + \sum_{k,r:s-s_r-k+1-1/p>0} \langle D^{k-1}_{\nu} f_r, e'_{kr} v \rangle = 0,$$
$$v \in (C^{\infty}(\overline{G}))^N : e'_{kr} v\big|_{\partial G} = 0, b' v\big|_{\partial G} = 0, k, r :$$
$$s - s_r - k + 1 - 1/p < 0, \Bigg\}$$

is the subspace of the space $H^{s-S,p}$.

In the special case of one equation with normal boundary conditions, this isomorphism was established in [9] (see also [10], Ch. III, §6, Subsec. 10, §5.5).

3.8

Let us show an example illustrated the using of the graph method.

Consider isomorphism (41) with the space $K^3_{s,p}$ of the form (42). Let $Y_{s,p}$ be a functional Banach space such that $(C^{\infty}(\overline{G}))^N \subset Y_{s,p} \subset H^{s-S,p}/\widetilde{M}^3_{s,p}$, and let the space $(C^{\infty}(\overline{G}))^N$ is dense in Y. For example, $Y_{s,p} = \prod_{r=1}^N H^{s-s_r+t,p}(G)/\widetilde{M}^3_{s+t,p}$, $t > 0$. We get that the set $Q_{s,p} = (A_{3sp})^{-1}(Y_{s,p} \times \prod_{h=1}^m B^{s-\sigma_h-1/p,p}(\partial G))$ is a linear, generally speaking, nonclosed subset of the set $\widetilde{H}^{T+s,p,(\tau)}$. However, in the graph norm $\|u\|_{\widetilde{H}_b^{T+s,p,(\tau)}} + \|lu\|_{Y_{s,p}} = \|u\|_{Q_{s,p}}$ the space $Q_{s,p}$ is a complete Banach space, and the operator A_{3sp} naturally establishes an isomorphism $Q_{s,p} \to Y_{s,p} \times \prod_{h=1}^m B^{s-\sigma_h-1/p,p}(\partial G)$.

In the special case of one equation with normal boundary conditions, this isomorphism was obtained by Lions and Magenes for $s - s_N < 0$ and $Y_{s,p} = L_p(G)$ (see, for details, [5], [1]).

References

[1] Ya.A. Roitberg, *Elliptic Boundary Value Problems in the Spaces of Distributions*, Kluwer Academic Publishers, Dordrecht/Boston/London, 1996, p. 427.

[2] Ya.A. Roitberg, *Theorems on homeomorphisms and Green's formula for general systems of elliptic boundary value problems with boundary conditions those are not normal*, Mat. Sb. **83** (125) (1970), no. 2(10), pp. 181–213.

[3] Ya.A. Roitberg and Z.G. Sheftel, *Theorems on homeomorphisms for elliptic systems with boundary conditions that are not normal*, Mat. Issledov. Kishinev **7** (1972), no. 2, pp. 143–157.

[4] Ya.A. Roitberg and Z.G. Sheftel, *Theorems on homeomorphisms for elliptic systems and its applications*, Mat. Sb. **78** (1969), no. 3, pp. 446–472.

[5] Ya.A. Roitberg, *Theorems on homeomorphisms realized by elliptic operators*, Dokl. Akad. Nauk SSSR, vol. **180** (1968), no. 3, pp. 542–545.

[6] Yu.V. Kostarchuk and Ya. A. Roitberg, *Theorems on isomorphisms for elliptic boundary value problems with boundary conditions, those are not normal*, Ukrain. Mat. J. vol. **25** (1972), no. 2, pp. 271–277.

[7] Ya.A. Roitberg, *Elliptic problems with nonhomogeneous boundary conditions and local increase of smoothness of generalized solutions up to the boundary*, Dokl. Akad. Nauk SSSR, vol. **157** (1964), no. 4, pp. 798–801.

[8] Ya.A. Roitberg, *On the values of generalized solutions of elliptic equations on the boundary of domain*, Mat. Sb. vol. **86** (1971), no. 2, pp. 248–267.

[9] Yu.M. Berezanskii, S. G. Krein and Ya. A. Roitberg, *Theorem on homeomorphisms and local increase of smoothness of solutions of elliptic equations up to the boundary*, Dokl. Akad. Nauk SSSR, vol. **148** (1963), no. 4, pp. 745–748.

[10] Yu.M. Berezanskii, *Expansions in Eigenfunctions of Self-Adjoind Operators*, Kiev, "Naukova Dumka", 1965, p. 800.

[11] I.Ya. Roitberg and Ya. A. Roitberg, *Green's formula and density of solutions of general parabolic boundary value problems in functional spaces on manifolds*, Differencial'nye Uravnenija **31** (1995), no. 8, pp. 1437–1444.

Chernigov State Pedagogical University
Department of Mathematical Analysis
Sverdlova str. 53
Chernigov, 250038
Ukraine
roitberg@elit.chernigov.ua

1991 AMS Classification: 35J40, 35J55.

Operator Theory:
Advances and Applications, Vol. 117
© 2000 Birkhäuser Verlag Basel/Switzerland

Sobolev's Problem in Complete Scale of Banach Spaces*

Yakov Roitberg, Valerii Los and Andrei Sklyarets

In the present work the theorems on complete collections of isomorphisms are proved for the Sobolev's boundary value problem both for one equation and for the system of equations. The same results are obtained also for parameter-elliptic and, therefore, for the Sobolev's parabolic problem. A number of applications is considered.

0

The Sobolev's problem was studied completely in the classes of sufficiently smooth functions (see [1]–[3] and the bibliography therein). Elliptic problems were studied in complete scales of Hilbert and Banach spaces in the works of Lions-Magenes, Yu.M. Berezanskii, S.G. Krein, Ya.A. Roitberg, and others. For such problems the theorems on complete collection of izomorphisms were obtained; the various applications of these theorems were considered (see [4], [5], and the bibliography therein). The Sobolev's problem in complete scales of Banach spaces was investigated in [6], [7]; it was assumed therein that the boundary expressions on $\mathbf{\Gamma}_k (k = 1, \dots, \overline{k})$ form the Dirichlet system, and the orders of boundary expressions on $\mathbf{\Gamma}_0$ are lower that the order of the equation.

In the present work we refuse from these restrictions. In addition, all the expressions are, generally speaking, pseudodifferential along ∂G. Here the solvability of the Sobolev's problem in complete scales of Banach spaces is obtained also for parameter-elliptic and parabolic problems. We consider also a number of applications.

1

Let $\mathbf{G} \subset \mathbf{R}^n$ be a bounded domain with the boundary $\partial\mathbf{G} = \mathbf{\Gamma}_0 \cup \mathbf{\Gamma}_1 \cup \dots \mathbf{\Gamma}_{\overline{k}} \in C^\infty$, and let $\mathbf{\Gamma}_0$ be an $(n-1)$-dimensional compact being an exterior boundary of $\mathbf{G}$. In addition, let $\mathbf{\Gamma}_j$ $(j = 1, \dots, \overline{k})$ denote an i_j-dimensional manifold without border situated inside $\mathbf{\Gamma}_0$, $0 \le i_j \le n-1$, and let $i'_j = n - i_j$ denote the codimension of $\mathbf{\Gamma}_j$. Assume that $\mathbf{\Gamma}_j \in C^\infty$ $(j = 0, \dots, \overline{k})$, and $\mathbf{\Gamma}_j \cap \mathbf{\Gamma}_k = \emptyset$ $(j \ne k)$.

*This work has been partially supported by Grant INTAS-94-2187.

Let us consider the Sobolev's problem

$$(1)\qquad L(x,D)u(x)=f(x)\quad (x\in\mathbf{G};\ \text{ord}\, L=2m),$$

$$(2)\qquad B_{j0}(x,D)u\big|_{\mathbf{\Gamma}_0}=\varphi_{j0}\quad (j=1,\dots,m;\ \text{ord}\, B_{j0}=q_{j0}),$$

$$(3)\quad B_{jk}(x,D)u\big|_{\mathbf{\Gamma}_k}=\varphi_{jk}\quad (k=1,\dots,\bar{k},\ j=1,\dots,m_k;\ \text{ord}\, B_{jk}=q_{jk}).$$

Let

$$(4)\qquad \begin{aligned} r&=\max\{2m,q_{10}+1,\dots,q_{m0}+1\},\\ q_k&=q_{1k}\geq\dots\geq q_{m_k,k}\quad (k=1,\dots,\bar{k}).\end{aligned}$$

Assume that in the neighborhood $\mathbf{G}_{0\delta}=\{x\in\mathbf{G}:\text{dist}(x,\mathbf{\Gamma}_0)<\delta\}$ (with sufficiently small $\delta>0$) the following representations are true:

$$(5)\qquad L(x,D)=\sum_{k=0}^{2m}L_k(x,D')D_\nu^k,$$

$$(6)\quad D_\nu^{j-1}L(x,D)\big|_{\mathbf{\Gamma}_0}=\sum_{k=0}^{2m+j-1}L_{kj}(x',D')D_\nu^k\quad (j=1,\dots,r-2m+1).$$

$$(7)\qquad B_{j0}(x,D)=\sum_{l=1}^{q_{j0}+1}B_{jl}(x,D')D_\nu^{l-1}\quad (j=1,\dots,m)$$

Here L_k, L_{kj} and B_{jl} are tangential differential (or pseudodifferntial) expressions of orders $2m-k$, $2m-k+j$, and $q_{j0}-l+1$, respectively. In the neighborhood

$$\mathbf{G}_{k\delta}=\{x\in\mathbf{G}:\text{dist}(x,\mathbf{\Gamma}_k)<\delta\}=\mathbf{\Gamma}_k\times\{|(y_1,\dots,y_{i'_k})|<\delta\}$$

of manyfold $\mathbf{\Gamma}_k$ one can write

$$(8)\qquad \begin{aligned} B_{jk}(x,D)&=\sum_{|\beta|\leq q_{jk}}T_{kj\beta}(t,D_t)D_y^\beta\\ (D_y^\beta&=D_{y_1}^{\beta_1}\cdots D_{y_{i'_k}}^{\beta_{i'_k}},\ x=(t,y),\ (t,0)\in\mathbf{\Gamma}_k,\ |y|=|(y_1,\dots,y_{i'_k})|;\\ j&=1,\dots,m_k;\ k=1,\dots,\bar{k}).\end{aligned}$$

For simplicity, the coefficients (the symbols) of all the expressions are assumed to be infinitely smooth in $\overline{\mathbf{G}}$. In what follows we set that problem (1)–(2) is elliptic in $\overline{\mathbf{G}}$; this means that the expression L is properly elliptic in $\overline{\mathbf{G}}$, and the expressions $\{B_{j0}\}$ satisfy the Lopatinskii conditions on $\mathbf{\Gamma}_0$ (see, for example, [4], [5]). Moreover, we assume that for every $k\in\{1,\dots,\bar{k}\}$ conditions (3) form a Douglis-Nirenberg elliptic system on $\mathbf{\Gamma}_k$ (see **5** below).

2

Let us introduce the notion of generalized solution of problem (1)–(3). By integration by parts we get

$$
(9)\qquad
\begin{aligned}
(Lu, v) &= (u, L^{+}v) - i\sum_{j=1}^{2m}\langle D_{\nu}^{j-1}u, M_j(x, D)v\rangle_{\Gamma_0} \quad (u, v \in C^{\infty}(\overline{\mathbf{G}})),\\
M_j v &= \sum_{k=j}^{2m} D_{\nu}^{k-j}L_k^{+}(x, D')v, \quad \operatorname{ord} M_j = 2m - j.
\end{aligned}
$$

Here and below $(\cdot,\cdot)$ and $\langle\cdot,\cdot\rangle_k$ denote the scalar products (or their extensions) in $L_2(\mathbf{G})$ and $L_2(\mathbf{\Gamma}_k)$, respectively. The expressions L^{+} and L_k^{+} are formally adjoint to L and L_k, respectively.

We identify the element $u \in C^{\infty}(\overline{\mathbf{G}})$ with the vector

$$
(10)\qquad
\begin{aligned}
u &= (u_0, u_1, \ldots, u_r, (u_{k\beta} : k = 1, \ldots, \overline{k}, |\beta| \le q_k))\\
(u_0 &= u|_{\overline{\mathbf{G}}}, u_j = D_{\nu}^{j-1}u|_{\mathbf{\Gamma}_0}\ (j = 1, \ldots, r), u_{k\beta} = D_y^{\beta}u|_{\mathbf{\Gamma}_k}),
\end{aligned}
$$

and the element $f \in C^{\infty}(\overline{\mathbf{G}})$ with the vector

$$
(11)\quad f = (f_0, \ldots, f_{r-2m}),\ f_0 = f|_{\overline{\mathbf{G}}},\ f_j = D_{\nu}^{j-1}f|_{\mathbf{\Gamma}_0}\ (1 \le j \le r - 2m).
$$

It follows from (5)–(9) that vector (10) $u \in C^{\infty}(\overline{\mathbf{G}})$ is a solution of problem (1)–(3) if the following relations hold:

$$
(12)\qquad (u_0, L^{+}v) - i\sum_{j=1}^{2m}\langle u_j, M_j v\rangle = (f_0, v) \quad (v \in C^{\infty}(\overline{\mathbf{G}})),
$$

$$
(13)\qquad \sum_{k=0}^{2m+j-1} L_{kj}(x', D')u_{k+1} = f_j \quad (j : 1 \le j \le r - 2m),
$$

$$
(14)\qquad B_{j0}u|_{\mathbf{\Gamma}_0} \equiv \sum_{l=1}^{q_{j0}+1} B_{jl}(x, D')u_l = \varphi_j \quad (j = 1, \ldots, m),
$$

$$
(15)\quad B_{jk}u|_{\mathbf{\Gamma}_k} \equiv \sum_{|\beta|\le q_{jk}} T_{kj\beta}(t, D_t)u_{k\beta} = \varphi_{jk} \quad (k = 1, \ldots, \overline{k};\ j = 1, \ldots m_k).
$$

Now let u and f be vectors (10), (11), where u_0 and f_0 are distributions in $\overline{\mathbf{G}}$, and $\{u_j, f_l\}$ and $\{u_{k\beta}\}$ are distributions in $\mathbf{\Gamma}_0$ and $\mathbf{\Gamma}_k$, respectively. If relations

(12)–(15) hold, then the vector u is called a generalized solution of problem (1)–(3). Relations (12)–(15) establish the mapping

$$\begin{aligned}\mathcal{A}: u \longmapsto F &= (f, \varphi_0, \varphi_1, \ldots, \varphi_{\overline{k}}),\ f = (f_0, \ldots, f_{r-2m}),\\ \varphi_0 &= (\varphi_{10}, \ldots, \varphi_{m0}),\ \varphi_k = (\varphi_{1k}, \ldots, \varphi_{m_k k}).\end{aligned} \tag{16}$$

In order to study this mapping and to precise the definition of a generalized solution, we introduce some convenient functional spaces.

3

Let $p, p' \in (1, \infty)$, $1/p + 1/p' = 1$, $s \in \mathbf{R}$. We denote by $H^{s,p}(\mathbf{R}^n)$ the spaces of Bessel potentials (see, for example, [5], [8]); $\| f, \mathbf{R}^n \|_{s,p}$ is the norm in $H^{s,p}(\mathbf{R}^n)$. The spaces $H^{s,p}(\mathbf{R}^n)$ and $H^{-s,p'}(\mathbf{R}^n)$ are dual to each other with respect to the extension of the scalar product in $L_2(\mathbf{R}^n)$. Denote by $H^{s,p}(\mathbf{G})$ ($s \geq 0$) the space of restrictions of elements of $H^{s,p}(\mathbf{R}^n)$ to $\mathbf{G}$ with the quotient space topology:

$$H^{s,p}(\mathbf{G}) = H^{s,p}(\mathbf{R}^n)/H^{s,p}_{C\mathbf{G}}(\mathbf{R}^n), \tag{17}$$

where $H^{s,p}_{C\mathbf{G}}(\mathbf{R}^n) = \{v \in H^{s,p}(\mathbf{R}^n) : \operatorname{supp} v \subset \mathbf{R}^n \backslash \mathbf{G}\}$ is a subspace of $H^{s,p}(\mathbf{R}^n)$. By $H^{-s,p}(\mathbf{G})$ ($s \geq 0$) we denote the space dual to $H^{s,p'}(\mathbf{G})$ with respect to the extension $(\cdot, \cdot)$ of the scalar product in $L_2(\mathbf{G})$. It follows from (17) (see [5], [8]) that $H^{-s,p}(\mathbf{G}) \simeq H^{-s,p}_{\overline{\mathbf{G}}}(\mathbf{R}^n) = \{f \in H^{-s,p}(\mathbf{R}^n) : \operatorname{supp} f \subset \overline{\mathbf{G}}\}$. Since for $s \geq 0$ there are not elements concentrated at Γ_k ($k = 1, \ldots, \overline{k}$) in $H^{s,p}(\mathbf{R}^n)$, the relation

$$H^{s,p}(\mathbf{G}) \simeq H^{s,p}(G \cup \mathbf{\Gamma}_1 \cup \ldots \cup \mathbf{\Gamma}_k) \quad (s \in \mathbf{R})$$

is true. The norm in $H^{s,p}(\mathbf{G})$ ($s \in \mathbf{R}$) is denoted by $\|u, G\|_{s,p}$.

We denote by $B^{s,p}(\mathbf{\Gamma}_k)$ ($k = 0, \ldots, \overline{k}$) the Besov space; $\langle\langle \varphi, \mathbf{\Gamma}_k \rangle\rangle_{s,p}$ is the norm in $B^{s,p}(\mathbf{\Gamma}_k)$. The spaces $B^{s,p}(\mathbf{\Gamma}_k)$ and $B^{-s,p'}(\mathbf{\Gamma}_k)$ are dual to each other with respect to the extension $\langle \cdot, \cdot \rangle_{\mathbf{\Gamma}_k}$ of the scalar product in $L_2(\mathbf{\Gamma}_k)$ (see [5], [8]).

Let

$$s \in \mathbf{R}, \quad s \neq j + \frac{i'_k}{p} \qquad (j = 1, \ldots, q_k, \quad k = 0, \ldots, \overline{k}, \quad q_0 = r). \tag{18}$$

By $\widetilde{H}^{s,p}$ we denote the completion of $C^\infty(\overline{\mathbf{G}})$ in the norm

$$\begin{aligned}|||u|||_{s,p} = \Bigg(&\|u\|^p_{s,p} + \sum_{j=1}^{r} \langle\langle D_\nu^{j-1} u, \mathbf{\Gamma}_0 \rangle\rangle^p_{s-j+1-1/p,p}\\ &+ \sum_{k=1}^{\overline{k}} \sum_{|\alpha| \leq q_k} \langle\langle D_y^\alpha u, \mathbf{\Gamma}_k \rangle\rangle^p_{s-|\alpha|-i'_k/p,p} \Bigg)^{1/p}.\end{aligned} \tag{19}$$

The space $\widetilde{H}^{s,p}$ is isometric to the closure $\mathcal{H}_0^{s,p}$ of the space of vectors $u \in C^\infty(\overline{\mathbf{G}})$ (10) in

$$
\begin{aligned}
\mathcal{H}^{s,p} &= H^{s,p}(\mathbf{G}) \times \prod_{j=1}^{r} B^{s-j+1-1/p,p}(\mathbf{\Gamma}_0) \\
&\times \prod_{k=1}^{\overline{k}} \prod_{|\alpha| \le q_k} B^{s-|\alpha|-i'_k/p,p}(\mathbf{\Gamma}_k).
\end{aligned} \tag{20}
$$

Therefore, the space $\widetilde{H}^{s,p}$ coincides with the subspace $\mathcal{H}_0^{s,p}$ of the space of the vectors

$$u = (u_0, u_1, \ldots, u_r, \{u_{\alpha k} : k = 1, \ldots, \overline{k};\ |\alpha| \le q_k\}) \in \mathcal{H}^{s,p}. \tag{21}$$

In addition, $u \in \mathcal{H}^{s,p}$ belongs to $\mathcal{H}_0^{s,p} \simeq \widetilde{H}^{s,p}$ if and only if

$$
\begin{aligned}
u_j &= D_\nu^{j-1} u_0\big|_{\mathbf{\Gamma}_0} \quad (\forall j : s - j + 1 - 1/p > 0), \\
u_{\alpha k} &= D_y^\alpha u_0\big|_{\mathbf{\Gamma}_k} \quad (\forall \alpha, k : s - |\alpha| - i'_k/p > 0)
\end{aligned} \tag{22}
$$

(cf. [5], [9]); if $s < 1/p$, then $\mathcal{H}_0^{s,p} = \mathcal{H}^{s,p} = \widetilde{H}^{s,p}$.

By $\widetilde{H}^{s,p,(r)}$ ([5], [9]) we denote the completion of $C^\infty(\overline{\mathbf{G}})$ in the norm

$$|||u|||_{s,p,(r)} = \left(\|u\|_{s,p}^p + \sum_{j=1}^{r} \langle\langle D_\nu^{j-1} u, \mathbf{\Gamma}_0 \rangle\rangle_{s-j+1-1/p,p}^p \right)^{1/p}. \tag{23}$$

The space $\widetilde{H}^{s,p,(r)}$ consists of vectors

$$(u_0, \ldots, u_r) \in H^{s,p}(\mathbf{G}) \times \prod_{j=1}^{r} B^{s-j+1-1/p,p}(\mathbf{\Gamma}_0)$$

such that

$$u_j = D_\nu^{j-1} u_0\big|_{\mathbf{\Gamma}_0} \quad (\forall j : s - j + 1 - 1/p > 0).$$

For the rest of values of s (see (18)) the spaces $\widetilde{H}^{s,p}$ and $\widetilde{H}^{s,p,(r)}$, and norms (19) and (23) are defined by complex interpolation.

Theorem 1 (cf. [5]–[7], [9]). *For any $s \in \mathbf{R}$ and $p \in (1, \infty)$ the closure $\mathcal{A} = \mathcal{A}_{s,p}$ of the mapping*

$$u \longmapsto (Lu|_{\overline{\mathbf{G}}},\ (D_\nu^{j-1} Lu|_{\mathbf{\Gamma}_0} \mid j : 1 \le j \le r - 2m),\ B_1 u|_{\mathbf{\Gamma}_0}, \ldots, B_m u|_{\mathbf{\Gamma}_0},$$

$$\{B_{jk} u|_{\mathbf{\Gamma}_k} : k = 1, \ldots, \overline{k},\ j = 1, \ldots, m_k\}) \qquad (u \in C^\infty(\overline{\mathbf{G}}))$$

acts continuously in the pair of spaces

$$\widetilde{H}^{s,p} \longrightarrow K^{s,p} := \widetilde{H}^{s-2m,p,(r-2m)} \times \prod_{j=1}^{m} B^{s-q_{j0}-1/p,p}(\mathbf{\Gamma}_0) \times \prod_{k=1}^{\overline{k}} \prod_{j=1}^{m_k} B^{s-q_{jk}-i'_k/p,p}(\mathbf{\Gamma}_k) \tag{24}$$

If $s_1 \leq s$, $p_1 \leq p$, *then the operator* $\mathcal{A} \equiv \mathcal{A}_{s_1,p_1}$ *is an extension of the operator* $\mathcal{A}_{s_2,p_2}$ *in continuity.*

Definition: An element $u \in \widetilde{H}^{s,p}$ is called a generalized solution of problem (1)–(3) if $\mathcal{A}_{s,p}u = F = (f, \varphi_{10}, \dots, \varphi_{m0}, \{\varphi_{jk}\}) \in K^{s,p}$.

Theorem 2 *An element* $u \in \widetilde{H}^{s,p}$ *is a generalized solution of problem* (1)–(3) *if and only if relations* (12)–(15) *hold.*

4

In $\overline{\mathbf{G}}$ we consider elliptic problem (1)–(2).

The following statements are true ([5], [9], [10]):

4.1

For any $s \in \mathbf{R}$ and $p \in (1, \infty)$ the closure $A_{s,p}$ of the mapping

$$u \longmapsto (Lu, B_1 u|_{\mathbf{\Gamma}_0}, \dots, B_m u|_{\mathbf{\Gamma}_0}) \quad (u \in C^\infty(\overline{\mathbf{G}}))$$

acts continuously in the pair of spaces

$$\widetilde{H}^{s,p,(r)} \longrightarrow K^{s-2m,p,(r-2m)} := \widetilde{H}^{s-2m,p,(r-2m)} \times \prod_{j=1}^{m} B^{s-q_{j0}-1/p,p}(\mathbf{\Gamma}_0).$$

4.2

The element $\widetilde{u} = (u_0, u_1, \dots, u_r) \in \widetilde{H}^{s,p,(r)}$ is called a generalized solution of problem (1)–(2) if $A_{s,p}\widetilde{u} = (f, \varphi) \in K^{s-2m,p,(r-2m)}$. Furthermore, $A_{s,p}\widetilde{u} = (f, \varphi)$ if and only if equalities (12)–(14) hold. Therefore, if the element $u \in \widetilde{H}^{s,p}$ (21) is a generalized solution of problem (1)–(3), then the element $\widetilde{u} = (u_0, u_1, \dots, u_r) \in \widetilde{H}^{s,p,(r)}$ is a generalized solution of problem (1)–(2).

4.3

The operator $A_{s,p}$ is Noetherian, and the kernel $\mathfrak{N}$ and the cokernel $\mathfrak{N}^*$ are finite-dimensional, does not depend on s and p and consist of infinitely smooth elements:

$$\begin{aligned} \mathfrak{N} &= \{u \in C^\infty(\overline{\mathbf{G}}) : Au = 0\}, \\ \mathfrak{N}^* &= \{V = (v_0, \{v_k : 1 \leq k \leq r - 2m\}, \{v_j : 1 \leq j \leq m\})\} \\ &\subset C^\infty(\overline{\mathbf{G}}) \times (C^\infty(\mathbf{\Gamma}_0))^{r-2m} \times (C^\infty(\mathbf{\Gamma}_0))^m. \end{aligned}$$

The problem

$$A_{s,p}\widetilde{u} = (f,\varphi) \in K^{s-2m,p,(r-2m)}, \quad f = (f_0, f_1, \dots, f_{r-2m}),$$
$$\varphi = (\varphi_1, \dots, \varphi_m)$$

is solvable in $\widetilde{H}^{s,p,(r)}$ if and only if the relation

$$[(f,\varphi), V] := (f_0, v_0) + \sum_{j=1}^{r-2m} \langle f_j, v_j \rangle_{\Gamma_0} + \sum_{k=1}^{m} \langle \varphi_k, v_k \rangle_{\Gamma_0} = 0 \quad (\forall V \in \mathfrak{N}^*) \tag{25}$$

is valid.

4.4

The restriction $A_1 = A_{1\,s,p}$ of the operator $A_{s,p}$ onto the subspace

$$P\widetilde{H}^{s,p,(r)} = \{(u_0, u_1, \dots, u_r) \in \widetilde{H}^{s,p,(r)} : (u_0, \mathfrak{N}) = 0\}$$

of the space $\widetilde{H}^{s,p,(r)}$ realizes the isomorphism

$$A_1 : P\widetilde{H}^{s,p,(r)} \longrightarrow Q^+ K^{s-2m,p,(r-2m)}, \tag{26}$$

where $Q^+ K^{s-2m,p,(r-2m)}$ is a subspace of the space $K^{s-2m,p,(r-2m)}$ which consist of all the elements (f,φ) satisfying relation (25).

5

Let us further formulate the assumption on the boundary expressions on Γ_k ($k = 1, \dots, \overline{k}$). Let $u \in \widetilde{H}^{s,p}$ be a generalized solution of problem (1)–(3). Then, in view of Theorem 2, the elements $\{u_{\alpha k}\}$ satisfy equalities (15). For any $k \in \{1, \dots, \overline{k}\}$ we set

$$\widetilde{T}_{kj\beta}(t, D_t) = \begin{cases} T_{kj\beta}(t, D_t) & \text{for } |\beta| \le q_{jk} \\ 0 & \text{for } q_{jk} < |\beta| \le q_k \end{cases}$$

where the numbers q_k are defined by formula (4). Then one can rewrite equalities (15) in the form

$$B_{jk}u|_{\Gamma_k} \equiv \sum_{|\beta| \le q_k} \widetilde{T}_{kj\beta}(t, D_t) u_{k\beta} = \varphi_{jk} \quad (j = 1, \dots, m_k, \ k = 1, \dots, \overline{k}), \tag{27}$$

or, in short,

$$\widetilde{T}_k(t, D_t) U_k = \varphi_k \quad (\widetilde{T}_k = (T_{kj\beta})_{\substack{j=1,\dots,m_k \\ |\beta| \le q_k}}; \ \varphi_k = (\varphi_{1k}, \dots, \varphi_{m_k k})) \tag{27'}$$

For any k the matrix $\left(\widetilde{T}_{kj\beta}(t, D_t)\right)$ has the structure of Douglis-Nirenberg. This means that if we associate the number $s_j = q_{jk} - q_k$ with the line with index j, and the number $t_\beta = q_k - |\beta|$ with the row with index β, then we obtain that $\operatorname{ord} \widetilde{T}_{kj\beta} \leq s_j + t_\beta$ for $s_j + t_\beta \geq 0$, and $\widetilde{T}_{kj\beta} \equiv 0$ for $s_j + t_\beta < 0$.

We assume that for any $k \in \{1, \dots, \overline{k}\}$ system (27) is elliptic in the Douglis-Nirenberg sense on $\mathbf{\Gamma}_k$. This means that the system is square, and

$$\det(\widetilde{T}^0_{kj\beta}(t, \sigma)) \neq 0 \quad (\sigma = (\sigma_1, \dots, \sigma_{i_k}) \neq 0, \ t \in \mathbf{\Gamma}_k),$$

where $\widetilde{T}^0_{kj\beta}$ is the principal symbol of the expression $\widetilde{T}_{kj\beta}(t, D_t)$.

For any $s \in \mathbf{R}$ and $p \in (1, \infty)$ the mapping

$$T_k = T_{k,s,p} : U \longmapsto T_k U$$

acts continuously in the pair of spaces

$$\begin{aligned} T_k &= T_{k,s,p} : B^{T+s,p}(\mathbf{\Gamma}_k) := \prod_{j=1}^{m_k} B^{t_j+s,p}(\mathbf{\Gamma}_k) \\ &\longrightarrow \prod_{j=1}^{m_k} B^{s-s_j,p}(\mathbf{\Gamma}_k) =: B^{s-S,p}(\mathbf{\Gamma}_k). \end{aligned} \tag{28}$$

The ellipticity of system (27) implies (see [11]) the operator T_k is Noetherian. It means that its range $\mathcal{R}(T_k)$ is closed in $B^{s-S,p}(\mathbf{\Gamma}_k)$, the kernel $\mathfrak{N}_k$ and the cokernel $\mathfrak{N}^*_k$ are finite-dimensional, does not depend on s and p, and consist of infinitely smooth elements:

$$\mathfrak{N}_k \subset (C^\infty(\mathbf{\Gamma}_k))^{m_k}, \quad \mathfrak{N}^*_k \subset (C^\infty(\mathbf{\Gamma}_k))^{m_k}.$$

The equation $T_k U = \varphi_k \in B^{s-S,p}$ has a solution $U \in B^{T+s,p}(\mathbf{\Gamma}_k)$ if and only if the equalities

$$(\varphi_k, V)_{\mathbf{\Gamma}_k} = 0 \qquad (\forall V \in \mathfrak{N}^*_k; k = 1, \dots, \overline{k}). \tag{29}$$

hold (here $(\cdot, \cdot)$ denotes the extension of the scalar product in $(L_2(\mathbf{\Gamma}_k))^{m_k}$). In addition, the cokernel $\mathfrak{N}^*_k$ of the operator T_k coincides with the kernel of the operator T^*_k. The restriction T^1_k of the operator T_k onto the subspace

$$PB^{T+s,p}(\mathbf{\Gamma}_k) = \{U \in B^{T+s,p}(\mathbf{\Gamma}_k) : (U, W)_{\mathbf{\Gamma}_k} = 0 \ (\forall W \in \mathfrak{N}_k)\}$$

of the space $B^{T+s,p}(\mathbf{\Gamma}_k)$ realizes an isomorphism

$$T^1_k : PB^{T+s,p}(\mathbf{\Gamma}_k) \longrightarrow Q^+ B^{s-S,p}(\mathbf{\Gamma}_k) \quad (s \in \mathbf{R}, \ p \in (1, \infty)). \tag{30}$$

Here $Q^+ B^{S-s,p}(\mathbf{\Gamma}_k) = \{\varphi \in B^{S-s,p}(\mathbf{\Gamma}_k) : (\varphi, V)_{\mathbf{\Gamma}_k} = 0 \ (\forall V \in \mathfrak{N}^*_k)\}$ is a subspace of the space $B^{S-s,p}(\mathbf{\Gamma}_k)$.

6

The following statement is true.

Theorem 3 *Assume that $s \in \mathbf{R}$, and $p \in (1, \infty)$. For the solvability of problem* (1)–(3) *in $\widetilde{H}^{s,p}$ it is necessary that relations* (25) *and* (29) *are valid; this condition is also sufficient if, in addition, the compatibility conditions (see* (34) *below) hold for $s > 1/p$.*

Proof: If the element $u \in \widetilde{H}^{s,p}$ (see (21)) is a solution of the equation

$$(31)\qquad \begin{aligned} \mathcal{A}u &= F = (f, \varphi_0, \varphi_1, \ldots, \varphi_{\overline{k}}) \in K^{s,p}(f = (f_0, \ldots, f_{r-2m}), \\ \varphi_0 &= (\varphi_{10}, \ldots, \varphi_{m0}), \quad \varphi_k = (\varphi_{1k}, \ldots, \varphi_{m_k,k})), \end{aligned}$$

then the vector $(u_1, \ldots, u_r)$ is a solution of problem (1)–(2) in $\overline{\mathbf{G}}$ (see **4.2**). Therefore, relations (25) are fulfilled (see **4.3**). Moreover, the vector

$$U_k = \{u_{\alpha k} : |\alpha| \le q_k\} \in B^{T+s,p}(\mathbf{\Gamma})$$

satisfies the equation

$$(32)\qquad T_k U_k = \varphi_k \qquad (k = 1, \ldots, \overline{k}).$$

Therefore, relations (29) are valid. Thus, the necessity is proved.

To prove the sufficiency we convert the arguments mentioned. Solving the problem (1)–(2) we find the element

$$\widetilde{u} = A_1^{-1}(f, \varphi) = (u_0, \ldots, u_r) \in P\widetilde{H}^{s,p,(r)}$$

(see (26)), such that $\widetilde{u} + w$ $(W \in \mathfrak{N})$ is the general solution of problem (1)–(2). The element $\widetilde{u} + w = (u_0 + w_0, \ldots, u_r + w_r)$ gives us the first $r + 1$ components of the solution required. We find the rest of the components by solving problem (32). The element

$$U_k + W_k \quad (U_k = T_k^{1-1}\varphi_k,\ W_k \in \mathfrak{N}_k)$$

is the general solution of problem (32). The vector

$$(33)\qquad (\widetilde{u} + w, \{U_k + W_k : k = 1, \ldots, \overline{k}\}) \in \widetilde{H}^{s,p}$$

is the general solution of equation (31). If $s > 1/p$, then $u_0 \in H^{s,p}(\mathbf{G})$ and, hence, there exist the traces $D_y^\alpha(u_0 + w_0)|_{\mathbf{\Gamma}_k}$ $(\forall \alpha, k : s - |\alpha| - i'_k/p > 0)$. The solution obtained above belongs to $\widetilde{H}^{s,p}$ if and only if the compatibility conditions

$$(34)\qquad u_{\alpha k} + w_{\alpha k} = D_y^\alpha(u_0 + w_0)|_{\mathbf{\Gamma}_k} \quad (\forall \alpha, k : s - |\alpha| - i'_k/p > 0).$$

hold. This completes the proof of the theorem. □

Remark: It follows from the proof that, for $s < 1/p$, if $(e_{10}, \dots, e_{t0})$ is the basis in $\mathfrak{N}$, and $(e_{1k}, \dots, e_{t_k,k})$ is the basis in $\mathfrak{N}_k$ $(k = 1, \dots, \bar{k})$, then the elements $(e_{10}, \dots, e_{t0}, e_{11}, \dots, e_{t_1,1}, \dots, e_{1\bar{k}}, \dots, e_{t_{\bar{k}},\bar{k}})$ forms the basis in the kernel $\mathfrak{N}(\mathcal{A})$ of the operator $\mathcal{A}$. The similar statement is valid also for the cokernel of the operator $\mathcal{A}$. Therefore

$$\operatorname{ind} \mathcal{A} = \operatorname{ind} A + \sum_{k=1}^{\bar{k}} \operatorname{ind} T_k. \tag{35}$$

7

Under the conditions $\bar{k} = 1$, $\Gamma_1 = \Gamma$, $i_1 = i$, $i_1' = i'$, and $s < 2m - i'/p'$, we consider problem (1)–(3). In this case the element $f_0 \in H^{s-2m,p}(\mathbf{G})$ may be concentrated at Γ. It turns out that we can add to f_0 the element f_0' concentrated at Γ such that the compatibility conditions hold automatically. The following theorem is true.

Theorem 4 *Let $1/p < s < 2m - i'/p$. Assume that $F \in K^{s,p}$ satisfies relations (25) and (29). Then one can add to f_0 the element f_0' concentrated at Γ such that the compatibility conditions hold automatically for the problem $\mathcal{A}u = \widetilde{F}$, $\widetilde{F} = F + (f_0', 0, \dots, 0)$, and this problem is solvable in $\widetilde{H}^{s,p}$.*

Note that the element f_0 coincides with the elements $f_0 + f_0'$ inside of G.

Let us show now the example of S. L. Sobolev ([1], p. 114). In the domain $\mathbf{G} = \{x \in \mathbf{R}^3 : 0 < |x| < 1\}$ we consider the problem

$$\Delta^2 u = 0 \ (\text{in } \mathbf{G}), \quad u|_{|x|=1} = \left.\frac{\partial u}{\partial \nu}\right|_{|x|=1} = 0, \quad u(0) = 1,$$

where Δ denotes the Laplace operator. The function $u = (1 - |x|)^2$ is the unique solution from $H^{2,2}(\mathbf{G})$ of the problem under consideration. Moreover, $u \notin H^{3,2}(\mathbf{G})$. It easy to verify that we have $\Delta^2 u = c\delta(0) \in H^{-3/2-\varepsilon,2}(\mathbf{G})$ in $\mathbf{G} \cup \{0\}$ (here $\delta(0)$ denotes the Dirac measure concentrated at the point 0, and $\varepsilon > 0$ may be chosen arbitrary small). Therefore $u \in H^{5/2-\varepsilon,2}(\mathbf{G})$. To obtain a smoother solution it is necessary to require that the conjugation condition $u(0) = 0$ holds. Then $u \equiv 0$.

8 Generalizations. Applications

8.1

As it was mentioned in the Sec. **1**, all the statements with the same proofs remain true if expressions (5)–(8) are pseudodifferential along ∂G and differential in directions normal to ∂G.

8.2

All these statements remain valid for the parameter-elliptic Sobolev's problem (c.f. [5, Ch. 9], [11]). In this case instead of the Noetherity property we have the unique solvability for large values of the parameter.

8.3

The solvability of the parabolic Sobolev's problems follows from **8.2** (c.f. [11]–[13]).

8.4

All the results with the same proofs remain true for the Sobolev's problem for Douglis-Nirenberg elliptic systems of order $(T, S) = (t_1, \ldots, t_N, s_1, \ldots, s_N)$ (c.f. [5, Ch. 10]).

8.5

Now we indicate some of possible applications of the results obtained above. The theorem on complete collection of isomorphisms enables us to construct and to study the Green's function for the Sobolev's problem (c.f. [5, §7.4]). This theorem gives us a possibility to study the local smoothness of the solutions up to $\boldsymbol{\Gamma}_0$, $\boldsymbol{\Gamma}_k$, to investigate the strongly degenerated elliptic Sobolev's problems, and to study the Sobolev's problem in the case where the right-hand sides have arbitrary large power singularities along the manifolds of different dimentions (c.f. [5], Ch. 8).

References

[1] S.L. Sobolev, *Some Applications of Functional Analysis in Mathematical Physics*, Izd. Leningrad Univ., Leningrad, 1950, p. 255.

[2] B.Yu. Sternin, *Elliptic and parabolic problems on manofolds whose boundary consists of components of various dimensions*, Trudy Mosk. Mat. Obsc. **15** (1966), 346–382.

[3] B.Yu. Sternin, *Relative elliptic theory and problem of S.L. Sobolev*, Dokl. Akad. Nauk SSSR **230** (1976), no. 2, 287–290.

[4] Yu.M. Berezanskii, *Expansions in Eigenfunctions of Self-Adjoind Operators*, Kiev, "Naukova Dumka", 1965, p. 800.

[5] Ya.A. Roitberg, *Elliptic Boundary Value Problems in the Spaces of Distributions*, Kluwer Acad. Publ., Dordrecht/Boston/London, 1996, p. 427

[6] Ya.A. Roitberg and A.V. Sklyarets, *Sobolev's problem in complete scales of Banach spaces*, Ukr. Math. J. **48** (1996), no. 11, 1555–1563.

[7] Ya.A. Roitberg, and A.V. Sklyarets, *Sobolev's problem in complete scales of Banach spaces*, Dokl. Ukrain. Akad. Nauk **48** (1996), no. 1.

[8] G. Grubb, *Partial diffrential problems in L_p spaces*, Comm. Part. Diff. Eq-s. **15**(3) (1990), 289–340.

[9] Ya.A. Roitberg, *On values of generalized solutions of elliptic equations on the boundary of domain*, Mat.Sb. **86**(128) (1971), no. 2(10), 246–267.

[10] Ya.A. Roitberg, *Theorems on homeomorphisms and Green's formula for general general elliptic boundary value problems for systems with boundary conditions those are not normal*, Mat.Sb. **83**(125) (1970), no. 2(10), 181–213.

[11] M.S. Agranovich and M.I. Vishik, *Elliptic problems with a parameter and parabolic problems of the general form*, Uspekhi Mat. Nauk **19** (1964), no. 3, 53–161.

[12] I.Ya. Roitberg and Ya.A. Roitberg, *Green's formula and density of solutions of general parabolic boundary value problems in functional spaces on manifolds*, Differencial'nye Uravnenija **31** (1995), no. 8, 1437–1444.

[13] S.D. Eidelman and N.V. Zhitarashu, *Parabolic Boundary Value Problems*, Operator Theory: Advanses and Appl. vol. **101**, Birkhäuser Verlag, 1998, p. 298.

Chernigov State Pedagogical University
Department of Mathematical Analysis
Sverdlova str., 53, Chernigov, 250038
Ukraine
roitberg@elit.chernigov.ua

1991 AMS Classification: 35J40, 35J55.

Operator Theory:
Advances and Applications, Vol. 117
© 2000 Birkhäuser Verlag Basel/Switzerland

On Simple Waves with Profiles in the form of some Special Functions-Chebyshov-Hermite, Mathieu, Whittaker-in Two-phase Media

Jarema Rushchitsky and Svitlana Rushchitska

As is well-known, M.G. Krein took always an interest in new problems of the mathematical physics. Some of his results are directly related to such problems. In our lecture, we propose a new problem from the general wave theory. The lecture deals with simple solitary waves in materials, their existence and description.

First of all, we would like to give the following three introductory notes.

Note 1 (About nonlinear waves).

Right away it is necessary to say that harmonic waves are studied during hundreds years. The first scientific observation of velocity of harmonic waves was carried out by Biot in 1809. That was a well-known in mechanics measurement of sound velocity in pig iron pipes of new Parisian water-line built at that time.

The classic linear theory of elastic waves was practically formed in the last century. Unfortunately, the development of nonlinear theory was late, and some aspects of this theory are creating just now. One of such directions is the subject of this article.

Nonlinearity in physics is a typical phenomenon. Such divisions of physics as celestial mechanics, hydrodynamics, theory of gravitation and a number of other topics are nonlinear at the bottom. Therefore physics has studied nonlinearity for a long time and in its multiform manifestations. That applies to nonlinear oscillations first of all. As long ago as the 1940s, a principle was formulated in the school of the Muscowite academician Mandelstamm which was called the "*Principle of isomorphism of generalities of nonlinear physics*" [1–3]. This principle states that in general the oscillational motion does not depend on the nature of a physical phenomenon. Since the understanding of an oscillation problem in each division of physics was different, Mandelstamm named the phenomenon of a transport of achieved understanding of oscillations from one division physics to other as "*Oscillation mutual aid*" [1–3].

That fact from the history turned out to be characteristic. Very similar situation arose in recent decades in the nonlinear wave theory. For example, achieved understanding of wave motion in nonlinear optics was transported to nonlinear physics of plasma, or from nonlinear radiophysics to nonlinear optics. The principle of mutual aid works in the nonlinear theory of waves. Strictly speaking, we would like to testify the fact that achieved in nonlinear optics and nonlinear acoustics understanding of problems of wave interactions must be transported to nonlinear mechanics of materials.

Nonlinear waves in physics was a traditional study. But such a regular investigation of nonlinear waves in solids began only in the 1960s. Those were the sound waves, and the basic publications belong to representatives of acoustics, and not of mechanics [4]. The encyclopedic book of Whitham [5] in the nonlinear wave field, and a number of highly professional books written on nonlinear waves have not accentuated the attention to waves in solids.

Such a situation is caused, perhaps, by the fact that these waves have some difficult specificity. Especially, on the initial and end stages, i.e. in the statement of a problem and in the commentary of results. Just these stages have circumvented attention in above cited books. Classic books on waves in solids give little or no information about nonlinear waves.

Note 2 (About simple waves).

The concept of simple waves arose in hydrodynamics. As a mathematical concept it corresponds to the solution of hydrodynamics equations which has a certain structure. As a physical concept it defines the wave of a certain form. Now, let us outline short historical aspects and present-day knowledge.

For the first time, the exact solution for plane waves in an incompressible fluid was obtained by Poisson in 1808. That solution had the form of a plane running wave which was called simple. The theory of simple waves was extended in the works of Stokes, Airy, and Earnshow. Riemann gave in 1860 the general solution of an one-dimensional problem of hydrodynamical equations in the case where a disturbance is plane and a state equation is an arbitrary functional relation between a pressure and a density.

For a plane case, hydromechanics equations in Euler coordinates and in conventional symbols are nonlinear and have the form

$$\rho_t + \rho v_x + v\rho_x = 0; \quad \rho v_t + \rho v v_x + p_x = 0. \tag{1}$$

The function

$$\sigma(\rho) = \int_{\rho_0}^{\rho} c(\rho)\frac{d\rho}{\rho}$$

was introduced by Riemann. Here ρ_0 is the density of a nondisturbed medium and $c(\rho) = \sqrt{\frac{dp}{d\rho}}$ is the sound velocity.

Equations (1) may be rewritten as

$$P_t + (c+v)P_x = 0; \quad Q_t + (c-v)Q_x = 0, \tag{2}$$

where $P = v + \sigma$, $Q = v - \sigma$. Then a solution of the system (2) has the form

$$v = \frac{1}{2}\left\{P\left(t - \frac{x}{c+v}\right) + Q\left(t + \frac{x}{c-v}\right)\right\} \tag{3}$$

Definition 1 The separately taken function P or Q is called a simple wave.

Definition 2 A simple wave is defined as such a wave process, whose main parameters are expressed by one of the parameters. For example, $\rho = \rho(u)$, $p = p(u)$ or $u = u(\rho)$, $\quad p = p(\rho)$.

Comment 1 The beauty of the Riemann solution (3) consists in the fact that it assumes a simple form of representation, and, so to speak, generalizes D'Alembert solution.

The most brilliant description of Riemann waves is given in [6]. In this description of simple Riemann waves the word "brilliant" is used often: The brilliant mathematical disclosure of Riemann - one of the outstanding mathematician of the middle 19 century - has become the basis of all subsequent works on nonlinear theory of plane sound waves. This disclosure equivalent to transformation of motion equations to a form which is wonderfully easy of use for a study of waves has led, in due time, produced to a beautiful level of the subject understanding.

On the other hand, the glittering display of the mathematical technique used for a carrying out of this initial transformation, has, for long time, impressed a hypnotizing effect on acoustics. This has led to a certain stagnation in nonlinear sound theory, which is associated with a universal conviction that all success in the subject understanding has depended on the initial mathematically brilliant transformation. Over many decades, this has prevented generalization of results to any other conditions of a wave propagation.

Lighthill formulates such a question - *may the evolution of a plane wave with an arbitrary amplitude in the absence of dissipation be predicted with the aim of simple physical reasoning in supposing of knowledge of linear acoustics only.* The Lighthill answer is yes. For an arbitrary position $x = x_1$ and time $t = t_1$, there exist some space interval around x_1 and some time interval around t_1. They both are so small that for x, t from these intervals the corresponding disturbances of u, p of values u_1, p_1 which they take at the point (x_1, t_1) are sufficiently small for describing their behavior by linear theory. Such a linear theory gives a general solution in the form of a plane D'Alembert wave in a reference system, in which $x - u_1 t$ is a space coordinate

$$p - p_1 = f(x - u_1 t - c_1 t) + g(x - u_1 t + c_1 t).$$

Riemann used the integral $P(p) = \int_{p_0}^{p} \frac{dp}{\rho_1 c_1}$ and obtained beautiful (according to Lighthill) results:

first Riemann result -

$u + P = \text{const}$ along the curve $C_+ : dx = (u + c)dt$;

second Riemann result -

$u - P = \text{const}$ along the curve $C_- : dx = (u - c)dt$.

The curves $C_\pm$ are characteristic space-time ones which run along a point (x_1, t_1).

Definition 3 A simple wave is defined as the domain where a constant value $u+P$ along any curve C_+ is equal to 0. That has as an outcome

$$u = P(p - p_0) = \int_0^{p-p_0} \frac{d(p - p_0)}{\rho c}.$$

The corresponding curves C_+ are straight lines along where the function u has a constant value. Generally speaking, it is different for different lines. According to Lighthill, the expression *simple wave* designates well an immediate generalization of a plane running wave notion from a linear theory to disturbances with a finite amplitude.

Comment 2 The principal difference of Riemann waves from linear D'Alembert ones is that their amplitude is arbitrary and that the phase velocity of a simple wave is nonlinearly dependent on the local velocity of the particles in a gas.

Some textbooks give different definitions for simple waves. For example, Sedov [3, 4] differs linear running waves and waves, which are a solution of equations (1). This system has no solutions depending only on $x \pm a_0 t$. But a solution may be found which is a plane wave and is a generalization of linear theory solutions of D'Alembert type $f(x \pm a_0 t)$. For these solutions, the velocity u is a function of density ρ only ($u = u(\rho)$, $\rho = \rho(x, t)$). Sedov calls such solutions as Riemann solutions, and motions corresponding to these solutions are called Riemann waves or simple waves.

The development of the theory of simple waves in electromagnetic media has been rather profound and has led even to an extreme definition of the simple wave. The definition is as follows:

Definition 4 [4]. A simple wave in a domain V of space (x, t) is an inconstant continuosly differentiable solution, all components of which are constant along lines of one-parametric family covering the entire domain V. The lines are described by the formula

$$l(\sigma): \quad x = c(\sigma)t + a(\sigma), \quad \sigma \in [\sigma_1; \sigma_2].$$

Also, the condition

$$\Delta(\sigma, t) = c_{,}(\sigma)t + a_{,}(\sigma) \neq 0, \quad \sigma \in [\sigma_1; \sigma_2].$$

should be fulfilled. The condition $\Delta(\sigma, t) = 0$ means that the family of lines $\{l(\sigma)\}$ has an envelope.

So, viewpoints of different authors on the simple wave have been demonstrated. Simple waves are defined and described by different words, but there are no special contradictions between them. These viewpoints rather supplement each other.

Note 3 (About microstructural theories of materials).

Microstructural theories of deformations have achieved the greatest development in the theory of composite materials. Their allmechanical and allphysical nature has displayed more tardy, and today such theories went from the scope of mechanics of composites. The first rather complete description of microstructural theories has been realized in the fundamental series of monographs on composite materials. The attempt to qualify microstructural theories has also been made in [4].

Definition 5 Composite materials are usually defined as materials consisting of several components with differing physical properties, as a rule, alternate many times in a space. The alternation method, the geometrical shape, and physical properties of the components determine the microstructure of the composite.

In real composites, the microstructure is, in the best case, close to periodic one. Moreover, they possess specifical imperfections (for example, delaminations), which are difficult to account in finding exact solutions.

Complexities in the analytical description of composites for exact problem statements have led to the necessity to construct continual, and by their own nature, approximate models, which, on the one hand, retain the existing physical properties of the system, and, at the same time, are rather simple to permit analytical solution of boundary value problems for constrained bodies.

The increased attention to composite materials during more than thirty years, has led very rapidly to diverse realizations of above-mentioned necessity in approximate approaches. A lot of approximate theories capable of accounting for the microstructure of materials, defining required macroscopic parameters, and solving problems of practical importance, have been proposed to date. The full review of such theories is presented in [4].

One of the most developed microstructural theories of materials is the theory, based on a mixture model. Now, let us formulate briefly the background of the microstructural theory of solid mixtures. Its basic advantage consists in the fact that it can be constructed similar to the classical theory of elasticity. This provides good possibilities for analytical solutions.

The hypothesis concerning the fact that the microstructure of a multi-component composite (we will later consider two-component composites as the most widely used composites) can be described using continua, whose material particles occupy simultaneously each geometric point of a domain and interact one with the other, is the basic hypothesis of mixture theory. These are the so-called interpenetrating and interacting continua. Each continuum is characterized by its own set of field characterictics of such thermodynamical functions as partial density $\rho_{\alpha\alpha}$, partial vector of the displacements $\vec{u}^{(\alpha)}$, and partial tensors of the stresses $\sigma_{ik}^{(\alpha)}$, strains $\epsilon_{ik}^{(\alpha)}$, and rotations $\omega_{ik}^{(\alpha)}$. Here and below, the Greek superscripts and Latin subscripts are equal to 1 and 2, and 1, 2, and 3, respectively.

Comment 3 In accordance with traditions of mechanics of heterogeneous media, the parameter is called *partial*, if it characterizes one phase only.

The basic equations of the theory of mixtures are derived, assuming the laws of mass, momentum, angular momentum, and energy conservation are valid for the mixture.

Practically the first wave which is studied in the framework of any theory is a plane wave. Therefore we will study plane waves, too. The definition of these waves is the same as in the classical theory. But a motion in the mixture is described already by two partial vectors of displacements. In plane waves, these vectors are sure not collinear, but the plane front of a wave must be the same for both vectors. Therefore, the representation of a plane wave is written as

$$\vec{u}^{(\alpha)}(x,t) = \vec{u}^{0(\alpha)} e^{i(\xi - \omega t)},$$

$\vec{u}^{0(\alpha)}$ are arbitrary constant vectors, $\xi = \vec{k} \cdot \vec{r}$; $\vec{r}$ is the radius-vector of the point $x \equiv (x_1, x_2, x_3)$. Let the plane wave propagate in the direction of coordinate axis Ox_1. Then it has the form

$$\vec{u}^{(\alpha)} = \{\vec{u}_k^{(\alpha)}(x_1, t)\}. \tag{4}$$

Let us choose the system of equations [3,4] which describes the propagation of longitudinal plane waves of the type of (4)

$$\rho_{\alpha\alpha} u_{tt}^{(\alpha)} - a_\alpha u_{xx}^{(\alpha)} - a_3 u_{xx}^{(\alpha)} - \beta(u^{(\alpha)} - u^{(\delta)}) = 0. \tag{5}$$

System (5) is a generalization of the classic Klein-Gordon equation from a nonlinear wave theory and has [4] a periodical solution in the form of harmonic dispersive waves

$$u_{long}^{(\alpha)}(x,t) = A_{0long}^{(\alpha)} e^{-(k_\alpha^{(long)} x - \omega t)} + l(k_{(\delta)}^{(long)}) A_{0long}^{(\delta)} e^{-(k_\delta^{(long)} x - \omega t)}. \tag{6}$$

In this solution, the wave numbers $k_\alpha^{(long)}$ are determined from the dispersion equation

$$\begin{aligned}
& M_1^{(long)} k^4 - 2M_2^{(long)} k^2 \omega^2 + M_3^{(long)} \omega^4 = 0; \\
& \quad M_1^{(long)} = a_1^{(long)} a_2^{(long)} - (a_3^{(long)})^2, \\
& \quad M_3^{(long)} = \rho_{11}\rho_{22} - (\rho_{11} + \rho_{22}) \left(\frac{\beta}{\omega^2} + \rho_{12} \right); \\
& 2M_2^{(long)} = a_1^{(long)} \rho_{22} + a_2^{(long)} \rho_{11} \\
& \qquad - (a_1^{(long)} + a_2^{(long)} + 2a_3^{(long)}) \left(\frac{\beta}{\omega^2} + \rho_{12} \right); \\
& a_n^{(long)} = \lambda_n + 2\mu_n
\end{aligned} \tag{7}$$

the coefficients of the amplitude distributions matrix are expressed by the algebraic formula

$$l(k_{(\delta)}^{(long)}) = \left\{ -\frac{a_\alpha^{(long)} (k_{(\alpha)}^{(long)})^2 + \beta - \rho_{\alpha\alpha}\omega^2}{a_3^{(long)} (k_{(\alpha)}^{(long)})^2 - \beta} \right\}^{(-1)^\alpha}.$$

So, the features of solution (6) (features of a plane longitudinal harmonic wave in two-phase elastic mixtures) are the following:

1) two modes exist simultaneously, they are distinguished by their wave numbers (the subscript α fixes the mode number);
2) both the modes are essentially dispersive waves;
3) one of the modes is filtered by the mixture, it is blocked for low frequencies, leading off with frequency

$$\omega_{cut} = \sqrt{\frac{\beta(\rho_{11} + \rho_{22})}{\rho_{11}\rho_{22}}},$$

which is called the blocking or cut off frequency;

4) the two modes with their own amplitudes propagate in each phase of the mixture, the amplitudes distribution matrix depends on frequency, that displays in that when changing frequency, the energy of waves is pumped from one mode to the other.

All the effects described above have the microstructural nature, and the waves are linear.

Let us return to basic system (5). Now, for this system, the universal D'Alembert solution cannot be written because the system has nonclassical terms $\pm\beta(u^{(\alpha)} - u^{(\delta)})$. If the function F of the phase $z = x - vt$ used usually in D'Alembert solution is not arbitrary and satisfies an equation

$$F''(z) = f(z)F(z) \qquad (f(z) \text{ is known function}),$$

then an approximate solution of D'Alembert type of system (5) may be obtained according to the classical procedure

$$u^{(\alpha)}(x,t) = A^{\alpha)}F^{(\alpha)}(z^{(\alpha)}) + m(z^{(\alpha)})F^{(\delta)}(z^{(\delta)})$$

This solution has a few new properties, and it may be treated as simple Riemann waves. First of all, the phase velocity v is not constant, it depends nonlinearly on the phase z. Secondly, the solution may be considered as follows: the initial pulse in the form of a function $F(x)$ propagates for some time in a form of the simple wave $F[x - v(z)t]$ and accumulates distortion. Later it distorted so that the approximate solution is not true.

We consider the following three cases of the initial pulse presentation [7–9]:

1. Chebyshov-Hermite functions $\psi_n(z)$ which satisfy the equation (Weber's equation)

$$w'' - [x^2/4 - (n + 1/2)]w = 0;$$

2. Mathieu functions $ce_n(x,q), se_n(x,q)$ which satisfy the equation

$$w'' - (p - 2q\cos 2x)w = 0;$$

3. Whittaker functions $W_{\kappa,\mu}(x)$ which satisfy the equation

$$w'' - (-1/4 + \kappa/x + (1 - 4\mu^2)/2x^2)w = 0.$$

For these cases, a simple wave will be exist with a certain restriction in time. This restriction has the form

$$1 + tv'(z) \simeq 1.$$

We will illustrate the general procedure on the following example. Let us select the initial pulse in the form of bell-shaped function, i.e. the Chebyshov-Hermite function of zero index

$$u^{(\alpha)}(x,0) = A^{(\alpha)}\psi_0(x). \tag{8}$$

With the exception of the small area near the cut-off frequency of the second mode, the dispersivity of a mixture is weak. Therefore the pulse will distort slightly, and the solution of the basic equation (5) may be sought in the form of bell-shaped function. It can be represented as

$$\begin{aligned} u^{(\alpha)}(x,0) &= A^{(\alpha)}\psi_0(z^{(\alpha)}) + p(z^{(\delta)})A^{(\delta)}\psi_0(z^{(\delta)}) \\ &= A^{(\alpha)}e^{-\frac{1}{2}(z^{(\alpha)})^2} + p(z^{(\delta)})A^{(\delta)}e^{-\frac{1}{2}(z^{(\delta)})^2}, \end{aligned} \tag{9}$$

$$\begin{aligned} p(z^{(\delta)}) &= \left\{ \frac{\lambda_\alpha + 2\mu_\alpha - v(z^{(\delta)})\rho_{\alpha\alpha} + \left(\frac{\beta}{1-(z^{(\alpha)})^2}\right)}{\lambda_3 + 2\mu_3 - \left(\frac{\beta}{1-(z^{(\alpha)})^2}\right)} \right\}^{(-1)^\alpha}, \\ v(z^{(\alpha)}) &= \{K_1 - (-1)^\alpha\sqrt{K_1^2 - K_2}\}^{1/2}, \\ K_1 &= \frac{\lambda_1 + 2\mu_1}{\rho_{11}} + \frac{\lambda_2 + 2\mu_2}{\rho_{22}} - \frac{\rho_{11} + \rho_{22}}{\rho_{11}\rho_{22}}\frac{\beta}{1 - (z^{(\alpha)})^2}, \\ K_2 &= \frac{\lambda_1 + 2\mu_1}{\rho_{11}}\frac{\lambda_2 + 2\mu_2}{\rho_{22}} - \frac{(\lambda_3 + 2\mu_3)^2}{\rho_{11}\rho_{22}} \\ &+ \frac{(\lambda_1 + 2\mu_1) + (\lambda_2 + 2\mu_2) + 2(\lambda_3 + 2\mu_3)}{\rho_{11}\rho_{22}}\frac{\beta}{1 - (z^{(\alpha)})^2}. \end{aligned} \tag{10}$$

In this solution, a phase acquires the number, since two phase velocities exist in the mixture.

The simplicity in the writing of solution (9) is achieved by two essential complications - nonlinear dependence of the phase velocity on a phase, and the approximate character of fulfilling the basic system. The first complication transforms the wave into a simple one. The second is associated with the procedure of calculating the first and second derivatives. If we neglect the denominators in the expressions for these derivatives, i.e., set the above restriction, then the solution is invalid in the region of variation in x and t, where the latter equations are not satisfied. The answer follows from analysis of plots of the phase velocities.

Comment 4 A certain analogy exists between an obtained condition and the van der Pol condition appearing in the slowly varying amplitudes method widely used in nonlinear physics. In the van der Pol method, it is supposed that in a harmonic solution the phase and amplitude vary with time, but so slowly, that this varying may be neglected on one period. In reality, the "accelerations" of amplitudes are neglected, and shorten equations are obtained. Here we use a similar procedure. The condition obtained supposes the existence of the time, during which the "accelerations" of a phase may be neglected, and only the "velocities" of phase should be taken into account.

The obtained representation of a solution in the form of simple wave (9) is easily generalized to the case of an arbitrary Chebyshov-Hermite function. Then a solution is of the form

$$
\begin{aligned}
u^{(\alpha)}(x,0) &= A^{(\alpha)}\psi_n(z^{(\alpha)}) + p(z^{(\delta)})A^{(\delta)}\psi_n(z^{(\delta)}), \\
p(z^{(\delta)}) &= \left\{ \frac{\lambda_\alpha + 2\mu_\alpha - v(z^{(\delta)})\rho_{\alpha\alpha} + (\frac{\beta}{1+2n-(z^{(\alpha)})^2})}{\lambda_3 + 2\mu_3 - (\frac{\beta}{1+2n-(z^{(\alpha)})^2})} \right\}^{(-1)^\alpha}.
\end{aligned}
\tag{11}
$$

Let us now compare solutions (11) and (6). Classical waves (6) are harmonic; this is manifested in the frequency of the phase velocity, and the coefficient β is in conformity with the level of dispersity. Solution (6) also describes the simultaneous existence of two modes distinguished by phase velocities. These basic properties noted before are retained even for both the bell-shaped simple waves (9) and the simple waves with more complicated profiles (11). The dispersivity in (6) is transformed into a nonlinear dependence between the phase velocity and the phase in (11). Waves (11) have all the indications of simple waves. Particularly, the characteristics for these waves are always the straight lines along which the wave amplitude $\mid u^{(\alpha)}(x,t) \mid$ is constant. Two modes are also retained here, and the difference between the phase velocities is significant. The same procedure of describing the amplitude distribution between phases is also retained.

The disagreements between solutions (6) and (11) are, however, rather important. Above all, equation (6) yields classical harmonic dispersive waves, whereas equation (11) produces aperiodic solitary simple waves.

That means that, in particular, the approximate solution in the form of a known bell-shaped signal describes by convenient functional writting, with the aim of a Chebyshov-Hermite function, the solitary wave with a one hump. Such a form is valid during some non-small time. The realized computer modelling showed that the simple wave had these properties, expressed by Chebyshov-Hermite function of any subscript. We would like to repeat that the obtained approximate solution is valid for a small initial period of wave propagation. That will be the moment comparing with the eternity reserved for a harmonic running wave (the latter comes from minus infinity and goes away to infinity).

The computer modeling showed that the time of propagation of simple waves studied is not small for some composite materials, and the distances on which new waves propagate without essential distortions are not small, too. Which distance and time intervals in comparing with are they small? Such comparison for the studied class of composites and for solitary simple waves gave positive results. The distance for a comparison for bell-shaped waves, for example, was chosen as the distance of the bottom of the "bell". Then the critical distance was equal to 20–30 choosen distances.

In part, this investigation was supported by Grant INTAS-93-189.

References

[1] L.I. Mandelstamm, Lectures on optics, theory of relativity and quantum mechanics (In Russian), Moscow, Nauka, 1972.

[2] Nonlinear waves. Propagation and interaction. (In Russian), Moscow, Nauka, 1981.

[3] J.J. Rushchitsky, Nonlinear waves in solid mixtures (review), Intern. Applied Mechanics **33**, no. 1 (1997), 1–35.

[4] J.J. Rushchitsky and S.I. Tsurpal, Waves in materials with a microstructure (In Ukrainian, with English contents), Kiev, S.P. Timoshenko Institute of Mechanics, 1998.

[5] G.B. Whitham, Linear and nonlinear waves, New York-London-Sydney, John Wiley & Sons, 1974.

[6] J. Lighthill, Waves in fluids, London-New York-Melbourne, Cambridge University Press, 1978.

[7] J. Kampe de Feriet, R. Campbell, G. Petiau and T. Vogel, Founctions de la physique mathematique, Paris, Centre national de la recherche scientifique, 1957.

[8] F.W.J. Olver, Asymptotics and special functions. New York-London, Academic Press, 1974.

[9] M.A. Lavrentiev and V.B. Shabat, Methods of the theory of functions of complex variable (In Russian), Moscow, Nauka, 1963.

Institute of Mechanics
National Academy of Sciences of Ukraine
Nesterov str. 3, Kiev, 252680
Ukraine
fax: +380–44–446-0319
rushch@imech.freenet.kiev.ua
National Technical University of Ukraine "KPI"
Kyiv

AMR 158A

Operator Theory:
Advances and Applications, Vol. 117
© 2000 Birkhäuser Verlag Basel/Switzerland

Inverse Spectral Problem Related to the N-wave Equation

Alexander Sakhnovich

To the memory of M.G. Krein

A skewselfadjoint linear system of differential equations that is auxiliary for the well known N-wave equation is considered. Direct and inverse spectral problems for this system are formulated and solved in terms of the generalized Weyl functions.

Introduction

We shall consider the system

$$\frac{dw(x,z)}{dx} = (izD - \zeta(x))w(x,z), \qquad x \geq 0, \quad w(0,z) = E_m, \tag{0.1}$$

$$D = \operatorname{diag}\{d_1, d_2, \dots, d_m\}, \quad d_1 > d_2 > \cdots > d_m > 0, \qquad E_m = \{\delta_{kj}\}_{k,j=1}^m, \tag{0.2}$$

where $w(x,z)$ and $\zeta(x)$ are $m \times m$ matrix functions, diag means diagonal matrix. System (0.1) is the auxiliary linear system for the well known N-wave equation

$$\begin{aligned}(Du_t - u_tD) &- (\widetilde{D}u_x - u_x\widetilde{D}) \\ &= (Du - uD)(\widetilde{D}u - u\widetilde{D}) - (\widetilde{D}u - u\widetilde{D})(Du - uD),\end{aligned}$$

where $\widetilde{D}$ is another diagonal matrix ($\widetilde{D} = \widetilde{D}^* = \operatorname{diag}\{\widetilde{d}_1, \widetilde{d}_2, \dots, \widetilde{d}_m\}$), u_t and u_x are the partial derivatives of the $m \times m$ matrix function $u(x,t)$. In case $D = j$, $\zeta = ijV$,

$$j = \begin{bmatrix} E_p & 0 \\ 0 & -E_p \end{bmatrix}, \qquad V(x) = \begin{bmatrix} 0 & v(x) \\ v^*(x) & 0 \end{bmatrix} \tag{0.3}$$

system (0.1) turns into the canonical system. Fundamental results for canonical systems were obtained by M.G. Krein [K1, K2].

Various references on the N-wave problem and its auxiliary system one can find in [AS]. An extensive amount of research on the scattering problem for system (0.1) with complex-valued entries d_k of D was done by R. Beals, R.R. Coifman, P. Deift and X. Zhou with coauthors (see [BDZ] and references therein). The unique solvability of the problem on a suitable dense set of scattering data was obtained, in particular. The case (0.2) of the positive D that is considered here is less general, and therefore more explicit description of the class of data for which the solution of the inverse problem exists and the close to the classical procedure for solving this problem are obtained below. Notice that generalizations of the Weyl functions are successfully used both in the inverse scattering and in the inverse spectral problems (see [L], [Y], [BC], [BDT], [DZ], [BDZ], [SaA1] and [GKS]). The inverse spectral problem in case $D = j, \zeta = iV$ (the so called skewselfadjoint case) was considered in [SaA1]. The inverse problem was formulated and solved in terms of the so called generalized Weyl functions. The exact solutions were obtained in [GKS] by the state space method.

In the following two sections we shall consider system (0.1) with diagonal D of the form (0.2) that has m different eigenvalues. We shall always assume in Sections 1 and 2 that ζ is "skewselfadjoint":

$$\zeta(x) = -\zeta^*(x). \tag{0.4}$$

Condition (0.4) (as well as the other reductions of the form (3.1) below) follows from the physical considerations connected with the N-wave problem. The definitions of the generalized Weyl function (denoted by the acronym GW-function) and inverse spectral problem (denoted by the acronym ISpP) are given in Section 1. The results on the existence, uniqueness and procedure of recovering the solution of ISpP are obtained. Some properties of the GW-functions we prove in Section 2. The ISpP in the case of the class of physically important reductions (3.1) is discussed in Section 3. Some results were stated and applications to the N-wave equation were given in [SaA2] and [SaA3].

1 Procedure of Recovering the Solution

Definition 1.1 A generalized Weyl function (GW-function) of system (0.1) is an analytic matrix function $\varphi(z) = \{\varphi_{kj}(z)\}_{k,j=1}^{m}$, satisfying for certain $M > 0$ and $r > 0$ and for all z from the domain $Z_M = \{z : \operatorname{Im} z < -M\}$ ($\operatorname{Im} z = (z-\overline{z})/(2i)$), the relations

$$\int_0^{\infty} [\exp(i\,\overline{z}\,Dx)]\varphi^*(z)w^*(x,z)w(x,z)\varphi(z)\exp[(-izD - rE_m)x]\,dx < \infty \tag{1.1}$$

and the normalization conditions

$$\begin{aligned} \varphi_{kj}(z) &\equiv 1 \text{ for } k = j, \\ \varphi_{kj}(z) &\equiv 0 \text{ for } k > j. \end{aligned} \tag{1.2}$$

Theorem 1.1 *Let the inequality*

$$\sup_{0<x<\infty} \|\zeta(x)\| \leq M_0 \tag{1.3}$$

be true. Then there is a GW*-function of system* (0.1) *and this* GW*-function is unique.*

At first the following lemma is proved

Lemma 1.1 *Let the condition* (1.3) *be valid. Then there exist a value* $M > 0$ *and a matrix function* φ *that is analytic in* Z_M, *such that equalities* (1.2) *are true,* $\sum_{k=1}^{m} |\varphi_{kj}(z)|^2 \leq 1$ *and for all* $l < \infty$ *we have*

$$\sup_{x \leq l,\ \mathrm{Im}\, z < -M} \|w(x, z)\varphi(z) \exp(-izDx)\| < \infty, \tag{1.4}$$

where w *satisfies* (0.1).

Proof: Let w satisfy (0.1). Put $\mu(x, z) = w^*(x, z)\, J_k\, w(x, z)$,

$$J_k = P_1 - P_2, \qquad P_1 = \begin{bmatrix} E_k & 0 \\ 0 & 0 \end{bmatrix}, \qquad P_2 = \begin{bmatrix} 0 & 0 \\ 0 & E_{m-k} \end{bmatrix}.$$

In view of (0.1) and (1.3) it is easy to estimate the derivative of μ

$$\begin{aligned} \frac{d}{dx} \mu(x, z) &= w^*(x, z)[J_k(izD - \zeta(x)) - (i\overline{z}D + \zeta^*(x))J_k]w(x, z) \\ &\geq w^*(x, z)[i(z - \overline{z})DJ_k - 2M_0 I_m]w(x, z). \end{aligned}$$

Fix now $\varepsilon > 0$ and put $M_k = 2M_0/(d_k - d_{k+1})$ and $\delta = \varepsilon(d_k - d_{k+1})$. Then, taking into account (0.2), for $z \in Z_{M_k+\varepsilon}$ we have

$$\begin{aligned} & i(z - \overline{z})DJ_k - 2M_0 I_m - [i(z - \overline{z})d_{k+1} + 2M_0 + \delta]J_k \\ &= i(z - \overline{z})(D - d_{k+1}I_m)J_k - 4M_0 P_1 - \delta(P_1 - P_2) \\ &\geq 2(M_k + \varepsilon)(d_k - d_{k+1})P_1 - 4M_0 P_1 - \delta(P_1 - P_2) \\ &= \varepsilon(d_k - d_{k+1})(P_1 + P_2). \end{aligned}$$

As $P_1 + P_2 = E_m$, it means that

$$i(z - \overline{z})DJ_k - 2M_0 I_m - [i(z - \overline{z})d_{k+1} + 2M_0 + \delta]J_k \geq \delta E_m.$$

This inequality and the estimate of $\frac{d}{dx} \mu(x, z)$ above infer

$$\begin{aligned} & \frac{d}{dx} [\mu(x, z) \exp\{[-i(z - \overline{z})d_{k+1} - 2M_0 - \delta]x\}] \\ & \geq \delta\, w^*(x, z)\, w(x, z) \exp\{[-i(z - \overline{z})d_{k+1} - 2M_0 - \delta]x\}. \end{aligned} \tag{1.5}$$

Formula (1.5) yields that the derivative in its left hand side is nonnegative. Therefore we obtain

$$\mu(x, z) \exp\{[-i(z - \overline{z})]d_{k+1} - 2M_0 - \delta]x\} \geq J_k. \tag{1.6}$$

As $\varepsilon \to 0$, we have $\delta \to 0$ and (1.6) infers the inequality

$$\mu(x, z) \exp\{[i(\overline{z} - z)d_{k+1} - 2M_0]x\} \geq J_k \qquad (z \in Z_{M_k}). \tag{1.7}$$

The analytic $m \times (m-k)$ matrix function Q_k will be called J_k-nonpositive if

$$Q_k^*(z)\, J_k\, Q_k(z) \le 0, \tag{1.8}$$

$$Q_k^*(z)\, Q_k(z) > 0 \quad (\operatorname{Im} z < -M_k). \tag{1.9}$$

By $[E_k, 0]$ and $[0, E_{m-k}]$ we denote $k \times m$ and $(m-k) \times m$ matrices with blocks E_k, 0 and 0, E_{m-k}, respectively. According to (1.7)–(1.9) the matrix function

$$\Phi_k(x,z) = [E_k, 0]\, w^{-1}(x,z)\, Q_k(z)\{[0, E_{m-k}]\, w^{-1}(x,z)\, Q_k(z)\}^{-1} \tag{1.10}$$

is well-defined and satisfies the inequality

$$[\, \Phi_k^*(z), E_{m-k}\,]\, J_k \begin{bmatrix} \Phi_k(z) \\ E_{m-k} \end{bmatrix} \le 0,$$

i. e.

$$\Phi_k^*(z)\, \Phi_k(z) \le E_{m-k}. \tag{1.11}$$

Here we took into account that

$$\begin{bmatrix} \Phi_k(z) \\ E_{m-k} \end{bmatrix} = w^{-1}(x,z) Q_k(x,z)\{[0, E_{m-k}]\, w^{-1}(x,z)\, Q_k(z)\}^{-1}.$$

In view of this equality and (1.8) we see also that

$$[\Phi_k^*(z), E_{m-k}]\mu(x,z) \begin{bmatrix} \Phi_k(z) \\ E_{m-k} \end{bmatrix} \le 0. \tag{1.12}$$

On the other hand formula (1.12) infers (1.8)–(1.10) with

$$Q_k(z) = w(x,z) \begin{bmatrix} \Phi_k(z) \\ E_{m-k} \end{bmatrix}.$$

So the set of matrix functions $\Phi_k(z)$ of the form (1.10) (with Q_k satisfying (1.8) and (1.9)) coincides with the set of matrix functions satisfying inequality (1.12).

Let us write down μ in the block form $\mu = \{\mu_{kj}\}_{k,j=1}^2$, where μ_{11} is a $k \times k$ matrix and μ_{22} is an $(m-k) \times (m-k)$ matrix. From (1.7) it follows that

$$\mu_{11} > 0, \qquad \mu_{22} - \mu_{12}^* \mu_{11}^{-1} \mu_{12} < 0. \tag{1.13}$$

Formula (1.12) is equivalent to the relation

$$[\Phi_k^* \mu_{11}^{1/2} + \mu_{12}^* \mu_{11}^{-1/2}][\mu_{11}^{1/2}\, \Phi_k + \mu_{11}^{-1/2}\, \mu_{12}] \le \mu_{12}^* \mu_{11}^{-1}\, \mu_{12} - \mu_{22}.$$

This relation is equivalent to the representation

$$\begin{gathered} \Phi_k(x,z) = \rho_l(x,z) u(x,z) \rho_r(x,z) - \mu_{11}^{-1}(x,z)\, \mu_{12}(x,z); \\ u^* u \le E_{m-k}, \quad \rho_l = \mu_{11}^{-1/2}, \quad \rho_r = (\mu_{12}^* \mu_{11}^{-1}\, \mu_{12} - \mu_{22})^{1/2}. \end{gathered} \tag{1.14}$$

Here $u(x,z)$ is a $k \times (m-k)$ matrix function.

The set of matrices $\Phi_k(x, z)$ will be denoted by $\mathcal{N}_k(x, z)$. As

$$\mathcal{N}_k(x_1, z) \subset \mathcal{N}_k(x_2, z), \qquad x_1 > x_2,$$

$$\sup\{\|\Phi_k(x, z)\| : \qquad \Phi_k \in \mathcal{N}_k(x, z)\} < \infty,$$

there is an analytic function

$$\widetilde{\Phi}_k(z) \in \bigcap_{x<\infty} \mathcal{N}_k(x, z) \qquad (\operatorname{Im} z < -M_k). \tag{1.15}$$

By [P] from (1.7) it follows

$$\mu^{-1}(x, z) \leq \exp\{[i(\overline{z} - z)d_{k+1} - 2M_0]x\}J_k,$$

which implies

$$\rho_r(x, z) \leq \exp\{(1/2)[i(z - \overline{z})d_{k+1} + 2M_0]x\}E_{m-k}. \tag{1.16}$$

From (1.6) and (1.14) we derive

$$\rho_l(x, z) \leq \exp\{(1/2)[i(\overline{z} - z)d_{k+1} - 2M_0 - \delta]x\}E_k. \tag{1.17}$$

Relations (1.14), (1.16) and (1.17) infer the uniqueness of $\widetilde{\Phi}_k$. Therefore (1.15) may be rewritten as

$$\widetilde{\Phi}_k(z) = \bigcap_{x<\infty} \mathcal{N}_k(x, z) \qquad (\operatorname{Im} z < -M_k). \tag{1.18}$$

As (1.14) yields (1.12), $\widetilde{\Phi}_k$ satisfies

$$[\widetilde{\Phi}_k^*(z), E_{m-k}]\,\mu(x, z) \begin{bmatrix} \widetilde{\Phi}_k(z) \\ E_{m-k} \end{bmatrix} \leq 0 \qquad (0 < x < \infty). \tag{1.19}$$

Introduce the function

$$\begin{aligned} f(x, z) \;=\; & [\widetilde{\Phi}_k^*(z), E_{m-k}]w^*(x, z)(E_m - J_k)w(x, z) \\ & [\widetilde{\Phi}_k^*(z), E_{m-k}]^* \exp[i(\overline{z} - z)\, d_{k+1}\, x]. \end{aligned}$$

From (0.1), (1.3) and (1.19) we get $f_x(x, z) \leq 2M_0 f(x, z)$. Hence $f(x, z) \leq 2\exp(2M_0\, x E_{m-k})$ and in view of (1.19) we conclude

$$\sup_{\operatorname{Im} z < -M_k,\ x \leq l} \left\| w(x, z) \begin{bmatrix} \widetilde{\Phi}_k(z) \\ E_{m-k} \end{bmatrix} \exp(-izd_{k+1}x) \right\| < \infty. \tag{1.20}$$

Equalities (1.2) and formula

$$\{\varphi_{p,k+1}(z)\}_{p=1}^{k} = \widetilde{\Phi}_k(z) \begin{bmatrix} 1 \\ 0 \\ \vdots \\ 0 \end{bmatrix} \tag{1.21}$$

define the matrix function φ. According to (1.11) and (1.20) this φ satisfies the conditions of the lemma with $M = \max_{1 \leq k \leq m}(M_k + \varepsilon_0)$ for any $\varepsilon_0 > 0$. □

Proof of Theorem 1.1: The k-th column of $\varphi(z)$ will be denoted by $\varphi_k(z)$. In view of (1.19) and (1.21) the relation

$$\varphi_{k+1}^*(z)\,\mu(x,z)\,\varphi_{k+1}(z) \le 0 \tag{1.22}$$

is true. By (1.5) and (1.22) for $r = 2M + \delta$ ($\delta = \varepsilon_0 \min_{1\le k\le m}(d_k - d_{k+1})$) the result

$$\begin{aligned}&\int_0^\infty \varphi_{k+1}^*(z)w^*(x,z)w(x,z)\varphi_{k+1}(z)\exp\{[i(\overline{z}-z)d_{k+1} - r]x\}\,dx\\ &\quad\le -\delta^{-1}\varphi_{k+1}^*(z)\,J_k\,\varphi_{k+1}(z) < \infty \qquad (\operatorname{Im} z < -M)\end{aligned} \tag{1.23}$$

is valid and the existence of a GW-function is proved. From (1.7) it follows that for any M and r there is a $z \in Z_M$ such that the inequality

$$\begin{aligned}&\int_0^u [E_k,0]w^*(x,z)w(x,z)[E_k,0]^*\exp\{[i(\overline{z}-z)d_{k+1} - r]x\}\,dx\\ &\quad\ge uE_k \quad (0 < u < \infty)\end{aligned} \tag{1.24}$$

is valid. In view of (1.24) we column by column derive the uniqueness of $\varphi(z)$. □

Remark: By easy algebraic transformations and induction on k, from (1.10), (1.18) and (1.19) we get

$$\widetilde{\Phi}_k(z) = [E_k, 0]\,\varphi(z)\begin{bmatrix}0\\E_{m-k}\end{bmatrix}\left\{[0, E_{m-k}]\,\varphi(z)\begin{bmatrix}0\\E_{m-k}\end{bmatrix}\right\}^{-1}. \tag{1.25}$$

As is obvious from the proof of Theorem 1.1, the GW-function $\varphi(z)$ satisfies (1.4). To generalize the inverse problem let us introduce a more general definition.

Definition 1.2 The inverse spectral problem (ISpP) for system (0.1) is the problem of recovering from an analytic matrix function $\varphi(z)$ a matrix function $\zeta(x) = -\zeta^*(x)$ ($\zeta_{kk} = 0$) such that (1.4) is valid and the inequalities

$$\sup_{0<x<l} \|\zeta(x)\| < \infty \tag{1.26}$$

are true for each $l < \infty$.

We shall denote by Ω the operator mapping the pair D and $\varphi(z)$ into ζ ($\Omega(D,\varphi) = \zeta$). We no longer assume that $\varphi(z)$ satisfies (1.2).

Theorem 1.2 *For any matrix function $\varphi(z)$ that is analytic and bounded in Z_M and has the property*

$$\int_{-\infty}^\infty [\varphi(z) - E_m]^*[\varphi(z) - E_m]\,d\lambda < \infty \quad (z = \lambda - i\eta, \quad \eta > M) \tag{1.27}$$

there is at most one solution of the ISpP.

We shall start with the construction of the transformation operator (see representation (1.28) of w below).

Lemma 1.2 *Relations* (0.1) *and* (1.26) *yield the representation*

$$w(x,z) = \exp(izDx) + \int_{ax}^{bx} [\exp(izu)] N(x,u)\, du, \tag{1.28}$$

where $a = d_m$, $b = d_1$, $\sup \|N(x,u)\| < \infty$ $(x \le l)$.

Proof: The "wave" function w may be represented in the form

$$w(x,z) = \sum_{k=0}^{\infty} \nu_k(x,z), \tag{1.29}$$

where

$$\begin{aligned} \nu_0(x,z) &= \exp(izDx), \\ \nu_k(x,z) &= -\int_0^x \{\exp[izD(x-u)]\}\, \zeta(u)\, \nu_{k-1}(u,z)\, du \quad (k>0). \end{aligned} \tag{1.30}$$

In the case $k = 1$ we have

$$\nu_1(x,z) = \int_{ax}^{bx} [\exp(izu)]\, N_1(x,u)\, du, \quad N_1 = \{N_{1pj}\}_{p,j=1}^m. \tag{1.31}$$

$$N_{1pj}(x,u) = \frac{1}{d_j - d_p} \zeta_{pj} \left(\frac{d_p x - u}{d_p - d_j} \right) \begin{cases} 1 & \text{for } j > p, \ d_j x < u \le d_p x, \\ -1 & \text{for } j < p, \ d_p x < u \le d_j x, \\ 0 & \text{otherwise.} \end{cases} \tag{1.32}$$

By induction we show that

$$\nu_k(x,z) = \int_{ax}^{bx} [\exp(izu)]\, N_k(x,u)\, du \quad (k \ge 1), \tag{1.33}$$

$$N_{kpj}(x,u) = -\int_{\chi}^{x} \sum_{r=1}^{m} \zeta_{pr}(v)\, N_{k-1,r,j}(v, d_p(v-x)+u)\, dv \quad (k \ge 2), \tag{1.34}$$

$$\begin{aligned} \chi &= (d_p x - u)(d_p - a)^{-1} \quad \text{for} \quad u \le d_p x, \\ \chi &= (d_p x - u)(d_p - b)^{-1} \quad \text{for} \quad u \ge d_p x. \end{aligned}$$

Here $N_k(x,u)$ is defined for $ax < u < bx$. From (1.32) and (1.34) we conclude $\|N_k(x,u)\| \le C^k x^{k-1}/(k-1)!$. Hence, in view of (1.29), (1.30) and (1.33) the statement of the lemma follows. □

The transformation operator (1.28) was constructed in [SaA4], another kind of transformation operator for the same system one can find in [M].

Proof of Theorem 1.2: Suppose there exist two solutions $\zeta_1 \in \Omega(D, \varphi)$ and $\zeta_2 \in \Omega(D, \varphi)$ of the ISpP. Let w_k $(k = 1, 2)$ satisfy the system of the form (0.1) with the potential ζ_k. Lemma 1.2 and formulas (1.4) and (1.27) yield [PW] the relations

$$\begin{aligned} & w_k(x, z)\, \varphi(z) \exp(-izDx) \\ (1.35) \qquad & = E_m + \int_0^\infty [\exp(-izu)]\, \gamma_k(x, u)\, du, \quad \operatorname{Im} z < -M, \\ & \int_0^\infty [\exp(-2Mu)]\, \gamma_k^*(x, u)\, \gamma_k(x, u)\, du < \infty. \end{aligned}$$

By (1.35) there is $\widehat{M} > M$ such that

$$(1.36) \qquad \sup_{\operatorname{Im} z \le -\widehat{M}} \|w_k(x, z) w_j^{-1}(x, z)\| < \infty \quad (k \neq j;\ k = 1, 2;\ j = 1, 2).$$

Taking into account (0.1), (0.2) and (0.4) we see that

$$(1.37) \qquad w^*(x, \overline{z}) = w^{-1}(x, z),$$

and therefore $w_1(x, \overline{z}) w_2^{-1}(x, \overline{z}) = [w_2(x, z) w_1^{-1}(x, z)]^*$. Using (1.36) we now obtain the boundedness of $w_1 w_2^{-1}$ in both halfplanes $\operatorname{Im} z \geq \widehat{M}$ and $\operatorname{Im} z \leq -\widehat{M}$:

$$\sup \|w_1(x, z)\, w_2^{-1}(x, z)\| < \infty \quad (|\operatorname{Im} z| \geq \widehat{M}).$$

In view of (0.1), (1.26) and (1.37) there exist values $c_1 > 0$ and $c_2 > 0$ such that $\sup \|w_1(x, z) w_2^{-1}(x, z)\| \leq c_1 e^{c_2 |z|}$ for all z in the complex plane $\mathbb{C}$. So the Phragmen-Lindelöf Theorem ([B], Chapter 1) may be applied to get the boundedness of $w_1 w_2^{-1}$ in the strip $|\operatorname{Im} z| < \widehat{M}$. As $w_1 w_2^{-1}$ is bounded in $|\operatorname{Im} z| < \widehat{M}$ and $|\operatorname{Im} z| \geq \widehat{M}$, it is bounded in $\mathbb{C}$. In this way the equality

$$(1.38) \qquad w_1(x, z) w_2^{-1}(x, z) \equiv \text{const}$$

is derived. From Lemma 1.2 it follows that

$$(1.39) \qquad \lim_{\lambda \to \infty} w_k(x, \lambda - i\eta) \exp[-i(\lambda - i\eta) D x] = E_m.$$

The relations (1.38) and (1.39) mean that $w_1 = w_2$. □

To prove the existence of the solutions of ISpP we need the stricter conditions

$$\begin{aligned} & \sup \|[\varphi(z) - E_m] z\| < \infty \quad (\operatorname{Im} z < -M), \\ (1.40) \qquad & [\varphi(z) - E_m - \alpha/z] z \in L^2_{m \times m}(-\infty, \infty) \\ & (z = \lambda - i\eta,\ \eta > M,\ -\infty < \lambda < \infty). \end{aligned}$$

Without loss of generality we can suppose that

$$(1.41) \qquad \det \varphi(z) \neq 0.$$

Theorem 1.3 *Let the analytic matrix function φ satisfy* (1.40). *Then a solution of the ISpP exists and is unique.*

Proof: The uniqueness follows from Theorem 1.2. The procedure of recovering the solution is started by introducing the matrix function

$$\Pi(x) = \frac{1}{2\pi i}\int_{-\infty}^{\infty}[\exp(izDx)][\varphi^{-1}(z)/z]\,d\lambda \quad (z=\lambda-i\eta,\ \eta>M,\ x\geq 0). \tag{1.42}$$

The right hand side of (1.42) is well-defined, as by (1.40) and (1.41) we have

$$\begin{aligned}&\sup\|[\varphi^{-1}(z)-E_m]z\|<\infty(\operatorname{Im} z<-M),\\ &[\varphi^{-1}(z)-E_m+\alpha/z]z\in L^2_{m\times m}(-\infty,\infty)\\ &(z=\lambda-i\eta,\quad \eta>M)\end{aligned} \tag{1.43}$$

and the integral in (1.42) is understood as the norm limit of the functions in $L^2_{m\times m}(0,l)$. So $\Pi(x)$ is defined on each interval $(0,l)$. We have

$$\frac{1}{2\pi i}\int_{-\infty}^{\infty}[\exp(izDx)]/z\,d\lambda\equiv E_m \quad (x\geq 0). \tag{1.44}$$

According to (1.42)–(1.44), $\Pi(x)$ is twice differentiable:

$$\begin{aligned}&\Pi(0)=E_m,\qquad \Pi'(0)=-iD\alpha;\\ &\exp(-MDx)\,\Pi'(x)\ \in\ L^2_{m\times m}(0,\infty),\\ &\exp(-MDx)\,\Pi''(x)\ \in\ L^2_{m\times m}(0,\infty).\end{aligned} \tag{1.45}$$

Let us now introduce the bounded linear operator

$$S_l f = D^{-1}f+\int_0^l s(x,u)f(u)\,du, \tag{1.46}$$

where $s(x,u)=\{s_{kj}(x,u)\}_{k,j=1}^m$; $0\leq x,u\leq l$;

$$s_{kj}(x,u) = \int_\gamma \theta_{kj}(v,u+d_kd_j^{-1}(v-x))\,dv + \begin{cases} d_k^{-1}\Pi'_{kj}(x-d_jd_k^{-1}u) & \text{for } u\leq d_kd_j^{-1}x,\\ d_j^{-1}\overline{\Pi}'_{jk}(u-d_kd_j^{-1}x) & \text{for } d_kd_j^{-1}x<u;\end{cases} \tag{1.47}$$

$\theta(x,u)=\{\theta_{kj}(x,u)\}_{k,j=1}^m=\Pi'(x)[\,\Pi'(u)\,]^*D^{-1}$, γ is the interval $[\max(0,x-d_jd_k^{-1}u),\ x]$, on $L^2_m(0,l)$. Sometimes we omit "l" in S_l and write just S.

The operator S satisfies the operator identity

$$AS-SA^*=i\,\Pi\,\Pi^*, \tag{1.48}$$

where the operator $A = iD\int_0^x \bullet \, du$ acts in $L^2_m(0,l)$ and Π acts from $\mathbb{C}^m$ into $L^2_m(0,l)$: $\Pi g = \Pi(x) g$. About the method of operator identities and applications for the ISpP for selfadjoint systems see [SaL1]–[SaL3] and references therein.

According to (1.48) the operator $-A^*$ is S-dissipative, i.e. $-i(AS - SA^*) \geq 0$. As A^* does not have any eigenvectors, it follows [Az] that $S \geq 0$. If now $Sf = 0$ by (1.48) we obtain $\Pi^* f = 0$. And with the help of (1.48) we prove by induction that $S(A^*)^k f = 0$ $(k \geq 0)$. Recall that A^* does not have any eigenvectors and hence it does not have finite dimensional invariant subspaces. So either $f = 0$ or $\operatorname{Ker} S$ containing all the vectors $(A^*)^k f$ proves infinite dimensional. On the other hand it can be seen from (1.46) that $\dim \operatorname{Ker} S < \infty$. Therefore $Sf = 0$ yields $f = 0$. Taking now into account also (1.46) and $S \geq 0$ we obtain

$$S_l \geq \varepsilon(l) E \qquad (\varepsilon > 0). \tag{1.49}$$

Introduce the following projection $\mathcal{P}_r$ from $L^2_m(0,l)$ on $L^2_m(0,r)$: $(\mathcal{P}_r f)(x) = f(x)$ $(0 \leq x \leq r \leq l)$. By (1.49) it is true that

$$S_r = \mathcal{P}_r \, S_l \, \mathcal{P}_r^* \geq \varepsilon(l) \, E. \tag{1.50}$$

From (1.45)–(1.47) and (1.50) one gets the representation [GK]

$$\begin{gathered} S^{-1} = V^* V, \quad V = D^{1/2} + \int_0^x V(x,u) \bullet du, \\ \int_0^l \int_0^x V^*(x,u) V(x,u) \, du \, dx < \infty. \end{gathered} \tag{1.51}$$

We shall now introduce the transfer matrix function

$$w_A(l,z) = E_m + iz\Pi^* S^{-1} (E - zA)^{-1} \Pi. \tag{1.52}$$

By [SaL1] we have

$$\begin{aligned} \frac{dw_A(r,z)}{dr} &= izH(r) \, w_A(r,z), \\ H(r) &= \frac{d}{dr} \int_0^r \Pi^*(v) \, [S_r^{-1} \mathcal{P}_r \Pi](v) \, dv. \end{aligned} \tag{1.53}$$

From (1.51) and (1.53) we get

$$H(r) = \widetilde{\beta}^*(r) \, \widetilde{\beta}(r), \tag{1.54}$$

where

$$\widetilde{\beta}(r) = (V\Pi)(r). \tag{1.55}$$

Let us show that the matrix function $\widetilde{\beta}(x)$ is differentiable and satisfies the equality

$$\widetilde{\beta}(x) \, \widetilde{\beta}^*(x) \equiv D. \tag{1.56}$$

Indeed, according to (1.48) and (1.51) we get the relation

$$VAV^{-1} - (V^*)^{-1}A^*V^* = i\,\widetilde{\beta}(x)\int_0^l \widetilde{\beta}^*(u)\bullet du,$$

i.e.,

$$VAV^{-1} = i\,\widetilde{\beta}(x)\int_0^x \widetilde{\beta}^*(u)\bullet du. \tag{1.57}$$

The operator V^{-1} admits representation

$$\Gamma = V^{-1} = D^{-1/2} + \int_0^x \Gamma(x,u)\bullet du. \tag{1.58}$$

Formula (1.57) may be rewritten as $A\Gamma = i\,\Gamma\widetilde{\beta}(x)\int_0^x \widetilde{\beta}^*(u)\ \bullet du$, which is equivalent to the equality

$$\begin{aligned} D^{1/2} + D\int_u^x \Gamma(v,u)\,dv &= D^{-1/2}\,\widetilde{\beta}(x)\,\widetilde{\beta}^*(u) \\ &+ \int_u^x \Gamma(x,v)\widetilde{\beta}(v)\,dv\,\widetilde{\beta}^*(u) \quad (x \geq u). \end{aligned} \tag{1.59}$$

As $V^{-1} = SV^*$ we can put

$$\begin{aligned} V^*(u,x) &= -S_u^{-1}s(x,u)D^{1/2} && (x \leq u), \\ \Gamma(x,u) &= s(x,u)D^{1/2} + \int_0^u s(x,v)\,V^*(u,v)\,dv && (x \geq u). \end{aligned} \tag{1.60}$$

Hence $\widetilde{\beta}(x)$ is continuous in x and the matrix functions $\int_u^x \Gamma(x,v)\widetilde{\beta}(v)\,dv$, $\int_u^x \Gamma(v,u)\ dv$ are continuous in x and u. In this way we see that (1.59) in valid pointwise. In particular, for $x = u$ we get (1.56). As $\Gamma\widetilde{\beta}(x) = \Pi(x)$, formula (1.59) is transformed into

$$D^{1/2} + D\int_u^x \Gamma(v,u)\,dv = \Pi(x)\widetilde{\beta}^*(u) - \int_0^u \Gamma(x,v)\widetilde{\beta}(v)\,dv\,\widetilde{\beta}^*(u).$$

Multiply both sides of the equality by $D^{-1}\widetilde{\beta}(u)$. In view of (1.56) we get

$$\begin{aligned} D^{-1/2}\widetilde{\beta}(u) &= \Pi(x) - \int_0^u \Gamma(x,v)\widetilde{\beta}(v)\,dv \\ &\quad - D\int_u^x \Gamma(v,u)\,dv\,D^{-1}\widetilde{\beta}(u). \end{aligned} \tag{1.61}$$

Differentiating both sides of (1.61) with respect to u and putting $u = x$ we obtain the result

$$\widetilde{\beta}'(x) = D^{1/2}[D\,\Gamma(x,x)D^{-1} - \Gamma(x,x)]\widetilde{\beta}(x). \tag{1.62}$$

So $\widetilde{\beta}(x)$ is absolutely continuous. Put

$$w(x,z) = D^{-1/2}\,\widetilde{\beta}(x)\,w_A(x,z). \tag{1.63}$$

According to (1.53), (1.54) and (1.56) the given function w satisfies (0.1), where

$$\zeta(x) = -\beta'(x)\,\beta^*(x), \qquad \beta(x) = D^{-1/2}\,\widetilde{\beta}(x). \tag{1.64}$$

(We take into account that $w(0,z) = D^{-1/2}\,\widetilde{\beta}(0) = \Pi(0) = E_m$.)

Let us show that the ζ constructed is the solution of ISpP. It is easy to see that

$$((E - zA)^{-1}\Pi)(x) = \Pi(x) + izD\int_0^x \{\exp[izD(x-u)]\}\Pi(u)\,du. \tag{1.65}$$

By (1.42) and (1.43) we have

$$izD\int_0^\infty [\exp(-izDu)]\Pi(u)\,du = \varphi^{-1}(z) \qquad (\operatorname{Im} z < -M). \tag{1.66}$$

Hence in view of (1.45) and (1.65) we conclude

$$\begin{aligned} ((E - zA)^{-1}\Pi)(x) &= \Pi(x) + [\exp(izDx)] \\ &\quad \left\{\varphi^{-1}(z) - ziD\int_x^\infty [\exp(-izDu)]\Pi(u)du\right\} \\ &= [\exp(izDx)]\varphi^{-1}(z) + (i/z)D^{-1} \\ &\quad \left\{\Pi'(x) + \int_x^\infty \exp[izD(x-u)]\Pi''(u)du\right\}. \end{aligned} \tag{1.67}$$

Thus it is true that

$$((E - zA)^{-1}\Pi)(x)\varphi(z)\exp(-izDl) = \exp[izD(x-l)] + q(z,x,l), \tag{1.68}$$

where for all $l_0 < \infty$, $\varepsilon > 0$

$$\sup \|q(z,x,l)z\| < \infty \qquad (\operatorname{Im} z \le -M - \varepsilon, \quad x \le l \le l_0). \tag{1.69}$$

From (1.48) one gets [SaL1]:

$$w_A^*(l,z)w_A(l,z) = E_m + i(z - \overline{z})\Pi^*(E - \overline{z}A^*)^{-1}S_l^{-1}(E - zA)^{-1}\Pi. \tag{1.70}$$

Taking into account (1.56), (1.63) and (1.68)–(1.70) we get

$$\sup_{\operatorname{Im} z < -M-\varepsilon,\ l \le l_0} \|w(l,z)\,\varphi(z)\exp(-izDl)\| < \infty. \tag{1.71}$$

Put in (1.71) $\varepsilon = 0$ and substitute l by x and l_0 by l. We obtain (1.4). It remains to add that by (1.56), (1.62) and (1.64) we have

$$\zeta(l) = [\Gamma(l,l) - D\,\Gamma(l,l)D^{-1}]D^{1/2}. \tag{1.72}$$

So $\zeta_{kk} = 0$. □

Remark: From (1.60) it follows that

$$\Gamma(l,l) = D^{-1}[S_l^{-1}s(x,l)](l)\, D^{1/2}. \tag{1.73}$$

Under the conditions of Theorem 1.3, equations (1.42), (1.46), (1.72) and (1.73) now uniquely define the mapping $\Omega(D,\varphi)$ and produce a procedure for solving ISpP.

2 GW-function

In the previous section we have shown that one obtains (1.4) from (1.1) and (1.3), i.e. the GW-function satisfies (1.4). Now we shall show that from (1.3) and (1.4) follows (1.1). This means that under condition (1.3) inequalities (1.1) and (1.4) are equivalent.

Theorem 2.1 *Let $\varphi(z)$ be normalized, i.e. satisfy* (1.2). *Let $\varphi(z)$ also satisfy* (1.4) *with a $w(x,z)$ that satisfies* (0.1), *where $\zeta(x)$ $(x \geq 0)$ is bounded in the matrix norm. Then $\varphi(z)$ is the* GW*-function of the system* (0.1) *and* (1.1) *is true.*

Proof: For each y there is $C(y)$ such that for all z with $\operatorname{Im} z < -C(y)$ the relation

$$\varphi_k^*(z)\,\mu(y,z)\,\varphi_k(z) \leq 0 \tag{2.1}$$

is true. (Recall that φ_k is the k-th column of φ.) Indeed, if (2.1) is not fulfilled, there is a sequence $\{z_p\}$ $(\operatorname{Im} z_p \to -\infty)$ such that inequalities $\varphi_k^*(z_p)\,\mu(y,z_p)\,\varphi_k(z_p) > 0$ are valid. By (1.5) we get also

$$\varphi_k^*(z_p)\,\mu(x,z_p)\,\varphi_k(z_p) > 0 \qquad (x \geq y). \tag{2.2}$$

From (0.1) and (2.2) it follows that

$$\begin{aligned} &\frac{d}{dx}\{\varphi_k^*(z_p)\,w^*(x,z_p)(E_m + J_k)w(x,z_p)\varphi_k(z_p) \\ &\quad \exp[i(\overline{z}_p - z_p)d_{k+1}\,x + 4M_0x]\} \\ &\quad \geq \varphi_k^*(z_p)w^*(x,z_p)w(x,z_p)\varphi_k(z_p)i\,(z_p - \overline{z}_p)(d_k - d_{k+1}) \\ &\quad \times\, \exp[\,i\,(\overline{z}_p - z_p)d_{k+1}\,x + 4M_0x] \qquad (M_0 = \sup\|\zeta(x)\|). \end{aligned} \tag{2.3}$$

Notice that

$$\frac{d}{dx}\,w^*(x,z)\,J_{k+1}\,w(x,z)\exp[\,i\,(\overline{z} - z)d_{k+1}x + 2M_0x] \geq 0 \tag{2.4}$$

for $z \in Z_{M_1}$ $(M_1 > M)$. As $\varphi_k^*(z_p)\,J_{k+1}\,\varphi_k(z_p) \geq 1$, in view of (2.3) and (2.4) we have

$$\begin{aligned} &\frac{d}{dx}\,\{\varphi_k^*(z_p)w^*(x,z_p)(E_m + J_k)w(x,z_p)\varphi_k(z_p) \\ &\quad \exp[i(\overline{z}_p - z_p)d_{k+1}x + 4M_0x]\} \\ &\quad \geq i(z_p - \overline{z}_p)(d_k - d_{k+1})\exp(2M_0x). \end{aligned}$$

Therefore we easily conclude

$$\begin{aligned}
&\varphi_k^*(y+l,z_p)w^*(y+l,z_p)w(y+l,z_p)\varphi_k(y+l,z_p)\\
(2.5)\quad &\exp[i(\overline{z}_p - z_p)d_{k+1}(y+l) + 4M_0(y+l)]\\
&\geq i(d_k - d_{k+1})[(z_p - \overline{z}_p)/(4M_0)]\{\exp[2M_0(y+l)] - \exp(2M_0 y)\}.
\end{aligned}$$

The right hand side of (2.5) tends to infinity as $p \to \infty$, which contradicts (1.4). Hence (2.1) is true.

By Theorem 1.1 under the conditions of Theorem 2.1 there exists a GW-function $\widehat{\varphi}(z)$ of system (0.1). Then taking into account of (1.14), (1.16), (1.17) and (2.5) we obtain

$$(2.6)\qquad \begin{aligned}&\|\widehat{\varphi}_k(z) - \varphi_k(z)\| \leq 2\\ &\exp\{[\operatorname{Im} z(d_k - d_{k+1})y/2] + M_0 y\} \quad (\operatorname{Im} z < -C(y)).\end{aligned}$$

The bounded analytic vector function $\varphi_k - \widehat{\varphi}_k$ admits [PW] the representation

$$(2.7)\qquad \widehat{\varphi}_k(z) - \varphi_k(z) = z\int_0^{\infty}[\exp(-izu)]g(u)du.$$

By (2.6) and (2.7) for each $r > 0$ the relation

$$\sup\left\|\int_0^r \{\exp[iz(r-u)]\}g(u)\,du\right\| < \infty$$

is true in the domain $Z_{C_1(r)}$ and therefore in the whole plane. So $g = 0$ and $\varphi_k = \widehat{\varphi}_k$. □

3 A Generalization

The considerations of Sections 1 and 2 remain valid (or valid after slight modifications) for system (0.1) with the reductions of the form

$$(3.1)\quad \zeta(x) = -B\zeta^*(x)B \quad (B = \operatorname{diag}\{b_1,\dots,b_m\},\ \ b_p = \pm 1 (1 \leq p \leq m)).$$

(The reduction (0.4) is obtained when $B = E_m$.) For $B \neq E_m$ Definition 1.1 remains valid and we change the condition $\zeta = -\zeta^*$ to (3.1) in the Definition 1.2 of the ISpP. Then the construction of the solution of the ISpP in the proof of Theorem 1.3 is modified in the following way. The operator identity (1.48) takes the form $AS - SA^* = \Pi B \Pi^*$ with Π defined by (1.42). Hence in (1.46) we have $S_l f = BD^{-1}f(x) + \int_0^l \{s_{kj}(x,u)\}_{k,j=1}^m f(u)du$, where the entries s_{kj} are defined by the modified formula (1.47) with $\theta(x,u) = \Pi'(x)B[\Pi'(u)]^* D^{-1}$ and terms $d_k^{-1} b_j \Pi'_{kj}$ and $d_j^{-1} b_k \overline{\Pi}'_{jk}$ instead of terms $d_k^{-1}\Pi'_{kj}$ and $d_j^{-1}\overline{\Pi}'_{jk}$, respectively. Modifying the formulas for S_l and $s = \{s_{kj}\}_{k,j=1}^m$ as described above and supposing in addition that the operators S_l are boundedly invertible for all l, we get the solution of the inverse problem by (1.72) with $\Gamma(l,l) = BD^{-1}(S_l^{-1}s(v,l))(l)D^{1/2}B$. For $B = E_m$ the invertibility of S follows automatically from the conditions of Theorem 1.3. Theorems 1.1 and 2.1 (and their proofs) are valid for systems (0.1) without reductions.

References

[AS] M.J. Ablowitz and H. Segur, *Solitons and the inverse scattering transform*, SIAM Stud. Appl. Math. **4**, Society for Industrial and Applied Mathematics (SIAM), Philadelphia, 1981.

[Az] T. Ya. Azizov, *Dissipative operators in Hilbert space with indefinite metrics*, Izv. Akad. Nauk SSSR Ser. Mat. **37** (1973), 639–662.

[BC] R. Beals and R.R. Coifman, *Scattering and inverse scattering for first order systems*, Comm. Pure Appl. Math. **37** (1984), 39–90.

[BDT] R. Beals, P. Deift and C. Tomei, *Direct and inverse scattering on the line*, Math. Surveys and Monographs **28**, AMS, 1988.

[BDZ] R. Beals, P. Deift and X. Zhou, *The inverse scattering transform on the line*. In: A.S. Fokas and V.E. Zakharov (Eds.), Important Developments in Soliton Theory. Springer Ser. Nonlinear Dynamics, Springer, Berlin, 1993, pp. 7–32.

[B] R.P. Boas, *Entire functions*. Academic Press, New York, 1954.

[DZ] P. Deift and X. Zhou, *Direct and inverse scattering on the line with arbitrary singularities*, Comm. Pure Appl. Math. **44** (1991), 485–533.

[GK] I. Gohberg and M.G. Krein, *Theory and applications of Volterra operators in Hilbert space*. Amer. Math. Soc. Transl. **24**, AMS, Providence, R.I., 1970.

[GKS] I. Gohberg, M.A. Kaashoek and A.L. Sakhnovich, *Pseudo-canonical systems with rational Weyl functions: explicit formulas and applications*, J. Diff. Eqs. **146** (1998), 375–398.

[K1] M.G. Krein, *Continuous analogues of propositions on polynomials orthogonal on the unit circle*, Dokl. Akad. Nauk SSSR **105** (1955), 637–640.

[K2] M.G. Krein, *Topics in differential and integral equations and operator theory*. OT **7**, Birkhäuser Verlag, 1983.

[L] Z.L. Leibenzon, *The inverse problem of spectral analysis for higher order ordinary differential operators*, Trudy Moskov. Mat. Obšč. **15** (1966), 70–144.

[M] M.M. Malamud, *Spectral analysis of Volterra operators and inverse problems for systems of ordinary operators*. SFB 288, Preprint no. 269, Berlin, 1997.

[P] V.P. Potapov, *The multiplicative structure of J-contractive matrix functions*, Amer. Math. Soc. Transl. **15** (1960), 131–243.

[PW] R. Paley and N. Wiener, *Fourier transforms in the complex domain*. Amer. Math. Soc., NY, 1934.

[SaA1] A.L. Sakhnovich, *A nonlinear Schrödinger equation on the semiaxis and a related inverse problem*, Ukrain. Math. J. **42**:3 (1990), 316–323.

[SaA2] A.L. Sakhnovich, *The N-wave problem on the semiaxis*, Russ. Math. Surveys **46**:4 (1991), 198–200.

[SaA3] A.L. Sakhnovich, *The N-wave problem on the semiaxis*, in: 16 All-Union school on the operator theory in functional spaces (lecture materials), Nydzni Novgorod (1992), 95–114.

[SaA4] A.L. Sakhnovich, *Spectral theory for the systems of differential equations and applications*. Thesis for secondary doctorship, Kiev, Institute of Mathematics, 1992.

[SaL1] L.A. Sakhnovich, *Factorisation problems and operator identities*, Russian Math. Surveys **41** (1986), 1–64.

[SaL2] L.A. Sakhnovich, *Interpolation theory and its applications*. Kluwer, Dordrecht, 1997.

[SaL3] L.A. Sakhnovich, *Spectral theory of canonical differential systems, method of operator identities*. OT, Birkhäuser Verlag, to appear.

[Y] V.A. Yurko, *An inverse problem for differential operators on the halfaxis*, Sov. Math. Iz. VUZ **35** (1991), no. 12, 67–74.

Branch of Hydroacoustics
Marine Institute of Hydrophysics
Academy of Sciences of Ukraine
270100 Odessa
Preobradzenskaya 3
Ukraine

AMS Subject Classification. 34L40, 47A10

Operator Theory:
Advances and Applications, Vol. 117
© 2000 Birkhäuser Verlag Basel/Switzerland

Degenerated Hyperbolic Approximations of the Wave Theory of Elastic Plates

Igor Selezov

A 3-D problem of elastodynamics is formulated for a layer as an initial boundary value (IBV) problem for hyperbolic equations. The problem is investigated to reduce this 3-D problem to a degenerated 2-D problem, i.e. to construct hyperbolically degenerated model with respect to a spatial coordinate so that the corresponding approximate problem would be also of a hyperbolic type. This is a problem of hyperbolic degeneration and it is solved here by means of the power series method and introduction of the rule to truncate the infinite systems. Several hyperbolic models as mathematical approximations for wave motion in elastic plates are obtained without any physical assumptions ordinarily used in the theory of elastic plates.

1 Introduction

It would be difficult to overestimate the contribution of Mark Krein to the development of mathematics in the USSR and throughout the world, as well as his influence on different areas of mathematics. Apart from the pure mathematics, Mark Krein was deeply interested in the problems of applied mathematics and mathematical physics. It is a characteristic feature of Mark Krein to combine ideas of functional analysis with the particular problems of mechanics (Krein and Langer, 1965; Gohberg and Krein, 1967). He made an essential contribution in the development of the method of solving the inverse problems of scattering (Krein, 1953, 1954) which have found wide applications, for example (Selezov, 1971; Selezov and Yakovlev, 1974).

This paper deals with degeneration of the original IBV problem with respect to a spatial coordinate with the purpose of new simpler models development. We consider the problem in a domain, one size of which is much lesser than the others so that it allows to construct power series expansions of desired functions and to reduce consequently the problem dimensionality. However, it is reached by the price of degeneration of the spectrum of original 3-D problem.

For example, consider the IBV problem in R^n for hyperbolic equations in the layer between two hypersurfaces $x^s \in [-h^s, h^s]$ where the coordinate hyperline x^s is orthogonal to hypersurfaces $x^s = -h^s$ and $x^s = h^s$. The thickness of this hyperlayer $2h^s$ is assumed to be much lesser than a characteristic in-hypersurface length l: $\xi = h^s/l \ll 1$. Therefore, we can expand the desired functions in the vicinity of the hypersurface $x^s = 0$ and as a result to decrease the dimensionality of the problem to one. In terms of the theory of operators it is the mapping of the operator P from R^n into R^{n-1} (Hörmander, 1955; Kythe, 1996). In this case operator P is not one-valued but multivalued rather because the selection of approximations in R^{n-1} can be conducted on the basis of different criteria.

In general, the construction of degenerated models of such a kind can be developed by using various methods. The most popular of them is the phenomenological approach-the method of hypotheses and averagings over the coordinate $x^s \in [-h^s, h^s]$. As far as analytical methods are concerned, the asymptotic approaches and the power series method can be noted. Asymptotic methods approximate the operator P as the expansion in small parameter ξ while the power series method represents the operator P in the vicinity $x^s = 0$. The analytical approaches find their beginning with the works of Lagrange (1781) who derived the shallow water wave equation on the basis of asymptotic expansions. The power series method finds its beginning with the works of Cauchy (1828) and Poisson (1829) derived the equations of plate vibration. In the theory of plates and shells the power series method has been used by Krauss (1929), Kilchevsky (1939), Selezov (1960, 1962, 1994) and others.

It should be noted that above-mentioned approaches originate new simpler models which, however, can be of different type compared to the original model. In this paper the problem of degeneration of the original IBV hyperbolic problem with a spatial coordinate is considered as a tool to construct the simpler hyperbolic models of lesser dimensionality, that is to realise hyperbolic degeneration.

The content of the paper can be characterized as follows. Expanding field functions in power series by a degenerated coordinate s with respect to a middle hypersurface yields a degenerate problem to determine the series coefficients depending now only on $n-1$ coordinates. Substituting these power series into hyperbolic PDE equations and boundary and initial conditions yields an infinite system of recurrence relationships. The reduction of this recurrence system can be conducted by different ways allowing to obtain different approximations, i.e. simplified models. The necessary condition is to obtain a closed system to determine new unknown functions. Our aim is to derive hyperbolic approximations degenerated with respect to coordinate s, i.e. to construct a mapping $R^n \to R^{n-1}$ satisfying the condition of limiting correctness to be of hyperbolic type.

The enough condition is announced for obtaining hyperbolic approximations: to keep in infinite systems all space-time differential operators up to given order. This is proved for the case R^3 considering the elastodynamic problem for the plane layer. In this case the combersome infinite systems are splitted in two independent systems, corresponding symmetric and asymmetric fields. As a result, we derive the hyperbolic approximations of the first order for symmetric and asymmetric fields both corresponding to known models, and new more exact hyperbolic approximations of the second order. The obtained approximations include as particular cases known models which don't satisfy the hyperbolity principle (finite velocity principle).

2 Degeneration in Small Parameters and Coordinates

In the Euclidean space R^n with coordinates $x^q (q = \overline{1, n})$ we consider a mathematical model presented by a finite system of partial differential equations, for which a boundary value problem is stated in the region $\Omega \times [0, X^m]$, $X^m > 0$ bounded with hypersurfaces $x^s = \pm h^s$, $h^s > 0$ (subscript s is fixed): $\Omega = \{x \in R^n | -\infty < (x^1, x^2, \dots, x^{s-1}, x^{s+1}, \dots, x^n) < \infty, -h^s \leq x^s \leq h^s\}$. Such a body will be called as the hyperlayer, or, simply, the layer or shell. The model is assumed to depend on a finite number ν of parameters ε_r, $r = \overline{1, \nu}$. Formally, such a model may be given as the system of k partial differential equations (PDEs) of p-th order with k unknowns u_i, $(i = \overline{1, k})$ and n arguments

$$F_i(x^1, \dots, x^n; u_1, \dots, u_k; u_{1,1}, \dots, u_{k,n}; \dots u_{1,\underbrace{1\dots1}_{P \text{ times}}}, \dots, u_{k,\underbrace{n\dots n}_{P \text{ times}}}; \varepsilon_1, \dots, \varepsilon_\nu) = P_i(x^1, \dots, x^n) \quad \text{in} \quad \Omega. \tag{1}$$

Subject the following system of boundary conditions on the hypersurfaces $x^s = -h^s, x^s = h^s$

$$f_j(x^1, \dots, x^n; u_1, \dots, u_k; u_{1,1}, \dots, u_{k,\underbrace{n\dots n}_{(P-1) \text{ times}}}; \varepsilon_1, \dots, \varepsilon_\nu)\Big|_{x^s = \pm h^s} = Q_j^\pm, \; j = \overline{1, (k \cdot p)}. \tag{2}$$

Here the subscript after "comma" denotes differentiation with respect to the corresponding coordinate, in general case $p \neq n$, F_l depends on all the possible partial derivatives up to the p-th order inclusively, the position of the hypersurface may depend on u_i and their derivatives. The solution of the boundary value problem (1)–(2) consists in finding the functions u_i transforming equations (1) into the identities and in choosing from a set of these functions such ones which satisfy the conditions (2).

Let us consider the particular case when the differential equations (1) and the boundary conditions (2) are presented as a sum of linear and nonlinear parts, where the linear operator L is of higher order p than the order p_1 of the nonlinear operator (Courant and Hilbert, 1962)

$$\begin{aligned} &a_{ilq}(x^1, \dots, x^n; \varepsilon_1, \dots, \varepsilon_\nu) \frac{\partial^q u_l}{\partial x^{1(\alpha_1)}, \dots, \partial x^{n(\alpha_n)}} + F_i\Bigg(x^1, \dots, x^n; \\ &u_1, \dots, u_k; \frac{\partial u_1}{\partial x^1}, \dots, \frac{\partial u_k}{\partial x^n}; \frac{\partial^2 u_1}{\partial x^1 \partial x^2}, \dots; \\ &\frac{\partial^{p1} u_1}{\partial x^{1(p1)}}, \dots, \frac{\partial^{p1} u_k}{\partial x^{n(p1)}}; \varepsilon_1, \dots, \varepsilon_\nu\Bigg) = P_i, \quad \text{in} \quad \Omega \quad i = \overline{1, k}, l = \overline{1, k}, \end{aligned} \tag{3}$$

$$q = \alpha_1 + \alpha_2 + \dots + \alpha_n, q = \overline{1, p}, \quad p_1 = \overline{1, (p-1)},$$

$$\begin{aligned} &\Bigg\{b_{jlq}(x^1, \dots, x^n; \varepsilon_1, \dots, \varepsilon_\nu) \frac{\partial^q u_l}{\partial x^{1(\alpha_1)} \dots \partial x^{n(\alpha n)}} + f_j\Bigg(x^1, \dots; \\ &u_1, \dots; \dots \frac{\partial^{p2} u_k}{\partial x^{n(p2)}}; \varepsilon_1, \dots, \varepsilon_\nu\Bigg)\Bigg\}\Bigg|_{x^s = \pm h^s} = Q_j^\pm, \end{aligned} \tag{4}$$

$$j = p \cdot k, q = \overline{1, (p-1)}, p_2 = \overline{1, (p-2)}, l = \overline{1, k}.$$

In contrast to the model of general kind (1), (2) the model (3), (4) can be c lassified by the type of PDEs . If the system of equations (3) is of a hyperbolic type, then a corresponding model will be called as a hyperbolic model. If some physical model corresponds to above model, then it will be called hyperbolic model too.

In the case of well-posed Cauchy problem for hyperbolic model in infinite region Ω, there exist solutions in the form of weak propagating discontinuities (discontinuities of derivatives of the highest order in differential operator). In this case the corresponding physical model describes the propagation of disturbances with finite velocities, that is the principle of finite velocity of propagation of dirturbances is satisfied (Selezov, 1969). And it is in full correspondence with an evidence that in real physical media or systems any disturbance propagates with a finite velocity determined by the properties of the medium or the system. The mathematical statement of the principle says that the solution of the Cauchy problem with completely supported initial data is finite with respect to spatial derivatives at each fixed value of the time coordinate (Kalashnikov, 1969).

It is necessary to note that in some cases the principal part of the operator of hyperbolic equation may not be responsible for energy transport, and a hyperbolic operator of more low order will here be taken as determining one. It is also known that at some boundary conditions the problem of hyperbolic system may not describe the propagation of disturbances, as a consequence of ill-posed Cauchy problem (Hersh, 1964).

It should be noted that most of known principles may, apparently, be included in a more general principle - the principle of causality, which connects phenomenon or change in the state of a system with the union of conditions generating this phenomenon (Bunge, 1962).

A mathematical model, which arises from an original model as degenerated in some parameters, will be called as the simplified (degenerated) mathematical model or the approximation.

Definition 1 The model following from the original model (3), (4) at $\varepsilon_r \to 0$ (or $x^s \to 0$) is called the degenerated model.

It is of interest to consider the cases when $\varepsilon_r \to \varepsilon^o$, $0 < \varepsilon \ll 1$ or $|x^s| \leq h^s$, $0 < h^s \ll 1$. Such models are close to degenerated ones. At that, due to the presence of small parameters ε_r or h_s a possibility arises to construct analytically a simplified model (approximation) replacing an initial model in the vicinity of the parameter of degeneration or the surface of degeneration. Such models will be further called as quasi-degenerated.

Definition 2 If the hyperbolic system of differential equations (3) under degeneration in parameter or coordinate remains hyperbolic one, then such a degeneration is called hyperbolic degeneration. In other case the degeneration will be called non-hyperbolic. In accordance with this definition we will distinguish hyperbolic models and corresponding hyperbolically degenerated or quasi-degenerated models.

It is evident that the construction of degenerated or quasi-degenerated in coordinate models leads to decreasing the dimension of problem. In the case of the degeneration in parameters the problem can be simplified considerably: decreasing the order of a system of differential equations, partial decomposition and so on. This problem can be considered as mapping of partial differential operator from R^n to R^{n-1}.

3 Hyperbolic Degeneration

Let us consider the functions $u_r(x)$ $(r = \overline{1,k})$ in n-dimensional Euclidean space R^n the point of which $(x) = x^1, \dots, x^n$, where $x^1 = t$ is the time coordinate. Consider the region $\Omega \subset R^n$ bounded hypersurfaces $x^s = \pm h^s$, $h^s > 0$ (index s is fixed): $\Omega = \{x \in R^n : -\infty < (x^1, x^2, \dots, x^{s-1}, x^{s+1}, \dots, x^{n-1}) < \infty,$ $x^n \geq 0, -h^s \leq x^s \leq h^s\}$. Consider a finite hyperbolic system of linear partial differential equations in R^n for which a boundary value problem or initial boundary value problem is stated with given real-valued data and real-valued solutions belonged C^k.

The differential equations and boundary conditions are written as follows (Kythe, 1996)

$$\begin{aligned} L_i &\equiv a_{ilq} \frac{\partial^q u_l}{\partial x^{1(\alpha_1)} \dots \partial x^{n(a_n)}} + F_i = P_i \quad \text{in} \quad \Omega,\ t \geq 0, \\ (i,l) &= \overline{1,k}, q = \alpha_1 + \alpha_2 + \cdots \alpha_n,\ q = \overline{1,p}, \end{aligned} \tag{5}$$

$$\begin{aligned} &\left\{ b_{jlq} \frac{\partial^q u_l}{\partial x^{1\alpha} \dots \partial x^{n(\alpha_n)}} + f_j \right\}_{x^s = \pm h^s} \\ &= Q_j^{\pm}, j = p \cdot k, q = \overline{1,(p-1)}, l = \overline{1,k}, \end{aligned} \tag{6}$$

where (5) is the system of k equations of the p-th order with k unknown functions to be determined as the solutions of this system satisfying the boundary conditions (6) and the initial conditions (not presented here), so that a well-posed statement is guaranteed. The first term in (5) is a principal part of operator the second term F_i is the remained part of operator. In (6) the term f_j is the operator of lower order than the first term. The summation convention over repeated subscripts is implied and subscripts preceded by a comma denote partial differentiation with respect to the corresponding coordinate. It is assumed that coefficients a_{ilq} and b_{jlq} are constant but they can depend on small parameter $\varepsilon \ll 1$.

The system (5) can be reduced to the system of equations of the first order (Misokhata, 1965; Bers et al., 1964)

$$\frac{\partial v}{\partial t} - A_\nu \frac{\partial v}{\partial x_\nu} - Bv = p, \quad \nu = 2, 3, \dots, n, \tag{7}$$

where v is the vector having m component $v_1(x,t), \ldots, v_m(x,t)$, $m > k$, p is the vector of similar kind, A and B are matrices of the order m.

It is assumed that the characteristic equation (without lowest terms)

$$det\,(\lambda I - A_\nu \xi_\nu) = 0 \tag{8}$$

has the real roots $\lambda_i(\xi)$ and matrix $A_\nu \xi_\nu$ is reduced to diagonal kind for each $\xi \in R^{n-1}$ that is the necessary condition is satisfied for system (3) to be of hyperbolic type.

Hyperbolicity is necessary but not enough condition for propagation of disturbances with finite velocities. Principle of finite velocity can be violated by boundary conditions including boundary and initial data, and variable coefficients corresponding to inhomogeneity and/or anisotropic properties of media. For example, Hersh (1964) has shown that there are boundary conditions for which the initial problem will be incorrect posed and hence there are no exist solutions with the finite velocities. We consider the boundary conditions for which the IBV problem is correct posed.

The original model (5), (6) is reduced to degenerated model when $\varepsilon \to 0$ or $\xi = h^s/l \to 0$. Our aim is to derive hyperbolic approximations, i.e. to construct a mapping of the reference space $R^n(\varepsilon)$ or $R^n(\xi)$ into the degenerated space, $R^n(\varepsilon) \to R^n$ or $R^n(\xi) \to R^{n-1}$ satisfying the condition of limiting correctness to be of hyperbolic type, that is the condition of the finite velocity of disturbance propagation (Hersh, 1964; Selezov, 1969). At that, in degenerated space instead of the functions $u(x)$ we have new functions $\hat{u}(x)$. In further, we shall consider the case of coordinate degeneration $\xi \to 0$.

Expanding field functions in power series by a degenerated coordinate s with respect to a middle hypersurface yields the degenerated problem to determine the series coefficients depending now only on $n - 1$ coordinates

$$u_i(t,x^2,\ldots,x^{n-1},x^n) = \sum_{k=1}^{\infty} u_{ik}(t,x^2,\ldots,x^{s-1},x^{s+1},\ldots,x^n)(x^s)^k. \tag{9}$$

Substituting (9) into PDE equations (5) and boundary conditions, (6) yields recurrence relationships from equations (5) and several systems of differential equations of infinite order from (6) in R^{n-1}. The next principal step is the reduction of these infinite systems which is possible in different ways keeping the terms according to different regulations. It is possible from recurrence relationships to express all u_{ik} in terms of minimal finite number of desired functions corresponding to a number of the systems of differential equations. Substituting these functions into reduced equations yields the resolving equations. It allows us to obtain different approximations, i.e. simplified models. Regulations of keeping the terms should be of such a kind that this system be of hyperbolic type. The necessary condition is to obtain a closed system to determine new unknown functions.

The sufficient condition of hyperbolic degeneration is announced as follows: in order to obtain the hyperbolic approximation it is enough to keep in infinite systems all space-time differential operators up to definite order.

4 Construction of Hyperbolic Approximations for Elastic Layer

We prove this for the case R^4 considering the elastodynamic problem for plane layer. The mathematical statement of the corresponding IBV problem in terms of displacements $\mathbf{u} = (u_1, u_2, u_3)$ is presented as follows: to find the vector-function $\mathbf{u} = \mathbf{u}(u_1, u_2, u_3, t)$ as a solution of equations in $\Omega \times [0, T]$, $T > 0$ (Selezov, 1994)

$$\nabla^2 u_k + (1 + \lambda/G)\partial_k(\nabla \cdot u) = \partial_{tt} u_k, \quad k = 1, 2, 3 \tag{10}$$

satisfying the boundary conditions

$$\begin{aligned} &\sigma_{33}|_{x_3=\xi/2} = q^+(x_1, x_2, t), \quad \sigma_{33}|_{x_3=-\xi/2} = q^-(x_1, x_2, t), \\ &\sigma_{3i}|_{x_3=\xi/2} = p_i^+(x_1, x_2, t), \quad \sigma_{3i}|_{x_3=-\xi/2} = p_i^-(x_1, x_2, t), \quad i = 1, 2 \end{aligned} \tag{11}$$

and the initial conditions

$$u_k|_{t=0} = 0, \quad \partial_t u_k|_{t=o} = 0, \quad k = 1, 2, 3. \tag{12}$$

Here dimensionless quantities are used according to the formulae (asteriscs are suppressed throughout)

$$\begin{aligned} (x_i^*, u_i^*) &= \frac{1}{l}(x_i, u_i), t^* = \frac{c_s}{l}t, c^* = \frac{c}{c_s}, \quad (\sigma_{kl}^*, q^*, p_i^*) \\ &= \frac{1}{G}(\sigma_{kl}, q, p), \frac{2h}{l} = \xi. \end{aligned}$$

In further such notations are used

$$\begin{aligned} \sigma_{kk} &= (\lambda/G)(\nabla \cdot u) + 2u_{k,k} \quad \text{(no summation over } k), \\ \sigma_{kp} &= u_{k,p} + u_{p,k} \quad (k, p = 1, 2, 3; k \neq p), \\ \nabla \cdot u &= u_{k,k} = e + u_{3,3}, \\ u_3 &= w, L_s = c_s^2\nabla^2 - \partial_{tt}, L_e = c_e^2\nabla^2 - \partial_{tt}, \\ c_s &= \sqrt{G/\rho}, c_e = \sqrt{(\lambda + 2G)/\rho}. \end{aligned} \tag{13}$$

The next essential step is connected with introducing the assumption about smallness of the value ξ and, consequently, the smallness of the transverse coordinate

$$(2h/l) = \xi \ll 1 \Longrightarrow |x_3/l| \sim O(\xi). \tag{14}$$

It follows from (14) that we consider enough smooth disturbances in the planar coordinates – x_1, x_2 and time t.

Components of the displacement vector are presented in the form of power series in x_3

$$u_i(x_1,x_2,x_3,t)=\sum_{\nu=o}^{\infty}u_{i\nu}(x_1,x_2,t)x_3^{\nu},\ i=1,2,3. \tag{15}$$

The functions $u_{i\nu}$ are assumed to be differentiated so many times as it is required, and all the derivatives of $u_{i\nu}$ are continuous, and series (15) converge uniformly. The convergence of these series depends only on the value ξ characterising changes of the fields in the coordinates x_1, x_2, t. The smoother these changes are, the greater is l and the smaller ξ and the faster series (15) are converging.

Substituting (15) into (10)–(12), after long transformations we obtain

$$\sum_{\nu=0}^{\infty}[(\nu+1)u_{1(\nu+1)}+w_{(\nu),1}]x_3^{\nu}|_{x3=\pm 1/2\xi}=p_1^{\pm}, \tag{16}$$

$$\sum_{\nu=0}^{\infty}[(\nu+1)u_{2(\nu+1)}+w_{(\nu),2}]x_3^{\nu}|_{x3=\pm 1/2\xi}=p_2^{\pm}, \tag{17}$$

$$\sum_{\nu=0}^{\infty}\left[\frac{\lambda}{G}u_{1(\nu),1}+\frac{\lambda}{G}u_{2(\nu),2}+(\nu+1)\left(2+\frac{\lambda}{G}\right)w_{(\nu+1)}\right]x_3^{\nu}|_{x3=\pm 1/2\xi}=q^{\pm}, \tag{18}$$

$$\begin{aligned}\sum_{\nu=0}^{\infty}\{&L_s u_{1(\nu)}+(1+\lambda/G)u_{1(\nu),11}+(\nu+2)(\nu+1)u_{1(\nu+2)}\\&+(1+\lambda/G)u_{2(\nu),12}+(\nu+1)(1+\lambda/G)w_{(\nu+1),1}\}x_3^{\nu}=0,\end{aligned} \tag{19}$$

$$\begin{aligned}\sum_{\nu=0}^{\infty}\{&(1+\lambda/G)u_{1(\nu),12}+L_s u_{2(\nu)}+(1+\lambda/G)u_{2(\nu),22}\\&+(\nu+2)(\nu+1)u_{2(\nu+2)}+(\nu+1)(1+\lambda/G)w_{(\nu+1),2}\}x_3^{\nu}=0,\end{aligned} \tag{20}$$

$$\begin{aligned}\sum_{\nu=0}^{\infty}\{&(\nu+1)(1+\lambda/G)u_{1(\nu+1),1}+(\nu+1)(1+\lambda/G)u_{2(\nu+1),2}\\&+L_s w_{(\nu)}+(\nu+2)(\nu+1)(2+\lambda/G)w_{(\nu+2)}\}x_3^{\nu}=0.\end{aligned} \tag{21}$$

Equations (16)–(18) can be satisfied by taking the coefficient at each power of x_3 to be zero. It can be noted that it is possible to differentiate these relations with respect to x_3 and to take $x_3=0$. These two approaches lead to the same result.

It is convenient to add equations (16)–(18) pairwise and then to substract them. Keeping this in mind, we obtain from (16)–(18) two independent infinite systems of differential equations, describing the symmetric and asymmetric vibrations of elastic layer. For symmetric field we have (hereafter s is a subscript of summation, L_s is the operator of shear waves (13))

$$\sum_{s=0}^{\infty}[(2s+2)u_{1(2s+2)} + w_{(2s+1),1}]2^{-(2s+1)}\xi^{2s+1} = \frac{1}{2}(p_1^+ - p_1^-),$$
$$\sum_{s=0}^{\infty}[(2s+2)u_{2(2s+2)} + w_{(2s+1),2}]2^{-(2s+1)}\xi^{2s+1} = \frac{1}{2}(p_2^+ - p_2^-),$$

$$\sum_{s=0}^{\infty}\left[\frac{\lambda}{G}u_{1(2s),1} + \frac{\lambda}{G}u_{2(2s),2} + (2s+1)(2+\lambda/G)w_{(2s+1)}\right]2^{-2s}\xi^{2s}$$
$$= \frac{1}{2}(q^+ + q^-),$$

(22)
$$L_s u_{1(2s)} + (1+\lambda/G)u_{1(2s),11} + (2s+2)(2s+1)u_{1(2s+2)}$$
$$+ (1+\lambda/G)u_{2(2s),12} + (2s+1)(1+\lambda/G)w_{(2s+1),1} = 0,$$

$$(1+\lambda/G)u_{1(2s),12} + L_s u_{2(2s)} + (1+\lambda/G)u_{2(2s),22} + (2s+2)$$
$$\times (2s+1)u_{2(2s+2)} + (2s+1)(1+\lambda/G)w_{(2s+1),2} = 0,$$
$$(2s+2)(1+\lambda/G)u_{1(2s+2),1} + (2s+2)(1+\lambda/G)u_{2(2s+2),2}$$
$$+ L_s w_{(2s+1)} + (2s+3)(2s+2)(2+\lambda/G)w_{(2s+3)} = 0,$$

for asymmetric field

$$\sum_{s=0}^{\infty}[(2s+1)u_{1(2s+1)} + w_{(2s),1}]2^{-2s}\xi^{2s} = \frac{1}{2}(p_1^+ + p_1^-),$$
$$\sum_{s=0}^{\infty}[(2s+1)u_{2(2s+1)} + w_{(2s),2}]2^{-2s}\xi^{2s} = \frac{1}{2}(p_2^+ + p_2^-),$$

(23)
$$\sum_{s=0}^{\infty}[\lambda/G u_{1(2s+1),1} + \lambda/G u_{2(2s+1),2} + (2s+2)(2+\lambda/G)w_{(2s+2)}]$$
$$\times 2^{-(2s+1)}\xi^{2s+1} = \frac{1}{2}(q^+ - q^-),$$
$$L_s u_{1(2s+1)} + (1+\lambda/G)u_{1(2s+1),11} + (2s+3)(2s+2)u_{1(2s+3)}$$
$$+ (1+\lambda/G)u_{1(2s+1),12} + L_s(2s+2)(1+\lambda/G)w_{(2s+2),1} = 0,$$
$$(1+\lambda/G)u_{2(2s+1),12} + L_s u_{2(2s+1)} + (1+\lambda/G)u_{2(2s+1),22} + (2s+3)$$
$$\times (2s+2)u_{2(2s+3)} + (2s+2)(1+\lambda/G)w_{(2s+2),2} = 0,$$
$$(2s+1)(1+\lambda/G)u_{1(2s+1),1} + (2s+1)(1+\lambda/G)u_{2(2s+1),2}$$
$$+ L_s w_{(2s)} + (2s+2)(2s+1)(2+\lambda/G)w_{(2s+2)} = 0.$$

The initial conditions after expanding the functions in the right-had sides into series are formulated on the basis of (12), they have the same form for every sought-for function as (12) and are not given here.

The systems (22) and (23) can be reduced to a more simple form if to introduce e by (13) instead of u_1 and u_2 and to carry out some transformations. Hence, the stated problem is reduced to two infinite systems, but the dimension of the problem is decreased by unit. Eventually, we have for symmetric field

$$e(x_1, x_2, x_3, t) = \sum_{s=0}^{\infty} e_{(2s)}(x_1, x_2, t)x_3^{2s}, \tag{24}$$

$$w(x_1, x_2, x_3, t) = \sum_{s=0}^{\infty} w_{(2s+1)}(x_1, x_2, t)x_3^{2s+1}, \tag{25}$$

$$\begin{aligned} &\sum_{s=0}^{\infty}\left[-\frac{1}{2s+1}L_e e_{(2s)} - \frac{\lambda}{G}\nabla^2 w_{(2s+1)}\right]2^{-(2s+1)}\xi^{2s+1} \\ &= \frac{\partial}{\partial x_1}\frac{1}{2}(p_1^+ - p_1^-) + \frac{\partial}{\partial x_2}\frac{1}{2}(p_2^+ - p_2^-), \end{aligned} \tag{26}$$

$$\sum_{s=0}^{\infty}[(\lambda/G)e_{(2s)} + (2s+1)(2+\lambda/G)w_{(2s+1)}]2^{-2s}\xi^{2s} = \frac{1}{2}(q^+ + q^-), \tag{27}$$

$$e_{(2s+2)} = \frac{-1}{(2s+1)(2s+2)}L_e e_{(2s)} - \frac{1}{2s+2}(1+\lambda/G)\nabla^2 w_{(2s+1)}, \tag{28}$$

$$\begin{aligned} w_{2s+3} = {} & \frac{1}{(2s+1)(2s+2)(2s+3)}\frac{1+\lambda/G}{2+\lambda/G}L_e e_{2s} \\ & + \frac{1}{(2s+2)(2s+3)(2+\lambda/G)} \times [(1+\lambda/G)^2\nabla^2 + L_s]w_{(2s+1)}, \end{aligned} \tag{29}$$

for asymmetric field

$$e(x_1, x_2, x_3, t) = \sum_{s=0}^{\infty} e_{(2s+1)}(x_1, x_2, t)x_3^{2s+1}, \tag{30}$$

$$w(x_1, x_2, x_3, t) = \sum_{s=0}^{\infty} w_{(2s)}(x_1, x_2, t)x_3^{2s}, \tag{31}$$

$$\begin{aligned} &\sum_{s=0}^{\infty}[(2s+1)e_{(2s+1)} + \nabla^2 w_{(2s)}]2^{-2s}\xi^{2s} \\ &= \frac{\partial}{\partial x_1}\frac{1}{2}(p_1^+ + p_1^-) + \frac{\partial}{\partial x_2}\frac{1}{2}(p_2^+ + p_2^-), \end{aligned} \tag{32}$$

$$\text{(33)} \quad \sum_{s=0}^{\infty}\left[-e_{(2s+1)} - \frac{1}{2s+1}L_s w_{(2s)}\right]2^{-(2s+1)}\xi^{2s+1} = \frac{1}{2}(q^+ - q^-),$$

$$\text{(34)} \quad \begin{aligned} w_{(2s+2)} &= -\frac{1}{2s+2}\frac{1+\lambda/G}{2+\lambda/G}e_{2s+1} \\ &\quad -\frac{1}{(2s+1)(2s+2)(2+\lambda/G)}L_s w_{(2s)}, \end{aligned}$$

$$\text{(35)} \quad \begin{aligned} e_{(2s+3)} &= \frac{1}{(2s+2)(2s+3)}\left[-L_e + \frac{(1+\lambda/G)}{(2+\lambda/G)}\nabla^2\right]e_{2s+1} \\ &\quad +\frac{1}{(2s+1)(2s+2)(2s+3)}\frac{1+\lambda/G}{2+\lambda/G}\nabla^2 + L_s w_{(2s)}. \end{aligned}$$

Infinite series (15) for $u_i (i = 1, 2, 3)$, the coefficients of which are determined from the closed systems of equations (24)–(28) and (29)–(33), give an exact solution of the problem. However, instead of three-dimension problem (16)–(21), we solve here two two-dimensional problems (24)–(29) and (30)–(35). Limiting the number of terms in series (24), (25) and (30), (31) leads to truncated finite systems (26), (27) and (32), (33) which are closed with taking into account the recurrence relationships (28), (29) and (34), (35). Therefore, this allows to obtain quasi-degenerated models. It is clear that from relations (24)–(35) various approximations can be obtained. Our task here is to construct quasi-degenerated in coordnate hyperbolic models on the basis of the sufficient condition announced in 3.

Statement: In the case of the second boundary value problem of elastodynamics, to construct hyperbolic approximation of k-th order, it is sufficient to keep all the terms up ξ^{2k+1} inclusively in truncated systems (26)–(27) and (32), (33).

The same result is has been established considering the problem of elastodynamics for a hollow circular cylinder (Selezov, 1962).

5 Simplified Models

The above-mentioned approach for the construction of hyperbolic approximations allows to obtain both the known models obtained earlier on the basis of phenomenelogical approach involving physical assumptions and the new more exact models.

In the case of symmetric deformation with respect to the middle surface we obtain from equations (24)–(29) (when $q_1, q_2 = 0$), as the first approximation the one-mode hyperbolic model, that is so-called equation of generalized plane stress state

$$\text{(36)} \quad (-\xi a_1{}'\nabla^2 + \xi a_2{}'\partial^2/\partial t^2)e_0 = d_1{}'$$

and as the second approximation – new more exact two-mode model

$$
(37)\quad
\begin{aligned}
&\left(-\xi a_1'\nabla^2 + \xi a_2'\frac{\partial^2}{\partial t^2} + \xi^3 b_1'\nabla^2\nabla^2 - \xi^3 b_2'\nabla^2\frac{\partial^2}{\partial t^2} + \xi^3 b_3'\frac{\partial^4}{\partial t^4}\right) e_0 \\
&= \left\{d_1' + \xi^2 d_2'\nabla^2 + \xi^2 d_3'\frac{\partial^2}{\partial t^2}\right\}\left(\frac{\partial}{\partial x_2}\frac{p_1 - p_2}{2} + \frac{\partial}{\partial x_1}\frac{p_3 - p_4}{2}\right).
\end{aligned}
$$

Coefficients in (36) and (37) are of the form

$$
(38)\quad
\begin{aligned}
a_1' &= \frac{4}{1+2\nu}, a_2' = 2\frac{1-\nu}{1-2\nu}, b_1' = \frac{8}{1-2\nu}\frac{1}{3!}, \\
b_2' &= \left(\frac{6}{1-2\nu} + 2\frac{2-\nu}{1-\nu} + 4\frac{1-\nu}{1-2\nu}\right)\frac{1}{3!}, b_3' = \frac{5-8\nu}{1-2\nu}\frac{1}{3!}.
\end{aligned}
$$

In the case of asimmetric deformations we obtain from (30)–(35) (when p_1, p_2, p_3, $p_4 = 0$) as the first approximation the two-mode hyperbolic model

$$
(39)\quad
\begin{aligned}
&\left\{\xi\frac{\partial^2}{\partial t^2} + \xi^3 a_1\nabla^2\nabla^2 - \xi^3 a_2\frac{\partial^2}{\partial t^2}\nabla^2 + \xi^3 a_3\frac{\partial^4}{\partial t^4}\right\} w_0 \\
&= \left\{1 - \xi^2 d_1\nabla^2 + \xi^2 d_2\frac{\partial^2}{\partial t^2}\right\}(q_1 - q_2).
\end{aligned}
$$

This equation is an extension of the classical one-mode parabolic equation predicted by the Kirchhoff theory of plates. The hyperbolic equation of transverse vibrations of beams has been obtained by Timoshenko (1921) on the basis of phenomenological approach. Extension of this theory to bending vibrations of plates has been developed by Ufland (1948) and Mindlin (1951). Both Timoshenko and Mindlin theories include an auxiliary correcting coefficient – shear coefficient, while equation (39) is free from this defect and it includes the exact analytical coefficients depending only on the Poisson ratio.

In the second approximation we obtain new three-mode hyperbolic model

$$
(40)\quad
\begin{aligned}
&\left\{\xi\frac{\partial^2}{\partial t^2} + \xi^3 a_1\nabla^2\nabla^2 - \xi^3 a_2\frac{\partial^2}{\partial t^2}\nabla^2 + \xi^3 a_3\frac{\partial^4}{\partial t^4} - \xi^5 b_1\nabla^2\nabla^2\nabla^2\right. \\
&\quad\left. + \xi^5 b_2\frac{\partial^2}{\partial t^2}\nabla^2\nabla^2 - \xi^5 b_3\frac{\partial^4}{\partial t^4}\nabla^2 + \xi^5 b_4\frac{\partial^6}{\partial t^6}\right\} w_0 \\
&= \left\{1 - \xi^2 d_1\nabla^2 + \xi^2 d_2\frac{\partial^2}{\partial t^2}\right. \\
&\quad\left. + \xi^4 d_3\nabla^2\nabla^2 - \xi^4 d_4\frac{\partial^2}{\partial t^2}\nabla^2 + \xi^4 d_5\frac{\partial^4}{\partial t^4}\right\}(q_1 - q_2).
\end{aligned}
$$

Coefficients in (39) and (40) are determined by formulae

$$
\begin{aligned}
a_1 &= \frac{1}{3}\frac{\lambda+G}{\lambda+2G}, a_2 = \frac{1}{3!}\frac{3\lambda+4G}{\lambda+2G}, a_3 = \frac{1}{4!}\frac{3\lambda+7G}{\lambda+2G},\\
b_1 &= \frac{1}{10\cdot 3!}\frac{\lambda+G}{\lambda+2G}, b_2 = \frac{1}{5!}\frac{4\lambda^2+14G\lambda+11G^2}{(\lambda+2G)^2},\\
b_3 &= \frac{1}{2^2\cdot 5!}\frac{9\lambda^2+39G\lambda+3G^2}{(\lambda+2G)^2}, b_4 = \frac{1}{2^4\cdot 5!}\frac{5\lambda^2+30G\lambda+41G^2}{(\lambda+2G)^2},\\
d_1 &= \frac{1}{8}\frac{3\lambda+4G}{\lambda+2G}, d_2 = \frac{1}{8}, d_3 = \frac{1}{2^4\cdot 4!}\frac{5\lambda^2+16G\lambda+12G^2}{(\lambda+2G)^2},\\
d_4 &= \frac{1}{8\cdot 4!}\frac{2\lambda^2+8G\lambda+7G^2}{(\lambda+2G)^2}, d_5 = \frac{1}{16\cdot 4!}.
\end{aligned}
\tag{41}
$$

Coefficients (41) also depend only on the Poisson ratio because λ and G are proportional to the Young modulus E which is cancelled.

6 Concluding Remarks

1. An approach is developed to construct the degenerated hyperbolic models as mathematical approximations of the original IBV problem for a hyperbolic system of equations. The approach is based on the power series method and the rule to keep in infinite system all spatial-time differential operators up to definite order.

2. Unlike known phenomenological approaches, for realization of this purely analytical approach it is not necessary to attract some physical prerequisites to construct the model.

3. The problem of elastodynamics for a plane layer is considered in detail on the basis of the approach developed. As a result, new hyperbolic models have been obtained which describe the propagation of disturbances with a finite velocity unlike known parabolic models of the theory of plates and shells describing the propagation of disturbances with the infinite velocity.

References

[1] L. Bers, J. Fritz and M. Schechter, *Partial differential equations*. Interscience, New York-London-Sydney, 1964.

[2] M. Bunge, *Causality*. Russian translated edition: ..., 1962.

[3] A.L. Cauchy, *Sur l'equilibre et le mouvement d'une lame solide*. Exercises Math. **3** (1828), 245–326.

[4] R. Courant and D. Hilbert, *Methods of mathematical physics*. vol. **2**. Interscience, New York-London, 1962.

[5] I.C. Gohberg and M.G. Krein, *The theory of Volterra operators in Hilbert space and its application* (In Russian). Moscow, Nauka, 1967.

[6] E.I. Grigoluk and I.T. Selezov, *Nonclassical theories of vibrations of bars, plates and shells* (In Russian) (Advances in Sciences and Engineering. Mechanics of Deforming Solids) Moscow, Acad. Sci. USSR, 1973, 272.

[7] R. Hersh, *Boundary conditions for equations of evolution.* Archive Ration. Mech. and Analysis **16** (1964), N4, 243–264.

[8] L. Hörmander, *On the theory of general partial differential operators.* Acta Mathematica **94** (1955), 161–248.

[9] A.S. Kalashnikov, *On the conception of finite velocity of propagating disturbances* (in Russian). Advances in Math. Sciences **34** (1979), N2, 199–200.

[10] N.A. Kilchevsky, *Extension of the modern theory of shells* (in Russian). Applied Mathematics and Mecanics (Prikladnaya mathematica i mechanica) **2** (1939), N4, 427–438.

[11] F. Krauss, *Über die Grundgleichungen der Elastizitätstheorie schwach deformierter Schalen.* Math. Annalen **101** (1929), N1, 61–92.

[12] M.G. Krein, *On some cases of effective determination of density of inhomogeneous string by its spectral function* (in Russian). Reports of Acad. Sci. USSR **43** (1953), N4, 617–620.

[13] M.G. Krein, *On the method of effective solving of inverse boundary value problem* (in Russia). Reports of Acad. Sci. USSR **44** (1954), N6, 987–990.

[14] M.G. Krein and H. Langer, *On some mathematical principles of the linear theory of damped oscillations of continuum* (in Russian). Proc. Int. Symp. in Tbilisi "Applications of the Theory of the Functions in Mechanics of Continuous Medium". vol. **2**, Moscow, Nauka, 1965.

[15] P.K. Kythe, *Fundamental solutions for differential operators and applications.* Birkhäuser Boston, 1996.

[16] J.L. Lagrange, *Memoire sur la theorie du mouvement des fluides.* Oeuvres, 1781, vol. **4**.

[17] R.D. Mindlin, *Influence of rotatory inertia and shear on flexural motions of isotropic, elastic plates.* J. Appl. Mech. **18** (1951), N1, 31–58.

[18] C. Misokhata, The theory of partial differential equations. 1965. Russian translated edition:. . . , 1977, 504 c.

[19] S.D. Poisson, *Mémoire sur l'equilibre et le mouvement des corps élastiques.* Mém. Acad. Roy. Sci. **8** (1829), 357–570.

[20] I. Selezov, *Investigation of transverse vibrations of plate* (in Ukrainian). Applied Mechanics (Prikladna mechanica) **6** (1960), N3, 319–327.

[21] I.T. Selezov, *On waves in a cylindrical shell* (in Russian). In: "Theory of Plates and Shells". Kiev, Publishing of Ukr. Sci., 1962, 240–253.

[22] I.T. Selezov, *Conception of hyperbolicity in the theory of controlled dynamic systems* (in Russian). In: Cybernetics and Computational Engineering. Kiev, Institute of Cybernetics, Ukr. Acad. Sci., 1969, N1, 131–137.

[23] I.T. Selezov, *On inverse problems of diagnosis of plasma inhomogeneities.* In: "Distributed Control of Processes in Continuum Media", Inst. of Cybernetics, Ukr. Acad. Sci., Kiev, 1971, 22–48.

[24] I.T. Selezov and V.V. Yakovlev, *Inverse problem of scattering of plane electromagnetic wave by plasma inhomogeneity* (in Russian). In: "Linear and Nonlinear Boundary Value Problems in Mathematical Physics", Kiev, Inst. of Mathematics., Ukr. Acad. Sci., 1974, 172–182.

[25] I.T. Selezov, *Modelling of wave and diffraction processes in continuous media*. Kiev, Naukova Dumka, 1989, p. 204.

[26] I.T. Selezov, *Hyperbolic models of wave propagation in bars, plates and shells* (in Russian). Mechanics of Solids. Translated from Izvestia Akademii Nauk Rossii. Mekhanika tverdogo tela, 1994, N2, 64–77.

[27] S.P. Timoshenko, *On the correction for shear of the differential equation for transverse vibrations of prismatic bar*. Phil. Mag. **41** (1921), Ser. 6, N245, 744–746.

[28] Ya.S. Ufland, *Wave propagation under transverse vibrations of rods and plates* (in Russian). Applied Mathematics and Mechanics (Prikladnaya mathematica i mechanica) **12** (1948), N3, 287–300.

Department of Wave Processes
Inst. of Hydromechanics
NAS of Ukraine
8/4 Sheliabov Str.
Kiev, 252057
Ukraine

Classification number: MSI N 35L30

Operator Theory:
Advances and Applications, Vol. 117
© 2000 Birkhäuser Verlag Basel/Switzerland

Elliptic Problems with a Shift in Complete Scales of Sobolev-Type Spaces

Zinovi G. Sheftel

General boundary value problems with a shift for elliptic equations were first studied in [1]; such problems arise for instance by studying certain steady-state oscillations. They are natural generalization of the well-known Carleman boundary value problems for analytic functions. In this contribution we prove the isomorphism theorem for such problems (i.e. the solvability theorem in complete scales of Sobolev-type spaces). In the author's previous paper [2] such theorem was obtained under additional assumption of normality of the boundary conditions. We consider also some applications of the isomorphism theorem, in particular, the local increasing of smoothness of generalized solutions and the existence and smoothness properties of the Green's function of elliptic problem with a shift.

1 Posing of the Problem and some Remarks

Let $G \subset \mathbb{R}^n$ be a bounded domain with boundary $\Gamma \in \mathbb{C}^\infty$ and let $\alpha : \Gamma \to \Gamma$ be a diffeomorphism satisfying the Carleman condition $\alpha(\alpha x) = x$ for any $x \in \Gamma$. Because of the smoothness of Γ the transformation α can be extended into the diffeomorphism of some neighborhood $U(\Gamma)$ in $\mathbb{R}^n$. The transformation α defines in a natural way the shift operator J which transforms any function u defined in $U(\Gamma)$ into the function Ju according to the formula $Ju(x) = u(\alpha x)$. In G a linear differential expression $L(x, D)$ with sufficiently smooth complex-valued coefficients is assigned, $\operatorname{ord} L = 2m$. On Γ linear differential expressions $B_{jr}(x, D)$ $(j = 1, \dots, 2m,\ r = 1, 2)$ with sufficiently smooth complex-valued coefficients are assigned, $\operatorname{ord} B_{jr} \leq 2m + \sigma_j$ (σ_j are given integers). We study the following boundary value problem with a shift in boundary condition:

$$Lu(x) = f(x) \qquad (x \in G), \tag{1.1}$$

$$B_j u := B_{j1}u(x) + JB_{j2}u(x) = \phi_j(x) \quad (x \in \Gamma;\ j = 1, \dots, 2m). \tag{1.2}$$

In addition we assume that the expressions B_{jr} satisfy the following natural condition.

Matching condition. The system of conditions (1.2) is invariant with respect to the replacement of x by αx. More precisely: after such replacement we obtain conditions equivalent to (1.2).

Definition: The problem (1.1), (1.2) is called elliptic [1] if the expression L is properly elliptic in $\overline{G}$ and the expressions B_{jr} satisfy certain assumptions of Lopatinski type.

A priori two cases are possible. **A)** The system of expressions B_{jr} is α-normal (see [2]). This means, roughly speaking, that this system is a part of a system equivalent in certain sense to the system of Cauchy data

$$\begin{pmatrix} D_{\nu,x}^{2m-1} & 0 \\ 0 & D_{\nu,\alpha x}^{2m-1} \\ \dots & \dots \\ 1 & 0 \\ 0 & 1 \end{pmatrix}$$

In this case, in particular, the integers σ_j are negative, i.e. $\operatorname{ord} B_{jr} < 2m$. This case was studied in [2]. Using the normality assumption we can deduce in this case the Green's formula, introduce the adjoint problem and use the transposing procedure. In this case the isomorphism theorem (see Sec. 3) is valid for both the given and adjoint problems. Some applications of these theorems were considered in [2], [3].

B) The expressions B_{jr} satisfy no additional assumptions. This case will be studied in the next sections.

2 Function Spaces

Now we introduce appropriate function spaces. We denote by $H^{s,p}(G)$ ($s \geq 0$, $1 < p < \infty$) the spaces of Bessel potentials, $H^{-s,p}(G)$ is the space dual to $H^{s,p'}(G)$ ($1/p + 1/p' = 1$) with respect to the extension $(\cdot,\cdot)$ of the inner product in $L_2(G)$, $\|\cdot\|_{s,p}$ is the norm in $H^{s,p}(G)$ ($s \in \mathbb{R}$). By $B^{s,p}(\Gamma)$ ($s \in \mathbb{R}$) we denote the Besov spaces; the spaces $B^{s,p}(\Gamma)$ and $B^{-s,p'}(\Gamma)$ are dual to each other with respect to the extension of the inner product in $L_2(\Gamma)$; $\langle\langle\cdot\rangle\rangle$ is the norm in $B^{s,p}(\Gamma)$. Let k be a fixed natural number; for any $s \in \mathbb{R}$, $s \neq l + 1/p$, $l = 0, \dots, k-1$ we denote by $\widetilde{H}^{s,p,k}(G)$ the completion of $C^\infty(\overline{G})$ with respect to the norm

$$|||u|||_{s,p,k} := \left(\|u\|_{s,p}^p + \sum_{j=1}^{k} \langle\langle D_\nu^{j-1} u \rangle\rangle_{s-j+1-1/p,p}^p \right)^{1/p},$$

where $D_\nu = i\partial/\partial\nu$, ν is the unit inner normal vector to Γ. In the case $s = l + 1/p$, $l = 0, \dots, k-1$, these spaces and the corresponding norms are defined by means of complex interpolation. If $s \geq k$ then the norms $|||u|||_{s,p,k}$ and $\|u\|_{s,p}$ are equivalent; therefore for $s \geq k$ we can consider that $\widetilde{H}^{s,p,k}(G) = H^{s,p}(G)$. The spaces $\widetilde{H}^{s,p,k}(G)$ were introduced by Roitberg in [4] and studied in detail in [5]; see also [6,7].

3 Theorem on Isomorphism

The space $\widetilde{H}^{s,p,k}(G)$, defined above, are very convenient for studying boundary value problems. Because of [5,7], for any $s \in \mathbb{R}$ the closure $\Lambda_{s,p}$ of the operator

$$u \longmapsto (Lu, B_1 u, \ldots, B_{2m} u) \quad (u \in C^\infty(\overline{G}))$$

acts continuously from $\widetilde{H}^{2m+s,p,2m+\sigma}(G)$ into the space

$$\widetilde{H}^{s,p,\sigma}(G,\Gamma) := \widetilde{H}^{s,p,\sigma}(G) \times \prod_{j=1}^{2m} B^{s-\sigma_j-1/p,p}(\Gamma), \tag{3.1}$$

where $\sigma = \max\{0, \sigma_1+1, \ldots, \sigma_{2m}+1\}$. It turns out that if problem (1.1), (1.2) is elliptic and has zero kernel and cokernel, then the operator $\Lambda_{s,p}$ for any $s \in \mathbb{R}$ and $1 < p < \infty$ realizes an isomorphism between the spaces $\widetilde{H}^{2m+s,p,2m+\sigma}(G)$ and $\widetilde{H}^{s,p,\sigma}(G,\Gamma)$ (see [3]). In the general case of arbitrary kernel and cokernel the following assertion is valid.

Theorem 3.1 *Let the problem* (1.1), (1.2) *be elliptic. Then for any* $s \in \mathbb{R}$ *and* $1 < p < \infty$ *the above defined operator* $\Lambda_{s,p}$ *is noetherian. In particular its kernel* N *and cokernel* N^* *are finite-dimensional, do not depend on* s *and* p *and consist of infinitely smooth functions.*

The proof of this theorem is based on the construction of regularizers.

If, in particular, $N = N^* = \{0\}$, then we obtain the above mentioned isomorphism between $\widetilde{H}^{2m+s,p,2m+\sigma}(G)$ and $\widetilde{H}^{s,p,\sigma}(G,\Gamma)$. In the case of non-zero kernel and cokernel a similar isomorphism can be obtained by using certain projectors.

4 Applications

4.1 Local Increasing of Smoothness of Generalized Solutions

The solution of the problem (1.1), (1.2) obtained by means of Theorem 3.1 in the case $s < 0$ is naturally called a generalized solution. The theorem on local increasing of smoothness is valid. This theorem states, roughly speaking, the following: let $u \in \widetilde{H}^{2m+s,p,2m+\sigma}(G)$ be a generalized solution of the problem (1.1), (1.2) and let G_0 be a subdomain of G adherent to the part $\Gamma_0 \subset \Gamma$; if $F = (f, \phi_1, \ldots, \phi_{2m})$ locally in G_0 up to Γ_0 pertains to $\widetilde{H}^{s_1,p_1,\sigma}$ $(s_1 \geq s,\ p_1 \geq p)$ then u is also more smooth, namely,

$$u \in \widetilde{H}^{2m+s_1,p_1,2m+\sigma}_{\mathrm{loc}}(G) \qquad \text{(c.f. [7]).}$$

4.2 Existence and Smoothness of Green's Function

Using the methods of the paper [8] and the results of the Section 3 we can establish the existence of Green's vector-function for the problem (1.1), (1.2); this means that there exist functions

$$R_0(x,y) \quad (x, y \in \overline{G}), \qquad R_j(x,y) \quad (x \in \overline{G},\ y \in \Gamma;\ j = 1, \ldots, 2m),$$

infinitely smooth for $x \neq y$, such that the solution of (1.1), (1.2) for $F = (f, \phi_1, \ldots, \phi_{2m}) \in \widetilde{H}^{s,p,\sigma}(G, \Gamma)$ with sufficiently large s can be represented in the form:

$$u(x) = (f, R_0(x, \cdot)) + \sum_{j=1}^{2m} \langle \phi_j, R_j(x, \cdot) \rangle.$$

The approach of the paper [8] also makes possible the investigation of regularity properties of this vector-function with respect to the totality of its variables and the estimate of its singularity for $x = y$.

Acknowledgement

The research was partially supported by the grant INTAS-94-2187.

References

[1] A.B. Antonevich, *On normal solvability of general boundary value problems with a shift for equations of elliptic type* (Russian). Proc. of second conference of mathematicians of Belorussia. Minsk, 1969, pp. 253–255.

[2] Z.G. Sheftel, *Green's formula for general elliptic boundary value problems with a shift and its applications* (Russian). Doclady Nation. Akad. Nauk Ukrainy, 1995, no. 12, pp. 22–25.

[3] Z.G. Sheftel, *On approximation of functions by solutions of elliptic problems with a shift* (Russian). Doklady Nation. Akad. Nauk Ukrainy, 1996, no. 6, pp. 13–16.

[4] Ya.A. Roitberg, *Elliptic problems with non-homogeneous boundary conditions and local increasing of smoothness of generelized solutions up to the boundary* (Russian). Doclady AN SSSR, vol. **157** (1964), no. 4, pp. 798–901.

[5] Ya.A. Roitberg, *On the boundary values of generalized solutions of elliptic equations* (Russian). Matem. Sbornik, vol. **36** (1971), no. 2, pp. 246–267.

[6] Yu.M. Berezansky, *Expansion in eigenfunctions of selfadjoint operators* (Russian). Naukova Dumka, Kiev, 1965, p. 800. English trans.: AMS Transl., vol. **17** (1968), Providence.

[7] Ya.A. Roitberg, *Elliptic boundary value problems in the spaces of distributions*. Kluwer Akademic Publishers, Dordrecht, 1996.

[8] Yu.M. Berezansky and Ya.A. Roitberg, *Theorem on homeomorphisms and Green's function for general elliptic boundary value problems* (Russian). Ukr. Matem. Zhurnal, vol. **19** (1967), no. 5, pp. 3–32.

Z. Sheftel
Pedagogical University
Sverdlova str. 53
250038 Chernigov
Ukraine

1991 AMS Classification: Primary 35R10; Secondary 35J40.

Operator Theory:
Advances and Applications, Vol. 117
© 2000 Birkhäuser Verlag Basel/Switzerland

On the Extremal Regularization of the Variational Inequalities with Multivalued Operators

*O.V. Solonoukha**

In this paper we propose the method of solving of the variational inequalities with multivalued operators by some extremal problems. These extremal problems consist the inclusion with parameter and the penalty function. As examples we consider the free boundary problems on $W_p^1(\Omega)$, $p \geq 3$.

1 Introduction

The variational inequalities theory has emerged as a powerful and effective technique to study a wide class of free boundary problems, equilibrium, nonlinear optimization problems, etc. The variational inequalities have a wide use spectrum in various branches of the pure mathematics and applied sciences, in particular, in economics and engineering. In many researches the object is described by variational inequalities. Wide class of models consist of essentially multivalued ones. For example, the family of control problems which described by variational inequalities, the minimax problems, nondifferential optimal problem etc. are essentially multivalued. For solving of such problems the algorithms which were created for variational inequalities with singlevalued operator may be not applicable. In this paper the method of extremal regularization is extended on multivalued problems subject to topological properties of the multivalued objects.

2 Formulation of Problem

Let X be a reflexive Banach space, X^* be its topological dual space, $\langle \cdot, \cdot \rangle$ be the duality pairing on $X \times X^*$, 2^{X^*} be the totality of all nonempty subset of the space X^*, $K \subset X$ be the closed convex set, A be a multivalued mapping with $\mathrm{Dom}(A) = \{y \in X : A(y) \neq \emptyset\} = X$, the upper and lower support functions which are associated to A are:

$$[A(y), \xi]_+ = \sup_{d \in A(y)} \langle d, \xi \rangle, \quad [A(y), \xi]_- = \inf_{d \in A(y)} \langle d, \xi \rangle.$$

And let $[[A(y)]]_+ = \sup_{d \in A(y)} \|d\|_{X^*}$, $[[\emptyset]]_+ = 0$. (See [11] for details).

*The present research work was partially supported by ISF Grant PSU061103.

We consider the following variational inequality with multivalued operator (VIMO) where $f \in X^*$:

$$(1) \qquad [A(y), \xi - y]_+ \geq \langle f, \xi - y \rangle \quad \forall \xi \in K.$$

We suppose that A is a strong operator, i.e. $\mathrm{Dom}(A) = X$ (by analog we can consider the problem with nonstrong operator if $K \subset \mathrm{Dom}(A)$ is the set of the full dimension).

We denote by $\overline{\mathrm{co}}A(y)$ the minimal convex closed set which contains $A(y)$, also we denote by graph $\overline{\mathrm{co}}A$ the graphic of the operator $\overline{\mathrm{co}}A$.

Definition: The mapping $A : \mathrm{Dom}(A) \to 2^{X^*}$ is

— **locally bounded**, if for arbitrary $y \in \overline{\mathrm{Dom}(A)}$ there exist $\varepsilon > 0$ and $M > 0$, such that $[[A(\xi)]]_+ \leq M$ for each $\xi \in \mathrm{Dom}(A)$ such that $\|\xi - y\|_X \leq \varepsilon$;

— **s-weakly locally bounded**, if for arbitrary $y \in \overline{\mathrm{Dom}(A)}$ and $\{y_n\} \subset \mathrm{Dom}(A)$, where $y_n \to y$ weakly in X, there exist the subsequence $\{y_{n_k}\}$ and $N > 0$ such that $[[A(y_{n_k})]]_+ \leq N$;

— **generalized pseudomonotone**, if for arbitrary $(y_n, w_n) \in$ graph $\overline{\mathrm{co}}A$, such that $y_n \to y$ weakly in X, $w_n \to w$ weakly in X^* and $\overline{\lim}_{n\to\infty}\langle w_n, y_n - y\rangle \leq 0$, we have that $w \in \overline{\mathrm{co}}A(y)$ and $\langle w_n, y_n \rangle \to \langle w, y \rangle$;

— **quasi-bounded**, if there exists $\zeta \in int\, \mathrm{Dom}(A)$ such that for each $(y, w) \in$ graph $\overline{\mathrm{co}}A$, which satisfy the estimates $\|y\|_X \leq k_1$ and $\langle w, y - \zeta \rangle \leq k_2$, we have $\|w\|_{X^*} \leq N < \infty$.

— **operator of lower semibounded variation**, if $\forall y_i \in X$ such that $\|y_i\|_X \leq R$, $i = 1, 2$ the inequality

$$(2) \qquad [A(y_1), y_1 - y_2]_- \geq [A(y_2), y_1 - y_2]_+ - C(R, \|y_1 - y_2\|'_X)$$

holds, where $\|\cdot\|'$ is the compact seminorm with respect to $\|\cdot\|$, the function $C : \mathbf{R}_+ \times \mathbf{R}_+ \to \overline{\mathbf{R}} := \mathbf{R} \cup \{+\infty\}$ is continuous, and $\forall h, R > 0$ we have $\frac{1}{\tau}C(R, \tau h) \to 0$ as $\tau \to 0$;

— **radially semicontinuous**, if $\forall y \in \mathrm{Dom}(A)$, $\xi \in X$ and $\tau \in (0, \tau_0]$ such that $y + \tau\xi \in \mathrm{Dom}(A)$ for some τ_0 we have that $\underline{\lim}_{\tau\to+0}[A(y + \tau\xi), -\xi]_+ \geq [A(y), -\xi]_-$;

— **coercive**, if there is $y_0 \in K$ such that $[A(y), y - y_0]_- \geq \gamma(\|y\|_X)\|y\|_X$, where $\gamma(s) \to +\infty$ as $s \to \infty$;

— **has the property (M)**, if for arbitrary $(y_n, w_n) \in$ graph $\overline{\mathrm{co}}A$, such that $y_n \to y$ weakly in X, $w_n \to w$ weakly in X^* and $\overline{\lim}_{n\to\infty}\langle w_n, y_n - y \rangle \leq 0$, we have that $w \in \overline{\mathrm{co}}A(y)$.

Definition: The function $F : X^* \times X \to \overline{\mathbf{R}}$ is **coercive**, if $F(v, y) \to +\infty$ as $\|y\|_X \to \infty$ and(or) $\|v\|_{X^*} \to \infty$.

3 Main Construction

Let the operator $A : X \to 2^{X^*}$ be generalized pseudomonotone and s-weakly locally bounded. Moreover let K be bounded or A be coercive. Then VIMO has at least one solution for each $f \in X^*$ (see [12]). Additionally, let A be quasi-bounded.

We construct the penalty function:

$$F(y, v) = \sup_{\xi \in B(y)} \langle v, y - \xi \rangle + \beta(y),$$

where y is the solution of the inclusion $v + f \in \overline{co}A(y)$,

$$B(y) = \begin{cases} \overline{co}(\{y\} \cup \{\xi \in K : \|\xi - y\|_X \leq 1\}), & \text{as } \inf_{\xi \in K} \|\xi - y\|_X \leq 1, \\ \emptyset, & \text{as } \inf_{\xi \in K} \|\xi - y\|_X > 1, \end{cases}$$

$\sup_{\xi \in \emptyset} \langle v, y - \xi \rangle \equiv +\infty$; $\beta(y) = I_K(y)$ is the indicator of the set K. Since $\sup_{\xi \in \emptyset} \langle v, y - \xi \rangle = +\infty$ and $\sup_{\xi \in B(y)} \langle v, y - \xi \rangle \geq \langle v, y - y \rangle = 0$ as $B(y) \neq \emptyset$, then the function F is lower bounded, $F(v, y) \geq 0$. Let us consider properties of this function.

Proposition 1 *Let the mapping $A : X \to 2^{X^*}$ be quasi-bounded. And let K be bounded or A be coercive. Then $F(v, y)$ is coercive.*

Proof: If K is bounded then $F(v, y)$ is coercive since $\sup_{\xi \in \emptyset} \langle v, y - \xi \rangle \equiv +\infty$. In another case for some auxiliary element $y_\lambda = y - C(y - y_0)$, $y_\lambda \in B(y)$ we have

$$\begin{aligned} F(v, y) &\geq \sup_{\xi \in B(y)} \langle v, y - \xi \rangle \geq \langle v, y - y_\lambda \rangle \geq \|y - y_0\|_X^{-1} [A(y) - f, y - y_0] - \\ &\geq \hat{C}(\gamma(\|y\|_X) - \|f\|_{X^*}) \end{aligned}$$

Thus $F(v, y) \to \infty$ as $\|y\|_X \to \infty$.

If the sequence $\{(v_i, y_i)\}$ minimizes the values of the function F then $\{y_i\}$ is bounded. Thus, it is sufficiently to show that on bounded set $B_{k_1} := \{y \in K : \|y\|_X \leq k_1\}$ for elements $v \in \overline{co}A(y) - f$ such that $F(v, y) \leq k_2$, we have that $\|v\|_{X^*} < \infty$. It follows because A is quasi-bounded and we have the estimate $\langle v, y - \xi \rangle \leq F(v, y) \leq k_2$. □

Now we can prove that the solutions sets for VIMO and for following problem

$$\overline{co}A(y) \ni f + v, \quad v \in X^*, \tag{3}$$

$$F(v, y) = \sup_{\xi \in B(y)} \langle v, y - \xi \rangle + \beta(y) \longrightarrow \inf \tag{4}$$

coincide.

Lemma 1 *Let the problems* (1) *and* (3)–(4) *have solutions. Then y is a solution of* (1) *if and only if there exists v such that (y, v) is a solution of* (3)–(4).

Proof: Let y be a solution of (1) . Then $\langle v, \xi - y\rangle \geq 0$ for arbitrary $\xi \in K$ and $\beta(y) = 0$, i.e. $F(v, y) = 0$. But $F(v, y) \geq 0$. Thus (y, v) is the solution of (3)–(4).

Let (y, v) be a solution of (3)–(4). Then $F(v, y) = 0$, $\sup_{\xi\in B}\langle v, y - \xi\rangle = 0$ and $I_K(y) = 0$. Thus, $y \in K$. Let us suppose that there exists $\xi \in K$ such that $\langle v, y - \xi\rangle > 0$. And let $\xi_\lambda = \lambda\xi + (1 - \lambda)y$, $\lambda \in (0, 1)$. Then $\langle v, y - \xi_\lambda\rangle = \langle v, y - \lambda\xi - (1 - \lambda)y\rangle = \langle v, \lambda y - \lambda\xi\rangle = \lambda\langle v, y - \xi\rangle$. Hence, the function $\langle v, y - \xi\rangle$ remains true its sign constant along each ray outgoing from y. Since K is convex, an arbitrary ray from $y \in K$ intersects with unit ball the center of which is at y If $\langle v, y - \xi\rangle > 0$ then there exists $\xi_\lambda \in K$ such that $\langle v, y - \xi_\lambda\rangle > 0$. We obtain the contradiction with $\langle v, y - \xi\rangle \leq 0$, y is the solution of (1). □

The problem (1) is solvable. Let us show that the regularized problem (3)–(4) is solvable too.

Theorem: *Let $K \subset X$ be a closed convex set, the mapping $A : X \to 2^{X^*}$ be generalized pseudomonotone, quasi-bounded and s-weakly locally bounded. Moreover, K be bounded or A be coercive. Then for all $f \in X^*$ the problem* (3)–(4) *has at least one solution.*

Proof: If A is coercive or K is bounded, then for each $v \in X^*$ there exists the solution of (3) (see [13]). The function F is coercive and lower bounded (see Proposition 1). Thus, there exists the sequence $\{(v_n, y_n)\}$ such that $F(v_n, y_n) \to 0$. Let us prove that each weakly convergent subsequence of $\{(v_n, y_n)\}$ converges to solution, i.e. if $\{(v_m, y_m)\} \subset \{(v_n, y_n)\}$, $v_m \to v$ weakly on X^*, $y_m \to y$ weakly on X, then $v \in \overline{\text{co}}A(y) - f$ and y is the solution of (1). As we shown in Proposition 1, the sequences $\{v_n\}$ and $\{y_n\}$ are bounded and the set $\{\xi_n\}$ (where $\xi_n = \text{argsup}\sup_{\xi\in B(y_n)}\langle v_n, y_n - \xi\rangle$) is bounded by constructing: $\xi_n \in B(y_n)$, i.e. $\|\xi_n - y_n\|_X \leq 1$. Thus, on reflexive Banach spaces there exist the weakly convergent subsequences: $(y_m, v_m) \to (y, v)$ weakly on $X \times X^*$, and $\overline{\lim}_{m\to\infty}\langle v_m, y_m - y\rangle \leq \lim_{m\to\infty} F(v_m, y_m) = 0$. And since A is generalized pseudomonotone, we obtain that $\langle v_n, y_n\rangle \to \langle v, y\rangle$ and $v + f \in \overline{\text{co}}A(y)$. Let us assume that $y \notin K$. But K is weakly closed, i.e. there exist the neighborhood $U_y \subset X \setminus K$ in weak topology X and the number N such that $\{y_m\}_{m\geq N} \subset U_y$, i.e. $\lim_{m\to\infty}\beta(y_m) = \beta(y) = 1$. Hence, y_m can not belong to minimizing sequence. Therefore, $\beta(y) = 0$. Let us verify the first summand:

$$\begin{aligned}\sup_{\xi\in B(y)}\langle v, y - \xi\rangle &= \langle v, y - \hat{\xi}\rangle = \sup_{\xi\in B(y)}\lim_n\langle v_n, y_n - \xi\rangle \\ &\leq \sup_{\xi\in B(y)}\lim_n \sup_{\zeta\in B(y_n)}\langle v_n, y_n - \zeta\rangle \leq \lim_n F(v_n, y_n) = 0.\end{aligned}$$

The last estimate is obvious if there exists $n_1(\hat{\xi})$ such that $\hat{\xi} \in B(y_n)$ for each $n \geq n_1$. Else there exists the sequence $\zeta_n \in B(y_n)$ such that $y - \zeta_n \to y - \hat{\xi}$ weakly on X and $\|y - \zeta_n\|_X \to \|y - \hat{\xi}\|_X \leq 1$, i.e. $y - \zeta_n \to y - \hat{\xi}$ on X. Hence,

$\lim_n \langle v_n, y_n - \zeta_n \rangle = \langle v, y - \hat{\xi} \rangle$. Thus, y belong to K and satisfies VIMO (1), $F(v, y) = 0$. □

This methods is applicable also if the mapping A is quasi-bounded, regular generalized pseudomonotone and coercive (see [3, 14] for details).

Remark 1 It is obviously, that we can use arbitrary function β which satisfies the following conditions:

i) $\beta : X \to \overline{\mathbf{R}_+}$ is weakly lower semicontinuous;

ii) $\beta(y) = 0$ if $y \in K$; $\beta(y) > 0$ if $y \notin K$;

iii) $\beta(y) \to \infty$ if dist (y, K)$\to \infty$.

Remark 2 If K is the cone with corner at y_0, then the variational inequality defines the complementary problem find the pair (y, d), where

$$y \in K(y_0),\ d \in K^*(y_0) \cap \overline{\mathrm{co}} A(y),$$
$$K^*(y_0) = \{w \in X^* : \langle w, \zeta - y_0 \rangle \geq 0 \forall \zeta \in K\}$$

Thus, y is the solution of VIMO if and only if $[A(y) - f, y - y_0]_+ \geq 0$ and we can modify the penalty function

$$F(v, y) = |\langle v, y - y_0 \rangle| + \beta(y).$$

4 Additional Properties of Operators of Lower Semibounded Variation

The important class of quasi-bounded, s-weakly locally bounded, generalized pseudomonotone mappings is the class of radially semicontinuous operators of lower semibounded variation (see [13]). In this section we constrict the set of parameter for regularization of VIMO with the operators of this class.

Let there exists the reflexive Banach space X_1 such that $X \subset X_1$ compactly and densely, $\|\cdot\|'_X = \|\cdot\|_{X_1}$. We need to study the properties of selectors of $\overline{\mathrm{co}} A$ which belong to X_1^*, X_1^* is dual to X_1.

Proposition 2 *Let $A : X \to 2^{X^*}$ be a coercive, radially semicontinuous operator of lower semibounded variation, y be the solution of* (1), *$f \in X_1^*$. Then $\overline{\mathrm{co}} A(y) \cap X_1^* \neq \emptyset$.*

Proof: Let $J : X_1 \to X_1^*$ be the duality. Because the space X_1 is reflexive Banach one, there exists the equivalent norm such that X_1 is a strongly convex space in sense of this norm, i.e. J is monotone ($C \equiv 0$ in (2)) and demicontinuous (continuous from strong topology of X_1 to weak topology of X_1^*), see [10]. We can show that there exists $\varepsilon_1 > 0$ such that for each $\varepsilon_1 \geq \varepsilon > 0$ the inclusion

$$J(y_\varepsilon - y) + \varepsilon \overline{\mathrm{co}} A(y_\varepsilon) \ni 0 \tag{5}$$

has at least one solution. The map J is generated by some function Ψ, where $\Psi : \mathbf{R}_+ \to \mathbf{R}_+$, $\Psi(0) = 0$ and $\Psi(s) \to \infty$ as $s \to \infty$:

$$J(y) = \{g \in X_1^* : \langle g, y\rangle = \|g\|_{X_1^*}\|y\|_{X_1} \text{ and } \|g\|_{X_1^*} = \Psi(\|y\|_{X_1})\}.$$

Thus, $J(\cdot - y) + \varepsilon A(\cdot)$ is coercive, since

$$\begin{aligned}[J(\xi - y) + \varepsilon A(\xi), \xi - y_0]_- &\geq \varepsilon\gamma(\|\xi\|_X) \\ \|\xi\|_X + \Phi(\|\xi - y\|_{X_1})\|\xi - y_0\|_{X_1} &\longrightarrow \infty\end{aligned}$$

as $\|\xi\|_X \to \infty$. Therefore, $J + \varepsilon A$ is a radially continuous operator of lower semibounded variation. Hence for all $\varepsilon > 0$ the inclusion (5) is solvable. Since A is coercive, the solution of (1) is bounded, $\|y\|_{V_1} \leq k_1$. Since $J(\cdot - y) + \varepsilon A(\cdot)$ is coercive, there exists a bounded solution of (5), $\|y_\varepsilon\|_{X_1} \leq k_2$. Hence, $\varepsilon[A(y_\varepsilon), y - y_\varepsilon]_+ \geq [J(y - y_\varepsilon), y - y_\varepsilon]_- = \|y - y_\varepsilon\|^2_{X_1}$. Moreover,

$$\begin{aligned}[A(y), y_\varepsilon - y]_+ &\leq [A(y_\varepsilon), y_\varepsilon - y]_- + C(R, \|y_\varepsilon - y\|_{X_1}) \\ &\leq -\varepsilon^{-1}\|y_\varepsilon - y\|^2_{X_1} + C(R, \|y_\varepsilon - y\|_{X_1}),\end{aligned}$$

i.e. $\varepsilon^{-1}\|y_\varepsilon - y\|^2_{X_1} \leq [A(y), y - y_\varepsilon]_- - C(R, \|y_\varepsilon - y\|_{X_1}) \leq \langle f, y - y_\varepsilon\rangle - C(R, \|y_\varepsilon - y\|_{X_1})$. We obtain that

$$\varepsilon^{-1}\|y_\varepsilon - y\|_{X_1} \leq \|f\|_{X_1} - \|y_\varepsilon - y\|^{-1}_{X_1}C(R, \|y_\varepsilon - y\|_{X_1})$$

If $\|y_\varepsilon - y\|_{X_1} \to 0$ as $\varepsilon \to 0$, then $\|y_\varepsilon - y\|^{-1}_{X_1}C(R, \|y_\varepsilon - y\|_{X_1}) \to 0$, and since $C(R; \tau)$ is continuous, we obtain that $\varepsilon^{-1}(y - y_\varepsilon)$ is bounded in X_1. Otherwise, we can choose the subsequences $\{\varepsilon_n\}$ such that $\|y_{\varepsilon_n} - y\|_{V_2} \geq C_2$. Since continuous functions C is bounded on the bounded interval $[C_2, C_1]$, there exists the estimate $C(R, \|y_\varepsilon - y\|_{V_2}) \leq C_3$, i.e.

$$\varepsilon^{-1}\|y_\varepsilon - y\|_{V_2} \leq \|f\|_{X_1^*} + \frac{C_3}{C_2} < \infty.$$

The sequence $\{y_\varepsilon\}$ is bounded. We obtained that $\overline{\text{co}}A(y_\varepsilon) \cap X_1^* \neq \emptyset$ and $y_\varepsilon \to y$ weakly in X_1. But A is s-weakly locally bounded. Thus, we can consider the subsequence $\overline{\text{co}}A(y_{\varepsilon_m}) \ni d_{\varepsilon_m} \to d$ weakly in X_1^*. Since $X \subset X_1$ compactly, then $\overline{\lim}_{y_{\varepsilon_m} \to y} \langle d_{\varepsilon_m}, y_{\varepsilon_m} - y\rangle = 0$. And by property (M) $\overline{\text{co}}A(y) \ni d$. □

Proposition 3 *Let $A : X \to 2^{X^*}$ be coercive, radially semicontinuous operator of lower semibounded variation. Then there exists the dense subset $\widetilde{X}$ of X such that $\overline{\text{co}}A(y) \cap X_1^* \neq \emptyset$ for arbitrary $y \in \widetilde{X}$.*

Proof: Let us consider the inequality on arbitrary convex closed set K:

$$[A(y), \xi - y]_+ \geq 0 \quad \forall\xi \in K.$$

For each K this inequality has at least one solution $y \in K$. Using the Proposition 2 we obtain that $\overline{\text{co}}A(y) \cap X_1^* \neq \emptyset$, $y \in K$. □

Thus, we can find all of the solutions of (1) if we would know all of the solutions of the following problem

$$\overline{\mathrm{co}}A(y) \ni f + v, \quad v \in X_1^*,$$
$$F(v) = \sup_{\xi \in B(y)} \langle v, y - \xi \rangle + \beta(y) \longrightarrow \inf .$$

In the Proposition 2 we proved that there exists the element v such that $v + f \in \overline{\mathrm{co}}A(y)$ and $F(v, y) = 0$, and in the Proposition 3 we proved that there exists the convergent to v in X_1^* subsequences.

The problem has at least one solutions as $f \in X_1^*$. If $f \in X^* \setminus X_1^*$ then we can modify the problem using the auxiliary mapping $\widetilde{A}(y) := \overline{\mathrm{co}}A(y) - f$:

$$\widetilde{A}(y) \ni v, \quad v \in X_1^*,$$
$$F(v, y) = \sup_{\xi \in B(y)} \langle v, y - \xi \rangle + \beta(y) \longrightarrow \inf .$$

5 Example

Let we consider the free-boundary problem on bounded set Ω with the regular boundary Γ, ν is a normal vector to Γ, $y \in W_p^1(\Omega)$, $p \geq 3$, $f \in W_q^{-1}(\Omega)$, $1/q + 1/p = 1$:

$$-\sum_{i,j=1}^{n} a_{ij}(x, y) \frac{\partial^2 y}{\partial x_i \partial x_j} = f, \tag{6}$$

$$y_{|\Gamma} \geq 0, \quad \frac{\partial y}{\partial \nu_A} = \sum_{i,j=1}^{n} a_{ij}(x, y) \frac{\partial y}{\partial x_j} \cos(x, \nu_i) \geq 0, \quad y \frac{\partial y}{\partial \nu_A} = 0, \tag{7}$$

where ν is the vector of normal. Let the functions a_{ij} satisfy the following conditions:

a) $a_{ij}(\cdot, y)$ are measurable for each $y \in \mathbf{R}^1$;

b) $a_{ij}(x, \cdot)$ are absolutely continuous at almost all $x \in \Omega$ and locally Lipschitz with respect to y almost everywhere;

c) for almost all $x \in \Omega$ and for each $y \in \mathbf{R}$

$$|a_{ij}(x, y)| \leq \alpha_{ij}^1(x) + \alpha_{ij}^2(x) y \quad \text{or} \quad |a_{ij}(x, y)| \leq \alpha_{ij}^1(x) + C|y|^{p-2}$$

where $\alpha_{ij}^1 \in L_{q'}(\Omega)$, $\alpha_{ij}^2 \in L_{q''}(\Omega)$, $\frac{1}{q'} + \frac{2}{p} = 1$, $\frac{1}{q''} + \frac{3}{p} = 1$, $C > 0$;

d) if at $x \in \Omega$ there exist the partial differentials $\frac{\partial}{\partial x_i} a_{ij}(x, y)$, then

$$\left| \frac{\partial}{\partial x_i} a_{ij}(x, y) \right| \leq \beta_{ij}^{1i}(x) + \beta_{ij}^{2i}(x) y \quad \text{or} \quad \left| \frac{\partial}{\partial x_i} a_{ij}(x, y) \right| \leq \beta_{ij}^{1i}(x) + C'|y|^{p-2}$$

where $\beta_{ij}^{1i} \in L_{q'}(\Omega)$, $\beta_{ij}^{2i} \in L_{q''}(\Omega)$, $C' > 0$;

e) $|a_{ij}(x, y)| \geq C(1 + |y|^{p-2})$ for sufficiently large $|y| >> 1$.

The analogous conditions are considered in [1, 4, 9].

Then the functions $a_{ij}(x, y)\xi_i\xi_j$ are integrable (by properties a)-)). Let

$$\partial_{x_i} a_{ij}(x, y) = \left\{ w | w\delta x_i \leq \overline{\lim_{z \to x, \lambda \to +0}} \frac{a_{ij}(z + \lambda\delta x_i, y) - a_{ij}(z, y)}{\lambda} \right\},$$

$$\partial_y a_{ij}(x, y) = \left\{ w | wh \leq \overline{\lim_{\zeta \to y, \lambda \to +0}} \frac{a_{ij}(x, \zeta + \lambda h) - a_{ij}(x, \zeta)}{\lambda} \right\}$$

be partial generalized gradients. By b) and by Rademarcher's theorem [3, 4] the functions $a_{ij}(\cdot, y) : \mathbf{R^n} \to \overline{\mathbf{R}}$ are differentiable almost everywhere. Let these functions $a_{ij}(\cdot, y)$ are not differentiable on the set $\Gamma^x_{a_{ij}} \subset \Omega$. Then

$$\partial_{x_i} a_{ij}(x, y) = \overline{\text{co}} \left\{ \lim_{x^m \to x} \frac{\partial}{\partial x_i} a_{ij}(x^m, y) : x^m \longrightarrow x,\ x^m \notin S_x \cup \Gamma^x_{a_{ij}} \right\},$$

where $S_x \subset \Omega$ is fixed set the Lebesgue measure of which is zero (see [1]). Let $\Gamma^y_{a_{ij}}$ be the set on which the functions $a_{ij}(x, \cdot) : \mathbf{R} \to \overline{\mathbf{R}}$ are not differentiable. Then

$$\partial_y a_{ij}(x, \hat{y}) = \overline{\text{co}} \left\{ \lim_{y^m \to \hat{y}} \frac{\partial}{\partial y} a_{ij}(x, y^m) : y^m \longrightarrow \hat{y},\ y^m \notin S_y \cup \Gamma^y_{a_{ij}} \right\}.$$

Hence in the generalized sense

$$\frac{\partial}{\partial x_i}\left(a_{ij}(x, y)\frac{\partial y}{\partial x_j}\right) - a_{ij}(x, y)\frac{\partial^2 y}{\partial x_i \partial x_j} \in \partial_{x_i} a_{ij}(x, y)\frac{\partial y}{\partial x_j} + \partial_y a_{ij}(x, y)\frac{\partial y}{\partial x_i}\frac{\partial y}{\partial x_j},$$

thus,

$$f \in -\frac{\partial}{\partial x_i}\left(a_{ij}(x, y)\frac{\partial y}{\partial x_j}\right) + \partial_{x_i} a_{ij}(x, y)\frac{\partial y}{\partial x_j} + \partial_y a_{ij}(x, y)\frac{\partial y}{\partial x_i}\frac{\partial y}{\partial x_j}. \tag{8}$$

The second and third composeds of right part are measurable [1], the first is integrable (by properties a)-)). ([1, 4]). Since

$$|\nabla a_{ij}(x^n, y)\xi_i\xi_j| \leq \beta_{ij}^1(x^n)y + \beta_{ij}^2(x^n)(\xi_i + \xi_j) + \beta_{ij}^3(x^n) \quad \text{a.e. } x^n \in S_x$$

by property d); the second summand of right part of (8) is integrable. For each $y \notin \Gamma^y_{a_{ij}}$ and $x \notin S_x$ $|\nabla_y a_{ij}(x, y)(h)| \leq \tilde{\alpha}_{ij}^1(x)h$ by property c). Hence, the third summand is integrable too.

Thus, we can consider the integral multivalued form on $W_p^1(\Omega)$:

$$
\begin{aligned}
a(y,\xi) &= \sum \int_\Omega a_{ij}(x,y)\frac{\partial y}{\partial x_j}\frac{\partial \xi}{\partial x_i}dx + \sum \int_\Omega \partial_{x_i} a_{ij}(x,y)\frac{\partial y}{\partial x_j}\xi dx \\
&+ \sum \int_\Omega \partial_y a_{ij}(x,y)\frac{\partial y}{\partial x_i}\frac{\partial y}{\partial x_j}\xi dx.
\end{aligned}
$$

We denote $a(y,\xi) = a_1(y,\xi) + \langle A_2(y),\xi\rangle + \langle A_3(y),\xi\rangle$. Then

$$
\begin{aligned}
&|a_1(y_1, y_1 - y_2) - a(y_2, y_1 - y_2)| \\
&\le \sum_{i,j=1}^n \int_\Omega \left| \left(a_{ij}(x,y_1)\frac{\partial y_1}{\partial x_j} - a_{ij}(x,y_2)\frac{\partial y_2}{\partial x_j}\right) \frac{\partial (y_1 - y_2)}{\partial x_i}\right| dx \\
&\le \sum_{i,j=1}^n \int_\Omega \left| a_{ij}(x,y_1)\frac{\partial y_1}{\partial x_j} - a_{ij}(x,y_2)\frac{\partial y_2}{\partial x_j}\right| \left| \frac{\partial (y_1 - y_2)}{\partial x_i}\right| dx \\
&\le K(R)\|y_1 - y_2\|_{L_p(\Omega)}
\end{aligned}
$$

because the functions $a_{ij}(x,\cdot)$ are continuous almost everywhere and them arguments are bounded: $\|y\|_{W_p^1(\Omega)} = \|y\|_{L_p(\Omega)} + \sum_{j=1}^n \|\frac{\partial y}{\partial x_j}\|_{L_p(\Omega)} \le R$. Hence, the first summand is generated by operator of lower semibounded variations A_1. Moreover, the mapping A_1 is radially semicontinuous:

$$
\begin{aligned}
&\lim_{\tau \to +0} \sum_{i,j=1}^n \int_\Omega a_{ij}(x, y+\tau h)\frac{\partial (y+\tau h)}{\partial x_j}\frac{\partial h}{\partial x_i}dx \\
&= \lim_{\tau \to +0} \sum_{i,j=1}^n \int_\Omega a_{ij}(x,y+\tau h)\frac{\partial y}{\partial x_j}\frac{\partial h}{\partial x_i}dx + \lim_{\tau\to +0} \tau \sum_{i,j=1}^n \int_\Omega \\
&\quad a_{ij}(x,y+\tau h)\frac{\partial h}{\partial x_j}\frac{\partial h}{\partial x_i}dx = \sum_{i,j=1}^n \int_\Omega a_{ij}(x,y)\frac{\partial y}{\partial x_j}\frac{\partial h}{\partial x_i}dx.
\end{aligned}
$$

Let us study the properties of the operator A_2.

$$
\begin{aligned}
&[A_2(y_1), y_1 - y_2]_- - [A_2(y_2), y_1 - y_2]_+ \\
&= \sum_{i,j=1}^n \inf_{w_{ijk}\in \partial_{x_i} a_{ij}(x,y_k)} \int_\Omega \left(w_{ij1}\frac{\partial y_1}{\partial x_j} - w_{ij2}\frac{\partial y_2}{\partial x_j}\right)(y_1 - y_2)dx \\
&\ge -\left(\int_\Omega \sum_{i,j=1}^n \inf_{w_{ijk}\in \partial_{x_i} a_{ij}(x,y_k)} \left| w_{ij1}\frac{\partial y_1}{\partial x_j} - w_{ij2}\frac{\partial y_2}{\partial x_j}\right|^q dx\right)^{1/q} \\
&\quad \times \left(\int_\Omega |y_1 - y_2|^p dx\right)^{1/p} \ge -C(R)\|y_1 - y_2\|_{L_p(\Omega)},
\end{aligned}
$$

because $\inf_{w_{ijk}\in\partial_{x_i}a_{ij}(x,y_k)} w_{ijk}$ are bounded by Lipschitz constant ([4]). Thus, A_2 is the operator of lower semibounded variations. Moreover, the mapping A_2 is radially semicontinuous because $a_{ij}(x,\cdot)$ are continuous almost everywhere.

It still to remain the properties of A_3. Analogously,

$$[A_3(y_1), y_1 - y_2]_- - [A_3(y_2), y_1 - y_2]_+ = \sum_{i,j=1}^{n} \inf_{w_{ijk}\in\partial_y a_{ij}(x,y_k)}$$

$$\int_\Omega \left(w_{ij1}\frac{\partial y_1}{\partial x_i}\frac{\partial y_1}{\partial x_j} - w_{ij2}\frac{\partial y_2}{\partial x_i}\frac{\partial y_2}{\partial x_j}\right)(y_1 - y_2)dx = \sum_{i,j=1}^{n} \inf_{w_{ijk}\in\partial_y a_{ij}(x,y_k)} \int_\Omega$$

$$\left((w_{ij1} - w_{ij2})(y_1 - y_2)\frac{\partial y_1}{\partial x_i}\frac{\partial y_1}{\partial x_j} + w_{ij2}(y_1 - y_2)\left(\frac{\partial y_1}{\partial x_i}\frac{\partial y_1}{\partial x_j} - \frac{\partial y_2}{\partial x_i}\frac{\partial y_2}{\partial x_j}\right)\right)dx$$

$$\geq \sum_{i,j=1}^{n} \inf_{w_{ij2}\in\partial_y a_{ij}(x,y_2)} \int_\Omega w_{ij2}(y_1 - y_2)\left(\frac{\partial y_1}{\partial x_i}\frac{\partial y_1}{\partial x_j} - \frac{\partial y_2}{\partial x_i}\frac{\partial y_2}{\partial x_j}\right)dx$$

$$\geq -\left(\int_\Omega \sum_{i,j=1}^{n} \inf_{w_{ij2}\in\partial_y a_{ij}(x,y_2)} \left| w_{ij2}\left(\frac{\partial y_1}{\partial x_j}\frac{\partial y_1}{\partial x_j} - \frac{\partial y_2}{\partial x_j}\frac{\partial y_2}{\partial x_j}\right)\right|^q dx\right)^{1/q}$$

$$\times\left(\int_\Omega |y_1 - y_2|^p dx\right)^{1/p} \geq -C(R)\|y_1 - y_2\|_{L_p(\Omega)}$$

and

$$\varliminf_{\tau\to+0} [A_3(y+\tau h), -h]_+ = \varliminf_{\tau\to+0}\ \varlimsup_{z\to y+\tau h,\lambda\to+0}$$

$$\sum_{i,j=1}^{n} \int_\Omega \frac{a_{ij}(x, z-\lambda h) - a_{ij}(x,z)}{\lambda}\frac{\partial y}{\partial x_i}\frac{\partial y}{\partial x_j}dx$$

$$\geq \varlimsup_{\lambda\to+0}\ \varliminf_{\tau\to+0}\ \varlimsup_{z\to y+\tau h} \sum_{i,j=1}^{n} \int_\Omega \frac{a_{ij}(x, z-\lambda h) - a_{ij}(x,z)}{\lambda}\frac{\partial y}{\partial x_i}\frac{\partial y}{\partial x_j}dx$$

$$= \varlimsup_{\lambda\to+0}\ \varlimsup_{z\to y} \sum_{i,j=1}^{n} \int_\Omega \frac{a_{ij}(x, z-\lambda h) - a_{ij}(x,z)}{\lambda}\frac{\partial y}{\partial x_i}\frac{\partial y}{\partial x_j}dx$$

$$= [A_3(y), -h]_+ \geq [A_3(y), -h]_-.$$

Hence, A_3 is radially semicontinuous operator of lower semibounded variations.

Thus, the operator $A = A_1 + A_2 + A_3$ is coercive by property e) and there exists at least one solution of the inequality

$$\sup_{d \in a(y,\xi - y)} d \geq \int_{\Omega} f(\xi - y)dx \qquad \forall \xi \in W_p^1(\Omega)$$

which we call the weak solution of (6)–(7) on $W_p^1(\Omega)$.

For solving of this inequality we use the proposed method:

$$-\frac{\partial}{\partial x_i}\left(a_{ij}(x,y)\frac{\partial y}{\partial x_j}\right) + \partial_{x_i} a_{ij}(x,y)\frac{\partial y}{\partial x_j} + \partial_y a_{ij}(x,y)\frac{\partial y}{\partial x_i}\frac{\partial y}{\partial x_j} \ni f \tag{9}$$

$$\frac{\partial y}{\partial \nu_A} = \sum_{i,j=1}^{n} a_{ij}(x,y)\frac{\partial y}{\partial x_j}\cos(x,\nu_i) = v, \tag{10}$$

$$F(v,y) = \langle v, |y|\rangle_\Gamma + C\|y^-\|_X \longrightarrow \inf, \tag{11}$$

where $v \in L_q(\Gamma)$, $v \geq 0$.

References

[1] J.-P. Aubin and H. Frankowska, *Set-Valued Analysis.* Birkhauser, 1990.

[2] V. Barbu, *Analysis and control of non-linear infinite dimensional systems.* Acad. Press, Inc., 1995.

[3] F.E. Browder and P. Hess, *Nonlinear Mappings of Monotone Type in Banach Spaces.* J. Func. Anal., vol. **11**, N2 (1972), pp. 251–294.

[4] F. Clarce, *Optimization and Nonsmooth Analysis.*, John Willey & Sons, Inc., 1983.

[5] Yu.A. Dubinsky, *Nonlinear elliptic and parabolic equations.* Itogi nauki i techniki: VINITI. Sovremennye problemy matematiki, vol. **9** (1976), pp. 5–130. (in Russian).

[6] I. Ekeland and R. Temam, *Convex Analysis and Variational Problems.* North-Holand Publ. Company, 1976.

[7] V.I. Ivanenko and V.S. Melnik, *Variational Methode in Control Problems for Distributed Systems.* Kiev: Naukova dumka (in Russian), 1988.

[8] N.N. Ladyzhenskaya and O.A. Uraltseva, *Linear and quasilinear elliptic type equations.* Moskou: Nauka (in Russian), 1973.

[9] G.I. Laptev, *First boundary-value problem for quasilinear elliptic equation with double degeneration.* Differential equation, vol. **30**, no. 6 (1994), pp. 1057–1068. (in Russian).

[10] J.-L. Lions, *Quelques Methodes de Resolution de Problemes aux Limities Non Lineaires.* Paris: Dunod, 1969.

[11] V.S. Mel'nik and O.V. Solonoukha, *On stationary variational inequality with multivalued operators.* ibernetica i systemny analis, no. 3 (1997), pp. 74–89. (in Russian).

[12] O.V. Solonoukha, *On Extremal Problem of the Variational Inequalities Regularization.* Applied Mathematics and Computer Science, vol. **6**, no. 4 (1996). pp. 733–752.

[13] O.V. Solonoukha, *On the Stationary Variational Inequalities with the Generalized Pseudomonotone Operators.* Method of functional analysis and Topology, vol. **3**, no. 4 (1997), pp. 81–95.

[14] O.V. Solonoukha, *On Extremal Regularization of Optimization Problems which are Described by Variational Inequalities with Multivalued Operators.* Kibernetika i vychislitelnaya technika, vol. **118** (1998) (in Russian).

O.V. Solonoukha
Department of Mathematical
Simulation of Economical Systems
National Technical University of Ukraine "KPI"
Pr. Pobedy, 37
Kiev, 252056
Ukraine

1991 Mathematics Subject Classification 47H04, 49J40, 35R35, 35R70

Operator Theory:
Advances and Applications, Vol. 117
© 2000 Birkhäuser Verlag Basel/Switzerland

Poly-Fock Spaces*

N.L. Vasilevski

Consider the space $L_2(\mathbb{C}^n, d\mu_n)$, where $d\mu_n$ is the Gaussian measure, and its Fock subspace $F^2(\mathbb{C}^n)$ consisting of all analytic (entire) functions in $\mathbb{C}^n$. We introduce the so-called true-poly-Fock spaces, and prove that $L_2(\mathbb{C}^n, d\mu_n)$ is the direct sum of the Fock and all true-poly-Fock spaces.

The structure of these spaces and connections between them are described. The orthogonal (Bargmann type) projections onto true-poly-Fock spaces are given.

1 Introduction

We will use the following standard notations: $z = x + iy = (z_1, \ldots, z_n) \in \mathbb{C}^n$, $\overline{z} = (\overline{z}_1, \ldots, \overline{z})$ with the usual notion of the complex conjugation, for $z, w \in \mathbb{C}^n$ $z \cdot w = z_1 w_1 + \cdots + z_n w_n$, $|z|^2 = |z_1|^2 + \cdots + |z_n|^2$ $(= z \cdot \overline{z})$. Denote by $d\mu_n(z)$ the following Gaussian measure over $\mathbb{C}^n$

$$d\mu_n(z) = \pi^{-n} e^{-z \cdot \overline{z}} dv(z),$$

where $dv(z) = dxdy$ is the usual Euclidean volume measure on $\mathbb{C}^n = \mathbb{R}^{2n}$.

Introduce the Hilbert space $L_2(\mathbb{C}^n, d\mu_n)$ of μ_n square-integrable functions on $\mathbb{C}^n$ with the following inner product

$$\langle f, \varphi \rangle = \int_{\mathbb{C}^n} f(z)\, \overline{\varphi(z)} d\mu_n(z).$$

The closed subspace $F^2(\mathbb{C}^n)$ of $L_2(\mathbb{C}^n, d\mu_n)$ consisting of all analytic (entire) functions is usually called the Fock [3, 4], or the Segal-Bargmann space [1, 6].

The orthogonal Bargmann projection

$$P_n : L_2(\mathbb{C}^n, d\mu_n) \longrightarrow F^2(\mathbb{C}^n)$$

is given by the formula [3]

$$(P_n\varphi)(z) = \int_{\mathbb{C}^n} \varphi(\zeta) e^{\overline{\zeta} \cdot z} d\mu_n(\zeta).$$

Note, that for each natural numbers k and l, with $n = k + l$, we have

$$\begin{aligned} L_2(\mathbb{C}^n, d\mu_n) &= L_2(\mathbb{C}^k, d\mu_k) \otimes L_2(\mathbb{C}^l, d\mu_l), \\ F^2(\mathbb{C}^n) &= F^2(\mathbb{C}^k) \otimes F^2(\mathbb{C}^l), \end{aligned}$$

*This work was partially supported by CONACYT Project 3114P-E9607, México.

and

$$P_n = P_k \otimes P_l.$$

Thus it is natural to start with the spaces over one-dimensional complex space $\mathbb{C}$. To have then results for the spaces over $\mathbb{C}^n$ one needs just "to tensor" the one-dimensional results.

In the article we answer the following questions:

- being a subspace of $L_2(\mathbb{C}^n, d\mu_n)$, how much room does the Fock space $F^2(\mathbb{C}^n)$ occupy inside $L_2(\mathbb{C}^n, d\mu_n)$,
- does there exist a direct sum decomposition of $L_2(\mathbb{C}^n, d\mu_n)$ onto analytic type spaces,
- what are the structure of these pieces of decomposition, and what are the connections between them,
- how do the orthogonal projections onto these pieces look like,
- does there exist an unitary operator *simultaneously* reducing all these pieces and corresponding projections to some simple transparent form?

It is fitting to note here that it was M.G. Krein himself, who first called my attention to the beauty of the Fock spaces and Berezin machinery, and kindly provided me with abundant information on these topics.

2 One-dimensional Case

Consider the space $L_2(\mathbb{C}, d\mu)$ of square-integrable functions on $\mathbb{C}$ with respect to the Gaussian measure

$$d\mu(z) = \pi^{-1} e^{-z\cdot\overline{z}} dv(z),$$

where $dv(z) = dxdy$ is the Euclidean volume measure on $\mathbb{C} = \mathbb{R}^2$, and its Fock subspace $F^2(\mathbb{C})$, consisting of all analytic in $\mathbb{C}$ functions.

Besides the Fock space $F^2(\mathbb{C})$, which can be defined alternatively as the closure of the set of all smooth functions satisfying the (Cauchy-Riemann) equation

$$\frac{\partial \varphi}{\partial \overline{z}} = 0,$$

introduce the poly-Fock spaces, i.e., given $k \in \mathbb{N}$, the k-Fock space $F_k^2(\mathbb{C})$ is the closure of the set of all smooth functions from $L_2(\mathbb{C}, d\mu)$ satisfying the equation

$$\left(\frac{\partial}{\partial \overline{z}}\right)^k \varphi = 0.$$

Introduce the unitary operator

$$U_1 : L_2(\mathbb{C}, d\mu) \longrightarrow L_2(\mathbb{R}^2) = L_2(\mathbb{R}^2, dxdy),$$

by the rule

$$(U_1\varphi)(z) = \pi^{-\frac{1}{2}} e^{-\frac{z\cdot\bar{z}}{2}} \varphi(z),$$

or

$$(U_1\varphi)(x, y) = \pi^{-\frac{1}{2}} e^{-\frac{x^2+y^2}{2}} \varphi(x + iy).$$

Then the image $F_k^{(1)} = U_1(F_k^2(\mathbb{C}))$ of the k-Fock space $F_k^2(\mathbb{C})$ is the set of all smooth functions in $L_2(\mathbb{R}^2)$ which satisfy the equation

$$\begin{aligned} D_k^{(1)} f &= U_1 \left(\frac{\partial}{\partial \bar{z}}\right)^k U_1^{-1} f = \left(\frac{\partial}{\partial \bar{z}} + \frac{z}{2}\right)^k f \\ &= \frac{1}{2^k} \left(\frac{\partial}{\partial x} + i\frac{\partial}{\partial y} + x + iy\right)^k f = 0. \end{aligned}$$

The unitary operator $U_2 = I \otimes F$, where

$$(F\psi)(y) = (2\pi)^{-1/2} \int_{\mathbb{R}} e^{-i\eta\cdot y} \psi(\eta)\, d\eta$$

is the Fourier transformation, maps isometrically the space

$$L_2(\mathbb{R}^2, dxdy) = L_2(\mathbb{R}, dx) \otimes L_2(\mathbb{R}, dy)$$

onto itself. The image $F_k^{(2)} = (I \otimes F)(F_k^{(1)})$ of the space $F_k^{(1)}$ under the mapping $I \otimes F$ is the closure of the set of all smooth functions in $L_2(\mathbb{R}^2)$ which satisfy the equation

$$\begin{aligned} D_k^{(2)} f &= (I \otimes F) D_k^{(1)} (I \otimes F^{-1}) f \\ &= \frac{1}{2^k} \left(\frac{\partial}{\partial x} - y + x - \frac{\partial}{\partial y}\right)^k f = 0. \end{aligned}$$

Finally introduce thc isomorphism

$$U_3 = U_3^* = U_3^{-1} : L_2(\mathbb{R}^2) \longrightarrow L_2(\mathbb{R}^2),$$

where

$$(U_3 f)(x, y) = f\left(\frac{1}{\sqrt{2}}(x + y), \frac{1}{\sqrt{2}}(x - y)\right).$$

The operator U_3 maps the space $F_k^{(2)}$ onto the space $F_k^{(3)}$, which is the closure of the set of the smooth functions satisfying the equation

$$D_k^{(3)} f = U_3 D_k^{(2)} U_3^{-1} f = 2^{-k/2} \left(\frac{\partial}{\partial y} + y\right)^k f = 0.$$

This last equation can be easily solved. Its general solution has the form

$$e^{-y^2/2}\sum_{j=0}^{k-1} y^j d_j(x),$$

where $d_j(x)$, $j = \overline{0, k-1}$, are function from $L_2(\mathbb{R}, dx)$. Or, rearranging the polynomials on y, we have

$$\sum_{j=0}^{k-1}(2^j j!\sqrt{\pi})^{-1/2} H_j(y)\, e^{-y^2/2} g_j(x),$$

where

$$H_j(y) = (-1)^j e^{y^2}\left(\frac{d}{dy}\right)^j e^{-y^2} := j!\sum_{m=0}^{[j/2]}\frac{(-1)^m(2y)^{j-2m}}{m!\,(j-2m)!} \tag{2.1}$$

is the Hermite polynomial of degree j (see, for example [2, 7]), and $g_j(x)$, $j = \overline{0, k-1}$, are function from $L_2(\mathbb{R}, dx)$.

The functions

$$h_j(y) = (2^j j!\sqrt{\pi})^{-1/2} H_j(y)\, e^{-y^2/2},\ j = 0, 1, 2, \ldots \tag{2.2}$$

form an orthonormal basis in $L_2(\mathbb{R})$. Denote by H_j the one-dimensional subspace of $L_2(\mathbb{R})$, generated by the function $h_j(y)$, and by Q_j the one-dimensional orthogonal projection of $L_2(\mathbb{R})$ onto H_j, which is given obviously by

$$(Q_j\psi)(y) = h_j(y)\int_{\mathbb{R}}\psi(\eta)\, h_j(\eta)\, d\eta. \tag{2.3}$$

Let

$$H_k^{\oplus} = \bigoplus_{j=o}^{k-1} H_j.$$

Now we have

$$F_k^{(3)} = L_2(\mathbb{R}) \otimes H_k^{\oplus}.$$

The above work leads up to the following theorem.

Theorem 2.1 *The unitary operator $U = U_3U_2U_1$ provides an isometric isomorphism of the space $L_2(\mathbb{C}, d\mu)$ onto the space $L_2(\mathbb{R}^2) = L_2(\mathbb{R}, dx) \otimes L_2(\mathbb{R}, dy)$, under which the k-Fock space $F_k^2(\mathbb{C})$ is mapped onto $L_2(\mathbb{R}) \otimes H_k^{\oplus}$*

$$U : F_k^2(\mathbb{C}) \longrightarrow L_2(\mathbb{R}) \otimes H_k^{\oplus}.$$

It is convenient to introduce the spaces

$$\begin{aligned} F^2_{(k)}(\mathbb{C}) &= F^2_k(\mathbb{C}) \ominus F^2_{k-1}(\mathbb{C}), \quad \text{for } k > 1, \\ F^2_{(1)}(\mathbb{C}) &= F^2_1(\mathbb{C}) = F^2(\mathbb{C}), \quad \text{for } k = 1. \end{aligned}$$

We will call the space $F^2_{(k)}(\mathbb{C})$ the *true-k-Fock space.* Introduce as well the true-k-Bargmann projection $P_{(k)}$ as the orthogonal projection on $L_2(\mathbb{C})$ with image $F^2_{(k)}(\mathbb{C})$.

Corollary 2.2 *Under the isometric isomorphism*

$$U : L_2(\mathbb{C}, d\mu) \longrightarrow L_2(\mathbb{R}, dx) \otimes L_2(\mathbb{R}, dy)$$

1. *the true-k-Fock space $F^2_{(k)}(\mathbb{C})$ is mapped onto $L_2(\mathbb{R}) \otimes H_{k-1}$*

$$U : F^2_{(k)}(\mathbb{C}) \longrightarrow L_2(\mathbb{R}) \otimes H_{k-1},$$

2. *the true-k-Bargmann projection $P_{(k)}$ is unitary equivalent to the following one*

$$U P_{(k)} U^{-1} = I \otimes Q_{k-1},$$

where the one-dimensional projection Q_{k-1} is given by (2.3).

Corollary 2.3 *For the Fock space $F^2(\mathbb{C})$ we have*

$$\begin{aligned} U : F^2(\mathbb{C}) &\longrightarrow L_2(\mathbb{R}) \otimes H_0, \\ U P U^{-1} &= I \otimes Q_0. \end{aligned}$$

Corollary 2.4 *We have the following decomposition of the space $L_2(\mathbb{C}, d\mu)$*

$$L_2(\mathbb{C}, d\mu) = \bigoplus_{k=1}^{\infty} F^2_{(k)}(\mathbb{C}).$$

All the true-k-Fock spaces $F^2_{(k)}(\mathbb{C})$ are isomorphic, one to each other, and each of them is naturally isomorphic to $L_2(\mathbb{R})$. Describe these isomorphisms. Given $k \in \mathbb{N}$, introduce the isometric imbedding

$$R_k : L_2(\mathbb{R}, dx) \longrightarrow L_2(\mathbb{R}, dx) \otimes L_2(\mathbb{R}, dy)$$

defined by

$$R_k : g(x) \longmapsto g(x) \cdot h_{k-1}(y),$$

where the function $h_{k-1}(y)$ is given by (2.2). The image of R_k is exactly $L_2(\mathbb{R}) \otimes H_{k-1}$. Then the adjoint operator

$$R_k^* : L_2(\mathbb{R}^2) \longrightarrow L_2(\mathbb{R})$$

is defined by

$$(R_k^* f)(x) = \int_{\mathbb{R}} f(x, y)\, h_{k-1}(y)\, dy.$$

And we have

$$R_k^* R_k = I \;:\; L_2(\mathbb{R}, dx) \longrightarrow L_2(\mathbb{R}, dx),$$
$$R_k R_k^* = I \otimes Q_k \;:\; L_2(\mathbb{R}^2) \longrightarrow L_2(\mathbb{R}) \otimes H_{k-1}.$$

Now, the operator $\widetilde{R}_k = R_k^* U$ maps the space $L_2(\mathbb{C}, d\mu)$ onto $L_2(\mathbb{R}, dx)$ and the restriction

$$\widetilde{R}_k|_{F^2_{(k)}(\mathbb{C})} : F^2_{(k)}(\mathbb{C}) \longrightarrow L_2(\mathbb{R}, dx)$$

is an isometric isomorphism.

The adjoint operator

$$\widetilde{R}_k^* = U^* R_k : L_2(\mathbb{R}, dx) \longrightarrow F^2_{(k)}(\mathbb{C}) \subset L_2(\mathbb{C}, d\mu)$$

maps isomorphically and isometrically the space $L_2(\mathbb{R}, dx)$ onto the true-k-Fock space $F^2_{(k)}(\mathbb{C})$.

Theorem 2.5 *The isomorphism*

$$\widetilde{R}_k^* : L_2(\mathbb{R}, dx) \longrightarrow F^2_{(k)}(\mathbb{C})$$

is given by the formula

$$\begin{aligned}(\widetilde{R}_k^* g)(z) &= (2^{k-1}(k-1)!\sqrt{\pi})^{-1/2} \int_{\mathbb{R}} e^{-(z^2+\eta^2)/2+\sqrt{2}z\eta} \\ &\quad H_{k-1}\left(\frac{z+\overline{z}}{\sqrt{2}} - \eta\right) g(\eta)\, d\eta,\end{aligned}$$

where $H_{k-1}(y)$ is the Hermite polynomial (2.1).

Proof: Calculate

$$\begin{aligned}(\widetilde{R}_k^* g)(z) &= (U_1^{-1} U_2^{-1} U_2^{-1} R_k g)(z) \\ &= U_1^{-1}(I \otimes F^{-1}) g\left(\frac{x+y}{\sqrt{2}}\right) h_{k-1}\left(\frac{x-y}{\sqrt{2}}\right) \\ &= \pi^{1/2} e^{(x^2+y^2)/2} (2\pi)^{-1/2} \int_{\mathbb{R}} e^{iy\eta} g\left(\frac{x+\eta}{\sqrt{2}}\right) h_{k-1}\left(\frac{x-\eta}{\sqrt{2}}\right) d\eta \\ &= (2^{k-1}(k-1)!\sqrt{\pi})^{-1/2} \int_{\mathbb{R}} e^{x^2/2+y^2/2+iy(\sqrt{2}x-\eta)-(\sqrt{2}x-\eta)^2/2} \\ &\quad g(\eta) H_{k-1}(\sqrt{2}x-\eta) d\eta \\ &= (2^{k-1}(k-1)!\sqrt{\pi})^{-1/2} \int_{\mathbb{R}} e^{-(z^2+\eta^2)/2+\sqrt{2}z\eta} \\ &\quad H_{k-1}\left(\frac{z+\overline{z}}{\sqrt{2}} - \eta\right) g(\eta) d\eta.\end{aligned}$$

□

Corollary 2.6 ([1]). *For the Fock space case the isomorphism*

$$\widetilde{R}^* = \widetilde{R}_1^* : L_2(\mathbb{R}, dx) \longrightarrow F^2(\mathbb{C})$$

is given by the formula

$$(\widetilde{R}^* g)(z) = \pi^{-1/4} \int_{\mathbb{R}} e^{-(z^2+\eta^2)/2+\sqrt{2}z\eta}\, g(\eta)\, d\eta.$$

Remark 2.7 The operators $\widetilde{R}_k$ and $\widetilde{R}_k^*$ provide the following decompositions of the true-k-Bargmann projection $P_{(k)}$ and of the identity operator on $L_2(\mathbb{R})$

$$P_{(k)} = \widetilde{R}_k^* \widetilde{R}_k \ : \ L_2(\mathbb{C}, d\mu) \longrightarrow F_{(k)}^2(\mathbb{C}),$$
$$I = \widetilde{R}_k \widetilde{R}_k^* \ : \ L_2(\mathbb{R}) \longrightarrow L_2(\mathbb{R}).$$

To describe the isomorphisms between the true-poly-Fock spaces $F_{(k)}^2(\mathbb{C})$ start with the spaces $F_{(k)} = L_2(\mathbb{R}) \otimes H_{k-1}$. In $L_2(\mathbb{R}, dy)$ introduce the creation and annihilation operators

$$a^\dagger = \frac{1}{\sqrt{2}}\left(-\frac{d}{dy} + y\right), \quad a = \frac{1}{\sqrt{2}}\left(\frac{d}{dy} + y\right),$$

and, in the space $L_2(\mathbb{R}^2) = L_2(\mathbb{R}, dx) \otimes L_2(\mathbb{R}, dy)$, the operators

$$\mathbf{a}^\dagger = I \otimes a^\dagger, \quad \mathbf{a} = I \otimes a.$$

It is known (see, for example, [7]), that on functions $h_k(y)$ of the form (2.2) we have

$$a^\dagger h_k(y) = \sqrt{k+1}\, h_{k+1}(y),$$
$$a h_k(y) = \sqrt{k}\, h_{k-1}(y).$$

Thus the operator

$$\frac{1}{\sqrt{k}} \mathbf{a}^\dagger|_{F_{(k)}} : F_{(k)} \longrightarrow F_{(k+1)}$$

is an isometric isomorphism, and the operator

$$\frac{1}{\sqrt{k}} \mathbf{a}|_{F_{(k+1)}} : F_{(k+1)} \longrightarrow F_{(k)}.$$

is its inverse. Iterating these isomorphisms we get the following lemma.

Lemma 2.8 *Given natural numbers $k < l$, the operator*

$$\sqrt{\frac{(k-1)!}{(l-1)!}} (\mathbf{a}^\dagger)^{l-k}|_{F_{(k)}} : F_{(k)} \longrightarrow F_{(l)}$$

is an isometric isomorphism, and the operator

$$\sqrt{\frac{(k-1)!}{(l-1)!}}\, \mathbf{a}^{l-k}|_{F_{(l)}} : F_{(l)} \longrightarrow F_{(k)}$$

is its inverse.

Introduce now the operators

$$\mathfrak{a}^{\dagger} = -\frac{\partial}{\partial z} + \overline{z}, \quad \mathfrak{a} = \frac{\partial}{\partial \overline{z}} \tag{2.4}$$

in the space $L_2(\mathbb{C}, d\mu)$.

Theorem 2.9 *Given natural numbers $k < l$, the isometric isomorphism between the true-poly-Fock spaces $F^2_{(k)}(\mathbb{C})$ and $F^2_{(l)}(\mathbb{C})$ is given by the operator*

$$\sqrt{\frac{(k-1)!}{(l-1)!}}(\mathfrak{a}^{\dagger})^{l-k}|_{F^2_{(k)}(\mathbb{C})} : F^2_{(k)}(\mathbb{C}) \longrightarrow F^2_{(l)}(\mathbb{C}),$$

and the operator

$$\sqrt{\frac{(k-1)!}{(l-1)!}}\mathfrak{a}^{l-k}|_{F^2_{(l)}(\mathbb{C})} : F^2_{(l)}(\mathbb{C}) \longrightarrow F^2_{(k)}(\mathbb{C})$$

gives the inverse isomorphism.

Proof: We have $F^2_{(k)}(\mathbb{C}) = U^{-1}(F_{(k)})$, for each $k \in \mathbb{N}$, and the direct calculation shows that

$$\mathfrak{a}^{\dagger} = U^{-1}\mathbf{a}^{\dagger}U, \quad \mathfrak{a} = U^{-1}\mathbf{a}U.$$

To finish the proof apply Lemma 2.8. □

Corollary 2.10 *The true-k-Fock space $F^2_{(k)}(\mathbb{C})$ consists of all the functions of the form*

$$\psi(z) = \psi(z, \overline{z}) = \sum_{m=0}^{k-1}(-1)^m \frac{\sqrt{(k-1)!}}{m!(k-1-m)!}\overline{z}^{k-1-m}\varphi^{(m)}(z), \tag{2.5}$$

where $\varphi(z) \in F^2(\mathbb{C})$, $\varphi^{(m)}$ is the m-th derivative of the (analytic) function φ, and

$$\|\psi\|_{F^2_{(k)}(\mathbb{C})} = \|\varphi\|_{F^2(\mathbb{C})}.$$

3 Basic Operators

Besides the operators (2.4) introduce

$$\overline{\mathfrak{a}}^{\dagger} = -\frac{\partial}{\partial \overline{z}} + z, \quad \overline{\mathfrak{a}} = \frac{\partial}{\partial z} \tag{3.1}$$

in the space $L_2(\mathbb{C}, d\mu)$.

Lemma 3.1 *Functions from $F^2_{(k)}(\mathbb{C})$ admit the following representation*

$$\begin{aligned}\psi(z,\overline{z}) &= \frac{1}{\sqrt{(k-1)!}}\langle f(\zeta), [(\mathrm{a}_z^{\dagger})^{k-1}e^{\zeta\overline{z}}]\rangle \\ &= \frac{1}{\sqrt{(k-1)!}}\int_{\mathbb{C}} f(\zeta)\left[\left(-\frac{\partial}{\partial z}+\overline{z}\right)^{k-1} e^{\overline{\zeta}z}\right] d\mu(\zeta),\end{aligned} \tag{3.2}$$

where $f \in L_2(\mathbb{C}, d\mu)$.

Proof: For each $m \in \mathbb{N}$ and $z \in \mathbb{C}$, the functions $\zeta^m e^{\zeta\overline{z}}$ and $\overline{\zeta}^m e^{\overline{\zeta}z}$ belong to $L_2(\mathbb{C}, d\mu)$. Really

$$\begin{aligned}\|\zeta^m e^{\zeta\overline{z}}\|_{L_2(\mathbb{C},d\mu)} &= \|\overline{\zeta}^m e^{\overline{\zeta}z}\|_{L_2(\mathbb{C},d\mu)} = \int_{\mathbb{C}} |\zeta|^{2m} e^{\zeta\overline{z}+\overline{\zeta}z} d\mu(\zeta) \\ &= \frac{1}{\pi}\int_{\mathbb{R}^2} (\xi^2+\eta^2)^m e^{-(\xi^2+\eta^2)+\xi x+\eta y} d\xi d\eta < \infty,\end{aligned}$$

where $\zeta = \xi + i\eta$, $z = x + iy$. To calculate the exact value of the norm see, for example, [5] formula 3.462.2.

Prove now that all functions of the form (3.2) belong to $F^2_{(k)}(\mathbb{C})$. We have

$$\begin{aligned}\left(-\frac{\partial}{\partial z}+\overline{z}\right)^{k-1} e^{\overline{\zeta}z} &= \sum_{m=0}^{k-1}(-1)^m \frac{(k-1)!}{m!(k-1-m)!}\overline{z}^{k-1-m}\overline{\zeta}^m e^{\overline{\zeta}z} \\ &= (\overline{z}-\overline{\zeta})^{k-1}e^{\overline{\zeta}z}.\end{aligned}$$

Thus the integral (3.2) exists for all $z \in \mathbb{C}$ as the scalar product of two $L_2(\mathbb{C}, d\mu)$ functions. Further,

$$\begin{aligned}\psi(z,\overline{z}) &= \frac{1}{\sqrt{(k-1)!}}\int_{\mathbb{C}} f(\zeta)\left[\left(-\frac{\partial}{\partial z}+\overline{z}\right)^{k-1} e^{\overline{\zeta}z}\right] d\mu(\zeta) \\ &= \frac{1}{\sqrt{(k-1)!}}\int_{\mathbb{C}} f(\zeta)\sum_{m=0}^{k-1}(-1)^m \frac{(k-1)!}{m!(k-1-m)!}\overline{z}^{k-1-m}\overline{\zeta}^m e^{\overline{\zeta}z} d\mu(\zeta) \\ &= \sum_{m=0}^{k-1}(-1)^m \frac{\sqrt{(k-1)!}}{m!(k-1-m)!}\overline{z}^{k-1-m}\varphi_m(z),\end{aligned}$$

where

$$\varphi_m(z) = \int_{\mathbb{C}} f(\zeta)\overline{\zeta}^m e^{\overline{\zeta}z} d\mu(\zeta), \quad m = \overline{0, k-1}.$$

The functions $\varphi_m(z)$ are well defined as the scalar product of two $L_2(\mathbb{C}, d\mu)$ functions, $\varphi_0(z) \in F^2(\mathbb{C})$, and standard arguments show that $\varphi_m(z) = \varphi_0^{(m)}(z)$. Thus by (2.5) the function $\psi(z, \overline{z})$ belongs to $F^2_{(k)}(\mathbb{C})$.

Finally, each function (2.5) can be represent in the form (3.2) with $f = \varphi$

$$\begin{aligned} \psi(z) = \psi(z, \overline{z}) &= \sum_{m=0}^{k-1} (-1)^m \frac{\sqrt{(k-1)!}}{m!(k-1-m)!} \overline{z}^{k-1-m} \varphi^{(m)}(z) \\ &= \frac{1}{\sqrt{(k-1)!}} \langle \varphi(\zeta), [(\overline{\mathrm{a}}_z^{\dagger})^{k-1} e^{\zeta \overline{z}}] \rangle. \end{aligned} \tag{3.3}$$

□

For each $k \in \mathbb{N}$ introduce the system of functions

$$q_z^{\{k\}}(\zeta) = \frac{1}{(k-1)!} (\mathrm{a}_\zeta^{\dagger})^{k-1} (\overline{\mathrm{a}}_z^{\dagger})^{k-1} e^{\zeta \overline{z}} \tag{3.4}$$

$$= \frac{1}{(k-1)!} \left(-\frac{\partial}{\partial \zeta} + \overline{\zeta}\right)^{k-1} \left(-\frac{\partial}{\partial \overline{z}} + z\right)^{k-1} e^{\zeta \overline{z}}, \quad \zeta \in \mathbb{C}, \tag{3.5}$$

parameterized by points $z \in \mathbb{C}$.

Lemma 3.2 *Functions $q_z^{\{k\}}(\zeta)$ belong to $F_{(k)}^2(\mathbb{C})$, have the form*

$$q_z^{\{k\}}(\zeta) = e^{\zeta \overline{z}} p_{k-1}((z - \zeta)(\overline{\zeta} - \overline{z})), \tag{3.6}$$

for some real coefficient polynomials $p_{k-1}(\lambda)$, and

$$q_\zeta^{\{k\}}(z) = \overline{q_z^{\{k\}}(\zeta)}.$$

Proof: Function

$$\begin{aligned} g(\zeta) &= \frac{1}{\sqrt{(k-1)!}} \left(-\frac{\partial}{\partial \overline{z}} + z\right)^{k-1} e^{\zeta \overline{z}} \\ &= \sum_{m=0}^{k-1} (-1)^m \frac{\sqrt{(k-1)!}}{m!(k-1-m)!} z^{k-1-m} \zeta^m e^{\zeta \overline{z}} = (z - \zeta)^{k-1} e^{\zeta \overline{z}} \end{aligned}$$

is analytic and belongs to $L_2(\mathbb{C}, d\mu)$, and thus belongs to $F^2(\mathbb{C})$. Now by Lemma 2.8 the function

$$q_z^{\{k\}}(\zeta) = \frac{1}{\sqrt{(k-1)!}} (\mathrm{a}^{\dagger})^{k-1} g(\zeta)$$

belongs to $F_{(k)}^2(\mathbb{C})$. The following equalities

$$\left(-\frac{\partial}{\partial \zeta} + \overline{\zeta}\right) \left(-\frac{\partial}{\partial \overline{z}} + z\right) e^{\zeta \overline{z}} = e^{\zeta \overline{z}}((z - \zeta)(\overline{\zeta} - \overline{z}) + 1),$$

$$\begin{aligned} &\left(-\frac{\partial}{\partial \zeta} + \overline{\zeta}\right) \left(-\frac{\partial}{\partial \overline{z}} + z\right) e^{\zeta \overline{z}} (z - \zeta)^m (\overline{\zeta} - \overline{z})^m \\ &= e^{\zeta \overline{z}}((z - \zeta)^{m+1} (\overline{\zeta} - \overline{z})^{m+1} + (2m+1)(z - \zeta)^m (\overline{\zeta} - \overline{z})^m \\ &\quad + m^2 (z - \zeta)^{m-1} (\overline{\zeta} - \overline{z})^{m-1} \end{aligned}$$

prove the second statement of Lemma. The last statement follows from the representation (3.6). □

Theorem 3.3 *The operator*

$$(P_{(k)}f)(z) = \langle f(\zeta), q_z^{\{k\}}(\zeta)\rangle = \int_{\mathbb{C}} f(\zeta) q_\zeta^{\{k\}}(z) d\mu(\zeta)$$

is the orthogonal Bargmann projection of $L_2(\mathbb{C}, d\mu)$ onto the true-k-Fock space $F^2_{(k)}(\mathbb{C})$.

Proof: Consider

$$L_2(\mathbb{C}, d\mu) = F^2_{(k)}(\mathbb{C}) \oplus (F^2_{(k)}(\mathbb{C}))^{\perp}.$$

By Lemma 3.6 $q_z^{\{k\}}(\zeta) \in F^2_{(k)}(\mathbb{C})$, thus for functions $f \in (F^2_{(k)}(\mathbb{C}))^{\perp}$ we have obviously

$$(P_{(k)}f)(z) = \langle f(\zeta), q_z^{\{k\}}(\zeta)\rangle = 0,$$

and thus $(F^2_{(k)}(\mathbb{C}))^{\perp} \subset \ker P_{(k)}$.

Consider now functions from $F^2_{(k)}(\mathbb{C})$. Using (2.5), Lemma 2.8, and (3.4), we have

$$\begin{aligned}(P_{(k)}\psi)(z) &= \langle \psi(\zeta), q_z^{\{k\}}(\zeta)\rangle \\ &= \left\langle \frac{1}{\sqrt{(k-1)!}}(\mathfrak{a}^{\dagger})^{k-1}\varphi(\zeta), \frac{1}{\sqrt{(k-1)!}}(\mathfrak{a}_\zeta^{\dagger})^{k-1} \right. \\ &\qquad \left. \frac{1}{\sqrt{(k-1)!}}(\overline{\mathfrak{a}}_z^{\dagger})^{k-1} e^{\zeta\overline{z}} \right\rangle \\ &= \left\langle \varphi(\zeta), \frac{1}{\sqrt{(k-1)!}}(\overline{\mathfrak{a}}_z^{\dagger})^{k-1} e^{\zeta\overline{z}} \right\rangle = \psi(z).\end{aligned}$$

Thus $P_{(k)}|_{(F^2_{(k)}(\mathbb{C}))^{\perp}} = 0$, and $P_{(k)}|_{F^2_{(k)}(\mathbb{C})} = I$. □

Corollary 3.4 *The system of functions $\{q_z^{\{k\}}(\zeta)\}_{z\in\mathbb{C}}$, given by (3.4), forms a system of coherent states on the true-k-Fock space $F^2_{(k)}(\mathbb{C})$.*

Return now to the operators (2.4) and (3.1). Operators from each pair

$$\mathfrak{a}^{\dagger} = -\frac{\partial}{\partial z} + \overline{z}, \quad \mathfrak{a} = \frac{\partial}{\partial \overline{z}},$$

or

$$\overline{\mathfrak{a}}^{\dagger} = -\frac{\partial}{\partial \overline{z}} + z, \quad \overline{\mathfrak{a}} = \frac{\partial}{\partial z} \tag{3.7}$$

are (formally) adjoint one to another in the space $L_2(\mathbb{C}, d\mu)$, and satisfy the commutation relations for the quantum mechanic creation and annihilation operators. But their action in the space

$$L_2(\mathbb{C}, d\mu) = \bigoplus_{k=1}^{\infty} F^2_{(k)}(\mathbb{C})$$

are quite different. According to Theorem 2.9 the operators from the first pair move the true-poly-Fock spaces, while each true-poly-Fock space is invariant with respect to the operators from the second pair. In the last case corresponding vacuum vector in the true-k-Fock space $F^2_{(k)}(\mathbb{C})$ has the form

$$\Phi_0^{(k)}(z) = \frac{1}{\sqrt{(k-1)!}} \overline{z}^{k-1}.$$

Remark 3.5 The analytic nature of the Fock space $F^2(\mathbb{C})$ implies that $\overline{\mathfrak{a}}^{\dagger}|_{F^2(\mathbb{C})} = z$, which coincides with the usual definition of the creation operator in the Fock space. But for true-k-Fock spaces $F^2_{(k)}(\mathbb{C})$ with $k > 1$ one has to consider the operators (3.7).

4 Spaces Over $\mathbb{C}^n$

Return now to the spaces over $\mathbb{C}^n$. Both the space $L_2(\mathbb{C}^n, d\mu_n)$ and its Fock subspace $F^2(\mathbb{C}^n)$ are the following n-times tensor products of the corresponding spaces over $\mathbb{C}$

$$\begin{aligned} L_2(\mathbb{C}^n, d\mu_n) &= L_2(\mathbb{C}, d\mu) \otimes \cdots \otimes L_2(\mathbb{C}, d\mu), \\ F^2(\mathbb{C}^n) &= F^2(\mathbb{C}) \otimes \cdots \otimes F^2(\mathbb{C}). \end{aligned} \tag{4.1}$$

For a multi-index $k = (k_1, k_2, \ldots, k_n)$, with $k_j \in \mathbb{N}$, $j = \overline{1,n}$, define as usual $|k| = k_1 + \cdots + k_n$, $k! = k_1! \cdot \ldots \cdot k_n!$, and $z^k = z_1^{k_1} \cdot \ldots \cdot z_n^{k_n}$, for $z = (z_1, \ldots, z_n) \in \mathbb{C}^n$.

Given multi-index $k = (k_1, k_2, \ldots, k_n)$, introduce the true-$k$-Fock space $F^2_{(k)}(\mathbb{C}^n)$ as the following tensor product of the true-poly-Fock spaces over $\mathbb{C}$

$$F^2_{(k)}(\mathbb{C}^n) = \bigotimes_{j=1}^{n} F^2_{(k_j)}(\mathbb{C}),$$

and denote by $P_{(k)}$ the orthogonal (Bargmann) projection of $L_2(\mathbb{C}^n, d\mu_n)$ onto the true-k-Fock space $F^2_{(k)}(\mathbb{C}^n)$.

For the multi-index $1 = (1, 1, \ldots, 1)$ we have

$$F^2_{(1)}(\mathbb{C}^n) = F^2_{(1)}(\mathbb{C}) \otimes \cdots \otimes F^2_{(1)}(\mathbb{C}) = F^2(\mathbb{C}) \otimes \cdots \otimes F^2(\mathbb{C}) = F^2(\mathbb{C}^n).$$

Preserve the notations for the following analogs of the previously used unitary operators:

$$U_1 : L_2(\mathbb{C}^n, d\mu_n) \longrightarrow L_2(\mathbb{R}^{2n}) = L_2(\mathbb{R}^{2n}, dxdy),$$

defined by the rule

$$(U_1\varphi)(z) = \pi^{-\frac{n}{2}} e^{-\frac{z\cdot\bar{z}}{2}} \varphi(z);$$

$$\begin{aligned} U_2 &= I \otimes F : L_2(\mathbb{R}^{2n}, dxdy) \\ &= L_2(\mathbb{R}^n, dx) \otimes L_2(\mathbb{R}^n, dy) \longrightarrow L_2(\mathbb{R}^n, dx) \otimes L_2(\mathbb{R}^n, dy), \end{aligned}$$

where

$$(F\psi)(y) = (2\pi)^{-n/2} \int_{\mathbb{R}^n} e^{-i\eta\cdot y} \psi(\eta) d\eta;$$

and

$$U_3 = U_3^* = U_3^{-1} : L_2(\mathbb{R}^{2n}) \longrightarrow L_2(\mathbb{R}^{2n}),$$

defined by the rule

$$(U_3 f)(x, y) = f\left(\frac{1}{\sqrt{2}}(x + y), \frac{1}{\sqrt{2}}(x - y)\right).$$

Given multi-index $k = (k_1, k_2, \ldots, k_n)$, introduce the function

$$\widetilde{h}_k(y) = \prod_{j=1}^{n} h_{k_j}(y_j),$$

where the functions $h_{k_j}(y_j)$ are given by (2.2).

The functions $\widetilde{h}_k(y)$, $|k| = \overline{0, \infty}$, form obviously an orthonormal basis in $L_2(\mathbb{R}^n)$. Denote by $\widetilde{H}_k$ the one-dimensional subspace of $L_2(\mathbb{R}^n)$ generated by the function $\widetilde{h}_k(y)$, then the operator

$$\widetilde{Q}_k = \bigotimes_{j=1}^{n} Q_{k_j},$$

where each Q_{k_j} is the one-dimensional projection (2.3) acting on the j-th multiple of the decomposition (4.2), is the one-dimensional orthogonal projection of $L_2(\mathbb{R}^n)$ onto $\widetilde{H}_k$, and has the following form

$$(\widetilde{Q}_k\psi)(y) = \widetilde{h}_k(y) \int_{\mathbb{R}^n} \psi(\eta) \widetilde{h}_k(\eta) d\eta.$$

Theorem 4.1 *The unitary operator $U = U_3U_2U_1$ provides an isometric isomorphism of the space $L_2(\mathbb{C}^n, d\mu_n)$ onto the space $L_2(\mathbb{R}^{2n}) = L_2(\mathbb{R}^n, dx) \otimes L_2(\mathbb{R}^n, dy)$, under which*

1. *the true-k-Fock space $F^2_{(k)}(\mathbb{C})$ is mapped onto $L_2(\mathbb{R}^n) \otimes \widetilde{H}_{k-1}$*

$$U : F^2_{(k)}(\mathbb{C}^n) \longrightarrow L_2(\mathbb{R}^n) \otimes \widetilde{H}_{k-1},$$

2. *the true-k-Bargmann projection $P_{(k)}$ is unitary equivalent to the following one*

$$U P_{(k)} U^{-1} = I \otimes \widetilde{Q}_{k-1},$$

where $k - 1 = (k_1 - 1, k_2 - 1, \ldots, k_n - 1)$.

The following direct sum decomposition of the space $L_2(\mathbb{C}^n, d\mu_n)$ holds

$$L_2(\mathbb{C}^n, d\mu_n) = \bigoplus_{|k|=n}^{\infty} F^2_{(k)}(\mathbb{C}^n).$$

Introduce now multidimensional analogs of the operators (2.4) and (3.1), i.e., the two pairs of the operators

$$\begin{aligned} \mathfrak{a}^\dagger &= \left(-\frac{\partial}{\partial z_1} + \overline{z}_1, -\frac{\partial}{\partial z_2} + \overline{z}_2, \ldots, -\frac{\partial}{\partial z_n} + \overline{z}_n\right), \\ \mathfrak{a} &= \left(\frac{\partial}{\partial \overline{z}_1}, \frac{\partial}{\partial \overline{z}_2}, \ldots, \frac{\partial}{\partial \overline{z}_n}\right) \end{aligned}$$

and

$$\begin{aligned} \overline{\mathfrak{a}}^\dagger &= \left(-\frac{\partial}{\partial \overline{z}_1} + z_1, -\frac{\partial}{\partial \overline{z}_2} + z_2, \ldots, -\frac{\partial}{\partial \overline{z}_n} + z_n\right), \\ \overline{\mathfrak{a}} &= \left(\frac{\partial}{\partial z_1}, \frac{\partial}{\partial z_2}, \ldots, \frac{\partial}{\partial z_n}\right), \end{aligned}$$

acting in the space $L_2(\mathbb{C}^n, d\mu_n)$.

Given multi-index $k = (k_1, k_2, \ldots, k_n)$, the system of functions

$$q_z^{(k)}(\zeta) = \frac{1}{(k-1)!} (\mathfrak{a}_\zeta^\dagger)^{k-1} (\overline{\mathfrak{a}}_z^\dagger)^{k-1} e^{\zeta \cdot \overline{z}}, \quad \zeta \in \mathbb{C}^n,$$

parameterized by points $z \in \mathbb{C}^n$, forms a system of coherent states in the true-k-Fock space $F^2_{(k)}(\mathbb{C}^n)$. The expressions $(\mathfrak{a}_\zeta^\dagger)^{k-1}$ and $(\overline{\mathfrak{a}}_z^\dagger)^{k-1}$ are understood in the usual multi-index sense, i.e., for example,

$$(\mathfrak{a}_\zeta^\dagger)^{k-1} = \bigotimes_{j=1}^{n} \left(-\frac{\partial}{\partial \overline{z}_j} + z_j\right)^{k_j - 1}.$$

Theorem 4.2 *All true-poly-Fock spaces are isomorphic, one to each other (and each of them is isomorphic to $L_2(\mathbb{R}^n)$). The isometric isomorphism between the Fock space $F^2(\mathbb{C}^n)$ and the true-k-Fock space $F^2_{(k)}(\mathbb{C}^n)$ is given by the operator*

$$\frac{1}{\sqrt{(k-1)!}}(\mathfrak{a}^\dagger)^k|_{F^2(\mathbb{C}^n)} : F^2(\mathbb{C}^n) \longrightarrow F^2_{(k)}(\mathbb{C}^n),$$

and the operator

$$\frac{1}{\sqrt{(k-1)!}}\mathfrak{a}^k|_{F^2_{(k)}(\mathbb{C}^n)} : F^2_{(k)}(\mathbb{C}^n) \longrightarrow F^2(\mathbb{C}^n)$$

gives the inverse isomorphism.

The true-k-Fock space $F^2_{(k)}(\mathbb{C}^n)$ consists of all the functions of the form

$$\psi(z) = \psi(z,\overline{z}) = \sum_{|m|\leq k-1} (-1)^{|m|} \frac{\sqrt{(k-1)!}}{m!(k-1-m)!} \overline{z}^{k-1-m} \partial^m \varphi(z),$$

where $\varphi(z) \in F^2(\mathbb{C})$,

$$\partial^m \varphi = \frac{\partial^{|m|}\varphi}{\partial z_1^{m_1} \partial z_2^{m_2} \ldots \partial z_n^{m_n}},$$

and

$$\|\psi\|_{F^2_{(k)}(\mathbb{C}^n)} = \|\varphi\|_{F^2(\mathbb{C}^n)}.$$

The operator

$$(P_{(k)}f)(z) = \langle f(\zeta), q_z^{\{k\}}(\zeta)\rangle = \int_{\mathbb{C}^n} f(\zeta) q_\zeta^{\{k\}}(z) d\mu_n(\zeta)$$

is the orthogonal Bargmann projection of $L_2(\mathbb{C}^n, d\mu_n)$ onto the true-k-Fock space $F^2_{(k)}(\mathbb{C}^n)$.

References

[1] V. Bargmann, *On a Hilbert space of analytic functions*. Comm. Pure Appl. Math. **3** (1961), 215–228.

[2] Harry Bateman and Arthur Erdélyi, *Higher transcendental functions*, vol. **2**. McGraw-Hill, 1954.

[3] F.A. Berezin, *Covariant and contravariant symbols of operators*. Math. USSR Izvestia **6** (1972), 1117–1151.

[4] V.A. Fock, *Konfigurationsraum und zweite Quantelung*. Z. Phys. **75** (1932), 622–647.

[5] I.S. Gradshteyn and I.M. Ryzhik, *Tables of Integrals, Series, and Products*. Academic Press, New York, 1980.

[6] I.E. Segal, Lectures at the Summer Seminar on Appl. Math. Boulder, Colorado, 1960.

[7] Sundaram Thangavelu, *Lectures on Hermitte and Laguerre expansions*. Princeton University Press, Preiceton, New Jersey, 1993.

Departamento de Matemáticas
Cinvestav del I.P.N.
Apartado Postal 14-740
07000 México, D.F.
México
nvasilev@math.cinvestav.mx

AMS Classification: 46E20, 46E22, 81S05

Operator Theory:
Advances and Applications, Vol. 117
© 2000 Birkhäuser Verlag Basel/Switzerland

The Diffraction of Elastic Shear Wave on the Circular Cylinder which is Situated in the Elastic Halfspace

N. Whitefield

Let's consider the hollow thick circular cylinder which is situated in the elastic half-space $y > 0, -\infty < x < \infty, -\infty < z < \infty$ in the condition of antiplane deformation on the distance l from the bottom. The position of it in the cylindrical coordinates, the axis of which coincided with cylinder's axis, is defined so: $r_1 < r < r_2, -\pi < \theta \leq \pi, -\infty < z < \infty$. On the external surface of cylinder the following conditions are fulfilled:

$$\begin{aligned} w(r_2+0,\theta,t) &= w(r_2-0,\theta,t) \\ \tau_{rz}(r_2+0,\theta,t) &= \tau_{rz}(r_2-0,\theta,t) \end{aligned} \tag{1.1}$$

The internal surface of cylinder is free from stresses:

$$\tau_{rz}(r_1+0,\theta,t) = 0 \tag{1.2}$$

The wave of longitudinal shear interacts with the cylinder in the moment $t = 0$:

$$w_p(r,\theta,t) = -(ct - r_2 - r\cos(\theta-\theta_0))H(ct - r_2 - r\cos(\theta-\theta_0)) \tag{1.3}$$

Here H is the Heaviside unite step function, θ_0 is the angle of incident wave, c is the velocity of wave propagation. The following condition is fulfilled on the bottom:

$$w(x,0,t) = 0 \tag{1.4}$$

The elastic state of cylinder under the incident wave impact is necessary to be defined.

The field of displacement and stresses inside the cylinder is represented as the superposition of discontinuous solutions [1] of wave equation:

$$T = T_1 + T_2 \tag{1.5}$$

here, $T_k = [w_k, \tau_{rz}^k]_T, k = 1,2$ are vectors, the components of which are the displacements and stresses which have discontinuities on the lines $r = r_k, -\pi <$

$\theta \leq \pi$ with the jumps:

$$
\begin{aligned}
w_k(r_k - 0, \theta, t) - w_k(r_k + 0, \theta, t) &= X_k(\theta, t) \\
\frac{\partial w_k(r_k - 0, \theta, t)}{\partial r} - \frac{\partial w_k(r_k + 0, \theta, t)}{\partial r} &= \Psi_k(\theta, t)
\end{aligned} \tag{1.6}
$$

In the Laplace transform space these discontinuous solutions for displacements are expressed by formulas:

$$
\overline{w_k}(r, \theta) = \frac{r_k}{\pi} \left\{ \sum_{n=-\infty}^{\infty} \overline{\Psi_k^n} h_{nk}(r, \sigma) e^{in\theta} - \sum_{n=-\infty}^{\infty} \overline{X_k^n} g_{nk}(r, \sigma) e^{in\theta} \right\} \tag{1.7}
$$

and for stresses:

$$
\overline{\tau_k}(r, \theta) = \frac{G r_k}{\pi} \left\{ \sum_{n=-\infty}^{\infty} \overline{\Psi_k^n} q_{nk}(r, \sigma) e^{in\theta} - \sum_{n=-\infty}^{\infty} \overline{X_k^n} y_{nk}(r, \sigma) e^{in\theta} \right\} \tag{1.8}
$$

here the line above the symbol define the Laplace transforms, G is the rigidity, p is the Laplace transformation parameter,

$$
\begin{aligned}
h_{nk}(r, \sigma) &= \left\{ \begin{array}{l} I_n(\sigma r) K_n(\sigma r_k), r < r_k \\ I_n(\sigma r_k) K_n(\sigma r), r > r_k \end{array} \right\}, \\
g_{nk}(r, \sigma) &= \left\{ \begin{array}{l} \sigma I_n(\sigma r) K_n'(\sigma r_k), r < r_k \\ \sigma I_n'(\sigma r_k) K_n(\sigma r), r > r_k \end{array} \right\} \\
q_{nk}(r, \sigma) &= \frac{\partial h_{nk}(r, \sigma)}{\partial r}, \; y_{nk} = \frac{\partial g_{nk}(r, \sigma)}{\partial r} \\
\left[\begin{array}{c} \overline{X_k^n} \\ \overline{\Psi_k^n} \end{array} \right] &= \int_0^{\infty} \int_{-\pi}^{\pi} \left[\begin{array}{c} X_k(\theta, t) \\ \Psi_k(\theta, t) \end{array} \right] e^{in\theta} e^{-pt} d\theta dt, \sigma = \frac{p}{c}
\end{aligned}
$$

The displacement field in the external medium is represented in the form:

$$
w = w_0 + w_p + w_* \tag{1.9}
$$

where w_0 is the reflected from the bottom wave:

$$
w_0(r, \theta, t) = -(ct - r_2 - r\cos(\theta + \theta_0)) H(ct - r_2 - r\cos(\theta + \theta_0)) \tag{1.10}
$$

and w_* is the wave, which is reflected from the cylinder.

The condition on the bottom (1.4) in terms of representation (1.9) will have the form:

$$
w_*(x + 0, \theta, t) = 0 \tag{1.11}
$$

For the searching of displacements which are created by the reflected from the cylinder wave the following method is applied: we supplement the half-space to the whole space and input in the lower half-space the cylinder $r_1 < r_- < r_2, -\pi \leq \theta_- < \pi, -\infty < z < \infty$. After these the potential of reflected by the cylinder waves is constructed in the form:

$$w_*(r,\theta,t) = w_*^+(r_+,\theta_+,t) + w_*^-(r_-,\theta_-,t) \tag{1.12}$$

here r_+, θ_+ are the coordinates of initial cylindrical system, r_-, θ_- are the cylindrical coordinates which are symmetric to the coordinate r_+, θ_+ relatively to the plane bound:

$$r = r_+, \theta = \theta_+, r_- = \sqrt{r_+^2 4l^2 + 4lr_+ \sin\theta_+}, \theta_- = \arccos\frac{r_+}{r_-}\cos\theta_+$$

$w_*^\pm$ are the discontinuous solutions of wave equation of (1.7) type with the jumps:

$$\langle w_*^\pm \rangle_{r_\pm = r_2} = \omega^\pm(\theta_\pm, t), \left\langle \frac{\partial w_*^\pm}{\partial r} \right\rangle_{r_\pm = r_2} = \gamma^\pm(\theta_\pm, t) \tag{1.13}$$

By direct checking we can sure that condition (1.11) is fulfilled automatically, if we take following:

$$\overline{\omega}^+(\theta) = \overline{\omega}^-(-\theta), \overline{\gamma}^+(\theta) = \overline{\gamma}^-(-\theta) \tag{1.14}$$

Than, according to (1.12) the presentation of reflected wave potential will be following:

$$\begin{aligned} \overline{w_*}(r_+,\theta_+) &= \frac{r_2}{\pi}\sum_{n=-\infty}^{\infty} \overline{\gamma_n^+}[h_{n2}^{(2)}(r^+,\sigma)e^{in\theta_+} + h_{n2}^{(2)}(r^-,\sigma)e^{in\theta_-}] \\ &\quad -\frac{r_2}{\pi}\sum_{n=-\infty}^{\infty} \overline{\omega_n^+}[g_{n2}^{(2)}(r^+,\sigma)e^{in\theta_+} + g_{n2}^{(2)}(r^-,\sigma)e^{in\theta_-}] \\ \left[\begin{matrix} \overline{\gamma_n^\pm} \\ \overline{\omega_n^\pm} \end{matrix}\right] &= \int_0^\infty \int_{-\pi}^{\pi} \left[\begin{matrix} \gamma^\pm(\theta,t) \\ \omega^\pm(\theta,t) \end{matrix}\right] e^{in\theta} e^{-pt} d\theta dt \end{aligned} \tag{1.15}$$

The formulas (1.15) and (1.5) define the displacement and stresses inside the cylinder and outside it, if we will know 6 jumps $X_j, \Psi_j (j = 1,2)$ and ω^+, γ^+. We can obtain them from the conditions (1.1), (1.2). But these conditions are not sufficient for their single value definition, so it is why we put the additional restrictions on the discontinuous solutions (1.15), (1.5). As function $\overline{w_*}(r,\theta)$ describes the displacements outside the cylinder, so we require from the discontinuous solution the fulfilling of equality:

$$\overline{w_*}(r_2 - 0, \theta) = 0 \tag{1.16}$$

After substitution (1.15) we obtain:

$$\omega^+(\theta_+) = \frac{r_2}{\pi}\sum_{n=-\infty}^{\infty}\overline{\omega_n^+}[g_{n2}^{(2)}(r^+,\sigma)e^{in\theta_+} + g_{n2}^{(2)}(r^-,\sigma)e^{in\theta_-}]_{r_\pm=r_2+0}$$

(1.17)

$$-\frac{r_2}{\pi}\sum_{n=-\infty}^{\infty}\overline{\gamma_n^+}[h_{n2}^{(2)}(r^+,\sigma)e^{in\theta_+} + h_{n2}^{(2)}(r^-,\sigma)e^{in\theta_-}]_{r_\pm=r_2+0}$$

By similar way, because of the discontinuous solutions (1.5) should define the displacements only inside the cylinder ($r_1 < r < r_2$) we require the fulfilling of equalities:

$$\overline{w_1}(r_1-0,\theta)=0, \overline{w_2}(r_2+0,\theta)=0 \quad (1.18)$$

From the last ones, after substitution (1.5) we find the following formulas for the jumps:

$$\overline{X_j^n} = \frac{h_{nj}}{g_{nj}}\overline{\Psi_j^n},\ j=1,2;\ h_{nj}=h_{nj}(r_j\pm 0),\ g_{nj}=g_{nj}(r_j\pm 0) \quad (1.19)$$

Now we should only to fulfill the conditions on the bound of the mediums (1.1), (1.2). For obtaining the formulas of the jumps $\overline{\omega_n^+}$, $\overline{\gamma_n^+}$ relation in these equalities the transfer from the boundary functions value to unknown jumps could be done. Then we solve the system of three equations. After the substitution of obtained formula in (1.17) we get the integral equation relatively to unknown jump $\overline{\omega}^+$:

$$\overline{\omega}^+(\theta_+) = \frac{r_2}{\pi}\left\{\int_{-\pi}^{\pi}\overline{\omega}^+(\nu)\overline{Q}(\theta_+-\nu)d\nu + \overline{F(\theta_+)}\right\} \quad (1.20)$$

After transfer from Laplace transforms and using the theorem of convolution the expression (1.20) will be following:

$$\overline{\omega}^+(\theta_+,t) = \frac{r_2}{\pi}\left\{\int_{-\pi}^{\pi}\int_0^{t+}\omega^+(\nu,\tau)Q(\theta_+-\nu,t-\tau)d\nu d\tau + F(\theta_+,t)\right\}$$

(1.21)

Function $Q(\theta_+-\nu,t-\tau)$ is the original of function $\overline{Q}(\theta_+-\nu)$ and is searched by the help of numerical inverse of Laplace transformation by the Ltr_2 algorithm. The solution of equation (1.21) is constructed numerically. That's why the time interval [0; T], during which we will research the interaction of cylinder with medium, is divided on the intervals $[\tau_k;\ \tau_{k+1}]$ with the step $h=\frac{T}{N}$, $\tau_k = k\frac{T}{N}$, $k=\overline{0,n}$.

The integrals by variable rechanged by the integral sum with the help of Simpson formula, the values of unknown functions in the sites of which a researched in the following representation:

$$\omega_n^+(\theta_+) = \sum_{m=-\infty}^{\infty}\omega_{nm}^+ e^{im\theta_+},\ \omega_n^+(\theta_+) = \omega(\theta_+,\tau_n) \quad (1.22)$$

As a result, we obtain the infinity system of linear algebraic equations relatively to ω_{nm}^{+} which are the unknown coefficients from the (1.22):

$$\omega_{nl}^{+} - \frac{r_2}{\pi}\left\{\sum_{k=1}^{n} A_k \sum_{m=-\infty}^{\infty} \omega_{km}^{+} B_{kmn}\right\} = F_{nl} \tag{1.23}$$

here A_k are the coefficients of Simpson quadrature formula,

$$\begin{aligned}
B_{kmn} &= \int_{-\pi}^{\pi} [R(\theta_+, t_n - \tau_k) + R(\theta_-, t_n - \tau_k)] e^{-il\theta_+} d\theta_+ \\
R(\theta, t) &= e^{im\theta}\left\{\sum_{j=0}^{2N} \frac{a_{j,N}}{1+\delta_j^2} L_j(2\gamma t)\right. \\
&\quad \left. - \int_0^{\infty} \xi N_m'\left(\frac{\xi}{c} r_2\right) J_m\left(\frac{\xi}{c} r_2\right) \cos \xi t d\xi\right\} \\
&\quad - e^{im\theta} \int_0^{\infty} \xi N_m'\left(\frac{\xi}{c} r_2\right) J_m\left(\frac{\xi}{c} r_2\right) \sin \xi t d\xi \\
F_{nl} &= \sum_{k=1}^{n} E_k \sum_{m=-\infty}^{\infty} [w_{km}^{0} + w_{km}^{p}] D_{kmn} \\
D_{kmn} &= \int_{-\pi}^{\pi} [U_m(\theta_+, t_n - \tau_k) + U_m(\theta_-, t_n - \tau_k)] e^{-il\theta_+} d\theta_+ \\
U_m(\theta, t) &= \frac{\rho}{2\mu} \int_0^{\infty} \xi^2 J_m\left(\frac{\xi}{c} r_2\right) \\
&\quad \left[J_m'\left(\frac{\xi}{c} r_1\right) \cos \xi t d\xi - N_m'\left(\frac{\xi}{c} r_1\right) \sin \xi t\right] d\xi e^{im\theta}
\end{aligned}$$

where L_j are the Laugerre polynomials, $a_{j,n}$ are expressed through the values of $\overline{Q}(\theta_+ - \nu)$ transform function. Then, system (1.23) is solved by the reduction method. For obtaining the accuracy $\varepsilon = 10^{-3}$ it is necessary to save the first 16 members of the set. The solution of system is obtained in the form of set of coefficients w_{km}^{+}. By the help of last ones we define the jump $\omega^{+}(\theta_+, t)$.

Depending of stresses τ_{rz} from time for different values of cylinder's distance from bottom is investigated. Dependence of stresses τ_{rz} at the frontal point $\theta = \frac{\pi}{2}$ for case when cylinder's thickness makes $\frac{h}{R} = 0.01$ for distance from the bottom $\frac{R}{l} = 0.9$ and $\frac{R}{l} = 0.01$ is investigated. Also comparison of these results with the values of stresses τ_{rz} at the point $\theta = \frac{\pi}{2}$ at the thickness $\frac{h}{R} = 0.1$ at the same values of cylinders distance is completed. Numerical calculation permit to get following conclusions. The nearer cylinder is situated to the bound ($\frac{R}{l} \to 1$) the higher stress's values and the more typical for it existence of points of maximum and minimum. That can be explained by the complicated diffraction process which takes place near bound. With moving off from the bound ($\frac{R}{l} \to 0$) values of stress decrease. Comparison of numerical results with obtained stresses values at the bottom absence shows that at the value of cylinder distance $\frac{R}{l} = 0.01$ influence of bottom became already inessential.

Reference

[1] G.Ya. Popov, *The Concentration of Elastic Stresses near Punches*, cuts, Thin Inclusions and Reinforcements. Nauka, Moscow, 1982.

Institute Mathematics
Economics and Mechanics
Odessa State University
270057 Dvoryanskaya str. 2
Odessa
Ukraine

Operator Theory:
Advances and Applications, Vol. 117
© 2000 Birkhäuser Verlag Basel/Switzerland

On M.G. Krein's Spectral Shift Function for Canonical Systems of Differential Equations

*V.A. Yavrian**

In this paper the spectral shift function (s.s.f.) for the pair of operators, generated by a canonical system of differential equations with two different boundary conditions, is studied. The main result is a connection between the s.s.f. and the spectral function. Also the inverse problem is studied: For a given s.s.f. we find a canonical system and investigate its uniqueness.

1 Introduction

Let $H(x)$, $x \in (0, \infty)$, be a complex, symmetric and nonnegative 2×2 matrix-function with entries $h_{ik}(x)$, $i, k = 1, 2$, which are locally integrable on $[0, \infty)$. We suppose that $\operatorname{tr} H(x) > 0$ for $x \in (0, \infty)$. Then without loss of generality we can assume that $\operatorname{tr} H(x) = 1$, $x \in (0, \infty)$.

Let $L^2(H; 0, l), 0 < l \leq \infty$, be the Hilbert space of measurable vector-functions $f(x) = (f_1(x), f_2(x))^T$ with the inner product

$$(f, g)_{H,l} = \int_0^l g(x)^* H(x) f(x) dx.$$

For short we write $L^2(H)$ instead of $L^2(H; 0, \infty)$ and $(f, g)_H$ instead of $(f, g)_{H,\infty}$. We say that f and g are equivalent on (a, b) and write $f(x) \sim g(x), x \in (a, b)$, if $H(x)(f(x) - g(x)) = 0$ a.e. on(a, b). Further, let

$$J = \begin{pmatrix} 0 & -1 \\ 1 & 0 \end{pmatrix}, \qquad \eta_\gamma = \begin{pmatrix} \cos \gamma \\ \sin \gamma \end{pmatrix}, \quad \text{for} \quad \gamma \in (-\infty, \infty).$$

An interval (a, b) is called H-indivisible (see [dB1],[Ka1]) if $H(x) = \eta_\gamma \eta_\gamma^*$ a.e. on (a, b); then the number γ is called the type of the H-indivisible interval (a, b). The H-indivisible interval (a, b) is called maximal H-indivisible if it is not properly contained in a larger H-indivisible interval. Note that for an H-indivisible interval (a, b) of type γ the relations $H(x)\eta_\gamma = \eta_\gamma$, $H(x)J\eta_\gamma = 0$, a.e. on (a, b), hold. In the following we shall always suppose that the whole interval $(0, l)$ is not H-indivisible.

Denote by $L_0^2(H; 0, l)$, $l < \infty$, the subspace of $L^2(H; 0, l)$, the elements of which are equivalent to a constant vector on each H-indivisible interval. For

*This research was made possible in part by Grant no. 93-0249 from INTAS.

$\alpha, \beta \in [-\frac{\pi}{2}, \frac{\pi}{2})$ we define the space $L_0^2(H; 0, l, \alpha, \beta)$ $(\subset L_0^2(H; 0, l))$ as follows: $L_0^2(H; 0, l, \alpha, \beta) = L_0^2(H; 0, l)$ if no interval of the forms $(0, b)$ and $(a, l))$ is H-indivisible of type $\alpha + \pi/2$ and $\beta + \pi/2$, respectively. If some interval $(0, b)((a, l)$, respectively) is H-indivisible of type $\alpha + \pi/2$ $(\beta + \pi/2$, respectively), then $L_0^2(H; 0, l, \alpha, \beta)$ consists of those elements of $L_0^2(H; 0, l)$, which are equivalent to zero on $(0, b)((a, l)$, respectively).

On $L^2(H; 0, l), l < \infty$, we consider the following operator V:

$$(Vf)(x) = \int_x^l JH(s)f(s)ds.$$

Let $L^2(0, l)$ denote the usual Hilbert space of 2-dimensional vector functions and define an operator $W : L^2(H; 0, l) \to L^2(0, l)$ by $(Wf)(x) = H^{1/2}(x)f(x)$. Evidently W is an isometry and $V = W^{-1}A_M W$, where A_M is the so-called model operator (see [GK2])

$$(A_M f)(x) = H^{1/2}(x)J \int_x^l H^{1/2}(s)f(s)ds.$$

Since A_M is a compact operator, V is also compact. It is easy to check that

$$\int_0^l JH(x)f(x)dx = \frac{1}{\sin(\beta - \alpha)}\{(f, \eta_\alpha)_H \eta_\beta - (f, \eta_\beta)_H \eta_\alpha\}, \quad \alpha \not\equiv \beta \pmod{\pi}. \tag{1.1}$$

For $f \in L^2(H; 0, l)$ define

$$(B_{l,\alpha,\beta} f)(x) = \int_x^l JH(s)f(s)ds - \frac{1}{\sin(\beta - \alpha)}(f, \eta_\alpha)_H\, \eta_\beta, \quad \alpha \not\equiv \beta \pmod{\pi}.$$

Obviously, $B_{l,\alpha,\beta}$ is a compact operator, and since

$$(B_{l,\alpha,\beta}^* f)(x) = -\int_0^x JH(s)f(s)ds - \frac{1}{\sin(\beta - \alpha)}(f, \eta_\beta)_H \eta_\alpha, \quad \alpha \not\equiv \beta \pmod{\pi},$$

it follows from (1.1) that $B_{l,\alpha,\beta} = B_{l,\alpha,\beta}^*$.

Theorem 1.1 *If $\alpha \neq \beta$ then* $(\ker B_{l,\alpha,\beta})^\perp = L_0^2(H; 0, l, \alpha, \beta)$.

The proof is easily obtained from the following lemma, see [Ka2].

Lemma 1.2 *Let $g(x)$ be a locally absolutely continuous vector function, such that $g(x) \sim 0$ on $[0, l)$ and*

$$Jg'(x) = H(x)f(x) \quad \text{a.e. on } (0, l), \; g(0) \in \operatorname{span}\{\eta_\alpha\}.$$

If $g(x) \neq 0$ a.e. on (a, b), then (a, b) is H-indivisible of some type $\gamma + \frac{\pi}{2}$ and

$$g(x) = \rho(x)\eta_\gamma, x \in (a, b),$$

where ρ is an absolutely continuous scalar function on $[a, b]$.

As in [Ka2] we define in $L_0^2(H; 0, l, \alpha, \beta)$ an operator $A_{l,\alpha,\beta}$: The element $g \in L_0^2(H; 0, l, \alpha, \beta)$ belongs to the domain $D(A_{l,\alpha,\beta})$ and the relation $f = A_{l,\alpha,\beta}g$ holds if and only if g is an absolutely continuous vector-function on $[0, l]$ and

$$Jg'(x) = H(x)f(x) \text{ a.e. on } (0, l), g(0) \in \operatorname{span}\{\eta_\alpha\}, \quad g(l) \in \operatorname{span}\{\eta_\beta\}.$$

If B_0 denotes the restriction of $B_{l,\alpha,\beta}$ to $L_0^2(H; 0, l, \alpha, \beta)$ and $\alpha \neq \beta$, then

$$A_{l,\alpha,\beta} = B_0^{-1}.$$

Denote by $L_0^2(H)$ the subspace of $L^2(H)$, whose elements are equivalent with a constant vector on each H-indivisible interval. If an interval of the form $(0, b)$, $b < +\infty$ is H-indivisible, denote by $L_1^2(H)$ the subspace of $L_0^2(H)$, whose elements are equivalent to zero on $(0, b)$. It is clear, that if (a, ∞) is H-indivisible and $f \in L_0^2(H)$ then $f(x) \sim 0$ on (a, ∞).

Now we define the space $L_0^2(H; \alpha)$, $\alpha \in [-\frac{\pi}{2}, \frac{\pi}{2})$ as follows: If there is no H-indivisible interval of the form $(0, b)$ of type $\alpha + \pi/2$, we set $L_0^2(H; \alpha) = L_0^2(H)$, otherwise $L_0^2(H; \alpha) = L_1^2(H)$.

In $L_0^2(H; \alpha)$ we consider the following operator A_α: $g \in D(A_\alpha)$ and $A_\alpha g = f$ if g is locally absolutely continuous on $[0, \infty)$ and

$$Jg'(x) = H(x)f(x) \text{ a.e. on } (0, \infty), \quad g(0) \in \operatorname{span}\{\eta_\alpha\}. \tag{1.2}$$

It follows from Lemma 1.2 that A_α is well-defined. If (l, ∞) is an H-indivisible interval of type $\beta + \pi/2$, then $L_0^2(H; \alpha) = L_0^2(H; 0, l, \alpha, \beta)$. For $g \in D(A_\alpha)$ it follows that $g(x) = c\eta_\beta$ if $x \geq l$, that is $g(l) \in \operatorname{span}\{\eta_\beta\}$ and, consequently, $A_\alpha = A_{l,\alpha,\beta}$. Conversely, given $H(x)$ for $x \in (0, l)$ with $\operatorname{tr} H(x) = 1$. If we set $H(x) = (J\eta_\beta)(J\eta_\beta)^*$ on $x \in (l, \infty)$ then $A_\alpha = A_{l,\alpha,\beta}$.

Denote by $D(\hat{A}_\alpha)$ the set of all functions $f \in D(A_\alpha)$ which are finite at ∞ (by which we mean that functions f vanish identically at ∞) and set

$$\hat{A}_\alpha g = A_\alpha g, \quad g \in D(\hat{A}_\alpha).$$

It is easy to see that $\hat{A}_\alpha$ is a symmetric operator. From (1.2) for $f = \hat{A}_\alpha g$ it follows that

$$g(x) = \int_x^\infty JH(s)f(s)ds, \text{ where } (f, \eta_\alpha)_H = 0.$$

We shall show that $D(\hat{A}_\alpha)$ is dense in $L_0^2(H;\alpha)$. Assume that $h \in L_0^2(H;\alpha)$ and $(g,h)_H = 0$ for every $g \in D(\hat{A}_\alpha)$. Then

$$\int_0^\infty h(x)^* H(x) \left(\int_x^\infty JH(s) f(s)\, ds \right) dx$$
$$= - \int_0^\infty \left(\int_0^s JH(x) h(x)\, dx \right)^* H(s) f(s)\, ds = 0$$

for every function $f \in L_0^2(H,\alpha)$ which is finite at ∞ and is such that $(f,\eta_\alpha)_H = 0$. Hence,

$$- \int_0^x JH(s)h(s)ds + c\eta_\alpha = g_1(x),$$

where c is a constant, and $g_1(x) \sim 0$ on $(0,\infty)$. This means that $g_1 \in D(A_\alpha)$ and $h = A_\alpha g_1 = 0$.

In the sequel the following condition will be used:

(*) *There is no H-indivisible interval of the form (l,∞).*

Lemma 1.3 *If the condition (*) is satisfied then*

$$\hat{A}_\alpha^* = A_\alpha.$$

Proof: From $(\hat{A}_\alpha f, g)_H = (f, \tilde{g})_H$, $f \in D(\hat{A}_\alpha)$, $g, \tilde{g} \in L_0^2(H;\alpha)$, it follows that

$$\begin{aligned} &\int_0^\infty g(x)^* H(x) f_1(x)\, dx \\ &\quad = - \int_0^\infty \left(\int_0^s JH(x)\tilde{g}(x)\, dx \right)^* H(s) f_1(s)\, ds, \end{aligned} \tag{1.3}$$

where $f_1 = \hat{A}_\alpha f$. For any $l' < \infty$,

$$\int_0^x JH(s)\tilde{g}(s)\, ds \in L_0^2(H;0,l'), \quad \eta_\alpha \in L_0^2(H;0,l').$$

The asumption (*) implies that (1.3) holds for every $f_1 \in L_0^2(H;0,l')$ such that $f_1(x) \sim 0$ on (l',∞) and $(f_1,\eta_\alpha)_H = 0$. Consequently, there exists a $c = c_g$, such that

$$g(x) \sim - \int_0^x JH(s)\tilde{g}(s)\, ds + c_g \eta_\alpha \text{ on } (0,\infty),$$

hence $g \in D(A_\alpha)$ and $\tilde{g} = A_\alpha g$. □

Let $\tilde{A}_\alpha$ denote the closure of $\hat{A}_\alpha$.

Theorem 1.4 *If the condition (*) is satisfied, at least one of the defect numbers of $\tilde{A}_\alpha$ is zero.*

Proof: We put $\alpha = 0$. Let $\varphi(x, \lambda)$ and $\psi(x, \lambda)$ be the solutions of the equation

$$Jy' = \lambda H(x)y, \quad x \in (0, \infty), \tag{1.4}$$

which satisfy the conditions

$$\varphi(0, \lambda) = \eta_0, \quad \psi(0, \lambda) = J\eta_0.$$

Consider an H-indivisible interval (a, b). It is easy to see that for fixed λ the function $\varphi(x, \lambda)$ is equivalent to a constant vector in (a, b) and, consequently, $\varphi(\cdot, \lambda) \in L^2(H)$ implies that $\varphi(\cdot, \lambda) \in L_0^2(H, 0)$. Thus, according to Lemma 1.3, for $\Im\lambda \neq 0$ there exists a $g \neq 0$ which satisfies

$$(\widetilde{A}_0^* - \lambda I)g = 0$$

if and only if $\varphi(\cdot, \lambda) \in L^2(H)$. Then $g(x) \sim c\varphi(x, \lambda), c \neq 0$. Assume that the defect index of $\widetilde{A}_0$ is (1, 1), that is $\varphi(\cdot, \lambda) \in L^2(H)$ if $\Im\lambda \neq 0$. Then the defect index of $\widetilde{A}_{\frac{\pi}{2}}$ is also (1, 1), and hence $\psi(\cdot, \lambda) \in L^2(H)$, $\Im\lambda \neq 0$. On $L^2(H)$ we consider the operator

$$(V_\lambda f)(x) = \int_0^x V_\lambda(x, s)H(s)f(s)ds, \quad \Im\lambda \neq 0,$$

where

$$V_\lambda(x, s) = \varphi(x, \lambda)\psi(s, \overline{\lambda})^* - \psi(x, \lambda)\varphi(s, \overline{\lambda})^*.$$

The operator V_λ is isometrically equivalent to the operator $\widetilde{V}_\lambda$, defined on $L^2(0, \infty)$ by the relation

$$(\widetilde{V}_\lambda f)(x) = H^{1/2}(x) \int_0^x V_\lambda(x, s)H^{1/2}(s)f(s)ds,$$

that is $V_\lambda = W^{-1}\widetilde{V}_\lambda W$. Since $H^{1/2}(\cdot)\varphi(\cdot, \lambda) \in L^2(0, \infty)$, $H(\cdot)^{1/2}\psi(\cdot\lambda) \in L^2(0, \infty)$, it follows that the elements of the matrix-kernel $V_\lambda(x, s)$ of the integral operator $\widetilde{V}_\lambda$ belong to the space $L^2((0, \infty) \times (0, \infty))$, and therefore $\widetilde{V}_\lambda$ and V_λ are Hilbert-Schmidt operators (see [GK1]).

Now, if $g = V_\lambda f$ then $g(0) = 0$ and

$$Jg'(x) = \lambda H(x)g(x) + H(x)f(x) \text{ a.e on } (0, \infty). \tag{1.5}$$

To verify (1.5) recall (see [GK2], [At]) that the fundamental matrix

$$U(x, \lambda) = \begin{pmatrix} \varphi_1(x, \lambda) & \psi_1(x, \lambda) \\ \varphi_2(x, \lambda) & \psi_2(x, \lambda) \end{pmatrix}$$

satisfies the identity

$$U(x, \lambda)JU(x, \overline{\lambda})^* = \psi(x, \lambda)\varphi(x, \overline{\lambda})^* - \varphi(x, \lambda)\psi(x, \overline{\lambda})^* = J.$$

Now we show that $f \sim 0$ if

$$(I + \lambda V_\lambda) f = 0. \tag{1.6}$$

The latter means that

$$f(x) + \lambda \int_0^x V_\lambda(x,s) f(s)\,ds = \widetilde{g}(x) \text{ a.e., } \widetilde{g}(x) \sim 0 \text{ on } (0,\infty).$$

Then from (1.5) it follows that $J(\widetilde{g}(x) - f(x))' = 0$ and whence $f(x) \sim 0$. Since for $\Im\lambda \neq 0$ the operator V_λ is compact, the inverse $(I + \lambda V_\lambda)^{-1}$ exists and is bounded. Let $f := (I + \lambda V_\lambda)^{-1}\varphi(\cdot,\lambda)$. Then

$$\lambda(V_\lambda f)(x) = -f(x) + \varphi(x,\lambda) + g(x) \text{ a.e., } g(x) \sim 0 \text{ on } (0,\infty).$$

Hence, in view of (1.5) we have that $J(f(x) - g(x))' = 0$, and it follows that $f(x) \sim \eta_0$ in $(0,\infty)$. Thus,

$$(I + \lambda V_\lambda)^{-1}\varphi(\cdot,\lambda) = \eta_0, \quad \Im\lambda \neq 0.$$

Similarly we get

$$(I + \lambda V_\lambda)^{-1}\psi(\cdot,\lambda) = J\eta_0, \quad \Im\lambda \neq 0.$$

Thus $\eta_0 \in L^2(H)$, $J\eta_0 \in L^2(H)$, which is impossible since $\operatorname{tr} H(x) = 1, x \in (0,\infty)$. The Theorem 1.4 is proved. □

In particular, for real H we have $\overline{\varphi(x,\lambda)} = \varphi(x,\overline{\lambda})$, whence the defect numbers of $\widetilde{A}_\alpha$ are zero and thus $\widetilde{A}_\alpha$ $(= A_\alpha)$ is self-adjoint. This fact was first established by L. de Branges [dB1]. If

$$H(x) = \begin{pmatrix} 1 & ih(x) \\ -ih(x) & 1 \end{pmatrix},$$

where $0 \leq h(x) \leq 1$, $1 - h(x) \in L^1(0,\infty)$, then it is easy to show that $\varphi(\cdot,\lambda) \in L^2(H)$ for $\Im\lambda < 0$, that is the defect numbers of $\widetilde{A}_0$ are 1 and 0.

2 The Spectral Expansion

We consider the problem

$$Jy'(x) = \lambda H(x) y(x) \quad \text{a.e. on } (0,\infty), \tag{2.1}$$

$$y(0) \in \operatorname{span}\{\eta_\alpha\}. \tag{2.2}$$

Let $\varphi_\alpha(x,\lambda)$ and $\psi_\alpha(x,\lambda)$ be the solutions of (2.1), which satisfy the initial conditions

$$\varphi_\alpha(0,\lambda) = \eta_\alpha, \quad \psi_\alpha(0,\lambda) = J\eta_\alpha.$$

For vector functions $f \in L^2(H)$ which are H-finite at ∞ (by which we mean that $f(x) \sim 0$ on some interval $(c_f, +\infty)$) we define

$$F(f, \lambda) = \int_0^\infty \varphi_\alpha(x, \overline{\lambda})^* H(x) f(x) dx. \tag{2.3}$$

The nondecreasing function $\sigma_\alpha(\lambda)$, $\lambda \in (-\infty, \infty)$, is called a spectral function of the problem (2.1), (2.2) if

$$\int_0^\infty f(x)^* H(x) f(x)\, dx = \int_{-\infty}^\infty |F(f, \lambda)|^2 d\sigma_\alpha(\lambda) \tag{2.4}$$

for every vector function $f \in L_0^2(H; \alpha)$ which is H-finite at ∞. The relation (2.4) is extended in the usual way to the whole space $L_0^2(H; \alpha)$.

Next we show the existence of a spectral function for the problem (2.1), (2.2) by the same method which was used in the case of the Sturm-Liouville problem in [CL]. First we consider the regular problem

$$Jy'(x) = \lambda H(x) y(x), \quad x \in (0, l), \tag{2.5}$$

$$y(0) \in \operatorname{span}\{\eta_\alpha\}, \quad y(l) \in \operatorname{span}\{\eta_\beta\}. \tag{2.6}$$

Let

$$\chi_{l,\alpha,\beta}(x, \lambda) = \psi_\alpha(x, \lambda) + m_{l,\alpha,\beta}(\lambda)\varphi_\alpha(x, \lambda)$$

be a solution of (2.5) which satisfies the boundary condition $\chi_{l,\alpha,\beta}(l, \lambda) \in$ span $\{\eta_\beta\}$. Then

$$m_{l,\alpha,\beta}(\lambda) = -\frac{\psi_{2,\alpha}(l, \lambda)\cos\beta - \psi_{1,\alpha}(l, \lambda)\sin\beta}{\varphi_{2,\alpha}(l, \lambda)\cos\beta - \varphi_{1,\alpha}(l, \lambda)\sin\beta}, \quad \Im\lambda \neq 0. \tag{2.7}$$

Note, that if y_μ and y_ν are solutions of (2.1) for $\lambda = \mu$ and $\lambda = \nu$, respectively, then (see [GK2], [At])

$$(\mu - \overline{\nu}) \int_0^l y_\nu(x)^* H(x) y_\mu(x)\, dx = y_\nu(x)^* J y_\mu(x)\Big|_0^l. \tag{2.8}$$

Applying (2.8) to $y_\lambda = \chi_{l,\alpha,\beta}(x, \lambda)$ and $\mu = \nu = \lambda$ and taking into account that

$$(J\chi_{l,\alpha,\beta}(l, \lambda), \chi_{l,\alpha,\beta}(l, \lambda)) = 0$$

we obtain the relation

$$\int_0^l \chi_{l,\alpha,\beta}(x, \lambda)^* H(x) \chi_{l,\alpha,\beta}(x, \lambda)\, dx = -\frac{\Im m_{l,\alpha,\beta}(\lambda)}{\Im\lambda}, \quad \Im\lambda \neq 0. \tag{2.9}$$

The eigenvalues and the eigenfunctions of the problem (2.5), (2.6) and of $A_{l,\alpha,\beta}$ coincide. Since $A_{l,\alpha,\beta}^{-1} = B_0$ it follows that the spectrum of $A_{l,\alpha,\beta}$ consists of a finite or countable number of eigenvalues λ_n with the only accumulation point ∞. Then the functions $\varphi_\alpha(\cdot,\lambda_n)(\in L_0^2(H;0,l,\alpha,\beta))$ are the corresponding eigenfunctions. The Hilbert-Schmidt theory implies, that for every $f\in L_0^2(H;0,l,\alpha,\beta)$ and with

$$F_l(f,\lambda)=\int_0^l \varphi_\alpha(x,\lambda)^* H(x)f(x)dx,$$

the relation

$$\int_0^l f(x)^* H(x)f(x)\,dx=\int_{-\infty}^{\infty}|F_l(f,\lambda)|^2\,d\sigma_{l,\alpha,\beta}(\lambda), \tag{2.10}$$

holds, where $\sigma_{l,\alpha,\beta}$ is a step function with jumps ρ_n at the points λ_n,

$$\rho_n^{-1}=\int_0^l \varphi_\alpha(x,\lambda_n)^* H(x)\varphi_\alpha(x,\lambda_n)\,dx,\quad n=1,2,\dots.$$

Note that $\varphi_\alpha(\cdot,\lambda_n)\in L_0^2(H;0,l,\alpha,\beta)$ implies that $L_0^2(H;0,l,\alpha,\beta)$ is the maximal subspace of $L^2(H)$, where the set of eigenfunctions of the boundary problem (2.5), (2.6) is complete.

If there is no H-indivisible interval of type $\alpha+\frac{\pi}{2}$ of the form $(0,d)$, then $\chi_{l,\alpha,\beta}$ belongs to $L_0^2(H;0,l,\alpha,\beta)$ and (2.10) holds for $f=\chi_{l,\alpha,\beta}$. In view of (2.8) we have

$$\begin{aligned}\int_0^l &\varphi_\alpha(x,\lambda_n)^* H(x)\chi_{l,\alpha,\beta}(x,\lambda)dx\\ &=-\frac{1}{\lambda-\lambda_n}\varphi_\alpha(0,\lambda_n)^* J\chi_{l,\alpha,\beta}(0,\lambda)=\frac{1}{\lambda-\lambda_n}.\end{aligned}$$

It follows from (2.9) and (2.10) that

$$-\frac{\Im m_{l,\alpha,\beta}(\lambda)}{\Im\lambda}=\int_{-\infty}^{\infty}\frac{d\sigma_{l,\alpha,\beta}(t)}{|t-\lambda|^2}.$$

Now let $(0,d)$, $d<l$, be a maximal H-indivisible interval of type $\alpha+\frac{\pi}{2}$. This means that $\chi_{l,\alpha,\beta}(x,\lambda)=J\eta_\alpha+(\lambda x+m_{l,\alpha,\beta}(\lambda))\eta_\alpha$ on $[0,d]$. We define

$$\widetilde{\chi}(x)=\begin{cases}\chi_{l,\alpha,\beta}(x,\lambda)-J\eta_\alpha & \text{if } 0\le x\le d,\\ \chi_{l,\alpha,\beta}(x,\lambda) & \text{if } x>d.\end{cases}$$

Since $\widetilde{\chi}\in L_0^2(H;0,l,\alpha\beta)$, in (2.10) we can choose $f=\widetilde{\chi}$. It follows that

$$\begin{aligned}\int_0^l \widetilde{\chi}(x)^* H(x)\widetilde{\chi}(x)\,dx &= \int_0^l \chi_{l,\alpha,\beta}(x,\lambda)^* H(x)\chi_{l,\alpha,\beta}(x,\lambda)\,dx-d,\\ &\int_0^l \varphi_\alpha(x,\lambda_n)^* H(x)\widetilde{\chi}(x)\,dx=\frac{1}{\lambda-\lambda_n}.\end{aligned}$$

Using (2.9) and (2.10) we obtain

$$-\frac{\Im m_{l,\alpha,\beta}(\lambda)}{\Im\lambda}=d+\int_{-\infty}^{\infty}\frac{d\sigma_{l,\alpha,\beta}(t)}{|t-\lambda|^2}.$$

Hence,

$$-m_{l,\alpha,\beta}(\lambda)=c+d\lambda+\int_{-\infty}^{\infty}\left(\frac{1}{t-\lambda}-\frac{t}{1+t^2}\right)d\sigma_{l,\alpha,\beta}(t),\quad \Im c=0,\Im\lambda\neq 0,\tag{2.11}$$

where d is the length of the maximal H-indivisible interval at zero of type $\alpha+\frac{\pi}{2}$. It follows from (2.7) that for fixed α and λ, $\Im\lambda\neq 0$, $m_{l,\alpha,\beta}(\lambda)$ maps the interval $[-\frac{\pi}{2},\frac{\pi}{2})$ onto a circle $C_{l,\alpha}(\lambda)$. If $K_{l,\alpha}(\lambda)$ and $r_{l,\alpha}(\lambda)$ denote the interior and the radius of $C_{l,\alpha}(\lambda)$, respectively, it follows from (2.9) that $K_{l',\alpha}(\lambda)\subset K_{l,\alpha}(\lambda)$ for $l<l'$. Using (2.7) we obtain

$$\begin{aligned} r_l(\lambda)&=\exp\left(2y\int_0^l\Im h_{12}(x)\,dx\right)\\ &\left\{2|y|\int_0^l\varphi_\alpha(x,\lambda)^*H(x)\varphi_\alpha(x,\lambda)dx\right\},\quad y=\Im\lambda.\end{aligned}\tag{2.12}$$

The relation (2.11) implies that

$$\overline{m_{l,\alpha,\beta}(\lambda)}=m_{l,\alpha,\beta}(\overline{\lambda})\tag{2.13}$$

and, consequently, $r_{l,\alpha}(\lambda)=r_{l,\alpha}(\overline{\lambda})$. Then from (2.12) we obtain

$$\begin{aligned} r_{l,\alpha}^{-2}(\lambda)&=4y^2\int_0^l\varphi_\alpha(x,\lambda)^*H(x)\varphi_\alpha(x,\lambda)dx\\ &\int_0^l\varphi_\alpha(x,\overline{\lambda})^*H(x)\varphi_\alpha(x,\overline{\lambda})\,dx,\quad y=\Im\lambda.\end{aligned}\tag{2.14}$$

This relation and Theorem 1.4 imply $\lim_{l\to\infty}r_{l,\alpha}(\lambda)=0$, $\Im\lambda\neq 0$. We mention that this is true also if the condition (*) is not satisfied. Therefore, $m_{l,\alpha,\beta}(\lambda)$ converges if $l\to\infty$; the function

$$m_\alpha(\lambda)=\lim_{l\to+\infty}m_{l,\alpha,\beta}(\lambda)$$

is analytic in the upper half plane. It is called the Weyl function of the problem (2.1), (2.2). Since $\Im m_\alpha(\lambda)/\Im\lambda>0$ if $\Im\lambda\neq 0$, $m_\alpha(\lambda)$ admits a unique representation (see [AG])

$$m_\alpha(\lambda)=\tilde{c}_\alpha+\tilde{d}_\alpha\lambda+\int_{-\infty}^{\infty}\left(\frac{1}{t-\lambda}-\frac{t}{1+t^2}\right)d\sigma_\alpha(t),\quad\Im\lambda\neq 0.\tag{2.15}$$

where $\tilde{c}_\alpha$ is real, $\tilde{d}_\alpha\geq 0$ and σ_α is a nondecreasing function on the real axis such that $\int_{-\infty}^{+\infty}(1+t^2)^{-1}\sigma_\alpha(t)<\infty$.

Theorem 2.1 *The function $\sigma_\alpha(\lambda)$ defined by (2.15) is the unique spectral function of the problem (2.1), (2.2), and the expansion formula*

$$f(x) = \int_{-\infty}^{\infty} F(f,\lambda)\varphi_\alpha(x,\lambda)d\sigma_\alpha(\lambda), \quad f \in L_0^2(H,\alpha), \tag{2.16}$$

holds.

As usual, the integral in the relation (2.16) is to be understood in the sense that $\|f_\Delta - f\|_H \to 0$ if $\Delta = (\mu, \nu) \to (-\infty, +\infty)$, where

$$f_\Delta(x) = \int_\Delta F(f,\lambda)\varphi_\alpha(x,\lambda)\,\sigma_\alpha(\lambda).$$

In the case of a real matrix function H Theorem 2.1 (including the definition of the space $L_0^2(H;\alpha)$) was first established by L. de Branges [dB1] by means of his theory of Hilbert spaces of entire functions. If H is complex the existence of a spectral function and the expansion (2.16) without the relation (2.15) were proved in [Ka1] using the method of directing functionals for linear relations (see [LT]). As we have mentioned already, here we applied the method, which was used in [CL] for Sturm-Liouville operators.

In the proof of Theorem 2.1 we use a simple lemma which we formulate now. Denote by (R) the class of all functions F which are analytic in the upper half plane and map the upper half plane into itself. Such a function F admits the representation (see [AG])

$$F(z) = \nu + \mu z + \int_{-\infty}^{\infty} \left(\frac{1}{t-z} - \frac{t}{1+t^2} \right) d\sigma(t), \tag{2.17}$$

where σ is a nondecreasing function, $\int_{-\infty}^{+\infty}(1+t^2)^{-1}dt < \infty$, $\Im \nu = 0$ and $\mu \geq 0$.

Lemma 2.2 *Let $F_n \in (R)$, $n \in N$, and let σ_n be the corresponding functions from (2.17), normalized such that $\sigma_n(t_0) = \sigma(t_0)$, where t_0 is a continuity point of σ. Then, if $\lim_{n\to\infty} F_n(z) = F(z)$, $Imz > 0$, it follows that $\lim_{n\to\infty} \sigma_n(t) = \sigma(t)$ for all t where the function σ is continuous.*

The proof of this lemma is left to the reader.

Proof of Theorem 2.1: First assume that $f \in D(\hat{A}_\alpha)$. Then

$$\begin{aligned} \lambda F(f,\lambda) &= \int_0^\infty (J\varphi'_\alpha(x,\bar{\lambda}))^* f(x)\,dx = \int_0^\infty \varphi_\alpha(x,\lambda)^* Jf'(x)\,dx \\ &= F(\hat{A}_\alpha f, \lambda), \end{aligned}$$

that is

$$F(\hat{A}_\alpha f, \lambda) = \lambda F(f,\lambda). \tag{2.18}$$

It follows from $f \in D(\hat{A}_\alpha)$ that $f \in D(A_{l,\alpha,\beta})$ and $\hat{A}_\alpha f = A_{l,\alpha,\beta} f$ for sufficiently large l. Therefore we can apply the identity (2.10) to f, and it follows that

$$\int_0^\infty f(x)^* H(x) f(x)\,dx = \int_{-\infty}^\infty |F(f,\lambda)|^2 d\sigma_{l,\alpha,\beta}(\lambda).$$

In view of (2.18) we obtain

$$\begin{aligned}
&\left| \int_0^\infty f(x)^* H(x) f(x)\,dx - \int_{-\mu}^{\mu} |F(f,\lambda)|^2 d\sigma_{l,\alpha,\beta}(\lambda) \right| \\
&\quad \le \frac{1}{\mu^2} \int_{|\lambda|>\mu} \lambda^2 |F(f,\lambda)|^2 d\sigma_{l,\alpha,\beta}(\lambda) \le \frac{1}{\mu^2} \int_{-\infty}^{\infty} \lambda^2 |F(f,\lambda)|^2 d\sigma_{l,\alpha,\beta}(\lambda) \\
&\quad = \frac{1}{\mu^2} \int_{-\infty}^{\infty} |F(A_{l,\alpha,\beta} f,\lambda)|^2 d\sigma_{l,\alpha,\beta}(\lambda) = \frac{1}{\mu^2} \|A_{l,\alpha,\beta} f\|_H^2.
\end{aligned}$$

It follows from Lemma 2.2 that $\sigma_{l,\alpha,\beta}(\lambda) \to \sigma_\alpha(\lambda)$ if $l \to \infty$, for all points λ at which σ_α is continuous. Assume that μ is such a continuity point of σ_α, then from the above inequality for $l \to \infty$ we obtain

$$\left| |f|_H^2 - \int_{-\mu}^{\mu} |F(f,\lambda)|^2 d\sigma_\alpha(\lambda) \right| \le \frac{1}{\mu^2} \|\hat{A}_\alpha f\|.$$

If $\mu \to +\infty$ it follows that (2.4) holds for every $f \in D(\hat{A}_\alpha)$, and hence also for all $f \in L_0^2(H;\alpha)$.

Next we prove (2.16). First we assume that (l,∞) is a maximal H-indivisible interval of type $\beta + \frac{\pi}{2}$. Then it is easy to see that

$$\begin{aligned}
\varphi_\alpha(x,\lambda) &= c(\lambda) J \eta_\beta + (\lambda x c(\lambda) + c_1(\lambda))\eta_\beta, \quad x \ge l, \\
\psi_\alpha(x,\lambda) &= \tilde{c}(\lambda) J \eta_\beta + (\lambda x \tilde{c}(\lambda) + \tilde{c}_1(\lambda))\eta_\beta, \quad x \ge l,
\end{aligned}$$

where $c(\lambda)$ and $\tilde{c}(\lambda)$ do not vanish for $\Im\lambda \neq 0$. For $b \ge l$ we have

$$m_{b,\alpha,\beta}(\lambda) = -\frac{(\psi_\alpha(b,\lambda), J\eta_\beta)}{(\varphi_\alpha(b,\lambda), J\eta_\beta)} = -\frac{\tilde{c}(\lambda)}{c(\lambda)}.$$

Thus, if (l,∞) is a maximal H-indivisible interval of type $\beta + \frac{\pi}{2}$, then

$$m_\alpha(\lambda) = m_{l,\alpha,\beta}(\lambda). \tag{2.19}$$

Hence $\sigma_\alpha(\lambda) = \sigma_{l,\alpha,\beta}(\lambda)$. In this case (2.16) is the Fourier-series expansion of f.

Now assume that there is no H-indivisible interval of the form (l,∞). Set

$$f_\Delta(x) = \int_\Delta F(f,\lambda)\varphi_\alpha(x,\lambda)\,d\sigma_\alpha(\lambda), \qquad \Delta = (\mu,\nu).$$

For every vector function $P_a \in L_0^2(H;\alpha)$, such that $P_a(x) \sim 0$ on (a,∞), we get

$$\int_0^a P_a(x)^* H(x) f_\Delta(x)dx = \int_\Delta F(f,\lambda)\overline{F(P_a,\lambda)}d\sigma_\alpha(\lambda). \tag{2.20}$$

Hence,

$$\begin{aligned}
&\left|\int_0^a P_a(x)^* H(x) f_\Delta(x)\,dx\right| = \\
&\le \left(\int_\Delta |F(f,\lambda)|^2 d\sigma_\alpha(\lambda)\right)^{1/2}\left(\int_\Delta |F(P_a,\lambda)|^2 d\sigma_\alpha(\lambda)\right)^{1/2} \\
&\le \left(\int_{-\infty}^{\infty} |F(f,\lambda)|^2 d\sigma_\alpha(\lambda)\right)^{1/2}\left(\int_{-\infty}^{\infty} |F(P_a,\lambda)|^2 d\sigma_\alpha(\lambda)\right)^{1/2} \\
&= \left(\int_0^{\infty} f(x)^* H(x) f(x)\,dx\right)^{1/2}\left(\int_0^a P_a(x)^* H(x) P_a(x)\,dx\right)^{1/2}.
\end{aligned}$$

Let $P_a(x) = f_\Delta(x)$ for $x < a$, and $P_a(x) \sim 0$ on (a,∞). Evidently, $P_a(x) \in L_0^2(H;\alpha)$. Thus, it follows that

$$\int_0^a f_\Delta(x) H(x) f_\Delta(x)\,dx \le \int_0^a f(x)^* H(x) f(x)\,dx,$$

and $f_\Delta \in L_0^2(H;\alpha)$. The relation (2.4) yields

$$\int_0^a P_a(x)^* H(x) f(x)\,dx = \int_{-\infty}^{\infty} F(f,\lambda)\overline{F(P_a,\lambda)}\,d\sigma_\alpha(\lambda).$$

In view of (2.20), with $\Delta^c = (-\infty,\infty)\backslash\Delta$ this implies

$$\int_0^a P_a(x)^* H(x)(f(x) - f_\Delta(x))\,dx = \int_{\Delta^c} F(f,\lambda)\overline{F(P_a,\lambda)}\,d\sigma_\alpha(\lambda). \tag{2.21}$$

Now let $P_a(x) = f(x) - f_\Delta(x)$ for $x < a$ and $P_a(x) \sim 0$ on (a,∞). ¿From (2.21) we find

$$\begin{aligned}
&\int_0^a (f(x) - f_\Delta(x))^* H(x)(f(x) - f_\Delta(x))\,dx \\
&\le \left(\int_{\Delta^c} |F(f,\lambda)|^2 d\sigma(\lambda)\right)^{1/2} \int_{-\infty}^{\infty} |F(P_a,\lambda)|^2 d\sigma_\alpha(\lambda) \\
&= \left(\int_{\Delta^c} |F(f,\lambda)|^2 d\sigma(\lambda)\right)^{1/2} \\
&\quad \left(\int_0^a (f(x) - f_\Delta(x))^* H(x)(f(x) - f_\Delta(x))\,dx\right)^{1/2},
\end{aligned}$$

whence it follows that

$$\int_0^a (f(x) - f_\Delta(x))^* H(x)(f(x) - f_\Delta(x))\,dx \le \int_{\Delta^c} |F(f,\lambda)|^2 d\sigma_\alpha(\lambda),$$

and

$$f - f_\Delta \longrightarrow 0 \quad \text{if} \quad \Delta \longrightarrow (-\infty, \infty).$$

The uniqueness of σ_α follows from Theorem 1.4. □

Remark: Let $\widetilde{H}(x)$ be real, $\widetilde{H}(x) \ge 0$ and $H(x) = \widetilde{H}(x) - ib\,(x)J$, where $\Im b(x) = 0$ and $\det H(x) - b^2(x) \ge 0$. Then the Weyl function of the problem (2.1), (2.2) remains the same if we replace $H(x)$ in (2.1) by $\widetilde{H}(x)$. Indeed, it is easy to verify that if $z(x,\lambda)$ is a solution of $Jz'(x) = \widetilde{H}(x)z(x)$, then $y(x,\lambda) = \exp\{-\lambda i \int_0^x b(s)\,ds\}z(x,\lambda)$ is a solution of $Jy'(x) = H(x)y(x)$. Then it follows from (2.7) that $m_{l,\alpha,\beta} = \widetilde{m}_{l,\alpha,\beta}$, hence $m_\alpha = \widetilde{m}_\alpha$. Theorem 2.1 implies that $\widetilde{\sigma}_\alpha(\lambda) = \sigma_\alpha(\lambda)$. It is also clear, that the $\widetilde{H}$- and H-indivisible intervals coincide. Consequently, $\widetilde{d}_\alpha$ in (2.15) is equal to the length of the maximal H-indivisible interval of type $\alpha + \pi/2$ ([dB2], [W]).

3 The Trace Formula

Let B and $\widetilde{B}$ be self-adjoint operators in a Hilbert space. Suppose that their difference is a nuclear operator. M.G. Krein proved in [K1] that there exists a real-valued function $\xi(t, \widetilde{B}, B)$, such that for every rational function Φ, which is regular on the spectra of B and $\widetilde{B}$ and has a pole of order ≤ 1 at ∞, the formula

$$\text{(3.1)} \qquad \operatorname{tr}(\Phi(\widetilde{B}) - \Phi(B)) = \int_{-\infty}^{\infty} \Phi'(t)\xi(t, \widetilde{B}, B)\,dt$$

holds. In particular, if $\Phi(\lambda) = \lambda$ from (3.1) it follows that

$$\text{(3.2)} \qquad \operatorname{tr}(\widetilde{B} - B) = \int_{-\infty}^{\infty} \xi(t, \widetilde{B}, B)\,dt,$$

and if $\Phi(\lambda) = (\lambda - z)^{-1}$, $\Im z \ne 0$, then

$$\text{(3.3)} \qquad \operatorname{tr}(R_z(\widetilde{B}) - R_z(B)) = -\int_{-\infty}^{\infty} \frac{1}{(t-z)^2}\xi(t, \widetilde{B}, B)\,dt, \quad \Im z \ne 0.$$

Here $R_z(T) = (T - z)^{-1}$ is the resolvent of T.

Note that $\xi(t, \widetilde{B}, B)$ is uniquely determined by (3.2) and (3.3). The formula (3.1) is called the trace formula and $\xi(t, \widetilde{B}, B)$ is the spectral shift function of the pair $B, \widetilde{B}$.

In [K2], [K3] M.G. Krein extended the formula (3.1) to a broader class of operators, namely, only the difference $R_z(\widetilde{B}) - R_z(B)$ was required to be nuclear for some z, $\Im z \neq 0$. In this case Φ must be regular at ∞, and then the relation (3.3) is still valid. However, in this case $\xi(t, \widetilde{B}, B)$ is not unique, it is determined only up to an additive constant.

In the following we consider the operators $\widetilde{A}_\alpha$ and $\widetilde{A}_\gamma$ from Section 1, which are generated by a canonical system and different boundary conditions at $x = 0$. For all $\alpha \in [-\frac{\pi}{2}, \frac{\pi}{2})$, with possible exception of one value of α, the operators $\widetilde{A}_\alpha$ are maximal symmetric extensions of A, where the operator A is generated by the boundary condition $g(0) = 0$. More precisely, we define A in $L_0^2(H)$ as follows. For a function g such that $g(x) = 0$ on $[c_g, +\infty)$, we say that $g \in D(A)$ and $f = Ag$, if g is locally absolutely continuous on $[0, \infty)$, $g(0) = 0$ and $Jg'(x) = H(x)f(x)$. It follows from Theorem 1.4 that the defect index of A is either $(1, 1)$ or $(1, 2)$. Consider two cases:

(I) There is no H-indivisible interval of the form $(0, b)$ for some $b > 0$.

(II) There exists a $b > 0$ such that the interval $(0, b)$ is H-indivisible.

In case (I) we have $A \subset \widetilde{A}_\gamma$ for each $\gamma \in [-\frac{\pi}{2}, \frac{\pi}{2})$ and $\overline{D(A)} = L_0^2(H)$. In case (II), if $H(x)\eta_\alpha = 0$ a.e. on $(0, b)$, we have $A \subset \widetilde{A}_\gamma$ for $\gamma \neq \alpha$, $D(A) \not\subset D(\widetilde{A}_\alpha)$ and $D(A) \subset L_1^2(H)$.

Obviously, the resolvents of $\widetilde{A}_\alpha$ and $\widetilde{A}_\gamma$ differ by a one-dimensional operator. Consequently, if they are self-adjoint operators the relations (3.3) and (3.1) (for appropriate Φ) hold. Here we establish a connection between $\xi(t, \alpha, \gamma) := \xi(t, A_\alpha, A_\gamma)$ and the spectral function $\sigma_\alpha(t)$ and also the Weyl function $m_\alpha(\lambda)$. In case of a finite interval $(0, l)$ we fix the boundary conditions at the right endpoint $x = l$. Let $\varphi = \varphi_0$, $\psi = \psi_0$ and

$$\chi(x, \lambda) = \psi(x, \lambda) + m_0(\lambda)\varphi(x, \lambda).$$

Lemma 3.1 *If the operator A_0 is self-adjoint, then*

$$\int_0^\infty \chi(x, \nu)^* H(x)\chi(x, \mu)\, dx = -\frac{m_0(\mu) - m_0(\overline{\nu})}{\mu - \overline{\nu}}, \qquad \Im\mu \neq 0,\ \Im\nu \neq 0. \tag{3.4}$$

Proof: It follows from (2.9) that $K_l(\lambda) = K_{l,0}(\lambda)$ is given by the inequality

$$\|\psi(\cdot, \lambda) + m\varphi(\cdot, \lambda)\|_{H,l}^2 \leq -\frac{\Im m}{\Im\lambda}, \qquad m \in K_l(\lambda),\ \Im\lambda \neq 0.$$

Since $r_l(\lambda) \to 0$ it follows that $\cap_{l>0} K_l(\lambda) = \{m_0(\lambda)\}$. This and the previous inequality imply that $\chi(\cdot, \lambda) \in L^2(H)$ and

$$\|\chi(\cdot, \lambda)\|_H^2 \leq -\frac{\Im m_0(\lambda)}{\Im\lambda}, \qquad \Im\lambda \neq 0.$$

According to the assumption of Lemma 3.1 it follows that $\varphi(\cdot,\lambda) \notin L^2(H)$ if $\Im\lambda \neq 0$. Consequently, from (2.14) we have

$$\begin{aligned}\|\chi_{l,0,\beta}(\cdot,\lambda) - \chi(\cdot,\lambda)\|_{H,l}^2 &\leq 4r_l^2(\lambda)\|\varphi(\cdot,\lambda)\|_{H,l}^2 \\ &= \frac{1}{(\Im\lambda)^2}\|\varphi(\cdot,\overline{\lambda})\|_{H,l}^{-2} \longrightarrow 0, \quad l \longrightarrow \infty.\end{aligned}$$

Since $\chi(\cdot,\lambda) \in L^2(H)$ we obtain

$$(\chi_{l,0,\beta}(\cdot,\mu), \chi_{l,0,\beta}(\cdot,\nu))_{H,l} \longrightarrow (\chi(\cdot,\mu), \chi(\cdot,\nu))_H, \quad l \longrightarrow \infty.$$

It follows from (2.8) that

$$(\chi_{l,0,\beta}(\cdot,\mu), \chi_{l,0,\beta}(\cdot,\nu))_{H,l} = -\frac{m_{l,0,\beta}(\mu) - m_{l,0,\beta}(\overline{\nu})}{\mu - \overline{\nu}}.$$

Passing to the limits leads to (3.4).

Now let us find an expression for the resolvent $R_\lambda(\hat{A}_\alpha)$ of $\hat{A}_\alpha$. If f and g are functions from $L_0^2(H;\alpha)$, which are finite at ∞ and $(\hat{A}_\alpha - \lambda I)g = f$ then

$$Jg'(x) = \lambda H(x)g(x) + H(x)f(x) \text{ a.e. on } (0,\infty), \quad g(0) \in \operatorname{span}\{\eta_\alpha\}.$$

We shall show that

$$\begin{aligned} g(x) &= (R_\lambda(\hat{A}_\alpha)f)(x) = -\frac{1}{\Delta_\alpha(\lambda)}\left\{\chi(x,\lambda)\int_0^x \varphi_\alpha(s,\overline{\lambda})^* H(s)f(s)\,ds\right. \\ &\qquad \left. + \varphi_\alpha(x,\lambda)\int_x^\infty \chi(s,\lambda)^* H(s)f(s)\,ds\right\}, \quad \Im\lambda \neq 0, \end{aligned} \tag{3.5}$$

where

$$\Delta_\alpha(\lambda) = \cos\alpha - m_0(\lambda)\sin\alpha, \tag{3.6}$$

$$(f, \varphi_\alpha(\cdot,\overline{\lambda}))_H = 0. \tag{3.7}$$

To this end we use the relation

$$V(x,\lambda)JV(x,\lambda)^* = \chi(x,\lambda)\varphi(x,\overline{\lambda})^* - \varphi(x,\lambda)\chi(x,\overline{\lambda})^* = \Delta_\alpha(\lambda)J,$$

with

$$V(x,\lambda) = \begin{pmatrix} \varphi_1(x,\lambda) & \chi_1(x,\lambda) \\ \varphi_2(x,\lambda) & \chi_2(x,\lambda)\end{pmatrix}, \quad \varphi = \begin{pmatrix}\varphi_1 \\ \varphi_2\end{pmatrix}, \quad \chi = \begin{pmatrix}\chi_1 \\ \chi_2\end{pmatrix}.$$

If the closure $\widetilde{A}_\alpha$ of $\hat{A}_\alpha$ is self-adjoint: $\widetilde{A}_\alpha^* = \widetilde{A}_\alpha = A_\alpha$, it follows that $\varphi_\alpha(\cdot,\lambda) \notin L_0^2(H;\alpha)$, $\Im\lambda \neq 0$. Thus, the set of functions $f \in L_0^2(H;\alpha)$, which are finite at ∞ and satisfy (3.7) is dense in $L_0^2(H;\alpha)$ and $R_\lambda(\hat{A}_\alpha)$ can be defined by continuity on the whole space $L_0^2(H;\alpha)$. So, in this case the resolvent of A_α is defined by (3.5), $g = R_\lambda(A_\alpha)f$.

In case (II), if $(0, b)$ is of type $\alpha + \pi/2$, the resolvents $R_\lambda(A_\gamma)$, $\gamma \neq \alpha$, are defined on $L_0^2(H)$, while $R_\lambda(A_\alpha)$ is defined on $L_1^2(H)$. In the sequel, we will set $R_\lambda(A_\alpha)f = 0$ for $f \in L_0^2(H) \ominus L_1^2(H)$. Obviously, this agrees with the relation (3.5).

Note that if an interval of the form (l, ∞) is H-indivisible of type $\beta + \pi/2$, then

$$R_\lambda(A_\alpha) = R_\lambda(A_{l,\alpha,\beta}). \tag{3.8}$$

In this case we also have that $\varphi_\alpha(\cdot, \lambda) \notin L_0^2(H;\alpha)$, since otherwise $\varphi_\alpha(l, \lambda) \in \operatorname{span}\{\eta_\alpha\}$, whence it is easy to conclude that $\varphi_\alpha(x, \lambda) = \eta_\alpha$ on $(0, \infty)$. The latter means that $(0, \infty)$ is H-indivisible of type $\alpha + \pi/2$, which we have excluded.

Since

$$-\frac{1}{\Delta_\alpha(\lambda)}\varphi_\alpha(x, \lambda) + \varphi(x, \lambda) = -\frac{\sin\alpha}{\Delta_\alpha(\lambda)}\chi(x, \lambda),$$

it follows from (3.5) that

$$R_\lambda(A_\alpha) - R_\lambda(A_0) = -\frac{\sin\alpha}{\Delta_\alpha(\lambda)}(\cdot, \chi(\cdot, \bar{\lambda}))_H \chi(\cdot, \lambda), \quad \Im\lambda \neq 0.$$

Hence,

$$\operatorname{tr}(R_\lambda(A_\alpha) - R_\lambda(A_0)) = -\frac{\sin\alpha}{\Delta_\alpha(\lambda)}(\chi(\cdot, \lambda), \chi(\cdot, \bar{\lambda}))_H.$$

Thus, from (3.4) and (3.6) we obtain that

$$\operatorname{tr}(R_\lambda(A_\alpha) - R_\lambda(A_0)) = -\frac{d}{d\lambda}\log\Delta_\alpha(\lambda), \quad \Im\lambda \neq 0. \tag{3.9}$$

Since $-\Im m_0(\lambda)/\Im\lambda \geq 0$, it follows that $\Im\Delta_\alpha(\lambda)/\Im\lambda \geq 0$ and $0 \leq \arg\Delta_\alpha(\lambda) \leq \pi$ for $\Im\lambda > 0$, $0 \leq \alpha < \frac{\pi}{2}$. Hence, if we choose the branch of the logarithm which is continuous on $(0, +\infty)$ and such that $\log 1 = 0$, then $\log\Delta_\alpha(\lambda)$ admits the representation

$$\log\Delta_\alpha(\lambda) = c_\alpha + \int_{-\infty}^{\infty}\left(\frac{1}{t-\lambda} - \frac{t}{1+t^2}\right)\xi(t, \alpha)\,dt, \quad \Im\lambda \neq 0, \tag{3.10}$$

where c_α is real and

$$\xi(t, \alpha) = \frac{1}{\pi}\arg\Delta_\alpha(t + i0) \quad \text{a.e. on } (-\infty, \infty). \tag{3.11}$$

It is clear, that (3.10) is also valid for $-\frac{\pi}{2} \leq \alpha < 0$, where $\xi(t, \alpha) = -\xi(t, -\alpha)$. It follows from (3.9) and (3.10) that

$$\begin{gathered}\operatorname{tr}(R_\lambda(A_\alpha) - R_\lambda(A_0)) = -\int_{-\infty}^{\infty}\frac{\xi(t, \alpha)}{(t-\lambda)^2}\,dt, \\ -\frac{\pi}{2} \leq \alpha < \frac{\pi}{2}, \quad \Im\lambda \neq 0.\end{gathered} \tag{3.12}$$

Let

$$\Delta_{\alpha,\gamma}(\lambda) := \frac{\Delta_\alpha(\lambda)}{\Delta_\gamma(\lambda)} = \frac{\cos\alpha - m_0(\lambda)\sin\alpha}{\cos\gamma - m_0(\lambda)\sin\gamma}, \quad \alpha, \gamma \in \left[-\frac{\pi}{2}, \frac{\pi}{2}\right), \quad \Im\lambda \neq 0. \tag{3.13}$$

From

$$m_\alpha(\lambda) = \frac{\sin\alpha + m_0(\lambda)\cos\alpha}{\cos\alpha - m_0(\lambda)\sin\alpha} \tag{3.14}$$

we obtain

$$\Delta_{\alpha,\gamma}(\lambda) = \cos(\alpha - \gamma) - m_\gamma(\lambda)\sin(\alpha - \gamma).$$

Then (3.10) implies the representation

$$\log \Delta_{\alpha,\gamma}(\lambda) = c_{\alpha,\gamma} + \int_{-\infty}^{\infty} \left(\frac{1}{t-\lambda} - \frac{t}{1+t^2}\right) \xi(t,\alpha,\gamma)\,dt, \quad \alpha, \gamma \in \left[-\frac{\pi}{2}, \frac{\pi}{2}\right), \tag{3.15}$$

where $c_{\alpha,\gamma}$ is real, $\xi(t,\alpha,0) = \xi(t,\alpha)$, $c_{\alpha,0} = c_\alpha$. Since $\Delta_{\alpha,\gamma}(\lambda) = \Delta_{\alpha,\beta}(\lambda)$ $\Delta_{\beta,\gamma}(\lambda)$ and $\Delta_{\alpha,\alpha}(\lambda) = 1$, we have

$$\xi(t,\alpha,\gamma) = \xi(t,\alpha,\beta) + \xi(t,\beta,\gamma), \quad \xi(t,\alpha,\gamma) = -\xi(t,\gamma,\alpha).$$

The relation (3.12) implies that

$$\operatorname{tr}(R_\lambda(A_\alpha) - R_\lambda(A_\gamma)) = -\int_{-\infty}^{\infty} \frac{\xi(t,\alpha,\gamma)}{(t-\lambda)^2}\,dt, \quad \alpha, \gamma \in \left[-\frac{\pi}{2}, \frac{\pi}{2}\right), \quad \Im\lambda \neq 0. \tag{3.16}$$

Thus, $\xi(t,\alpha,\gamma)$ is the spectral shift function of the pair A_α, A_γ. Since $\Im\Delta_{\alpha,\gamma}(\lambda)/\Im\lambda \geq 0$ for $-\frac{\pi}{2} \leq \gamma < \alpha < \frac{\pi}{2}$, it follows that

$$0 \leq \xi(t,\alpha,\gamma) \leq 1. \tag{3.17}$$

¿From (3.14) we obtain

$$-\int_\alpha^\gamma m_\mu(\lambda)\,d\mu = \log\frac{\Delta_\alpha(\lambda)}{\Delta_\gamma(\lambda)} = \log\Delta_{\alpha,\gamma}(\lambda). \tag{3.18}$$

Let $\alpha = \alpha_0^{(n)} < \alpha_1^{(n)} < \cdots < \alpha_{k_n}^{(n)} = \gamma$, $n = 1, 2, \ldots$, be a sequence of partitions of the interval $[\alpha, \gamma]$ and

$$d_n = \max\{|\alpha_i^{(n)} - \alpha_{i-1}^{(n)}| : 1 \leq i \leq k_n\} \longrightarrow 0, \quad n \longrightarrow \infty.$$

Then the relation (3.18) means that for $n \to \infty$ and $\Im\lambda \neq 0$

$$F_n(\lambda) = -\sum_{i=1}^{k_n} m_{\alpha_i^{(n)}}(\lambda)(\alpha_i^{(n)} - \alpha_{i-1}^{(n)}) \longrightarrow \log \Delta_{\alpha,\gamma}(\lambda).$$

Now assume that $\sigma_\alpha(\lambda)$ in (2.15) is normalized by $\sigma_\alpha(0) = 0$. Using (2.15), Lemma 2.2 and (3.15) we have

$$\sum_{i=1}^{k_n} \sigma_{\alpha_i^{(n)}}(t)(\alpha_i^{(n)} - \alpha_{i-1}^{(n)}) \longrightarrow \int_0^t \xi(\tau, \alpha, \gamma)\, d\tau,$$

that is

$$\int_\alpha^\gamma \sigma_\mu(t)\, d\mu = \int_0^t \xi(\tau, \alpha, \gamma)\, d\tau, \quad \alpha, \gamma \in \left[-\frac{\pi}{2}, \frac{\pi}{2}\right), \ \sigma_\mu(0) = 0. \tag{3.19}$$

Thus, we have proved the following theorem. □

Theorem 3.2 *The spectral shift function $\xi(t, \alpha, \gamma)$ of the pair A_α, A_γ, $-\frac{\pi}{2} \leq \alpha$, $\gamma < \frac{\pi}{2}$, is given by the relation*

$$\xi(t, \alpha, \gamma) = \frac{1}{\pi} \lim_{\tau \to +0} \arg \Delta_{\alpha,\gamma}(t + i\tau). \tag{3.20}$$

If the spectral shift function σ_α is normalized by $\sigma_\alpha(0) = 0$ then for t fixed $\sigma_\alpha(t)$ is Riemann integrable with respect to α and the relation (3.19) *holds.*

Remark: In the case of a finite interval $(0, l)$, $l < +\infty$, we consider operators $A_{l,\alpha,\beta}$, $A_{l,\gamma,\beta}$, i.e. the boundary conditions which determine the operators coincide at the right endpoint and differ at the left endpoint. Let H be continued on (l, ∞) by $H(x) = J\eta_\beta(J\eta_\beta)^*$, and let A_α, A_γ correspond to the continued H. Then it follows from (2.19) and (3.8) that

$$\sigma_{l,\alpha,\beta}(\lambda) = \sigma_l(\lambda), \ \xi(t, A_{l,\alpha,\beta}, A_{l,\gamma,\beta}) = \xi(t, \alpha, \gamma).$$

Thus, Theorem 3.2 includes also the case of finite interval.

Let

$$\Omega = \begin{pmatrix} \omega_1 & \omega \\ \omega & \omega_2 \end{pmatrix} \neq 0, \ \omega_1 + \omega_2 = 1,$$

be a real nonnegative matrix and

$$\beta(x) = \frac{1}{x} \int_0^x \|H(s) - \Omega\|\, ds, \quad d = \sqrt{\det \Omega}$$

It was proved in [Y1] that if $d \neq 0$ and

$$\beta(x) = o(1), \ x \longrightarrow 0, \tag{3.21}$$

then

$$m_\alpha(i\tau) = -\frac{\omega(\alpha) + id}{\omega_1(\alpha)} + o(1),\ \tau \longrightarrow +\infty, \tag{3.22}$$

where

$$\begin{aligned} \omega_1(\alpha) &= \omega_1 \cos^2\alpha + 2\omega \sin\alpha \cos\alpha + \omega_2 \sin^2\alpha \\ \omega(\alpha) &= \omega \cos^2\alpha + (\omega_2 - \omega_1)\sin\alpha\cos\alpha - \omega\sin^2\alpha. \end{aligned}$$

Obviously, we have $\omega_1(\alpha) > 0$ for $d \neq 0$.

Theorem 3.3 *If $d \neq 0$ and $-\frac{\pi}{2} \leq \gamma < \alpha < \frac{\pi}{2}$, then under the condition* (3.21) *it holds*

$$\int_{-\infty}^{\infty} \frac{\xi(t,\alpha,\gamma)}{1+|t|}\, dt = \infty, \quad .$$

Proof. Assume that $(1+|t|)^{-1}\xi(t,\alpha,\gamma) \in L^1(-\infty,\infty)$. In (3.15) set $\lambda = i\tau$ and let $\tau \to \infty$. Then the right-hand side of (3.15) tends to a real limit, while the left-hand side in view of (3.22) and $d \neq 0$, tends to a nonreal limit. This contradiction proves the theorem.

Now let us consider the case $d = 0$. Without loss of generality we suppose that $\omega = \omega_2 = 0$, $\omega_1 = 1$. If

$$\beta(x) = o\left(\left(\log\frac{1}{x}\right)^{-\nu}\right),\ x \longrightarrow 0,\ \nu > 4, \tag{3.23}$$

we have([Y1]):

$$\begin{gathered} m_\alpha(i\tau) = \tan\alpha + o((\log\tau)^{-\delta}),\ \tau \longrightarrow \infty, \\ \alpha \in \left(-\frac{\pi}{2}, \frac{\pi}{2}\right),\ \delta \in \left(1, \frac{\nu}{4}\right), \end{gathered} \tag{3.24}$$

$$m_{-\frac{\pi}{2}}(i\tau) \longrightarrow \infty, \quad \tau \longrightarrow \infty. \tag{3.25}$$

From (3.24) it follows (see [KK1]) that

$$-m_\alpha(\lambda) = -\tan\alpha + \int_{-\infty}^{\infty} \frac{d\sigma_\alpha(t)}{t-\lambda}, \quad \alpha \in \left(-\frac{\pi}{2}, \frac{\pi}{2}\right). \tag{3.26}$$

Theorem 3.4 *Suppose that $\omega = \omega_2 = 0$, $\omega_1 = 1$ and let* (3.23) *be satisfied. Then for $-\frac{\pi}{2} < \gamma < \alpha < \frac{\pi}{2}$ the relations*

$$\int_{-\infty}^{\infty} \frac{\xi(t,\alpha,\gamma)}{1+|t|}\, dt < +\infty, \tag{3.27}$$

$$\int_{-\infty}^{\infty} \frac{\xi(t,\alpha,-\frac{\pi}{2})}{1+|t|}\, dt = \infty \tag{3.28}$$

hold. Moreover, if $\Im\lambda \neq 0$ *and* $\alpha, \gamma \in (-\frac{\pi}{2}, \frac{\pi}{2})$ *then*

$$\log\left(1 + (\tan\alpha - \tan\gamma)\cos^2\gamma \int_{-\infty}^{\infty} \frac{d\sigma_\gamma(t)}{t-\lambda}\right) = \int_{-\infty}^{\infty} \frac{\xi(t,\alpha,\gamma)}{t-\lambda}\,dt. \tag{3.29}$$

Proof: Let

$$\widetilde{\Delta}_{\alpha,\gamma}(\lambda) = \frac{\cos\gamma}{\cos\alpha}\Delta_{\alpha,\gamma}(\lambda), \quad \alpha,\gamma \in \left(-\frac{\pi}{2}, \frac{\pi}{2}\right).$$

The relation (3.15) yields

$$\begin{aligned} &\log\widetilde{\Delta}_{\alpha,\gamma}(\lambda) \\ &= \widetilde{c}_{\alpha,\gamma} + \int_{-\infty}^{\infty}\left(\frac{1}{t-\lambda} - \frac{t}{1+t^2}\right)\xi(t,\alpha,\gamma)\,dt, \quad \Im\widetilde{c}_{\alpha,\gamma} = 0. \end{aligned} \tag{3.30}$$

In view of (3.24), from

$$\widetilde{\Delta}_{\alpha,\gamma}(\lambda) = \frac{1 - m_0(\lambda)\tan\alpha}{1 - m_0(\lambda)\tan\gamma}$$

it follows that $\widetilde{\Delta}_{\alpha,\gamma}(i\tau) = 1 + o((\log\tau)^{-\delta})$, $\tau \to +\infty$ and

$$\log\widetilde{\Delta}_{\alpha,\gamma}(i\tau) = o((\log\tau)^{-\delta}), \ \tau \longrightarrow +\infty \quad \delta \in \left(1, \frac{\nu}{4}\right). \tag{3.31}$$

Hence

$$\int_1^{\infty} \frac{\Im\log\widetilde{\Delta}_{\alpha,\gamma}(i\tau)}{\tau}\,d\tau < \infty,$$

and using Kac's theorem (see [KK1]), (3.30) and (3.31) we find that the relation (3.27) holds and

$$\log\widetilde{\Delta}_{\alpha,\gamma}(\lambda) = \int_{-\infty}^{\infty} \frac{\xi(t,\alpha,\gamma)}{t-\lambda}\,dt.$$

From this relation, (3.26) and the identity

$$\widetilde{\Delta}_{\alpha,\gamma}(\lambda) = 1 + \cos^2\gamma(\tan\alpha - \tan\gamma)(-m_\gamma(\lambda) + \tan\gamma),$$

we get (3.29).

Now let us check (3.28). From (3.15) we have

$$\begin{aligned} &\log(-\sin\alpha - m_{-\frac{\pi}{2}}(\lambda)\cos\alpha) \\ &= c_{\alpha,-\frac{\pi}{2}} + \int_{-\infty}^{\infty}\left(\frac{1}{t-\lambda} - \frac{t}{1+t^2}\right)\xi\left(t,\alpha,-\frac{\pi}{2}\right)dt. \end{aligned} \tag{3.32}$$

If $(1+|t|)^{-1}\xi(t,\alpha,-\frac{\pi}{2}) \in L^1(-\infty,\infty)$, the right-hand side on the last equality should tend to a finite limit for $\lambda = i\tau$, $\tau \to \infty$, while the left-hand side, in view of (3.25), tends to ∞. This contradiction completes the proof. □

Note that if A_α is semi-bounded from below, under the conditions $\omega = \omega_2 = 0$, $\omega_1 = 1$ and (3.23) it follows from (3.27) that

$$\xi(t,\alpha,\gamma) = 0, \quad t < \min\{\sigma(A_\alpha) \cup \sigma(A_\gamma)\}, \; \alpha,\gamma \in \left(-\frac{\pi}{2},\frac{\pi}{2}\right).$$

Let us suppose that the spectrum $\sigma(A_\alpha)$ of A_α consists of a finite or countable number of eigenvalues $\lambda_n(A_\alpha)$, $n = 1,2,\ldots$. Then m_α is a meromorphic function with poles $\lambda_n(\alpha)$. Consequently, if $-\frac{\pi}{2} \le \gamma < \alpha < \frac{\pi}{2}$, $\Delta_{\alpha,\gamma(\lambda)}$ is meromorphic and maps the upper half plane $\Im\lambda > 0$ into itself. It follows that its zeros and poles must alternate. On the other hand, taking into account (3.13) and (3.14) the poles and the zeros of $\Delta_{\alpha,\gamma}(\lambda)$ coincide with $\lambda_n(\gamma)$ and $\lambda_n(\alpha)$, respectively. Therefore the sequences $\lambda_n(\gamma)$ and $\lambda_n(\alpha)$ are alternating. Thus, for $-\frac{\pi}{2} \le \gamma < \alpha < \frac{\pi}{2}$ the argument principle and (3.20) imply that

$$\xi(t,\alpha,\gamma) = \begin{cases} 1, & \text{if } \lambda_n(\gamma) < t < \lambda_n(\alpha), \\ 0, & \text{if } \lambda_n(\alpha) < t < \lambda_{n+1}(\gamma). \end{cases} \tag{3.33}$$

If the sequences $\lambda_n(\alpha)$ and $\lambda_n(\gamma)$ are bounded from below, then there are two possibilities: Either $\lambda_1(\gamma) < \lambda_1(\alpha) < \lambda_2(\gamma) < \cdots$ or $\lambda_0(\alpha) < \lambda_1(\gamma) < \lambda_1(\alpha) < \cdots$. In the first case $\xi(t,\alpha,\gamma) = 0$ if $t < \lambda_1(\gamma)$ and in the second case $\xi(t,\alpha,\gamma) = 1$ if $\lambda < \lambda_0(\alpha)$.

For Sturm-Liouville operators the relations (3.29) and (3.19) were given in [Y2]. The relations (3.15) and (3.29) together with (3.33) yield formulas, which in case of discrete spectrum establish a connection between the eigenvalues of the problem (2.1), (2.2) with different boundary conditions at zero and the spectral function. For Sturm-Liouville operators formulas of such type were found by Levitan and Gasimov [LG].

For Sturm-Liouville and Dirac operators the asymptotic behavior of the spectral function is known, see [M], [LS], [Y3]. This makes it possible to obtain from (3.19) formulas for the regularized trace of the difference of two Sturm-Liouville ([LG], [Y2]) or Dirac operators with different boundary conditions at zero.

4 The Inverse Problem

In this section we suppose that H is real, see the Remark in Section 2. We cite the following basic result of L. de Branges [dB2], [dB3], in the form refined by Winkler [W].

Theorem 4.1 *For each $F \in (R)$ and $\alpha \in [-\frac{\pi}{2},\frac{\pi}{2})$ there exists a unique real matrix function H on* $(0,\infty)$ *as above, such that F coincides with the Weyl function m_α of the problem* (2.1), (2.2).

In [KY] for a pair of self-adjoint extensions of a simple symmetric nonnegative operator A with defect 1 the spectral shift function was studied, and it was also shown that this pair of operators (as well as the operator A) can be restored uniquely up to unitary equivalence from the spectral shift function.

The operator A considered in this paper is not necessarily semi-bounded. Moreover, in the case (II) $\overline{D(A)} \neq L_0^2(H)$. Therefore the results of [KY] are not applicable in this case.

Let $0 \leq \xi(t) \leq 1$, $\xi(t) \not\equiv const$, be any measurable function on $(0, \infty)$. The previous results imply that for any $-\frac{\pi}{2} \leq \gamma < \alpha < \frac{\pi}{2}$, there exists a real $H(x)$, $x \in (0, \infty)$, as above such that

$$\xi(t) = \xi(t, H, \alpha, \gamma).$$

In fact, let

$$-m_\gamma(\lambda) = \frac{ce^{F(\lambda)} - \cos(\alpha - \gamma)}{\sin(\alpha - \gamma)},$$

where $c > 0$ and

$$F(\lambda) = \int_{-\infty}^{\infty} \left(\frac{1}{t - \lambda} - \frac{t}{1 + t^2} \right) \xi(t)\, dt, \quad \Im\lambda \neq 0.$$

Since $0 < \arg F(\lambda) < \pi$ for $\Im\lambda > 0$, we have $\Im m_\gamma(\lambda) < 0$ for $\Im\lambda > 0$. Because of de Branges's theorem there exists an $H(x)$, $x \in (0, \infty)$, such that $-m_\gamma(\lambda)$ coincides with the Weyl function of the corresponding canonical system with the initial condition $y(0) \in \operatorname{span}\{\eta_\gamma\}$. Then in view of (3.15) it follows that $\xi(t) = \xi(t, H, \alpha, \gamma)$.

In particular, from (3.33) we find that for any given alternating sequences $\{\lambda_n\}$, $\{\mu_n\}$ and $\beta, \gamma \in [-\frac{\pi}{2}, \frac{\pi}{2})$ there exists an H such that these sequences coincide with the eigenvalues of the problem (2.1), (2.2) for $\alpha = \beta$ and $\alpha = \gamma$, respectively. Since $c > 0$ is arbitrary, it follows that the matrix function H corresponding to a given function ξ is not unique. Therefore the following problem arises: Find additional conditions under which H will be uniquely determined.

Theorem 4.2 *Let $H_1(x)$ and $H_2(x)$ satisfy the condition* (3.21) *for the same Ω.*

(a) *If $d \neq 0$ and $\alpha_1, \alpha_2, \gamma \in [-\frac{\pi}{2}, \frac{\pi}{2})$, $\alpha_1, \alpha_2 \neq \gamma$, then the relation*

$$\xi(t, H_1, \alpha_1, \gamma) = \xi(t, H_2, \alpha_2, \gamma) \text{ for } t \in (0, \infty), \text{ a.e.},$$

implies $\alpha_1 = \alpha_2$ and $H_1(x) = H_2(x)$ a.e. on $(0, \infty)$.

(b) *If $d = 0$, $\alpha \neq \gamma$, $\alpha, \gamma \in [-\frac{\pi}{2}, \frac{\pi}{2})$ and $\Omega\eta_\alpha \neq 0$ and $\Omega\eta_\gamma \neq 0$, then the relation*

$$\xi(t, H_1, \alpha, \gamma) = \xi(t, H_2, \alpha, \gamma) \text{ for } t \in (0, \infty), \text{ a.e.},$$

implies $H_1(x) = H_2(x)$ a.e. on $(0, \infty)$.

Proof: (a) We have to show that under the assumption (3.21) for given Ω and γ the number α and the function m_γ are uniquely determined by $\xi(t) := \xi(t, H, \alpha, \gamma) = \xi(t, \alpha, \gamma)$. We can put $\gamma = -\frac{\pi}{2}$. In view of (3.22), we have

$$\lim_{y\to+\infty} m_{-\frac{\pi}{2}}(iy) = \frac{\omega - id}{\omega_2}.$$

Hence, the right-hand side in (3.32) converges for $y \to +\infty$ and

$$\begin{aligned}&\log\left(-\sin\alpha - \frac{\omega - id}{\omega_2}\cos\alpha\right)\\ &= c_{\alpha,-\frac{\pi}{2}} + \lim_{y\to+\infty}\int_{-\infty}^{\infty}\left(\frac{1}{t-\lambda} - \frac{t}{1+t^2}\right)\xi(t)\,dt.\end{aligned} \tag{4.1}$$

Taking the imaginary parts of both sides of (4.1) and using $\Im c_{\alpha,-\frac{\pi}{2}} = 0$ it follows that

$$\tan c = -\frac{d\cos\alpha}{\omega_2\sin\alpha + \omega\cos\alpha}, \text{ where } c = \lim_{y\to+\infty}\int_{-\infty}^{\infty}\frac{y\xi(t)}{t^2+y^2}\,dt.$$

Since $d \neq 0$, we can choose a unique $\alpha \in (-\frac{\pi}{2}, \frac{\pi}{2})$, and from (4.1) we find $c_{\alpha,-\frac{\pi}{2}}$. Then $m_{-\frac{\pi}{2}}(\lambda)$ will be uniquely determined from (3.32). Hence, in view of the theorem of de Branges we can uniquely determine $H(x)$.

(b) Let $\omega_1 \neq 0$. It was proved in [Y1] that

$$\lim_{y\to\infty} m_0(iy) = -\frac{\omega}{\omega_1}.$$

Thus, (3.13) implies that $\Delta_{\alpha,\gamma}(iy)$ tends to

$$\lim_{y\to\infty}\Delta_{\alpha,\gamma}(iy) = \frac{\omega_1\cos\alpha + \omega\sin\alpha}{\omega_1\cos\gamma + \omega\sin\gamma} = a \neq 0.$$

Passing to the limits in (3.15) we obtain

$$\log a = c_{\alpha,\gamma} + \lim_{y\to\infty}\int_{-\infty}^{\infty}\left(\frac{1}{t-iy} - \frac{t}{1+t^2}\right)\xi(t)\,dt,$$

whence we find $c_{\alpha,\gamma}$ and in view of (3.15) also $m_\gamma(\lambda)$.

The proof in the case $\omega_1 = 0$ is similar. □

References

[AG] N.I. Ahiezer and I.M. Glazman, *The theory of linear operators in Hilbert space* (Russian), Nauka, Moscow, 1961.

[At] F.V. Atkinson, *Discrete and continuous boundary problems*, Academic Press, New York, 1964.

[CL] E.A. Coddington and N. Levinson, *Theory of ordinary differential equations*, New York, 1955.

[dB1] L. de Branges, *Some Hilbert spaces of entire functions. III*, Trans. Amer. Math. Soc. **100** (1961), 73–115.

[dB2] L. de Branges, *Some Hilbert spaces of entire functions. II*, Trans. Amer. Math. Soc. **99** (1961), 118–152.

[dB3] L. de Branges, *Some Hilbert spaces of entire functions. IV*, Trans. Amer. Math. Soc. **105** (1962), 43–83.

[GK1] I.C. Gohberg and M.G. Krein, *Introduction in the theory of linear nonself-adjoint operators* (Russian), Nauka, Moscow, 1965.

[GK2] I.C. Gohberg and M.G. Krein, *Theory and applications of Volterra operators in Hilbert spaces* (Russian), Nauka, Moscow, 1967.

[Ka1] I.S. Kac, *Linear relations, generated by a canonical differential equation* (Russian), Functional Anal. i Priložen. **17** (1983), 86–87.

[Ka2] I.S. Kac, *Linear relations, generated by a canonical differential equation on an interval with a regular endpoint and expansibility in eigenfunctions* (Russian), Odessa, 1984.

[KK1] I.S. Kac and M.G. Krein, *R-functions-analytic functions mapping the upper half-plane into itself*, Amer. Math. Soc. Transl. (2) **103** (1974), 1–17.

[K1] M.G. Krein, *On the trace formula in perturbation theory* (Russian), Math. Sb. **33** (1953), 597–626.

[K2] M.G. Krein, *On the perturbation determinant and the trace formula for unitary and self-adjoint operators* (Russian), Dokl. Akad. Nauk SSSR **144** (1962), 268–271.

[K3] M.G. Krein, *About some new investigations in perturbation theory for self-adjoint operators* (Russian), First Summer School of mathematics, Kanev, 1963.

[KY] M.G. Krein and V.A. Yavrian, *On the spectral shift function arising from the perturbation of a positive operator* (Russian), J. Operator Theory **6** (1981), 155–191.

[LT] H. Langer and B. *Textorius, A generalization of M.G. Krein's method of directing functionals to linear relations*, Proc. Roy. Soc. Edinburgh **81A** (1978), 237–246.

[LG] B.M. Levitan and Y.S. Gasimov, *Determination of the differential equation by two spectrums* (Russian), Usp. matem. nauk. **19** (1964), 3–63.

[LS] B.M. Levitan and I.S. Sargsjan, *Introduction to spectral theory of self-adjoint ordinary differential operators*, Trans. of Math. Monographs **39** (Amer. Math. Soc., Providence, 1975).

[M] V.A. Marchenko, *Spectral theory of Sturm-Lioville operators* (Russian), "Naukova dumka", Kiev, 1972.

[Y1] V.A. Yavrian, The asymptotics of the Weyl function for canonical system of differential equations, Journal of Contemporary Mathematical Analysis (Armenian Academy of Sciences), vol. **33**, no. 4, 1998.

[Y2] V.A. Javrian (Yavrian), *On the regularized trace of the difference between two singular Sturm-Lioville operators*, Soviet Math. Dokl. **7** (1966), 888–891.

[Y3] V.A. Yavrian, *On the asymptotics of the spectral matrix-function of canonical system of differential equations* (Russian), Dokl. Akad. Nauk Armenian SSR **56** (1973), 129–134.

[W] H. Winkler, *The inverse spectral problem for canonical systems*, Integr. Equat. Oper. Th. **22** (1995), 360–374.

V.A. Yavrian
Department of Mathematics
Yerevan State University
Alek Manukian str. 1
Yerevan 375049
Armenia

AMS Subject Classification: 34A55, 34B20, 34L10, 47A75

Table of Contents of Volume II

Zeitfracht Medien GmbH
Ferdinand-Jühlke-Straße 7
99095 Erfurt, Deutschland
produktsicherheit@kolibri360.de